Food Engineering Principles and Practices

Syed S. H. Rizvi

Food Engineering Principles and Practices

A One-Semester Course

 Springer

Syed S. H. Rizvi
Institute of Food Science
Cornell University
Ithaca, NY, USA

ISBN 978-3-031-34125-0   ISBN 978-3-031-34123-6   (eBook)
https://doi.org/10.1007/978-3-031-34123-6

This Springer imprint is published by the registered company Springer Nature Switzerland AG
The registered company address is: Gewerbestrasse 11, 6330 Cham, Switzerland

To
Zahra, Ali and Raza

"Scientists study the world as it is, engineers create the world that never has been."
—*Theodore von Kármán*

Preface

Some forms of the basic unit operation of food processing and preservation have been practiced by *homo sapiens* from well before the time of recorded history. Today, it is indeed difficult to visualize a food processing operation in which one or more of the phenomena of transport processes, thermodynamics, reaction kinetics, and other related principles do not apply. Attempts to quantitatively analyze them are confronted with formidable challenges due to the limited availability of data on the physical and engineering properties of food and food components. This lack of technical information is counterbalanced by a vast experience based on empirical observations and techniques. Developments of the scientific approach in the quantification of physical and biological factors for application in the design and analysis of equipment and processes are of recent origin and have evolved into a new discipline of food engineering through the recognition that fundamental understanding of the underlying principles of unit operations of importance in the food industry must be capable of more meaningful analysis if the modern technologies are to become even more efficient in their utilization of resources.

Younger generations of food scientists are expected to have a thorough and better understanding of the food engineering principles to become professionals of full stature dealing with the problems of the food industry as opposed to becoming mere practitioners of a specific commodity. The engineering background of direct interest to such professionals should essentially include the theories of equilibrium (or thermodynamic) and non-equilibrium (or transport) processes. Once the fundamental understanding is firmly acquired, the various other applied topics of interest to food processing can be built up. This can be realized though only from a thorough understanding of the principles of engineering that emphasizes analytical reasoning, problem-solving, and intellectual rigor.

Recognizing the need of students enrolled in food science curricula, this text has been prepared. It is based on a one-semester course on concepts, principles, and some practices of food engineering that I have taught for many years. The main objective in doing this has been to enhance teaching since there is no one text currently available, which covers the material needed in a one-semester course. The text has been organized around the principles of transport processes rather than the detailed analysis of a particular application. Because of this approach new equipment and systems will not render obsolete the knowledge gained from this study but prepare the students well for later courses on unit operations and food processing they are required to take during undergraduate degree programs in food science and engineering.

It has been generally observed that teaching food engineering in one course often results in a survey-type approach with a minimal grounding in engineering principles. The topics and subject matter included in the text have been developed with a focus on the fundamentals of engineering and their applications in food and biomaterial processing. This should lead to a sound, satisfying, and useful grasp of the subject. It is also reasonable to expect engineering students of today to be bilingual and proficient in the use of both the System International (SI) and Engineering system of units. Although the former system is extensively used, examples of solutions and problems at the end of each chapter are given in both sets of units. The first three chapters concentrate on reviewing the basic knowledge of physics and chemistry and introducing the engineering language needed to eliminate confusion going forward.

Zeal and enthusiasm are necessary ingredients for successful mastery of anything. Before the student can solve specific problems, principles must be clearly understood. And the most effective and efficient way to learn those principles is to work on many, many problems. A correct theoretical solution to a problem is the most elegant and least expensive route to progress. Students are encouraged to be persistent since learning is a result of interest and effort, repeated over and over. The thrill of engineering comes from using it to understand processes we deal with every day and to invent new ones.

Very often errors in fact, omissions and misprints inevitably creep into an undertaking of this type. Since the text is based on "preliminary teaching notes" at best, any comments regarding errors, scope, and organization will be a big help in preparing the next edition and will appreciate having them brought to my attention. In this text, an attempt has been made to do justice to mathematical treatment and its applications to real life. Suggestions to do further justice to both are indeed welcome. I will have achieved my objective in publishing this text if it helps our students perform more effectively and creates interest in food engineering.

I thank the many graduate students who have helped me teach the course over the past many years and developed many of the problems included in the text. I also thank the faculty, staff, and administration of Cornell University for their support. It has also been a pleasure to work with the team at Springer Nature, especially Sofia Costa. Their patience and support during the preparation of the manuscript are sincerely appreciated.

Ithaca, NY, USA   Syed S. H. Rizvi

The original version of the book has been revised. A correction to this book can be found at https://doi.org/10.1007/978-981-99-0026-8_14

Contents

Abbreviations

Nomenclature

A	Area (m^2, ft^2)
a, b, c	Constants or coefficients
a	Acceleration (m/s^2, ft/s^2)
B	Bulk modulus (Pa, psi)
C_p	Specific heat (heat capacity) at constant pressure [J/(kg·K), BTU/(lb_m·°F)]
C_v	Specific heat (heat capacity) at constant volume [J/(kg·K), BTU/(lb_m·°F)]
D	Mass diffusivity, diffusion coefficient (m^2/s, ft^2/h)
d	Diameter (cm, in)
de	Equivalent diameter (cm, in)
df	Degree of freedom
E	Energy (kJ, BTU, cal), Young's modulus (Pa, psi)
E_i	Internal energy of a system relative to a reference state (kJ, BTU)
E_T	Total energy of the system (kJ, BTU)
E_k	Kinetic energy (kJ, BTU).
E_p	Potential energy (kJ, BTU).
Eff	Efficiency (%)
F	Force (N, lb_f), Farad
f	Fanning friction factor (dimensionless)
G	Shear modulus or modulus of rigidity (Pa, psi)
GRe	Generalized Reynolds number (dimensionless)
g	Acceleration due to gravity (9.81 m/s^2 or 32.17ft/s^2 at sea level).
g_c	Gravitational constant
H	Enthalpy of a system relative to a reference state (kJ, BTU)
h	Depth or height (m)
h_c	Convective heat-transfer coefficient (((J/s)(m^2-K) or (BTU/hr)(m^2-K))
I	Current (amp)
ID (i.d.)	Inside diameter (m, in.)
K	Consistency coefficient in power law model (Pa-s^n), Kelvin
k	Thermal conductivity ((J/s)(m-K) or (BTU/hr)(m-K))

k_f	Friction loss coefficient (dimensionless)
M_i	Molarity of solute i (moles of solute/liter of solution)
MC	Moisture content
M	Molecular weight
m	Mass (kg, lb_m)
$\dot{m}$	Mass flow rate (kg/s, lb_m/h)
N	Newton, Normality (number of equivalents of solute/liter of solution)
Nu	Nusselt number, dimensionless
OD (o.d.)	Outside diameter (m, in.)
n	Number of moles (mol, mole), flow behavior index in power law model
P	Pressure (Pa)
P_A	Partial pressure of species A in a mixture (N/m^2, lb_f/in^2), $= y_A P_T$.
$P^*_A\,(T)$	Vapor pressure of pure A at temperature T (N/m^2, lb_f/in^2).
P_T	Total pressure of a system (N/m^2, lb_f/in^2).
P_c	Critical pressure (N/m^2, lb_f/in^2)
Per	Perimeter (m, ft)
P	Power (N·m/s)
q	Total heat transferred to or from a system (kJ, BTU), heat content
$\dot{q}$	Heat flow rate (W, BTU/h)
q'	Heat flux (W/m^2, BTU/h-ft^2)
R	Gas constant [kJ/(mol-K)] or [l.- atm/mol-K]
Re	Reynolds number (dimensionless)
R_f	Fouling resistance
R_i	Resistance
r_h	Hydraulic radius (m, ft)
SG	Specific gravity (dimensionless)
T	Temperature (K or C or F or R)
T_m, T_b, T_c	Melting point temperature, boiling point temperature, and critical temperature
t	Time (s, h)
U	Overall heat-transfer coefficient (W/m^2.K, $BTU/h.ft^2$.F)
V	Volume (m^3, ft^3), Volt
V_T	Total volume of a mixture (m^3, ft^3)
V_i	Partial volume of i in a mixture (m^3, ft^3)
v	Velocity (m/s, ft/s)
w	Work transferred to or from a system (kJ, BTU)
W	Weight (N, lb_f)
X, Y	Variables
x, y	Mass fraction or mole fraction of a species in a mixture. (Also used as subscripts to identify different species. In liquid vapor systems, x usually denotes fraction in the liquid and y denotes fraction in the vapor)
z	Compressibility factor of a gas

Greek Symbols

α Thermal diffusivity ($k/\rho C_\mathrm{p}$, m^2/s·, or ft^2/h), kinetic energy correction factor

β Coefficient of volumetric expansion, a constant

ε Strain, Surface roughness

ε_o Porosity

$\dot{\gamma}$ Shear rate (s^{-1})

η Viscosity (Pa·s, cP)

η_a Apparent viscosity (Pa·s, cP)

η_0 Zero shear viscosity (Pa·s)

ν Kinematic viscosity, momentum diffusivity (m^2/h, ft^2/h), Poisson's ratio

ρ Density (kg/m^3, lb$_\mathrm{m}$/ft^3)

τ Shear stress (Pa, lb$_\mathrm{f}$/in^2)

τ_o Yield stress (Pa, lb$_\mathrm{f}$/in^2)

ω Angular velocity (rad/s)

ω_r Revolutions per second

Δ Difference, e.g., $X_\mathrm{final} - X_\mathrm{initial}$, where X is any system property

Special Superscript and Symbols

$\left(\text{e.g.,} \dot{m}, \dot{V}, \dot{q}\right)$ Rate, such as mass flow rate, volumetric flow rate, heat-transfer rate, etc.

$- \left(\text{e.g.,} \overline{V}, \overline{H}\right)$ Specific property, property per unit mass like specific volume, specific enthalpy

$<\,>$ (e.g., $<v>$)   Average such as average velocity

$\wedge$ (e.g., $\widehat{E}_i$)   Molar property such as molar internal energy

$\Delta\widehat{H}_\mathrm{m}, \Delta\widehat{H}_\mathrm{v}$ Molar enthalpy of melting (fusion) and vaporization, respectively (kJ/mol)

$[=]$ Read as "has the units of."

Greek Letters: CAP/Lower

A α alpha (AL-fuh)

B β beta (BAY-tuh)

Γ γ gamma (GAM-uh)

Δ δ delta (DEL-tuh)

E ε epsilon (EP-sil-on)

ζ Z   zeta (ZEY-tuh)

H η eta (AY-tuh)

Θ θ theta (THAT-tuh)

I ι iota (eye-OH-tuh)

K κ kappa (KAP-uh)

Λ λ lambda (LAM-duh)

M μ mu (MYOO)

N ν nu (NOO)

Ξ ξ xi (KS-EYE)

O o   omicron (OM-i-KRON)
Π π   pi (PIE)
P ρ   rho (ROW)
Σ σ   sigma (SIG-muh)
T τ   tau (TAU)
Y υ   upsilon (OOP-si-LON)
Φ ϕ   phi (FEE)
Χχ   chi (K-EYE)
Ψ ψ   sy (SIGH)
Ω ω   omega (oh-MAY-guh)

Prefixes in Common Use Before Units

Prefix	Symbol	Multiple	Prefix	Symbol	Multiple
atto	a	10^{-18}	deca	da	10^1
femto	f	10^{-15}	hecto	h	10^2
pico	p	10^{-12}	kilo	k	10^3
nano	n	10^{-9}	mega	M	10^6
micro	μ	10^{-6}	giga	G	10^9
milli	m	10^{-3}	tera	T	10^{12}
centi	c	10^{-2}	peta	P	10^{15}
deci	d	10^{-1}	exa	E	10^{18}

Basic Concepts and Material Properties

1

This chapter provides a review of important concepts that are needed to formulate, solve, and interpret food systems and manufacturing processes. They also provide a good foundation for learning more advanced scientific and engineering concepts and principles. Mastering these fundamental principles and their applications is essential for long-term successes both in and out of the classroom.

1.1 Dimensions and Units: Base Units

At the core of engineering is the ability to measure fundamental physical quantities that are each defined by a dimension and their magnitude expressed by appropriate units. Units of measure are used to quantify phenomena.

In science and engineering, there are seven basic dimensions used to measure properties: length, mass, time, temperature, unit amount of a substance, electric charge (current), and luminous intensity. The magnitude of each of these dimensions is given in units relative to some arbitrary standards. The three major unit systems used in science and engineering to express the seven **base units** are the Systeme International d'Units (SI), the centimeter-gram-second (CGS) system, and the American Engineering System, which is based on Imperial and US customary measurement systems (Table 1.1).

Although the International Treaty of the Meter was signed in Paris on May 20, 1875, by seventeen countries, including the USA, many different types of units continue to be in use in many parts of the world. The SI is the recommended and most used system of units of measurement worldwide, but the universal adoption of SI units is unlikely to happen for many years. This is because much of the historic and useful data of interest and utility exist in traditional units, and some of them continue to be in use today, such as barrel for volume, slug for mass, degrees Brix for sugar concentration, etc. The literature is full of disasters related to unit conversion errors, including NASA's loss of a $125 million Mars orbiter due to

S. S. H. Rizvi, *Food Engineering Principles and Practices*,
https://doi.org/10.1007/978-3-031-34123-6_1

Table 1.1 Base units

Dimension	SI unit	CGS unit	American Engineering Units
Length	Meter (m)	Centimeter (cm)	Inch (in); foot (ft)
Mass	Kilogram (kg)	Gram (g)	Pound (lb_m)
Time	Second (s)	Second (s)	Minute (min); hour (h)
Temperature	Kelvin (K)	Kelvin (K); Celsius (°C)	Fahrenheit (°F); Rankine (°R)
Unit amount of substance	Mole (mol)	g-mole	lb-mole
Electric current	Ampere (A)	Ampere (A)	Ampere (A)
Luminous intensity	Candela (cd)	Candela (cd)	Candela (cd)

confusion in the units of measurement—one team used SI, while another used American Engineering units for a key spacecraft operation. It thus becomes necessary for engineers to have familiarity with the SI, imperial, and other customary units and be able to interconvert and use data available in different sets of units.

Length is a measure of the distance between two points. The SI base unit for length is the **meter**, which is the distance traveled by light in a vacuum during 1/299792458 of a second.

Mass is a measure of the quantity of material. The SI unit used for mass is the **kilogram (kg).** A cylinder of platinum-iridium alloy kept at the International Bureau of Weights and Measures in France defines the kilogram. This is the only unit standard that is not reproducible in a laboratory.

Although mass is related to weight, they are distinctly different quantities. Weight measures the downward force exerted on an object by a gravitational field. On the moon, an object weighs less than on earth because the moon has 1/6 the gravity of earth. In an orbiting satellite, an object weighs nothing because the gravitational field is canceled by the centrifugal force. Since the weight of an object changes with circumstances, weight cannot be considered to be a base unit.

Time measures the ordering of events by comparing them to some regular periodic phenomena, such as the orbiting of the earth around the sun, the rotation of the earth on its axis, the swinging of a pendulum, or the vibration of an atom. The **second (s)** is the SI unit and is defined as the duration of 9,192,631,770 periods of radiation corresponding to the transition between two hyperfine levels of the ground state of the cesium-133 atom.

Temperature is related to the constant random motion of molecules. At high temperatures, the molecules move more rapidly and collide more frequently than at low temperatures. Thus, a substance having molecules with a relatively high average kinetic energy (the energy of motion) will be at a higher temperature than a substance having molecules with a low average kinetic energy.

The triple point of water is used to define Kelvin, the SI unit for thermodynamic temperature. A value of 273.16 K is assigned to the triple point of water, and a value of 0 K is assigned to absolute zero. The absolute zero is the temperature at which molecular motion is stopped and all disorder disappears. **Kelvin** is the fraction 1/273.16 of the thermodynamic temperature of the triple point of water.

The atomic or molecular mass of a chemical expressed in grams is called a *mole (mol)* in SI units, often written more correctly as **gram-mole**. One mole contains exactly 6.023×10^{23} (known as Avogadro's number) entities (atoms, molecules, electrons, particles, or groups of particles) of the substance.

The electric current is said to flow when charged particles (usually electrons) move through a conductor. The current is measured by counting the number of charges that pass a defined point each second. One coulomb is 6.281×10^{18} charged particles. If one coulomb of protons passes from right to left past a defined point in one second, then a current of one *ampere* flows through the wire.

Consider that when an electric current passes through a wire, a magnetic field surrounds the wire. Then, an ampere is the constant current that, if maintained in two straight parallel conductors of infinite length of negligible circular cross-section and placed 1 m apart in a vacuum, would produce a force of 2×10^{-7} Newton per meter of length.

Luminous Intensity is a measure of the brightness of light. The **candela (cd)** is the luminous intensity, in each direction, of a source that emits monochromatic radiation of a frequency 540×10^{12} cycles per second (540 Hz) and that has a radiant intensity of 1/683 watts per steradian (solid angle). The steradian (sr) is the unit of the solid angular measure, and a sphere has 4 pi or approximately 12.57 steradians.

1.2   Derived Units

The base units can be used to derive all other units of the physical measure. Table 1.2 summarizes several derived units commonly used in the processing and engineering of materials, including foods. A number of these are discussed in more detail below.

1.2.1   Force

Force is defined as a push or pull acting on a body. Force (F) when applied to a body causes the body to accelerate if no counteracting force is present. According to Newton's second law of motion, force is directly related to mass and acceleration as:

$$F = ma \tag{1.1}$$

Imagine a mass of 1 kg sitting on a large frictionless surface. Now imagine that the mass is pushed with enough force that the mass is moving at a velocity of 1 m/s at the end of 1 s. If at the end of 2 s the mass is moving at a velocity of 2 m/s, the accelerating of the mass is 1 meter per second or 1 m/s^2. The force required to accelerate 1 kg of mass by 1 m/s^2 is called 1 Newton (1 N). Note that force is a vector quantity.

Table 1.2 SI-derived units with their names

Quantity	Name	Symbol	Expression in terms of other units
Frequency	Hertz	Hz	s^{-1}
Force	Newton	N	$kg \cdot m/s^2$
Pressure, stress	Pascal	Pa	N/m^2
Energy, work, quantity of heat	Joule	J	$N \cdot m$
Power, radiant flux	Watt	W	J/s
Electric charge, electric quantity	Coulomb	C	$A \cdot s$
Electric potential, potential difference, electromotive force	Volt	V	W/A
Capacitance	Farad	F	C/V
Electric resistance	Ohm	Ω	V/A
Conductance	Siemens	S	A/V
Magnetic flux	Weber	Wb	$V \cdot s$
Magnetic flux density	Tesla	T	Wb/m^2
Inductance	Henry	H	Wb/A
Luminous flux	Lumen	lm	$cd \cdot sr$
Illuminance	Lux	lx	lm/m^2
Celsius temperature	Degree Celsius	°C	K

1.2.2 Weight

Weight is the downward force exerted on the mass of an object by the local gravitational field. So, weight will disappear in the absence of gravity, but mass will not. Also, since force equals mass times acceleration, it follows that the gravitational field can be measured in terms of the acceleration it imparts to a falling object. The symbol g is used for acceleration due to gravity and near the surface of the earth, it is 9.81 m/s^2. Weight (W), therefore, is defined as:

$$W = mg \tag{1.2}$$

Example 1

What is the weight of a 2.00 kg block of cheese?

Solution: $W = mg = (2.00 \ kg)\left(\frac{9.81 \ m}{s^2}\right) = 19.6 \ \frac{kg - m}{s^2} = 19.6 \ N$ ◀

In the SI and CGS unit systems, weight and mass use different units (Table 1.3). In the American Engineering System, however, the pound is used to refer to both mass and force. To distinguish between pound-mass and pound-force, the subscripts m and f are used. The terms pound-mass and pound-force have been defined so that a 1 lb$_m$ object will have a weight of 1 lb$_f$ on the surface of the earth. For this to be

Table 1.3 Units of mass and weight

Unit system	Mass	Acceleration (g)	Weight
SI	Kilogram (kg)	9.81 m/s^2	Newton (N)
CGS	Gram (g)	981 cm/s^2	Dyne (dyn)
American engineering	Pound-mass (lb$_m$)	32.17 ft/s^2	Pound-force (lb$_f$)

possible, a gravitational constant g_c has been introduced. The value of g_c has the same magnitude as g, but the units are different.

$$g_c = 32.17 \frac{lb_m - ft}{lb_f - s^2} \tag{1.3a}$$

To find the weight in lb$_f$ of an object whose mass is given in lb$_m$, replace Eq. (1.2) with Eq. (1.3b) below:

$$W = \frac{m\,g}{g_c} \tag{1.3b}$$

Example 2

A quart of milk has a mass of 2.00 lb$_m$. What is its weight?

Solution: $W = m\left(\frac{g}{g_c}\right) = (2.00\ lb_m)\left(\frac{32.17\ \frac{ft}{s^2}}{32.17\ \frac{lb_m \cdot ft}{lb_f \cdot s^2}}\right) = 2.00\ lb_f$ ◄

This is a lot of work to convert 2.00 lb$_m$ to 2.00 lb$_f$. However, on the surface of the moon, the acceleration due to gravity is one-sixth that on Earth. Since the value of g_c is constant, the weight of an object on the moon will correspondingly be less than that on earth.

Check Your Understanding

1. Use Eq. (1.3b) to verify that a tub of ice cream with a mass of 8 lb$_m$ has a weight of 8 lb$_f$. Include all units and show that they cancel correctly.
2. Verify that the tub of ice cream weighs 1.32 lb$_f$ on the moon where the acceleration due to gravity is 5.32 ft/s^2. ◄

1.2.3 Pressure

The basic definition of pressure is the force per unit area exerted by a fluid (gas or liquid). Within any fluid, molecules are constantly moving at random speeds in random directions. When molecules collide with the side of the container, they exert a force against the side. The average force exerted per unit area is what we call pressure (Fig. 1.1):

$$\text{Pressure } (P) = \frac{F}{A} \tag{1.4}$$

The magnitude of this force depends on the number of molecules striking a unit area in a unit of time and on the average kinetic energy of those molecules. In reality, a system under pressure contains stored energy per unit volume, as shown below:

$$P = \frac{\text{Force} \times \text{Distance}}{\text{Area} \times \text{Distance}} = \frac{\text{Work (Energy)}}{\text{Volume}} \tag{1.5a}$$

In SI units, this energy density may be expressed as J/m^3, which is equal to Pascal (Pa), the unit of pressure. The term **normal stress** instead of pressure is used for solids.

A. **Standard Atmospheric Pressure:** The Earth is surrounded by a blanket of air. Because of its weight, this air exerts a pressure on all surfaces. At sea level on the Earth's surface, this pressure averages 101.4 kilopascal (kPa) or 14.7 lb_f/in^2 (also written psi for "pounds per square inch") or 1.01 bar (bar is a non-SI unit but currently accepted for use with the International System). A pressure of this magnitude is called a **standard atmospheric pressure** (atm). Atmospheric pressure decreases as altitude increases. Another commonly used method to express pressure is in terms of a "head" of a particular fluid. A head of 760 mm or 29.92 in Hg or 33.9 ft. H_2O is equivalent to a pressure of 1 atm. With these few factors, it is possible to convert any pressure from one unit to another. For example, to convert 28.6 psi to mm Hg, it may be written as:

$$(28.6 \text{ psi}) \left(\frac{760 \frac{\text{mmHg}}{\text{atm}}}{14.7 \frac{\text{psi}}{\text{atm}}} \right) = (28.6 \text{ psi}) \left(51.7 \frac{\text{mmHg}}{\text{psi}} \right) = 1479 \text{ mmHg}$$

B. **Absolute and Gauge Pressures:** In engineering, there are two pressure types: absolute and gauge pressures, differentiated by their measurement datum. **Absolute** pressure is measured in relation to absolute vacuum as its reference point. **Gauge** pressure uses local atmospheric pressure as its reference point. It represents the difference between absolute pressure and atmospheric pressure:

Fig. 1.1 Pressure definition

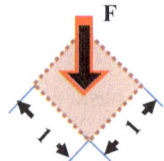

$$\text{Gauge Pressure} = \text{Absolute Pressure} - \text{Atmospheric Pressure} \qquad (1.5b)$$

Gauge pressure is used in common routine practices (e.g., pressure cooking, automobile tire, and blood pressure measurements), while absolute pressure is used for scientific observations and calculations. Given the varying nature of atmospheric pressure due to the weather and altitude, gauge pressure is less precise as opposed to absolute pressure, which is always definite. To distinguish between the two, a modifier in parentheses, such as "kPa (gauge)" or "kPa (absolute)," or units with appended "g" for gauge or "a" for absolute, such as "kPag," "kPaa" or "Barg," "Bara" are used. In the American engineering system of units, an absolute pressure is abbreviated as "psia," meaning "pounds per square inch, absolute," while a gauge is written as "psig," meaning "pounds per square inch, gauge." For use in equations of state such as the ideal gas law and other thermodynamic functions, pressure, like temperature, must also be expressed in terms of absolute values.

In some applications, knowledge of only the pressure difference between two locations or systems is needed, and it is often expressed as **differential** pressure.

C. **Vacuum:** It refers to any absolute pressure that is less than atmospheric. It can be expressed as a negative gauge pressure. Alternatively, the minus sign may be dropped and refer to the vacuum of the system, i.e., vacuum is the negative gauge pressure.

$$\text{Vacuum} = \text{Atmospheric Pressure} - \text{Absolute Pressure} \qquad (1.5c)$$

A perfect vacuum would have 0 absolute pressure or -1 atmosphere gauge pressure or one atmosphere of vacuum.

Example 3

Suppose we reduce the pressure inside a can to 10 lb_f/in^2 absolute. The gauge pressure will be $10 - 14.7 = -4.7$ lb_f/in^2. This is the same as the 4.7 lb_f/in^2 of vacuum. ◄

The various modes of pressure are illustrated in Fig. 1.2. Note the changeable nature of atmospheric pressure, which serves as the reference line for gauge pressure and may render gauge pressure less accurate.

1.2.4 Work, Energy, and Power

Work (w) is done if a force or portion of a force is applied to an object and the object is moved a measurable distance (s) in the direction of the applied force. The quantity of work done is:

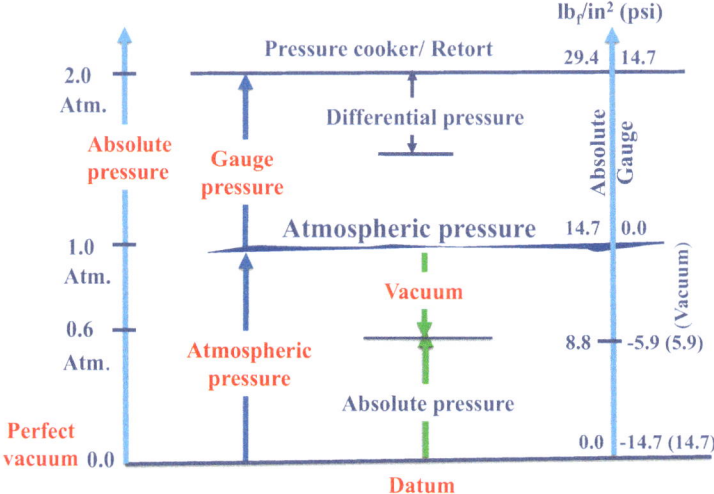

Fig. 1.2 Various modes of expressing pressure

$$w = F \cdot s \tag{1.6a}$$

When a force is exerted on an object at an angle to the horizontal, only a part of the force contributes to a displacement. Consider the force of a chain pulling an object rightward at an angle θ to move it to the right, Fig. 1.3. It is only the horizontal component of the pull force in the chain that causes the object to be displaced to the right. The horizontal component of the force is found by multiplying the force F by the cosine of the angle between F and s.

If a force is applied and the object does not move, no work is done. This is the case when you push on a large boulder, and it will not budge. If, for example, you push on an object and cause it to slide across the floor, you are exerting a force over a distance and hence doing work. If you push on an immovable block of granite, you may get tired, but if the block does not move, you are doing no work. Similarly, if an object is moving without an applied force, no work is done. For example, a satellite moves through space with no applied force.

$$w = F \cos(\theta) \cdot s \tag{1.6b}$$

Fig. 1.3 Work by component
of force and displacement

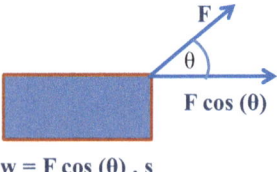

The SI unit for work is the Joule (J), which is the amount of work done by a force of 1 Newton acting over a distance of 1 m in the direction of displacement. A Joule is a Newton-meter or kg-m^2/s^2.

For a general case where force and displacement vary, it becomes:

$$w = \int_0^s F \cdot ds \qquad (1.6c)$$

which is the area under the force–displacement curve, as illustrated in Fig. 1.4.

Example 4

If a crate of apples is pushed with a force of 100 N and the crate moves 2 m, how much work is done?
   Solution: $w = F \cdot s = (100 \ N)(2 \ m) = 200 \ N \cdot m = 200 \ J$ ◄

Check Your Understanding

3. A crane lifts 100 kg of cheese 6 m off the floor. Verify that the crane does 5884 J of work on the cheese. Show all units and verify that they cancel correctly. ◄

Energy (E) is defined as the ability to do work or transfer heat. Equating energy with work may seem obvious, while equating energy with heat is a more difficult concept to grasp. Take the example of trying to push the large boulder. No work is done, but you will feel warmer for the effort put forth. The result is that energy in the form of heat is transferred from the body to the cooler surrounding air. Had you been able to move the boulder slightly, you would have expended energy in the form of work and heat. The SI unit for energy is the Joule (J). Calorie with a capital C is used to express the energy content of food, but there is often confusion between "calorie" with a small c and "food calorie or Calorie." The latter is a kilocalorie, a thousand small calories. In many parts of the world, food labels explicitly use kilocalories to avoid confusion. Many countries use kilojoules, a metric unit, on their food labels. One Calorie is equivalent to 4.184 kilojoules.

Power is a measure of the rate of work or energy expenditure. The SI unit for power is the Watt (W), which is a Joule per second (J/s). For historical reasons,

Fig. 1.4 Force-displacement graph for work

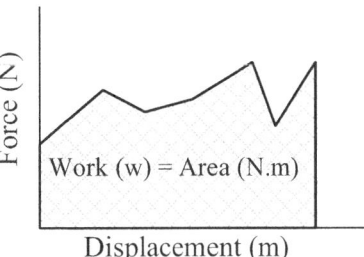

Force (N)

Work (w) = Area (N.m)

Displacement (m)

horsepower is often used to describe power. One horsepower is approximately equal to 750 W.

Example 5

If a 10-kg crate of apples is pushed and every second the crate moves 0.6 m, what is the power expended?

Solution: $Power = \frac{w}{t} = \frac{(10*9.8 \ N)(0.6 \ m)}{1 \ s} = 58.8 \ \frac{J}{s} = 58.8 \ W$ ◄

As indicated above, the definition of power is work done per unit time. Noting the definition of work as force × displacement, the expression for power becomes (force × displacement)/time. However, since velocity is displacement/time, power can also be written as

$$Power = (Force) \times (Velocity)$$

Check Your Understanding

Verify that the power required to lift 100 kg mass of cheese 6 m off the floor in 40 s is 147 W. Show all units and verify that they cancel correctly. ◄

1.3   Quantity of Materials

When a substance is dissolved in a solvent or added to a mixture, the amount of substance relative to the total volume or mass of the solution or mixture must be specified, as described below.

1.3.1   Concentration Based on Mass

Below is a description of the number of different ways that concentration based on mass can be expressed.

A. **Mass per unit volume** is the mass of a substance per unit volume of the total solution or mixture.

Example 6

If 20.2 g of sucrose is dissolved in enough water to make up 2 L of solution, what is the mass per unit volume concentration of sucrose?

Solution: $Concentration = \frac{Mass \ of \ substance}{Total \ volume \ of \ mixture} = \frac{20.2 \ g}{2 \ L} = 10.1 \ \frac{g}{L}$ ◄

B. **The mass fraction** is the mass of a substance per unit of mass of the total mixture.

$$\text{Mass Fraction(MF)} = \frac{\text{Mass of substance}}{\text{Total mass of mixture}} \tag{1.7}$$

There are two ways of expressing the mass fraction:

Mass Fraction, Wet, or "As-is" Basis (MF$_{wb}$): If a mixture has n components with masses of $m_1, m_2, \ldots, m_n$, then the mass fraction of component k on a *wet or "as-is" basis* is: expressed as:

$$\text{MF}_{wb} = \frac{m_k}{m_1 + m_2 + \cdots + m_n} = \frac{m_k}{\Sigma_{i=1}^{n} m_i} \tag{1.8}$$

Mass Fraction, Dry Basis (MF$_{db}$): The concentration of a component in a mixture based on the dry matter that would remain if all the water was removed. If m_n is m_{water}, then:

$$\text{MF}_{db} = \frac{m_k}{m_1 + m_2 + \cdots + m_{n-1}} = \frac{m_k}{\Sigma_{i=1}^{n-1} m_i} \tag{1.9}$$

Mass fraction is often multiplied by 100 and presented as a percent of the total mass. This representation of the concentration is termed **the mass percent**. It can be either on a wet or "as-is" basis or on a dry basis.

The mass percent of water in foods is also called **moisture content** and may be abbreviated as MC$_{wb}$ and MC$_{db}$ on wet or "as-is" and dry bases, respectively. To reiterate it, these may be simply computed as follows:

$$\text{MC}_{db} = \frac{\text{Mass of water}}{\text{Mass of solids}} \text{ and } \text{MC}_{wb} = \frac{\text{Mass of water}}{\text{Mass of product}} \tag{1.10}$$

The product in Eq. (1.10) means solids + moisture. To convert them into % MC$_{wb}$ and % MC$_{db}$, simply multiply by 100.

It is important to note that the generally used term "moisture content" invariably refers to wet basis moisture content.

Check Your Understanding

4. Show that: $\text{MC}_{db} = \frac{\text{MC}_{wb}}{1 - \text{MC}_{wb}}$ ◀

C. **Parts per million (ppm)** is the mass of a component per million units of mass of mixture or solution. This method of expressing quantity is used for components that are present in very low concentrations.

Table 1.4 Composition of a Cheddar cheese block

Component	Mass (g)
Water	37.4
Protein	26.2
Fat	32.5
Carbohydrate	2.1
Ash	3.7
Riboflavin	0.5×10^{-3}
Total	101.9

$$\text{ppm} = \left(\frac{\text{Mass of component}}{\text{Mass of mixture}}\right) 10^6 \qquad (1.11)$$

where both masses are expressed in the same units.

Example 7

A block of Cheddar cheese has the composition given in Table 1.4. Compute the mass fraction of water and moisture content of the product on wet and dry bases. What is the concentration of riboflavin in ppm?

Solution:

$$MF_{wb} = \frac{37.4 \ g \ water}{101.9 \ g \ total} = 0.367 \quad MF_{db} = \frac{37.4 \ g \ water}{101.9 \ g \ total - 37.4 \ g \ water} = 0.58$$

The mass percent water or moisture content is then 36.7% wet basis (MC_{wb}) or 58.0% dry basis (MC_{db}).

The concentration of riboflavin is $(0.5 \times 10^{-3})(10^6)/101.9 = 4.9$ ppm. ◀

Check Your Understanding

5. A batch of sugar solution has a mass of 24.82 kg. It contains 16.03 kg sucrose, 2.40 kg dextrose, and 0.002 kg preservative. The rest of the mass is water. Verify that the solution is:
   (a) 9.6% mass percent dextrose, wet basis, (b) 13.0% mass percent dextrose, dry basis, and (c) 81 ppm preservative. ◀

1.3.2   Concentration Based on Moles

As discussed previously, a quantity of a substance whose weight in grams equals its molecular weight is called a gram-mole (gram-mol) and abbreviated as a mole or mol in SI units. For instance, 32 g oxygen is 1 gram-mol. Similarly, a weight in kilograms equal to its molecular weight is called a kg-mole (kg-mol) and is used when dealing with large quantities of materials in industrial settings. In the American engineering system of units, the pound-mole (lb-mol) is similarly defined. Rather than express concentrations in terms of mass units, concentration can also be expressed in terms of mole units. This is particularly useful when chemical reactions are involved.

A. **Molarity** is defined as the number of moles of solute per liter of solution. The unit is given the symbol M.

$$\text{Molarity} = \left(\frac{\text{Number of moles of solute}}{\text{Liters of solution}} \right) \qquad (1.12)$$

Example 8

What is the molarity of a salt solution consisting of 29.25 g NaCl dissolved in water and brought to a volume of 1 L?

   Solution: The moles of NaCl can be obtained from its molecular weight (58.5). Then, the molarity can be computed as:

$$\text{Molarity} = \frac{(29.25 \ g \ NaCl)\left(\frac{1 \ mol \ NaCl}{58.5 \ g \ NaCl}\right)}{1 \ L} = 0.5 \ M \ NaCl$$

The solution is said to be a 0.5 molar NaCl solution. ◀

B. **Molality** is defined as the number of moles of material per kilogram of solvent:

$$\text{Molality} = \frac{\text{Number of solute moles}}{\text{Kilogram of solvent}} \qquad (1.13)$$

Example 9

What is the molality of a sugar solution consisting of 45 g glucose dissolved in 2 kg of water?

   Solution: The moles of glucose can be obtained from its molecular weight (180). Then, the molality can be computed as:

$$\text{Molality} = \frac{(45 \ g \ C_6H_{12}O_6)\left(\frac{1 \ mol \ C_6H_{12}O_6}{180 \ g \ C_6H_{12}O_6}\right)}{2 \ kg \ water} = 0.125 \ m \ C_6H_{12}O_6$$

The solution is a 0.125 molal glucose solution. ◀

C. **The mole fraction** is a ratio of the number of moles of a component to the total number of moles of all components in a solution or mixture. If there are a total of c components in a mixture with n_1 moles of component 1, n_2 moles of component 2, and n_c moles of component c, then the mole fraction of component k is:

$$\text{Mole fraction of component } k = \frac{n_k}{n_1 + n_2 + \cdots + n_c} = \frac{n_k}{\Sigma_{i=1}^{c} n_i} \qquad (1.14)$$

When referring to solids and liquids, the symbol x_i is used to represent the mole fraction of component i. When referring to gases, the symbol y_i is used to represent the mole fraction of component i. The mole fraction may also be presented as a percent of the total moles and then the concentration is termed **the mole percent**.

Example 10

A particular air sample contains 21% O_2, 78% N_2, and 1.0% CO_2 by weight. What is the mole fraction of O_2?

   __Solution__: The major constituents of air are O_2, N_2, and CO_2, and the molecular weights of these constituents are 32, 28, and 44, respectively. A basis of 1 kg of air will be used to compute the mole fraction of O_2.

$$(1\,kg\,air)\left(\frac{0.21\,kg\,O_2}{1\,kg\,air}\right)\left(\frac{1\,kg\text{-}mol\,O_2}{32\,kg\,O_2}\right) = 0.00656\,kg\text{-}mol\,O_2$$

$$(1\,kg\,air)\left(\frac{0.78\,kg\,N_2}{1\,kg\,air}\right)\left(\frac{1\,kg\text{-}mol\,N_2}{28\,kg\,N_2}\right) = 0.02786\,kg\text{-}mol\,N_2$$

$$(1\,kg\,air)\left(\frac{0.01\,kg\,CO_2}{1\,kg\,air}\right)\left(\frac{1\,kg\text{-}mol\,CO_2}{44\,kg\,CO_2}\right) = 0.00023\,kg\text{-}mol\,CO_2$$

Mole fraction O_2

$$= \frac{0.00656\,kg - mol\,O_2}{0.00656\,kg - mol\,O_2 + 0.02786\,kg - mol\,N_2 + 0.00023\,kg - mol\,CO_2}$$

$$= 0.18 \frac{kg - mol\,O_2}{kg - mol\,air}$$

The mole percent of O_2 in the air sample is 18%. ◄

D. **Molar Volume** is defined as the volume occupied by 1 mole of a substance. Some possible ways of expressing this concentration are m^3/mol, ft^3/mol, or L/mol. Recall that at standard temperature and pressure (0 °C and 1 atm), an ideal gas has a molar volume of 22.415 L/mol.

1.4   Density

Density is a measure of the mass per unit volume of a substance. It is commonly abbreviated with the Greek letter ρ (rho).

$$\text{Density } (\rho) = \frac{\text{Mass (m)}}{\text{Volume (V)}} \tag{1.15}$$

Typical units of density are kg/m^3, g/cm^3, and lb_m/ft^3. At 4 °C, the density of water is 1000 kg/m^3, 1 g/cm^3, or 62.4 lb_m/ft^3. These values of density of water are often needed in food engineering computations and should be committed to memory.

The density of solids and liquids varies only slightly with changes in temperature and pressure, and they are considered "incompressible." On the other hand, the densities of gases vary greatly with changes in temperature and pressure. Increases in pressure result in decreases in volume and, therefore, increases in density. Increases in temperature result in increases in volume and, therefore, decreases in density. The values for some selected products are shown in Table 1.5.

The density of a mixture with no volume change on mixing may be approximated by averaging the densities as shown below:

$$\rho_{ave} = \sum x_i \rho_i \tag{1.16}$$

Table 1.5 Typical density of selected items

	g/cm^3
Gases	
Air (0 °C, 1 atm)	1.29×10^{-3}
Carbon dioxide	1.98×10^{-3}
Liquids	
Alcohol, ethyl	0.79
Water, sea	1.03
Milk	1.03
Mercury	13.69
Solids	
Wood	~0.55
Glass	2.60
Steel	7.80
Food components	
Fat	~0.96
Proteins	~1.30
Carbohydrates	~1.50
Sucrose	~1.60
Fiber	~1.85
Salt	~2.20

In food systems, density is a highly variable property. It changes considerably with the processing techniques used in manufacturing operations, where porosity and void volumes are often manipulated to attain some desirable mechanical attributes, such as texture, in the final product. Additionally, for example, instantization via agglomeration of dry powder particles improves the rate of their solubilization, but it also reduces the density of the powder. Food density is critical for packaging, as food products are typically packaged according to volume but sold by weight. A product's density must be accurately quantified to ensure that packages contain the same quantity of food and meet the legal requirements. There are three types of densities that are used to characterize a product (Fig. 1.5).

1.4.1   True (or Solid) Density and Apparent Density

The true density or just density is the mass of a material divided by its volume, excluding any trapped air or other foreign items. For most foods, it ranges from 0.95 to 1.6. Density of materials with regular geometries may be determined from their characteristic dimensions and mass. Very often, it is not easy to determine a material's true density, especially when it is a composite material or has been transformed into various products such as dehydrated, baked, and extruded foods. In such cases, apparent density in lieu of true density is often used. The apparent density may be estimated by grinding the material enough to ensure that no air cells remain and then sieving through a 250 μm (60-mesh) screen. A graduated cylinder is then filled to the line with the ground samples, and the edge of the cylinder is tapped repeatedly until the sample is completely settled. The weight and volume of the sample in the cylinder are used to calculate the apparent density.

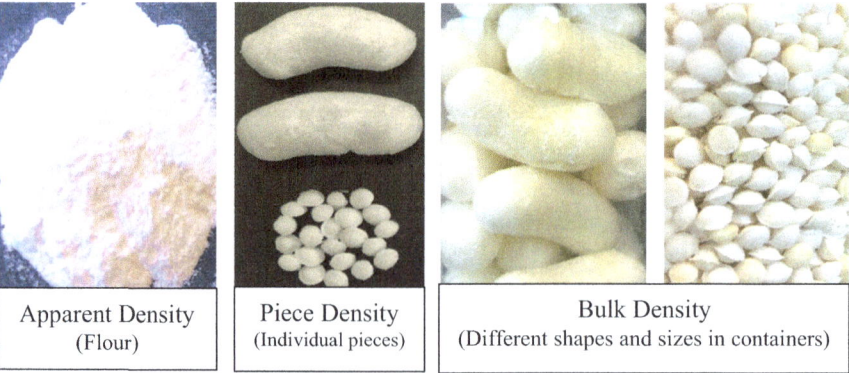

Apparent Density	Piece Density	Bulk Density
(Flour)	(Individual pieces)	(Different shapes and sizes in containers)

Fig. 1.5 Types of densities for characterization of food materials

1.4.2 Piece or Particle Density

Piece density is defined as the apparent density of a piece or a particle that may contain pores internally. It is computed as:

$$\rho_{particle} = \frac{\text{Mass of particle (piece)}}{\text{Volume of particle (piece)}} \quad (1.17)$$

If the samples have uniform shapes, such as cylindrical/rectangular or spherical shapes, the dimensions of the samples may be measured, and the volume can be calculated using the standard equations. For samples that are not of uniform shape, the volume can be estimated using the solid or liquid displacement methods. A simple rapeseed (or glass beads or sand) displacement is frequently used for solid particle density measurements. A 25 or 50 mL graduated cylinder is first filled to the line with the rapeseeds. The rapeseeds are then temporarily placed into a holding container. A known mass of the sample under study is placed in the graduated cylinder, the rapeseeds are poured back into the graduated cylinder to the top line, and the volume of the remaining rapeseed is measured. The volume of the sample is calculated by subtracting the volume of rapeseeds from the volume of the graduated cylinder. From knowledge of the mass and volume, density is then computed.

1.4.3 Bulk Density

Bulk density is defined as the mass of particles occupying a unit volume and thus accounts for the interparticle void space not occupied by the food. It is not an intrinsic property of a food; it varies depending on how the food is processed and the size and shape of the particles and their density. The bulk density of granular foods such as powders is usually reported both as "freely settled" (or "poured" density) and "tapped" density obtained after a specified tapping, which causes compaction. Examples of the bulk density of some food products are shown in Table 1.6.

Table 1.6 Typical bulk density of selected products

Product	g/cm^3
Coffee, beans (green)	0.67
Coffee, ground	0.40
Corn, ear	0.45
Corn, shelled	0.72
Cabbage	0.50
Carrots	0.55
Potatoes	0.67
Onions	0.65
Milk, dry powder	0.32
Sugar, granulated	0.80

The volume not occupied by the food is the void volume and is reported as porosity (ε_0) and is calculated as follows:

$$\text{Porosity} (\varepsilon_0) = \frac{\text{True or apparent density} - \text{Bulk desity}}{\text{True or apparent density}} = 1 - \frac{\text{Bulk density}}{\text{True or apparent density}} \quad (1.18)$$

Some porosity (air cells) exists in almost all foods. These cells may exist in three forms: closed cells (the space is enclosed from all sides), blind cells (one end is open), and open cells (interconnected void spaces). Porosity and density are the most common structural properties used to describe the qualities of many foods. Additionally, they are needed in the design of processing and handling equipment, aeration, storage, and transportation.

1.4.4   Specific Volume

The specific volume is defined as the volume per unit mass and is simply the reciprocal of density.

$$\text{Specific Volume} \left(\overline{V} \right) = \frac{\text{Volume}}{\text{Mass}} = \frac{1}{\rho} \quad (1.19)$$

Typical units of specific volume are m^3/kg, cm^3/g, and ft^3/lb_m. The specific volume of water at 4 °C is 0.001 m^3/kg, 1 cm^3/g, or 0.016 ft^3/lb_m. Steam and refrigerant tables typically report specific volumes rather than density. The specific volume values are useful in many real-life applications. For example, the density of liquid water at 100 °C is approximately 1.0 kg/L, but when converted into saturated steam, its density decreases to 0.0006 while its specific volume increases to 16,673 L/kg. The latter number is much easier to work with and conceptualize processes that involve low-density fluids like gas and steam.

1.4.5   Specific Gravity or Relative Density

It is often useful to express the density of a material relative to a reference material at a specified temperature. This ratio of the densities is termed the specific gravity or relative density.

$$\text{Specifig Gravity (SG)} = \frac{\text{Density of material}}{\text{Density of reference material}} = \frac{\rho}{\rho_{ref}} \quad (1.20)$$

For solids and liquids, the reference material is usually water at 4 °C because water is so common that most people have a fair idea of its density: "1 cm^3 has a mass of gram." A substance with a specific gravity of <1 will float in water. The reference for gases is usually air. Since specific gravity is a relative measure, it is **dimensionless**. It is used to determine the density of an unknown substance from the known density of the reference material.

To be precise, it is necessary to represent the specific gravity with the following notation: $[Cd(OH)_2] = 4.79_{4^0}^{15^0}$. This means that cadmium hydroxide at 15 °C is 4.79 times as dense as water at 4 °C.

Specific gravity measurements also provide information about the concentration of solutions of materials in a noninvasive way and are used in a wide variety of industries. In the food industry, the specific gravity of milk, fruit juices, brines, sugar solutions, and other liquids is commonly measured using a **hydrometer** (Fig. 1.6), based on Archimedes' principle, which states that an object such as a hydrometer suspended in a fluid will sink until it displaces its own weight. A hydrometer is made of glass and consists of a cylindrical stem and a bulb weighted with lead shot, acting as a ballast, to make it float upright in a liquid. In denser fluids, it will float higher (displacing less fluid needed to equal its weight), and in less dense fluids, it floats lower. A hydrometer is thus calibrated to float to different heights in liquids of different specific gravity (or density) and provides a direct reading of the specific gravity (or sometime density) of the liquid. Hydrometers are calibrated to different scales for different uses. A **lactometer** is a hydrometer used for the measurement of milk specific gravity. The refractive index is a material specific property which is easy to measure. The refractive index of a solution is usually directly proportional to its concentration and thus it is utilized to characterize sugar and water, water and alcohol, and other soluble solids containing liquids. The accuracy of the measured

Fig. 1.6 A typical hydrometer

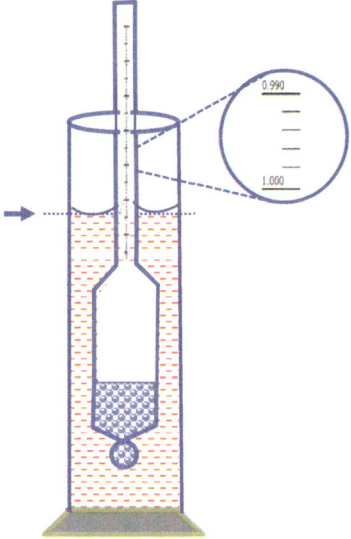

value decreases as mixtures become more complex but still good enough for quality control and process evaluation. A **Brix refractometer,** introduced in 1897 by Adolf F. Brix, is extensively used in fruit juice, wine making and related industries to measure sugar content. It is calibrated to read in degrees of Brix (°Bx), which is based on percentage by weight of sugar in the solution (e.g., 1 °Bx = one gram sugar per 100 grams of solution) at 20 °C. The measurement is done by placing a few drops of the solution on a prism, and the angle of refraction is calibrated to directly read the concentration of sugar in the liquid. The Brix scale is frequently used to express the sugar content of many kinds of beverages like fruit juices, wines, jelly and jams and jellies, honey, etc. Handheld refractometers are the most common because they are compact, simple to use, low maintenance and provide instant results.

1.5 Classification of Matter

Anything that occupies space and has mass and occupies space is known as matter. It can exist in three forms or phases:

1.5.1 Gas

A gas is a fluid that does not have a definite volume and expands to fill its container. A gas will not form a top surface. Gases are compressible, with volumes being approximately proportional to their temperature and to the inverse of their pressure. A gas at a temperature and pressure near its liquid phase is generally called a **vapor,** e.g., steam. A gas may be considered a superheated vapor far away from its liquid phase. Thus, atmospheric air consisting primarily of nitrogen and oxygen is considered a gas since it is far removed from the liquid phase of both its primary constituents.

The gaseous state of matter that contains electrically charged particles is called **plasma** and is thought of as a fourth state of matter.

1.5.2 Liquid

A liquid is a fluid that flows, has a definite volume, and takes the shape of its container. When placed in a container, the liquid tends to form an almost flat top surface (low surface tension). In general, liquids are nearly incompressible, and large increases in pressure result in very small changes in volume.

A **fluid** may be either a gas or a liquid that does not resist distortion of shape because its intermolecular cohesive forces are not strong enough. Therefore, a fluid flows under the slightest stress and assumes the shape of the container in which it is placed.

1.5.3 Solid

A solid is rigid and possesses a definite shape, unlike a gas or liquid. The atoms or molecules making up a solid are locked into place by intermolecular forces and possess a fixed volume.

Because of their low compressibility, liquid and solid are collectively termed **condensed matter**. A matter may undergo phase transitions based on the heat content of the system. Figure 1.7 shows these phase transitions and the nomenclature used to describe them. Some of the more useful thermophysical properties of gases, liquids, and solids are given in Appendix A.2.

1.6   Selected Properties of Gases

As one of the three classical states of matter, gases are unique in that they do not possess any definite shape or volume but fill all the space around them. Gases are integral parts of foods and drinks for several reasons. Nitrogen, oxygen, and carbon dioxide alone or in various combinations are used to extend food shelf life. The aeration or "leavening" of bread dough is achieved by yeast-generated carbon dioxide. It also plays an important role in determining the taste of beer by changing the pH value via the formation of carbonic acid. All gases exhibit some common behaviors that are governed by a few laws. Understanding these laws at a macroscopic level is helpful in handling them correctly in a variety of physical and chemical processes used in food manufacturing and packaging operations.

1.6.1 Ideal Gas

An ideal gas is defined as one in which the volume occupied by the molecules is negligible compared to the space between them and in which there is negligible attraction or repulsion between molecules. At high temperatures and low pressures, most gases obey these laws closely. Common gases do so well at room temperature and atmospheric pressure. At higher pressures and lower temperatures, the molecules of gas come closer together, occupy a more significant portion of the space, and begin to interact with each other. This leads first to deviations from the gas law, then to liquefaction or solidification.

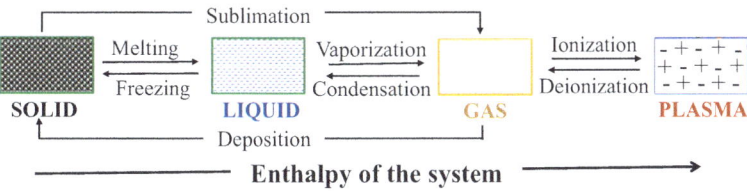

Fig. 1.7 States of matter and their transitions

A. **Boyle's Law:** In 1662, Robert Boyle found that at a constant temperature, the volume of a fixed quantity of gas was inversely proportional to its pressure. In other words, if you double the pressure of a quantity of gas, it will occupy half the volume. This relationship is known as Boyle's law and can be expressed as follows:

$$V \propto \frac{1}{P} \text{ at constant } T \quad V = K\frac{1}{P} \tag{1.21}$$

where V = volume, P = pressure, and K is a proportionality constant that depends on the units used, the nature of the gas, and the temperature. If a quantity of gas has volume V_1 at pressure P_1 and temperature T, then the volume at any other pressure P_2 at the same temperature can be obtained from the equation:

$$\frac{P_1}{P_2} = \frac{V_2}{V_1} \tag{1.22}$$

Check Your Understanding

6. *A cylinder contains 2 L of air at 25 °C and 305 kPa pressure. A piston reduces the volume to 1.5 L, still at 25 °C. Verify that its pressure is now 407 kPa. Verify that this is just over 4 atmospheres.* ◄

B. **Charles' Law:** In 1787, Charles studied the effect of temperature change on hydrogen, oxygen, carbon dioxide, and air and noticed that when he heated the gas from 0 °C to 90 °C at constant pressure, each gas expanded by the same amount, i.e.

$$\frac{dV/V}{dT} = \text{constant} = \beta = 2.1 * 10^{-4} \left(\frac{cm^3/cm^3}{°C} = \frac{1}{°C} \right) \tag{1.23}$$

where this constant is independent of the gas studied. He called this constant (β) the coefficient of thermal expansion.

C. **Gay-Lussac's Law:** In 1802, Joseph Gay-Lussac found that this relationship held for all gases and presented the following analysis. From Eq. (1.23):

$$\beta = \frac{\Delta V/V_0}{\Delta T} = \left(\frac{1}{V_0}\right)\left(\frac{\Delta V}{\Delta T}\right) = \left(\frac{1}{V_0}\right)\left(\frac{V - V_0}{T - T_0}\right) \tag{1.24}$$

where T_0 is the initial temperature (0 °C in the experiments), T is the final temperature, V_0 is the volume at T_0, V is the volume at T, and β is the coefficient of thermal expansion.

Solving for V, the relationship becomes:

$$V = V_0 + V_0\beta(T - T_0) \tag{1.25}$$

From the experiments, Gay-Lussac estimated the value of β to be 1/250. Subsequent work has shown that for "ideal" gases, the coefficient is actually 1/273.15. These observations can be summarized with the equation:

$$V = V_0 + V_0\beta T = V_0 + \frac{V_0}{273.15}T \text{ (at constant pressure)} \tag{1.26}$$

Figure 1.8 illustrates this principle. The solid lines labeled P_1 and P_2 represent the observed volume-temperature relationships for the same gas at two different pressures. The lines intercept the 0 °C line at values we have called V_0. The slope of each line is $V_0/273.15$. If you project the lines backward, they reach 0 volume at -273.15 °C.

D. **The Kelvin Scale:** These observations led to the defining of a new temperature scale with degrees of the same size as the Celsius scale but with 0 equal to -273.15 °C as in Fig. 1.9.

Fig. 1.8 Volume versus temperature

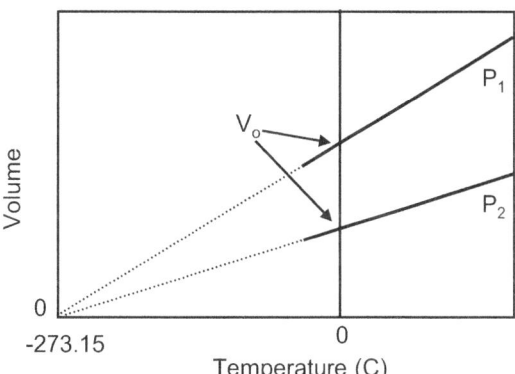

Fig. 1.9 The Kelvin (absolute) scale

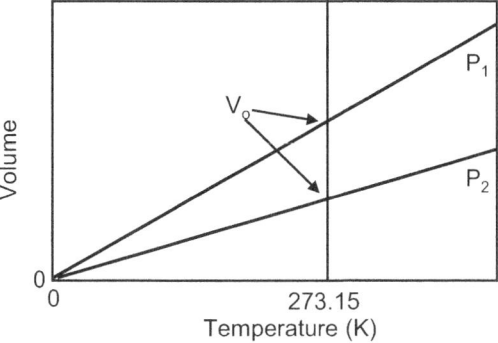

Using this new scale, now called the *Kelvin scale* or absolute scale, the temperature-volume relationship simplifies to:

$$V = \frac{V_0}{273.15} T \tag{1.27}$$

Since, for a fixed quantity and pressure of gas, $V_0/273.15$ is a constant, this equation states that the volume of a fixed quantity of gas is proportional to its temperature on the Kelvin scale. This equation implies that at 0 K, the volume of gas would become 0. In fact, all real gases liquefy or solidify before this temperature, so the 0 volume is never reached.

If a quantity of gas has volume V_1 at pressure P_1 and temperature T_1, then the volume at any other temperature T_2 at the same pressure can be obtained from the equation:

$$\frac{T_1}{T_2} = \frac{V_1}{V_2} \tag{1.28}$$

Note that because of the direct relationship between pressure and volume, the order of the subscripts is the same in both fractions.

E. **The Rankine Scale:** In the American engineering system, the absolute temperature scale is called the *Rankine scale*. It is defined to have the same units as the Fahrenheit scale but with 0 equal to $-459.7\ °F$.

Check Your Understanding

7. Twenty liters of CO_2 at 30 °C and 50 kPa pressure was heated to 150 °C. Verify that if the pressure remains unchanged, the gas must now occupy 27.9 L. ◄

F. **Avogadro's Law:** In 1811, Avogadro proposed that, at the same temperature and pressure, equal volumes of any gas contained the same number of molecules. He found that 1 g-mole of a gas contained 6.02×10^{23} molecules. This number is now called *Avogadro's number*. If a gas is at standard temperature and pressure (STP = 0 °C and 1 atm, an arbitrarily chosen reference point), then 1 g-mole of any gas occupies 22.4 L. One kg-mole occupies $22.4\ m^3$. One lb-mole at the STP will occupy $359\ ft^3$.

Example 11

What is the density of methane gas at 20 °C and 22 psi?
   Solution: *Methane gas has the formula CH_4, so its molecular weight is:*

$$1(12.0) + 4(1.0) = 16.0$$

Thus, by Avogadro's law, 16 grams of methane will occupy 22.4 L at STP. At 20 °C, Gay-Lussac's law tells us that its volume is

$$V = (22.4L)(273 + 20)/273 = 24.0 \, L$$

At 22 psi, Boyle's law tells us that the volume becomes

$$V = (22.4L)(14.7 \, psi)/22psi = 16.0 \, L$$

The density is, therefore, $\rho = \frac{16.0 \, g}{16.0 \, L} = 1.0 \, g/L$ ◄

G. **The Ideal Gas Law:** The preceding laws can be summarized as follows:

Charles' law: Volume $\propto$ Temperature
Boyle's/Gay-Lussac's law: Volume $\propto \frac{1}{Pressure}$
Avogadro's law: Volume $\propto$ Number of moles

These laws can be combined into a single equation, and with a constant, it becomes:

$$V = R\frac{nT}{P} \text{ or } PV = nRT \tag{1.29}$$

where R is a proportionality constant whose value depends on the units of the variables in the equation, but which is independent of the nature of the gas. For a single mole, this equation is usually expressed in the form:

$$P\widehat{V} = RT \tag{1.30}$$

where $\widehat{V}$ is the molar volume (V/n) and solving for P:

$$P = R\frac{T}{\widehat{V}} \tag{1.31}$$

Figure 1.10 is a three-dimensional plot of Eq. (1.31) and graphically displays the ideal gas law. Solving Eq. (1.31) for R gives:

$$R = \frac{PV}{nT} \tag{1.32}$$

Equation (1.32) indicates that the ratio PV/nT is always a constant. For a fixed quantity of gas n is a constant, so it may also be written as:

$$nR = \frac{PV}{T} = \text{constant} \tag{1.33}$$

Fig. 1.10 The ideal gas law

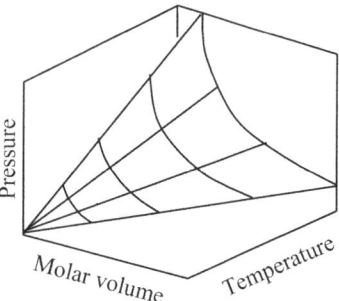

Thus, if a fixed quantity of gas has volume $= V_1$ at pressure $= P_1$ and temperature $= T_1$, then the volume at any other temperature and pressure can be obtained from the equation:

$$\frac{P_1 V_1}{T_1} = \frac{P_2 V_2}{T_2} \tag{1.34}$$

Example 12

A quantity of O_2 gas at 3.00 atmospheres pressure and 20 °C occupied 50 ft^3. What is its volume at 1.00 atmospheres and 50 °C?
  Solution: *Change the temperatures to the Kelvin scale thus:*

$$20 + 273.15 = 293.15 \text{ K, and } 50 + 273.15 = 323.15 \text{ K}$$

Solve Eq. (1.34) for V_2: $\frac{P_1 V_1}{T_1} = \frac{P_2 V_2}{T_2}$ *or* $V_2 = \frac{P_1 V_1}{T_1}\left(\frac{T_2}{P_2}\right)$
Now substitute the given values to get: $V_2 = \frac{(3\ atm)(50\ ft^3)}{(293.15\ K)}\left(\frac{323.15\ K}{1\ atm}\right) = 165.35\ ft^3$

◀

Check Your Understanding

8. Nitrogen gas (4 ft^3) at 58 °F and 2.8 psig is heated to 78 °F at 12.8 psig. Using the Rankine scale, verify that it now occupies 2.64 ft^3. ◀

H. **Other Forms of the Ideal Gas Law:** In the ideal gas law, the mole (n) is equal to mass divided by molecular weight, so it may be written as:

$$PV = nRT = \frac{m}{M} RT \tag{1.35}$$

where m = mass and M = molecular weight. On rearranging it becomes:

$$PM = \frac{m}{V} RT \tag{1.36}$$

However, mass divided by volume is density, so we can now write:

$$PM = \rho RT \tag{1.37}$$

where ρ = density. For an ideal gas, we can calculate its density from its molecular weight (M) as follows:

$$\text{Density} = \rho = \frac{PM}{RT} \tag{1.38}$$

Conversely, we can calculate the molecular weight of a gas as follows:

$$\text{Molecular Weight} = M = \frac{\rho RT}{P} \tag{1.39}$$

Check Your Understanding

9. Verify that at 1 atmosphere pressure and 20 °C, oxygen has a density of 1.33 g/L. (Verify both the quantity and the units.)
10. 0.178 lb. of an unknown gas is placed in a 1.8 ft^3 container. Its temperature is 75.1 °F, and its pressure is 20.2 psia. Verify that it could be nitrogen. ◀

I. **Evaluating the Gas Constant:** The proportionality constant R in the combined gas law is of great importance, so we need to determine its value and interpret its meaning. The value of R depends on the units of P, V, and T. Avogadro showed that at $T = 0\,°C$ (273.15 K) and $P = 1$ atm, 1 g mole of any gas occupies 22.415 L. Substituting these values into the equation, we obtain:

$$R = \frac{PV}{nT} = \frac{(1 \text{ atm})(22.415 \text{ L})}{(1 \text{ g} - \text{mole})(273.15 \text{ K})} = 0.08206 \frac{\text{L} - \text{atm}}{\text{mole} - \text{K}} \tag{1.40}$$

Interpreting the Gas Constant: You may be familiar with the calorie as the quantity of heat needed to raise 1 g of water by 1 °C (or Kelvin). R plays a similar, although not identical, role. It can be viewed as the quantity of energy needed to raise a mole of gas by 1°. To see why this is so, note that in Eq. (1.40), the numerator is pressure times volume. However, pressure = force/area or force/length^2 and volume = length^3. Substituting these for PV, we obtain:

Table 1.7 Values for the gas constant (R)

| Units of work or energy | | Units of | Units of | | |
Pressure	Volume	quantity	temperature	Value of R	Units of R
Atmosphere	Liter	gm-mole	Kelvin	0.08206	liter atm/g-mole K
Atmosphere	cm^3	gm-mole	Kelvin	82.057	cm^3 atm/g-mole K
Atmosphere	m^3	kg-mole	Kelvin	0.08206	m^3 atm/kg-mole K
Pascal	m^3	kg-mole	Kelvin	8314.34	m^3 Pa/kg-mole K
lb_f/ft^2	ft^3	lb-mole	Rankine	1545.3	ft lb/lb-mole R
Joule		kg-mole	Kelvin	8314.34	J/kg-mole K
Btu		lb-mole	Rankine	1.9872	Btu/lb-mole R
ft lb_f		lb-mole	Rankine	1545.3	ft lb_f/lb-mole R

$$PV = \frac{\text{Force}}{\text{Length}^2} * \text{Length}^3 = \text{Force} * \text{Length} = \text{Work (Energy)} \qquad (1.41a)$$

Thus, PV is a measure of work performed by expanding gas. It is also a measure of the energy needed to perform that work, so we can write:

$$R = \frac{PV}{nT} = \frac{\text{Energy}}{\text{Moles} * \text{Degrees}} \qquad (1.41b)$$

In other words, R is the energy needed to raise the temperature of 1 M of gas by one degree. Of course, the actual value of R will depend on the units selected for P, V, T, and n. Table 1.7 lists some values of R, and values in other units can be similarly computed.

1.6.2 Gas Mixtures

In many situations, mixtures of gases need to be handled. Air is one mixture consisting of oxygen, nitrogen, carbon dioxide, and a few minor gases. Fruits are often stored in modified atmospheres to slow their respiration processes to extend shelf life and reduce spoilage. At times, it may be necessary to deal with a mixture of water vapor and one or more other gases. Two laws, Dalton's and Amagat's, that are useful in dealing with mixtures of gases are presented here.

Dalton's and Amagat's laws are both valid when the mixture and each of its components behave as ideal gases. Air at normal temperature and pressure is reasonably ideal, as is each of its components. At higher pressures and lower temperatures, air ceases to be ideal. When a mixture is not ideal, it is still possible that one law or the other may apply but not both.

A. **Partial Pressure:** Imagine a 10-L tank containing nitrogen gas at 79.1 kPa pressure, a second 10-L tank containing oxygen gas at 19.2 kPa pressure, and a 10-L third tank containing carbon dioxide at 3.0 kPa as illustrated in Fig. 1.11. If the contents of all three tanks are pumped completely into a single 10-L tank, the resulting mixture will have a pressure of 101.3 kPa. Although the three gases are now thoroughly mixed, it is still convenient to imagine that each is exerting its own individual pressure, just as it did in its own tank. We call these pressures partial pressures.

B. **B. Dalton's Law of Partial Pressures:** Dalton's law states that the total pressure exerted by a gaseous mixture is equal to the sum of the partial pressures, i.e.,

$$P_T = P_1 + P_2 + P_3 + \cdots + P_n \text{ or } P_T = \Sigma_{i=1}^{n} P_i \tag{1.42}$$

where P_T is the total pressure of the mixture and P_1, P_2, P_3, etc., are the partial pressures of the gases in the mixture. In other words, partial pressures are additive.

Example 13

For the mixture of nitrogen, oxygen, and carbon dioxide described above, we can write:

$$P_T = P_{N2} + P_{O2} + P_{CO2} = 79.1 + 19.2 + 3.0 = 101.3 \text{ kPa} \qquad \blacktriangleleft$$

C. **Moles and Pressures of a Mixture:** When the ideal gas law applies to each component of the mixture, we can state that:

$$P_1 = \frac{n_1 RT}{V_T}, P_2 = \frac{n_2 RT}{V_T}, P_3 = \frac{n_3 RT}{V_T}, \ldots, P_i = \frac{n_i RT}{V_T}, \ldots \tag{1.43}$$

where each term of the equations in (1.42) refers to one of the components. Substituting these equations into Dalton's law, we have Eq. (1.44):

Fig. 1.11 Partial pressures in gas mixtures

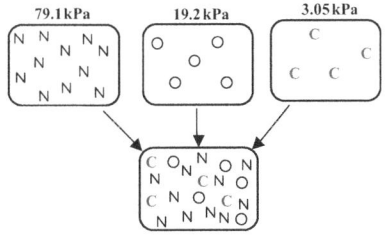

Total Pressure =101.3 kPa

$$P_T = \frac{n_1 RT}{V_T} + \frac{n_2 RT}{V_T} + \frac{n_3 RT}{V_T} + \ldots \tag{1.44}$$

This equation tells us that the total pressure of a mixture can be computed by applying the ideal gas law to each component and adding the results. However, R, T, and V_T are the same for each component, so we can factor RT/V_T out of the terms of Eq. (1.44); thus,

$$P_T = (n_1 + n_2 + n_3 + \ldots) \left(\frac{RT}{V_T} \right) \tag{1.45}$$

which states that the total pressure can be obtained by applying the ideal gas law to the sum of the moles of the components.

Now, select one of the components, namely, component i, and select its equation from (1.43). Divide the selected equation by Eq. (1.45):

$$\frac{P_i}{P_T} = \frac{\frac{n_i RT}{V_T}}{(n_1 + n_2 + n_3 + \ldots) \left(\frac{RT}{V_T} \right)} \tag{1.46}$$

and solve for P_i, the partial pressure of component i is then:

$$P_i = \left(\frac{n_i}{n_1 + n_2 + n_3 + \ldots} \right) P_T = \left(\frac{n_i}{\Sigma n_i} \right) P_T = y_i P_T \tag{1.47}$$

where $y_i = \frac{n_i}{\Sigma n_i} = \frac{n_i}{n_1 + n_2 + n_3 + \ldots} = $ mole fraction of gas i.

When the ideal gas law applies, the partial pressure of each gas is equal to the mole fraction of that gas times the total pressure of the mixture. Rearranging Eq. (1.47) gives:

$$y_i = \frac{P_i}{P_T} \tag{1.48}$$

If there is a device that measures partial pressure, we can use that device and Eq. (1.48) to determine the mole fraction of a component of a gas mixture.

D. **Partial Volume:** Imagine a 3-L tank of oxygen, an 11-L tank of nitrogen, and a 2-L tank of carbon dioxide, all at the same pressure, say 2 atmospheres, as shown in Fig. 1.12. Now, completely pump the contents of all three tanks into a single tank that holds 16 L. The pressure will still be 2 atmospheres. Although thoroughly mixed, one can imagine that each gas still occupies a portion of the total volume. This portion is called the partial volume of the gas.

Fig. 1.12 Partial volumes in gas mixtures

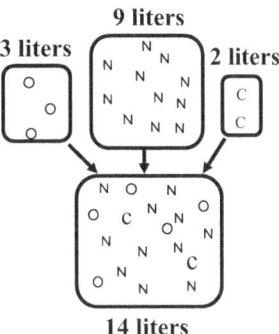

E. **Amagat's Law of Partial Volume** states that the total volume occupied by a mixture of gases at a constant pressure is equal to the sum of their partial volumes:

$$V_T = V_1 + V_2 + V_3 + \ldots + V_n = \Sigma_{i=1}^{n} V_i \tag{1.49}$$

where V_T = the total volume of the mixture and V_i = the partial volume of the ith gas in the mixture. Thus, partial volumes like partial pressures are also additive.

F. **Moles and Volumes of a Mixture:** If the components of a mixture of gases behave ideally, one can write the ideal gas law for the mixture and for each component of a mixture, thus:

$$PV_T = n_T RT, \text{ and } PV_1 = n_1 RT, PV_2 = n_2 RT, PV_3 = n_3 RT, \ldots \tag{1.50}$$

Add these equations for each component:

$$PV_T = P(V_1 + V_2 + V_3 + \ldots) = RT(n_1 + n_2 + n_3 + \ldots) \tag{1.51}$$

Now, if one component and its equation are selected from Eq. (1.50) and then divided by Eq. (1.51), it becomes:

$$\frac{PV_i}{P(V_1 + V_2 + V_3 + \ldots)} = \frac{n_i RT}{RT(n_1 + n_2 + n_3 + \ldots)} \tag{1.52}$$

Using Amagat's law, $(V_1 + V_2 + V_3 + \ldots)$ is then replaced with V_T. Since Ps cancel on the left and RTs cancel on the right, the result is:

$$\frac{V_i}{V_T} = \frac{n_i}{n_1 + n_2 + n_3 + \ldots} \tag{1.53}$$

Furthermore, $\frac{n_i}{n_1 + n_2 + n_3 + \ldots}$ = mole fraction of gas $i = y_i$, so

$$V_i = y_i V_T \tag{1.54}$$

This equation signifies that the partial volume of a gas in a mixture is proportional to the mole fraction of that gas. If the total volume of a gas mixture and the volume occupied by the ith component of the gas can be measured, the mole fraction can be computed as follows:

$$\text{Mole fractions} = y_i = \frac{V_i}{V_T} \tag{1.55}$$

This indicates that if 20% of the molecules in a mixture are oxygen, then 20% of the volume of the mixture is occupied by oxygen.

G. **Average Molecular Weight of a Gas Mixture:** Consider a system such as air that consists of a mixture of gases. Although such a system consists of molecules such as oxygen, nitrogen, and carbon dioxide with different molecular weights, it is sometimes convenient to think of the system as containing just one type of molecule, say "air" molecules. If the weight and chemical composition of the system do not change, such a simplified view of the system simplifies many problems. Taking this view, one can then assign these molecules an average molecular weight computed with the following equation:

$$M_{\text{average}} = y_1 M_1 + y_2 M_2 + y_3 M_3 + \ldots + y_n M_n = \Sigma_{i=1}^{n} y_i M_i \tag{1.56}$$

where y_i is the mole fraction of the ith gas and M_i is the molecular weight of that gas.

Example 14

(a) *Calculate the average molecular weight of air, assuming the following composition:* $[O_2] = 21$ *mole%,* $[N_2] = 79$ *mole%.*

	Oxygen :	0.21×32	$= 6.72$
Solution:	*Nitrogen :*	0.79×28	$= 22.18$
	Average molecular weight		$= 28.90$

(b) *Calculate the average molecular weight of the following modified atmosphere used in food packaging:* $[O_2] = 7.7$ *mole%,* $[CO_2] = 13.1$ *mole%,* $[N_2] = 79.2$ *mole%.*

	Oxygen :	0.077×32	$= 2.46$
Solution:	*Carbon dioxide :*	0.131×44	$= 5.76$ ◀
	Nitrogen	0.792×28	$= 22.18$
	Average molecular weight		$= 30.40$

H. **Density of a Gas Mixture:** If the composition of a gas mixture is known in either mole, weight or volume units, the density of the mixture at a particular temperature and pressure can be calculated.

Example 15

Compute the density in lb_m/ft^3 at 20.0 in Hg and 30 °C of a mixture of hydrogen and oxygen that is 12.1% H_2 by weight.

Solution: Take 1 lb of the mixture as the basis for the calculations.

$$[H_2] = \frac{\left(0.121 \frac{lb_m H_2}{lb_m Mix}\right)}{\left(2 \frac{lb_m H_2}{lb-mole\ H_2}\right)} = 0.0605 \frac{lb\text{-}mole\ H_2}{lb_m\ mixture}$$

$$[O_2] = \frac{\left(0.879 \frac{lb_m O_2}{lb_m Mix}\right)}{\left(32 \frac{lb_m O_2}{lb-mole\ O_2}\right)} = 0.0275 \frac{lb\text{-}mole\ O_2}{lb_m\ mixture}$$

$$Total\ molal\ quantity = 0.0605 + 0.0275 = 0.0880 \frac{lb-mole\ gas}{lb_m\ mixture}$$

The volume of 1 g-mole of gas at STP (29.9 in Hg, 273 K) = 22.41 L. The volume of a lb-mole is thus:

$$\left(22.41 \frac{liter}{gm\text{-}mole}\right) \left(453.6 \frac{gm\text{-}mole}{lb\text{-}mole}\right) \left(\frac{1ft^3}{28.315\ liters}\right) = 359 \frac{ft^3}{lb\text{-}mole}$$

Since 1 pound of this mixture contains 0.0833 lb-moles, its volume at STP is

$$\left(0.0880 \frac{lb\text{-}moles}{lb_m\ mixture}\right) \left(359 \frac{ft^3}{lb\text{-}mole}\right) = 31.6 \frac{ft^3}{lb_m\ mixture}$$

The volume at 20.0 in Hg and 30 °C is

$$V = \left(31.6 ft^3\right) \left(\frac{303\ K}{273\ K}\right) \left(\frac{29.9\ in\ Hg}{20.0\ in\ Hg}\right) = 52.4 ft^3$$

Since this volume is occupied by 1 lb_m of the mixture, its density is

$$\rho = \frac{(1\ lb_m)}{(52.4 ft^3)} = 0.0191 \frac{lb_m}{ft^3}$$

◄

Example 16

Air is assumed to contain 79.0 mole% of N_2 and 21.0 mole% of O_2. Compute its density at 70 °F and 742 mm Hg.

   Solution: Take 1 gram-mole as the basis for the calculations.

$$Mass\ O_2 = (0.21\,gmoles)\left(\frac{32\,g}{gmole}\right) = 6.72\ g$$

$$Mass\ N_2 = (0.79\,g)\left(\frac{28\,g}{gmole}\right) = 22.12\ g$$

$$Total\ mass = 6.72\,g + 22.12\ g = 28.84\ g$$

   The volume at STP = 22.41 L. The volume and density at 742 mm Hg and 70 °F is

$$V = (22.41\,L)\frac{(760\,mmHg)}{(742\,mmHg)}\frac{(530\,°R)}{(459.7\,°R)} = 26.44\ L$$

$$\rho = \frac{28.84\,g}{26.44\,L} = 1.07\ \frac{g}{L} \qquad \blacktriangleleft$$

Check Your Understanding

11. A hermetically sealed food package has a volume of 1000 cm^3 and contains atmospheric air (21% O_2, 79% N_2) at 20 °C and a total pressure of 760 mm Hg. Calculate each of the following:
    (a) The partial pressure of O_2 and N_2 in the package.
    (b) The partial volume of O_2 and N_2 in the package.
    (c) The total pressure of the package if it was to become totally anaerobic assuming the package is completely rigid.
    (d) The volume it would occupy if it were anaerobic, and the packaging was completely flexible. ◄

1.6.3 Real Gases and Compressibility Factor

In a previous section, the ideal gas law was discussed that related the volume, pressure, and temperature of a gas only when certain assumptions hold:

1. This law assumes that the molecules in a gas occupy negligible volume compared to the space between molecules as in Fig. 1.13a. When a gas is compressed, only the space between molecules is reduced in volume, not the space occupied by the

Fig. 1.13 Ideal gas law
assumptions

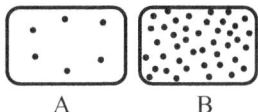

A   B

molecules themselves. If the molecules occupy a significant percentage of the gas as in Fig. 1.13b, the compression will not be proportional to pressure changes.

2. This law also assumes that there are no forces of attraction or repulsion between molecules. As molecules move closer together, they begin to affect each other in ways that increase or decrease the distances between them and hence change the volume of the gas.

Both assumptions hold for gases at very low pressures when molecules are far apart. For most gases they apply reasonably well, up to and slightly above atmospheric pressure. They also hold for gases at moderately high temperatures.

To study deviation from ideal behavior, we note that the ideal gas equation implies that for a fixed mass of gas (n = constant) at a fixed temperature (T = constant), the nRT product will be a constant. If the law is obeyed, the PV product must also be a constant equal to nRT at any pressure. Thus, the ratio:

$$z = \frac{(PV)_{actual}}{nRT} \qquad (1.57)$$

will be equal to 1.000 for an ideal gas.

For a nonideal gas, the PV product will change with pressure, and z will deviate from 1.00. A z value of 1.5, for example, means that the actual PV product of the real gas is 50% higher than that predicted by the ideal gas law. This implies that at a given pressure, the volume will be 50% greater than that predicted by the ideal gas law. In other words, the gas will be less compressible than predicted. Conversely, a z value less than 1.00 indicates a smaller volume than expected and a gas that is more compressible than predicted by the ideal gas law. Table 1.8 shows the extent of deviations from the ideal gas law for 3 common gases.

Figure 1.14 is a plot of the data in Table 1.8. Note that as the pressure approaches 0, the behavior of all gases approaches an ideal value (z = 1.000). At higher pressures, one can see that each gas behaves differently. The PV product of hydrogen increases linearly with pressure. This suggests that the deviation is due primarily to the fact that as increased pressure pushes the molecules closer together, they occupy a greater percent of the volume of the gas. On the other hand, nitrogen and carbon dioxide show an initial decrease in the PV product followed by an increase. Apparently, as the molecules of these gases are pushed closer together, they interact more strongly than hydrogen molecules. At lower temperatures, helium and neon are similar to hydrogen in their behavior. However, as temperatures increase, they behave increasingly like nitrogen and carbon dioxide. To account for the deviation in compressibility of real gases, the ideal gas law is modified as:

Table 1.8 Deviation from the ideal gas law

Pressure in atmospheres	z (PV product relative to ideal)		
	Hydrogen at 0 °C	Nitrogen at 0 °C	Carbon dioxide at 40 °C
1	1.000	1.000	1.000
50	1.033	0.985	0.741
100	1.064	0.985	0.270
200	1.134	1.037	0.409
400	1.277	1.256	0.718
800	1.566	1.796	1.299

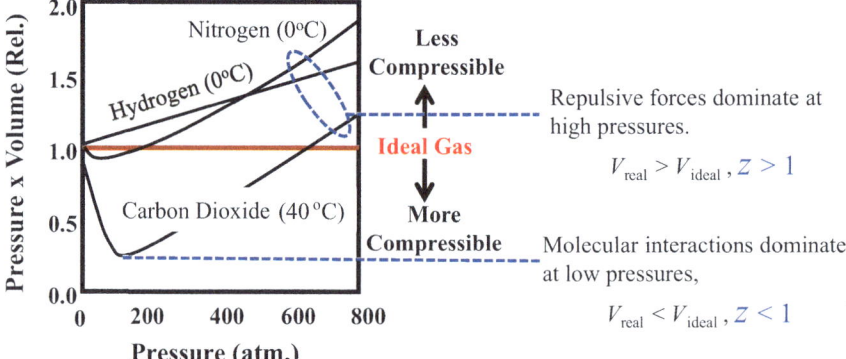

Fig. 1.14 Deviations from the ideal gas law

$$PV = znRT \tag{1.58}$$

The compressibility factor (z), a dimensionless number, is a very useful thermodynamic property that adjusts the ideal gas law to account for the nonideality of real gases. In general, the closer a gas is to a phase change (i.e., low temperatures or high pressures), the higher its deviation from ideal behavior.

The compressibility factor for specific gases is available from generalized compressibility charts that plot z as a function of reduced parameters based on the law of corresponding states. This law, formulated by Johannes van der Waals in 1873, states that any pure gas at the same reduced temperature (Tr), reduced pressure (Pr), or reduced volume (Vr) should have the same compressibility factor. The pressure, temperature, and volume are normalized with the critical values and are called the reduced parameters defined as:

$$T_r = \frac{T}{T_c}, P_r = \frac{P}{P_c}, V_r = \frac{V}{V_c} \tag{1.59}$$

where T_c, P_c, and V_c are the critical temperature, critical pressure, and critical volume of a gas, respectively. T_c is the temperature above which a gas cannot be liquefied, P_c is the pressure required to liquefy a gas at its critical temperature, and V_c is the volume occupied by a gas at its T_c and P_c. These parameters define the critical point of a fluid above which liquid and gas phases of a fluid become indistinguishable. These can be determined from the table of critical point data of various substances available in the literature. All fluids normalized in this manner exhibit similar nonideal gas behavior within a few percent and thus are plotted on a single chart. Figure 1.15 is a graph of a generalized compressibility factor at low pressure and has an accuracy within 4–6 percent for z values of 0.3–0.6. From a knowledge of any two of the three reduced parameters. The compressibility of a gas can be read from the graph.

The compressibility factor provides an easy way to correct the volume of a gas calculated using the ideal gas law, since at the same temperature and pressure, it may be written as $V_{real} = z \, V_{ideal}$. There are many uses of the compressibility factor. Since gases are highly compressible, knowledge of the z value permits an estimate of how large masses of a gas can be compressed in small containers, such as the cooking/grilling gas that is used for preparing foods or compressed oxygen cylinders used in hospitals.

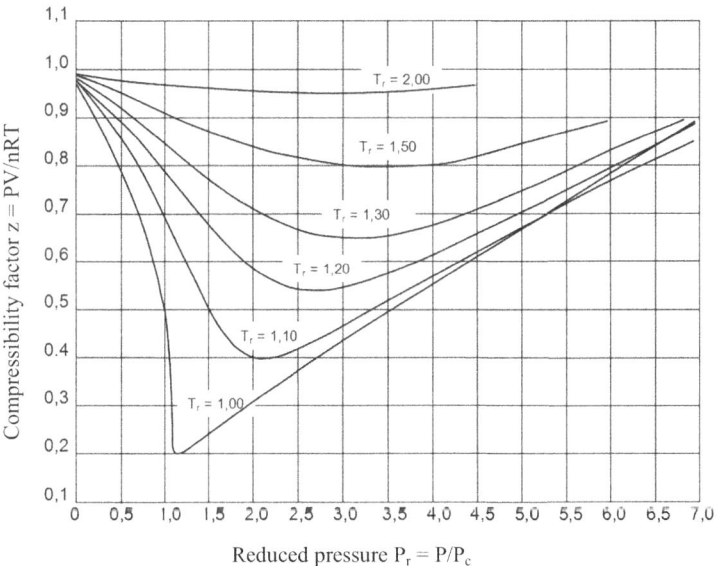

Fig. 1.15 Generalized compressibility chart. Source: Daniele Pugliesi, CC BY-SA 3.0, via Wikimedia Commons

Example 17

Estimate the volume occupied by a kilogram of superheated steam at 380 °C and 30 bar using the compressibility factor and the steam table. What is the percent difference between the two values? For water, T_c = 374 °C and P_c = 218 bar

Solution: T_r = T/T_c = (380.0 + 273.17)/(374.0 + 273.17) = 1.00; P_r = P/P_c = (30)/218 = 0.14

From Fig. 1.15, z = 0.96, and the volume of 1 kg water vapor can be calculated as:

$$V = (z\, n\, RT)/P$$
$$= [(0.96)\,(1\ kg/18\ kg/kg\text{-}mole)(8.31\ kJ/kg\text{-}mole\text{-}K)(653.17\ K)]$$
$$/[(30\ bar)]\,[(1\ bar/10^5\ N/m^2)(10^3 N \cdot m)/1\ kJ)] = 0.102\ m^3$$

From the steam table, Appendix A.4, at 3000 kPa (30 bar) and 420 °C, V = 0.103 m^3 for 1 kg of superheated steam. The difference between the two values is less than 1%. ◄

1.6.4 The Solubility of Gases and Henry's Law

Gases dissolve in liquids (most commonly water) to form solutions and establish equilibrium described as:

$$O_2\,(g) \leftrightarrow O_2(aq.) \tag{1.60}$$

The equilibrium solubility depends on temperature, the partial pressure of the gas over the liquid, the nature of the solvent, and the nature of the gas. The normal composition of air is 79% nitrogen, 21% oxygen, and trace amounts of others. However, the composition of air dissolved in water is 37% oxygen and 61% nitrogen due to the greater solubility of oxygen in water than nitrogen. This is significant for the survival of aquatic life; fish will not survive in a fish tank filled with unaerated distilled water. The solubility of carbon dioxide is even higher in water, in which more than 99% exists as the dissolved gas and less than 1% as carbonic acid (H_2CO_3), a week acid known to stimulate taste receptors in the mouth. Carbonic acid partly dissociates to give H^+, HCO_3^-, and CO_3^{2-} and lowers the pH to approximately 3–4.

The solubility of gases decreases with an increase in temperature because the enthalpy change of the dissolution of most gases is negative (exothermic reaction). This loss in solubility due to temperature effects may be observed during heating of water when air bubbles start appearing on the sides of the pan well before boiling. Almost all the dissolved gases disappear at the boiling point of water, making water taste very bland. On the other hand, the solubility of gases increases with increasing pressure. This happens because the concentration of molecules in the gas phase

Table 1.9 Henry's law constants for gas-water systems at 25 °C

Gas	k_H (Pa/(mol/L))	k_H (atm/(mol/L))
O_2	74.7×10^6	756.7
N_2	155.0×10^6	1600.0
CO_2	2.9×10^6	29.8
NH_3	5.7×10^6	56.9

increases with increasing pressure, and thus, the concentration of dissolved gas molecules in the solution at equilibrium is also higher at higher pressures. In 1800 J·W. Henry formalized this relationship and proposed an empirical law that stated that the concentration of a gas in a solution is directly proportional to the partial pressure of that gas above the solution. Now known as Henry's law, with the proportionality constant called the Henry's law constant, it is written as:

$$P_i = k_H C \text{ or } C = (1/k_H) P_i = k_H' P_i \quad (1.61)$$

where P_i = is the partial pressure of the gas, C is its concentration, and k_H or k_H' is Henry's law constant. Note that concentration can also be expressed as the mole fraction (x_i) of the gas dissolved in the liquid, and in that case, the above equation is written as:

$$P_i = k_H x_i \text{ or } x_i = (1/k_H) P_i = k_H' P_i \quad (1.62)$$

The units of k_H depend on the units selected for pressure and concentration, and its value for gas-water systems is shown in Table 1.9. Note that k_H' is simply the inverse of k_H. Since both are referred to as Henry's law constant, care must be exercised in their use. When multiplied by the partial pressure of the gas above the solution, k_H' gives the molar solubility of the gas. Thus, carbon dioxide at one atmospheric pressure and 25 °C would have a molar solubility of (1/29.8) mol/L or 33.56 mmol/L (1.07 g/L).

Remember, the value for Henry's law constant is the same for the same gas-solvent system at a specific temperature.

Check Your Understanding

12. Show that the equilibrium solubility of oxygen in water at 25 °C and one atmosphere is 8.9 mg/L and compare it with the solubility of carbon dioxide. ◄

Carbonated beverages contain dissolved carbon dioxide under pressure. Their effervescence and fizz are due to carbon dioxide coming out of solution when the container is opened, and pressure is released. The drink ultimately goes flat when most of the carbon dioxide is lost. The process is accelerated by warming of the liquid. For fizzy drinks commonly known as sparkling water, club soda, soda water, or fizzy water, 5–8 g carbon dioxide gas per liter is infused in water under pressure. In colas and related drinks, it ranges up to 3.5–7 g/L, and in fruity drinks, it is much

Table 1.10 Carbon dioxide limits for selected beverages

	Volume of CO_2/volume of drink (STP)	
Product	Lower	Upper
Draft porter, stout	1.8	2.4
Bottled ales	2.2	2.6
European lagers	2.4	2.6
American ales and lagers	2.4	2.8

lower. The final value varies based on the type of drink, manufacturer, packaging material, etc. At 25 °C, the internal pressure in a can of soda is approximately 379 kPa (55 psig) and approximately 117 kPa (17 psig) at 4 °C. Frequently, the degree of carbonation is expressed as "volumes," and one volume, for example, means 1 L of carbon dioxide per liter of drink, which translates to 1.96 g/L. The carbon dioxide limits for some products are shown in Table 1.10.

It is important to note that as an empirical correlation, Henry's law works well when the pressures and concentrations are low, and for this reason, solutions that obey it are often called ideal dilute solutions. Deviations and nonideal behavior of solutions are corrected through the introduction of an activity coefficient. For more information on how to estimate the activity coefficient, interested readers should consult advanced textbooks on phase equilibria.

1.7 Selected Properties of Liquids

In food manufacturing operations, many processes that are utilized to transform raw food materials into consumable products involve handling them in liquid states such as milk, juice, beer, wine, oil, and many others. Many separation, concentration, and purification processes are also liquid based. Some of the most critical liquid properties of utility are discussed below.

1.7.1 Surface Tension

Surface tension is the property of the surface of a liquid in contact with air. If the interface is between liquid–liquid, liquid–solid, or solid–air, it is called "interfacial tension." Surface tension arises due to asymmetries in the intermolecular forces between molecules of a liquid at the surface. As shown in Fig. 1.16, molecules on the surface have no molecules above and thus exert stronger net attractive forces upon neighboring molecules on and below the surface and pull them inward. This causes the surface to contract and develop tension, which may be called "surface tension." It is due to the surface tension that small objects that are denser than water such as insects, needles, paper clips, etc. can float on the surface of water. It is also because of the surface tension that droplets of water assume spherical shapes and minimize their surface area. When two immiscible liquids, like water and oil, in a salad

Fig. 1.16 Surface tension effects

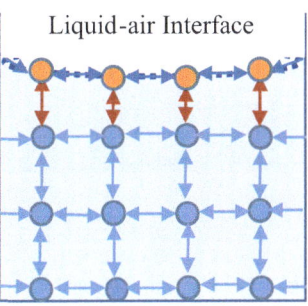

Liquid-air Interface

dressing are brought together, they do not mix and minimize their interfacial energy by minimizing the interfacial contact area and thus maintaining their separate identities. To make a stable emulsion system, a surfactant such as phospholipids with both hydrophilic and lipophilic ends is added, which orients between the two liquids, lowers the interfacial tension, and makes the two liquids disperse in each other. Similarly, disinfectants are formulated to have low surface tension to easily spread out on the surface where bacteria reside and kill them. Soaps and detergents lower the surface tension of water so that it can readily soak into pores and crevices of soiled surfaces and ease their cleaning. It is the surface tension effect that causes air-dried fruits and vegetables to lose their shapes.

The measure of the surface tension of a liquid is the amount of force required to break its surface over a linear distance of 1 m. It may also be considered as work required per unit area to break the surface. It can be expressed in the SI units of J/m^2 or N/m, but the cgs unit of dynes/cm is very commonly used. Water, with its strong intermolecular hydrogen bonding, has one of the highest surface tension values of 0.072 J/m^2 or 72 mN/m (72 dynes/cm) at 25 °C compared to any other liquid. Liquids with weak intermolecular forces, such as organic solvents, have much lower surface tensions. The surface tension of mercury, a metallic element that is liquid under standard atmospheric conditions, is very high (0.46 J/m^2) because of stronger bonds between the mercury atoms.

A liquid resting on a plane solid surface experiences the constraints of three surface energies (see Fig. 1.17) and is identified below:

- the liquid and vapor interfacial free energy, γ_{LG}
- the solid and liquid interfacial free energy, $\gamma_{SL,}$ and
- the solid and vapor interfacial free energy, γ_{SG}

The angle, known as the **contact angle** (θ_c), a liquid creates with a solid surface, is a function of the intermolecular cohesive forces of similar molecules and adhesive forces between dissimilar molecules operating at the interface (Fig. 1.17a). When cohesive forces are weaker than adhesive forces, the molecules tend to interact more with solid molecules, which results in a smaller contact angle, such as water in a glass capillary (water molecules are more strongly attracted to silicon dioxide in glass than to other water molecules). The interplay of adhesive and cohesive forces is

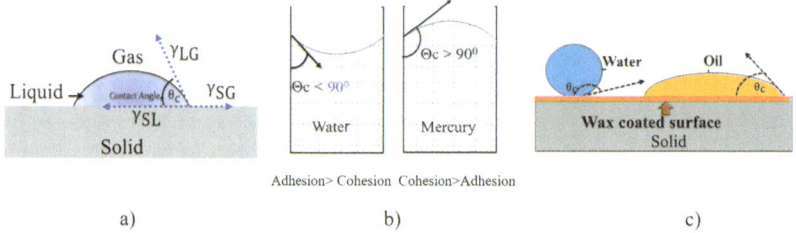

Fig. 1.17 Contact angle and wetting of surface. (**a**) Solid–liquid–gas contact. (**b**) Contact angle and meniscus. (**c**) Wetting of a surface

the opposite for larger contact angles, such as mercury in a capillary (Fig. 1.17b). A liquid with a low contact angle is thus pulled into the surface by capillary forces and is the reason why porous surfaces soak up liquids quickly. Dry food particles are examples of porous media that imbibe water via capillary forces. The contact angle plays an important role in determining the wetting ability of a surface. A completely wetting liquid forms a contact angle of zero with the surface, while a surface repelling a liquid has a contact angle close to 180° (Fig. 1.17c). In general, in a solid–liquid system, when the contact angle (θ_c) is between 0° and 90°, the surface will be covered by the liquid; between >90° and <180°, the solid surface will have good nonstick characteristics; and near 180°, the surface will be perfectly nonwetted, and the liquid will form beads on the surface.

The surface tension of a fluid decreases with increasing temperature and becomes zero at its critical temperature, which has significant bearing on the use of supercritical fluids for the extraction and separation of molecular mixtures, including food components. This will be discussed in more detail in the next chapter.

Knowledge of surface tension is of great interest for applications in numerous areas, including food product development, packaging, coating, painting, biomedical and pharmaceutical, cleaning, etc., especially in the design of emulsifiers, foaming agents, detergents, and wetting agents.

1.7.2 Colligative Properties of Solutions

A solution is a homogenous mixture of two or more components. The solute is the component present in the smaller amount, while the solvent is the component present in the larger amount.

Colligative properties are properties that result from adding solute to a solvent and depend only on the **number** (concentration) of solute particles in solution and not on the **type or identity** of the solutes. For different solvents, the effect will indeed be different. What this means is that a 0.10 molar solution of table salt and a 0.10 molar solution of sucrose in water would have the same effect. It is also obvious that higher the concentration of solutes, the more these properties will change. The colligative properties exhibited by a solution include:

A. **Vapor Pressure Lowering**: The vapor pressure of a pure solvent is the pressure of its vapor in equilibrium with the liquid phase at a constant temperature and is a measure of its escaping tendencies. When added to a solvent, the solute molecules interact with the solvent molecules, and the interacting solvent molecules become less likely to vaporize than if the solute molecules were not present. This leads to a decrease in the vapor pressure of the solvent. The vapor pressure of a solution is always lower than that of a pure solvent. Its value is given by **Raoult's law,** a phenomenological law that assumes ideal behavior with no interactive forces between molecules. It states that the vapor pressure exerted by a solvent in an **ideal solution** with a nonvolatile solute is equal to the mole fraction of the solvent in the solution times the saturation vapor pressure of the pure solvent at the solution temperature. Mathematically, it is stated as:

$$P_1 = x_1 P^* \qquad (1.63\text{a})$$

where P_1 = vapor pressure of the solvent in the solution, P^* = saturation pressure of the pure solvent, and x_1 = mole fraction of the solvent in the solution.

If the solution contains only one solute, then $x_1 + x_2 = 1$, where x_2 is the mole fraction of solute, and Eq. (1.63a) may be written as:

$$P_1 = (1 - x_2)P^* \ \text{ or } \ \Delta P = P^* - P_1 = x_2\,P^* \qquad (1.63\text{b})$$

A combination of Raoult's law and Dalton's law (Eq. 1.48) shows that for the vapor phase pressure of an ideal solution, the following relationship will apply:

$$P_1 = y_1 P_\text{T} = x_1 P^* \qquad (1.64)$$

where y_1 = mole fraction of component 1 in the gas phase and P_T is the total pressure above the solution.

In the case of solutions that do not behave ideally, an activity coefficient (γ) is used to address their nonideality. This is done by inserting the solvent activity (γ_1) on the right side of Eq. (1.63a) or the solute (γ_2) in Eq. (1.63b). The activity coefficients may be determined by either Van Laar's or Margules' equations. The binary constants needed for these equations for several systems are available in standard handbooks.

Figure 1.18 shows the phase diagram of a pure solvent such as water and its solution as a function of temperature and pressure. The solid lines represent the combinations of temperature and pressure at which the pure solvent in solid–liquid, liquid–vapor, and vapor–solid phases is in equilibrium. Each point on these lines therefore describes the freezing point, boiling point, and sublimation point of the pure solvent at that temperature and pressure. The dotted lines in the figure describe the properties of a solution obtained by dissolving a solute in the solvent (water). The solid and dotted lines BC show the vapor pressure line of the pure solvent and solution, respectively. The dotted line lies below the solid line because the vapor pressure of the solvent escaping from the solution at any given temperature is lower

Fig. 1.18 Phase diagrams of
solvent and solution

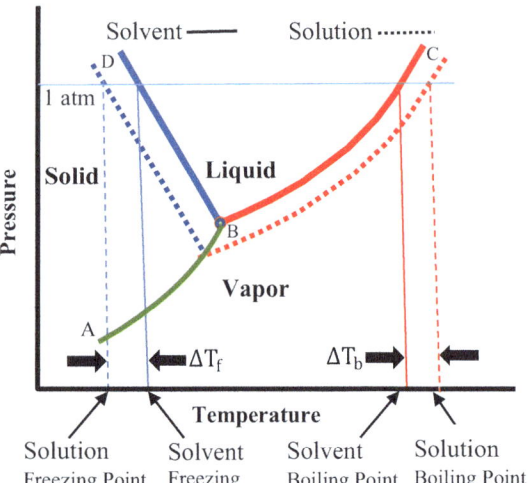

than the vapor pressure of the pure solvent and thus the boiling point is elevated. Similarly, the solid and dotted lines BD describe the freezing point depression of the solution.

Check Your Understanding

13. Why does a sugary soda spill on the floor remain sticky for a long time, but water spill evaporates away much sooner? ◄

B. **Boiling Point Elevation and Freezing Point Depression:** As a result of lowering the vapor pressure of a solvent when a solute is added, the boiling point of the solution is elevated. The magnitude of the change in the boiling point (ΔT_b) is proportional to the mole fraction of the solute (x_2). For a nonelectrolyte solute that remains intact when dissolved in a solvent, it is given as:

$$\Delta T_b = K_b x_2 \tag{1.65}$$

where $\Delta T_b = T_b - T_b^*$ and T_b = boiling point of the solution and T_b^* = boiling point of the pure solvent.

In dilute solutions, the mole fraction of the solute is proportional to the molality (m_2) of the solution, and the change in the boiling point that occurs is often written as follows:

$$\Delta T_b = K_b m_2 \tag{1.66a}$$

where K_b is a proportionality constant known as the **molal boiling point elevation constant** (equal to the change in boiling point for a 1-molal solution of a nonvolatile solute) for the solvent and m_2 is the molality of the solute in mol solute/kg solvent. The constant K_b is the property of a particular solvent and can be obtained from the following thermodynamic function:

$$K_b = \frac{R\, T_b^2 M_1}{\Delta \overline{H}_v} \qquad (1.66b)$$

where T_b is the boiling point of pure solvent, M_1 is the molecular weight of solvent, and $\Delta \overline{H}_v$ is the molar enthalpy of evaporation for solvent. For solutes that dissociate into ions when dissolved in solvent to make electrically conducting solution, van't Hoff in 1884 introduced another term into the boiling point elevation expressions as follows:

$$\Delta T_b = i\, K_b m_2 \qquad (1.67)$$

where i is the number of dissociated particles per mole of solute and is called the **van't Hoff factor.** A similar thing happens to the freezing point (or melting point) of a solvent when a solute is added to the solvent, and the following expression is used to calculate the depression in the freezing point of a solvent when nonvolatile solutes are added.

$$\Delta T_f = -K_f\, m_2 \qquad (1.68a)$$

where ΔT_f is the freezing point depression , $\Delta T_f = T_f - T_f^*$, T_f = freezing point of the solution, and T_f^* = freezing point of the pure solvent. K_f is the **molal freezing point depression constant** for the solvent. A negative sign indicates that the freezing point of the solvent decreases when a solute is added. Like K_b, the constant K_f is also the property of a particular solvent, given by the following thermodynamic function:

$$K_f = \frac{R\, T_f^2 M_1}{\Delta \overline{H}_f} \qquad (1.68b)$$

where T_f is the freezing point of pure solvent, M_1 is the molecular weight of solvent and $\Delta \overline{H}_f$ is the molar enthalpy of fusion for solvent. For electrolytes, analogous to the boiling point behavior, Eq. (1.68a) becomes:

$$\Delta T_f = -i\, K_f\, m_2 \qquad (1.69)$$

The K_b and K_f values of some common solvents are listed in Table 1.11.

Several interesting applications of the above principles may be observed in our daily life. For example, making ice cream by using an ice and salt mixture to depress the freezing point below the freezing point of the ice cream mix (milk, sugar, cream) is an age-old practice still in use in many parts of the world. Another example is the common antifreeze used in car radiators. It is a solution of ethylene glycol and water that serves to both drop the freezing and raise the

Table 1.11 Molal boiling point elevation and freezing point depression constants of selected liquids

Solvent	Boiling point at 1 atm (°C)	K_b (°C/m)	Freezing point at 1 atm (°C)	K_f (°C/m)
Acetic acid	117.9	2.93	16.6	3.90
Chloroform	61.2	3.88	−63.5	4.90
Ethanol	79.0	1.19	−114.6	2.00
Water	100	0.52	0	1.86

boiling point. This prevents the liquid in the cooling system from freezing during the winter months and reduces the risk of the liquid boiling over when the engine becomes too hot.

C. **Osmotic Pressure:** It is the last member of the quartet of colligative properties and refers to a pressure caused by a difference in the concentrations of solutes between solutions. A solution having the same, lower, or higher osmotic pressure than some other solution is known as isotonic, hypotonic, or hypertonic solution, respectively. Shown below in Fig. 1.19a are two compartments filled with solvents containing differing concentrations of solutes, which are separated by a semipermeable or differentially permeable membrane. Being semipermeable, the membrane is essentially invisible to the solvent molecules, so they freely diffuse from the high concentration region to the low-concentration region until equilibrium is achieved. This flow of solvent constitutes osmotic flow, or osmosis, and the pressure required to achieve osmotic equilibrium is called the osmotic pressure (Π). The osmotic pressure may also be considered the pressure required to stop osmosis.

The osmotic pressure of a dilute ("ideal") solution may be estimated using the van't Hoff equation, resembling the ideal gas equation:

$$\Pi = nRT/V \tag{1.70}$$

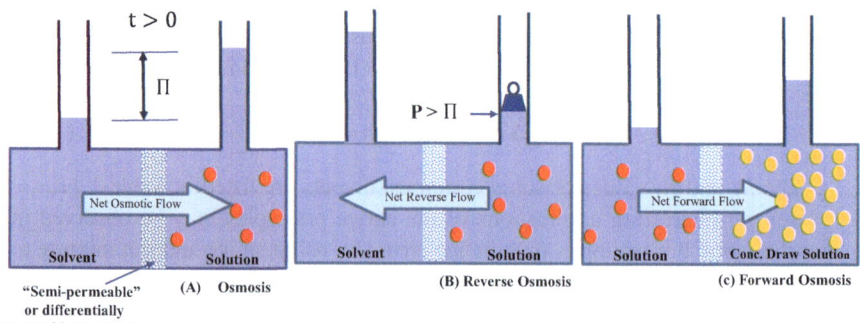

Fig. 1.19 Osmosis, reverse osmosis, and forward osmosis

Table 1.12 Osmotic pressure of selected liquid foods

Food	Concentration	Osmotic pressure (kPa)
Apple juice	15% total solids	2070
Coffee extract	28% total solids	3450
Grape juice	16% total solids	2070
Milk	9% solid-not-fat	690
Whey	6% total solids	690

Source: Adapted from Cheryan [1]

where n = moles of solute, V = volume of solution, R = gas constant, and T = absolute temp. Note that n/V represents the molarity of the solution. This contrasts with the need for molality to compute changes in boiling and freezing points of a solution.

Check Your Understanding

14. Show that sea water with a salt concentration of 1.1 moles/L at 25 °C will exert an osmotic pressure of 26.9 atm. ◄

For a more reliable estimate over a wider range of solutions, the following Gibb's equation has been found to be more realistic:

$$\Pi = - \frac{RT \ln x_A}{V_m} \tag{1.71}$$

where x_A is the mole fraction and V_m is the molar volume of pure solvent. The osmotic pressures of a few foods are listed in Table 1.12 below.

Reverse osmosis (RO) is a process commonly used to purify water and concentrates of food ingredients. It is based on osmotic pressure principles. The solution of interest is placed in a chamber separated by a differentially permeable membrane, and a pressure greater than the osmotic pressure of the solution is applied (Fig. 1.19b). The membrane allows the solvent molecules to pass through but not the solute particles. As a result, the solution becomes more concentrated in solutes, and almost pure solvent collects on the other side of the membrane. This reverse osmosis principle is commonly employed on large commercial scales to continuously desalinate brackish water and concentrate solutes from various food streams.

The drying of foods, the use of sugar in making jams and jellies and salts in curing meat are some of the most common ancient techniques of food preservation. The removal of water and enhancement of the concentrations of salts or sugar create a hypertonic environment that takes water out of microorganisms by osmosis. This loss of water severely interferes with cell function, retards their growth, and eventually leads to bacterial death due to dehydration. Osmotic dehydration involving the immersion of foods (fish, vegetables, fruits and meat) in osmotic solutions of salts, alcohols, or concentrated sugars is also used to some extent to dehydrate the foods. A derivative technology called **forward osmosis (FO)** is used to separate water from dissolved solids by employing a draw solution of higher osmotic pressure across the membrane than the feedstock (Fig. 1.19c).

1.8   Selected Properties of Solids

Solids fall into two main categories: crystalline and amorphous. In crystalline solids, the atoms and molecules making it up fare arranged in a regular, well-defined pattern. In amorphous solids, such as glass, atoms and molecules are positioned irregularly. In food manufacturing, a number of foods are handled in their solid state, such as fruits, vegetables, grains, meat, cheeses, and many kinds of powders and granular materials. Some of the more important behavioral properties of solids are described below.

1.8.1   Force–Deformation Behavior

The force–deformation characteristics are necessary to understand and quantify the behavior of materials during their manufacturing and uses. Shaping food materials by sheeting, rolling, extrusion, and other related operations and mastication of foods in the mouth require one or more of the basic modes of deformation and flow. The study of the deformation behavior of materials is called rheology, and a good understanding of the basic concepts of rheology is required in the manufacture and production of many foods that exhibit complex flow characteristics.

Any applied force per unit area that tends to deform (strain) a body is called **stress,** and the deformation produced by the stress is known as **strain**. Stress has the same units as pressure, and pressure is one special type of stress. Stress may be applied to produce tension (to lengthen an object, called tensile stress), compression (to shorten an object, called compressive stress), and shear (acts parallel to a surface, called shear stress), illustrated in Fig. 1.20. It is important to note the nomenclature and the differences among them, as they all show a change in shape.

When force is applied to a material in tension or compression (Fig. 1.20a, b), it is uniformly distributed over its cross-section and results in volume change. The ratio of the normal (perpendicular) force (F) to the cross-sectional area (A) is called the **normal or true stress.** On the other hand, when the applied force acts tangentially to the surface of a material element and produces shape change (Fig. 1.20c), it is called **shear stress**.

Stress is symbolized by letter σ and has the units of force per unit area, like pressure:

$$\text{Stress } (\sigma) = \frac{F}{A} \tag{1.72}$$

Since the area changes during deformation, it should be the actual or true, instantaneous area supporting the force. In most cases, the change in the area is

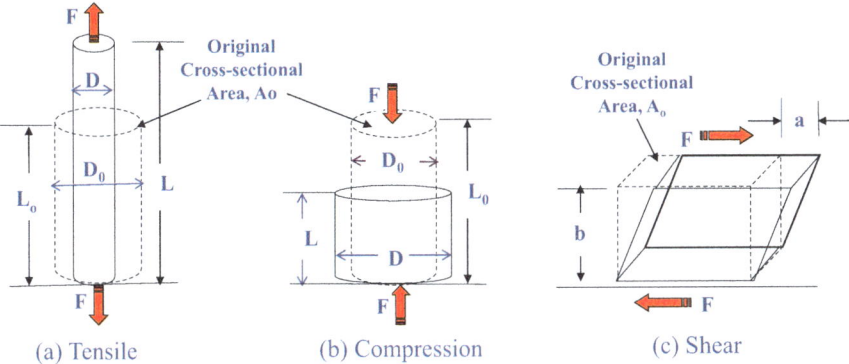

Fig. 1.20 Typical force-deformation behavior

small and an **engineering stress**, defined as the ratio of the applied normal force to the original area (A_o), is used:

$$\text{Engineering Stress } (\sigma) = \frac{F}{A_0} \tag{1.73}$$

The degree of deformation a material undergoes during tension or compression, Fig. 1.20a, b, is called **Hencky or true strain (ε),** and to account for the incremental change in the specimen length, it is defined as:

$$\text{True strain } (\varepsilon) = \int_{L_o}^{L} \frac{dL}{L} = \ln\left(\frac{L}{L_o}\right) \tag{1.74}$$

For practical purposes, both for compression and tension, the following expression, known as **Cauchy or engineering strain**, is commonly used:

$$\text{Engineering strain } (\varepsilon) = \left(\frac{L - L_0}{L_o}\right) = \frac{\Delta L}{L_o} \tag{1.75}$$

As may be observed from the above expressions, the value of strain is positive in tensile deformation and negative in compression.

Two other aspects of solid deformation are creep and relaxation. Creep is defined as the increase in deformation of a body under a constant stress, while relaxation is the decay of stress over time under constant strain.

The region where the relationship between the applied stress and strain of a material remains linear is called the **elastic region,** and the material is said to be ideal and obeys **Hooke's law**. In this region, the applied energy is stored in the material, and when the external force is removed, the original form is fully recovered. The ratio of stress to strain in this region is called the modulus of elasticity or Young's modulus (E), defined as:

$$E = \frac{\sigma}{\varepsilon} \tag{1.76}$$

Many materials, including many metals, polymers, and biomaterials, follow the above relationship. Eventually, beyond some stress values, elastic materials reach a point where they may stretch permanently. This kind of change is called **plastic deformation.**

A typical solid food material's response to compression is shown in Fig. 1.21a as a stress–strain curve. Following a linear elastic region, the dip in the curve shows that the force exceeded the maximum value the specimen could withstand and where permanent deformation or plastic flow occurred. This is called **the yield point.** The maximum amount of stress a material can withstand determines its strength or hardness. With the ever-increasing importance of food material properties, both in manufacturing and in eating, the concepts of yield strength and ultimate strength (the maximum stress before rupture) are important to master. The fracture point represents the brittleness of a material, which along with the Young's modulus is related to the crispiness, as shown in Fig. 1.21b. For example, fried potato chips with a higher modulus of elasticity will be crispier than those with a lower modulus, although the breaking (fracture) stress may be the same.

The compression or elongation of a material in one direction results in its expansion or contraction in the perpendicular direction. For example, when a sample of material is stretched in one direction, it tends to become thinner in the other direction, or when a sample is compressed in one direction, it tends to become thicker in the lateral direction, as illustrated in Fig. 1.22. The ratio of the lateral (transverse) strain to longitudinal strain under uniaxial stress in the longitudinal direction is known as **Poisson's ratio** (named after Siméon Denis Poisson), usually represented as a lowercase Greek nu (ν).

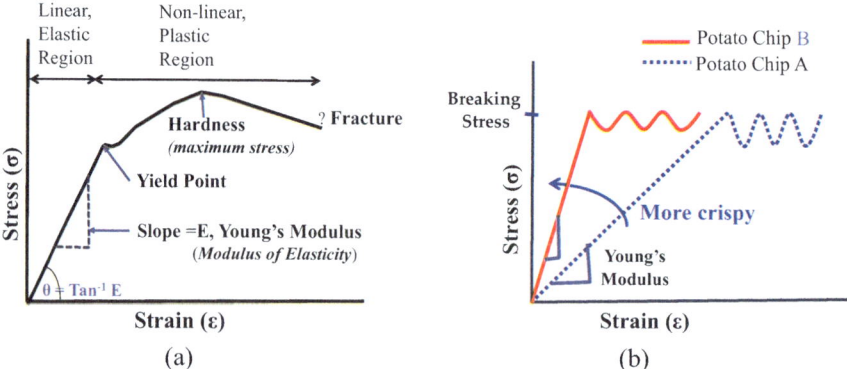

Fig. 1.21   (**a**) Typical stress-strain curve of a biomaterial. (**b**) Mechanical properties of potato chips

Fig. 1.22 Poisson's effect

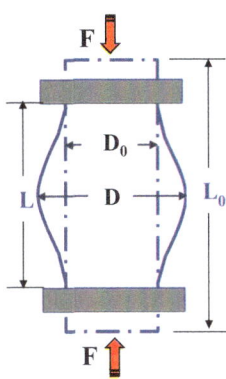

$$\nu = -\frac{\varepsilon_{\text{Trans}}}{\varepsilon_{\text{Longit}}} \qquad (1.77)$$

The above Poisson's ratio contains a minus sign so that normal materials have a positive ratio. Poisson's ratio generally varies from 0 to approximately 0.5; harder materials have lower values than softer materials. Rubber and stainless steel have values of ~0.5 and ~0.3, respectively. Referring to Fig. 1.22, Poisson's ratio can be estimated as follows:

$$\nu = -\frac{(D - D_0)/D_o}{(L - L_0)/L_o} \qquad (1.78)$$

For high-moisture foods, ν approaches 0.5. For example, for food gel and apple, the value has been reported to range between 0.4 and 0.5.

Viscoelastic materials are known to show time-dependent strain under constant stress. Such materials, when frozen quickly, behave harder with a higher modulus of elasticity and low Poisson's ratio, which often leads to freeze cracking.

Under shear conditions, the applied force is parallel to the surface. The front view of Fig. 1.20c is shown in Fig. 1.23. Based on the original area (A_0), the shear stress is defined as:

$$\text{Shear Stress } (\tau) = \frac{F}{A_0} \qquad (1.79)$$

And the shear strain is given as:

Fig. 1.23 Shear stress and shear strain

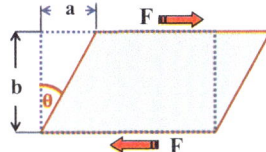

$$\text{Shear strain } (\gamma) = \frac{a}{b} = \tan \theta \qquad (1.80)$$

The **Shear Modulus or Modulus of Rigidity (G)** indicates a material's response to shearing strains and is defined as the ratio of shear stress to the displacement per unit sample length (shear strain) and has the units of pressure:

$$G = \frac{\tau}{\gamma} \qquad (1.81)$$

The shear modulus describes how far an object's shape changes when its volume remains constant. The bigger the shear modulus the more rigid is the material.

The **bulk modulus (B)** and its reciprocal, called compressibility (k), are also important properties of materials. Most materials decrease in volume when exposed to a uniform external pressure from all sides, as illustrated below (Fig. 1.24).

The ratio of the applied, uniform pressure to the resulting volumetric strain (change in volume/initial volume) of a material is called the bulk modulus (B):

$$B = -\frac{\Delta P}{\Delta V/V_i} \qquad (1.82)$$

where V_i is the initial volume and ΔV is the change in volume due to the change in pressure ΔP. The bulk modulus is thus a measure of the material's resistance to compression. It may also be considered as the applied pressure divided by the fractional change in volume. Compressibility ($1/B$), on the other hand, indicates the fractional change in volume per unit increase in pressure. Alternatively, since volume is inversely proportional to density, bulk modulus can also be written in terms of density, as follows:

$$B = \frac{\Delta P}{\Delta \rho/\rho_i} \qquad (1.83)$$

where ρ is the density of the material. A large bulk modulus indicates a relatively incompressible material. The bulk modulus of water is approximately 2.2×10^3 MPa, while for air, it is approximately 0.14 MPa. Air is thus over four orders of magnitude more compressible than water. For each atmospheric increase in pressure, the volume of water would decrease by 45.4 parts per million. Stainless steel has a bulk modulus of 160 GPa, approximately 70 times less compressible than water.

Fig. 1.24 Compression and bulk modulus

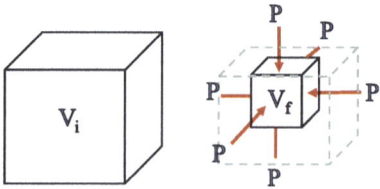

Example 18

When water freezes, it expands by approximately 9%. Compute the pressure increase that would occur inside a bottle filled with water if the water is allowed to freeze. The bulk modulus of water is 2.2×10^9 N/m².

Solution:

The initial volume V_i increases by 9%. Thus, $\Delta V/V_i = 9/100 = 0.09$ and $\Delta p = 0.09 * 2.2 \times 10^9 = 198$ MPa ◄

1.8.2   Food Particles: Shape, Size, and Flowability

Food products and ingredients often come in the form of many small pieces (or particles) of various sizes, ranging from emulsions (1–100 µm), colloids and suspensions (<1–50 µm) to powders (<1000 µm). Examples of such items include various types of flours, powdered drink mixes, ground seasonings and spices (salt, pepper, paprika, etc.), breakfast cereals, snack foods, salad dressings, pellets, etc. Table 1.13 lists the approximate particle sizes of some common items. The shape and size distributions of these particles play an important role in the processing packaging, handling and quality control of ingredients and formulations as well as in the appearance, mouthfeel, texture, taste, solubility, stability, processability, and functionality of the finished products.

A. **Particle Shape**: The shape of food particles is an important quality parameter. The bulk density and flowability of powders may change significantly depending on the shape of the particles. The shape of sugar crystals can affect the mouthfeel of chocolate. Regularity and uniformity are important quality factors in many products. Size reduction operations, such as cutting and grinding, are known to modify a food product's particle shape distribution and affect its properties. There are several quantitative means of expressing the shape of a particle. One simple method is to compare a particle to a sphere of the same volume. The ratio of the sphere's surface area to the particle's surface area is known as **sphericity**. The sphericity of a perfect sphere is 1; that of any other shape is less than 1.

Table 1.13 Approximate particle sizes of some common items

Item	Size range (µm)
Black pepper (table ground)	400–500
Coffee (ground)	5–500
Flour (wheat)	25–300
Ginger (ground)	25–50
Milk powder (spray dried)	30–100
Salt (table)	100–300
Sugar (granulated)	200–500
Sugar (icing)	20–25

B. **Particle Size**: In the handling of many products, it is important to quantify the size of food particles for several reasons, including the following:

- Smaller particles have larger surface-area-to-volume ratios.
- Particle size affects texture in foods made by combining many smaller particles into a larger mass, such as meat patties (made from ground meat), or foods consisting of particles suspended in a continuous phase, such as chocolate (cocoa solids and sugar suspended in cocoa butter) and milk (fat globules and proteins dispersed in water). Smaller particles give food products a smoother texture, while food containing larger particles may seem grainy. The smaller the ice crystal in ice cream is, the less grainy the mouthfeel.
- Particle size can affect the stability of emulsions, such as peanut butter and salad dressings. Larger particles are more prone to settling (for reasons that will be discussed in a later chapter) and causing phase separation, which is undesirable in products intended to serve as a single, homogeneous phase.

The size of a particle may be defined in a few different ways. Regular-shaped particles can easily be defined by a small number of dimensions—one for spheres and cubes (radius or side length), two for cylinders (radius and height), three for prisms (side lengths), and so on, Fig. 1.25.

However, food particles rarely have regular shapes; as such, several methods have been developed for defining the size of an irregularly shaped particle as simply and with as few dimensions as possible. Equivalent diameter is the most common method by which an irregular particle's size may be defined. Generally, a particle's equivalent diameter is defined as the diameter of the sphere that would behave like the particle under the relevant conditions. A few types of equivalent diameters are listed in Table 1.14.

Fig. 1.25 Standard shapes of particles and critical dimensions

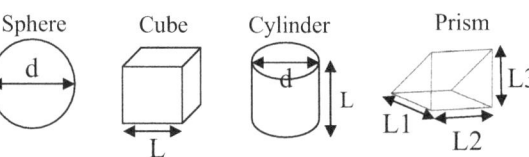

Table 1.14 Common types of equivalent diameter

Name	Equivalent behavior
Sieve diameter	Passes through same-sized sieve opening
Surface diameter	Has same surface area
Volume diameter	Has same volume
Surface-volume (Sauter) dia.	Has same surface/volume ratio
Stokes diameter	Has same density; falls at same stokes terminal velocity in a given fluid

Several techniques ranging from traditional sieving and sedimentation to modern analytical approaches including image analysis, light scattering, and electrical sensing are used to determine particle sizes. Sieves and screen are usually used for particles that are greater than approximately 50 μm and are classified by mesh sizes. Mesh sizes has historically been defined as the number of square openings per linear inch in a screen, Fig. 1.26, but due to the variability in the width of the wires in the mesh, different mesh sizes for the same mesh numbers have emerged.

The US Standard Sieve Series and Tyler Mesh size are very commonly used to classify particle sizes, Appendix A.3. The sieve size number or mesh size is defined as the square aperture or opening through which the particles can go through. The openings of successive sieves vary by a factor of √2.

Size Distribution and Mean Particle Size Many food products and ingredients consist of particles of many different sizes. These particulate materials may be characterized by determining their particle size distributions and mean particle sizes.

A food material's particle size distribution refers to the fraction of the total particle population that falls within a specified size range. This may be determined by sifting or microscopic examination of the material in question, among other methods. A material's particle size distribution may be easily determined via sieve analysis. A known mass of the material is poured onto a stack of nested mesh sieves (screens), in which the size of the sieve openings decreases from top to bottom (Fig. 1.27). The stack of sieves is then mechanically shaken or vibrated until all the material has settled. The material accumulated on each sieve is then weighed and recorded. This method yields the material's mass particle size distribution. Particle size distribution may also be defined, as mentioned earlier, on a number (i.e., the number of particles), volume, surface, or surface/volume basis. Depending on which type of particle size distribution is used, the mean particle size may be different.

A powder that passes through a 100-mesh screen will have particle sizes less than 149 microns and could range from submicron to 149 microns. To obtain a powder that was closer to the desired size range, two mesh sizes are specified, as for example −40 + 100 mesh. This powder would have particles that are smaller than 40 mesh (400 micron) but larger than 100 mesh (149 micron).

Fig. 1.26 Mesh number and sieve opening size

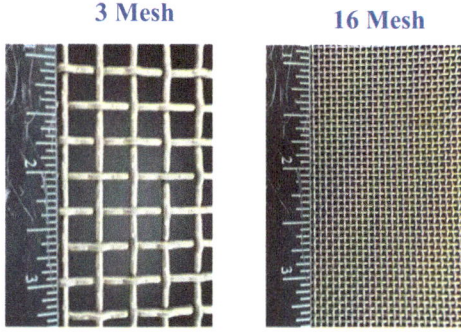

3 Mesh   16 Mesh

Fig. 1.27 Nested sieves
stacked on a shaker

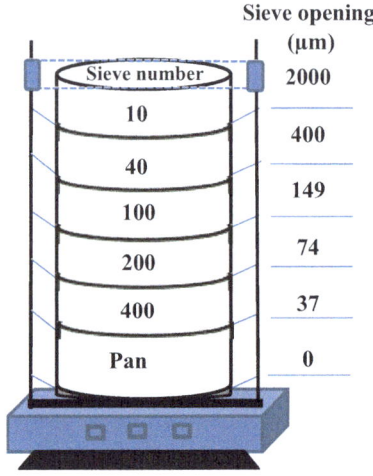

C. **Flowability and Angle of Repose**

The shape and size of a powder particle influence its flow behavior, and the measurement of the angle of repose is one of the commonly used methods for testing powder flowability. The angle of repose is defined as the angle that a loose powder makes when poured onto a surface. It is well known that a powder poured onto a surface forms a heap and that both the repose angle and the heap shape depend on powder flow properties. It is observed that a cohesive powder forms an irregular heap and does not flow well, while a non-cohesive powder forms a regular conical heap and has better flowability. The angle of repose is measured by passing a powder sample through a funnel and heaping it in the form of a cone on a flat surface. The height of the cone and the radius of the base are measured, and the angle of repose is computed by Tan^{-1} of height/radius (Fig. 1.28). The following empirical classification based on repose angle is used as a guide to assess a powder's flowability (Table 1.15).

The angle of repose depends on the particle size and surface roughness (coefficient of friction). It tends to increase with the decrease in particle size, and thus, very fine and sticky particles have a higher angle of repose. Highly flowable materials exhibit a low angle of repose.

Fig. 1.28 Angle of repose of
powders

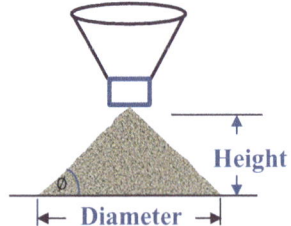

Table 1.15 Classification of flowability of powder based on repose angle [2]

Description	Repose angle
Very free flowing	$<30°$
Free flowing	30–38°
Fair to passable flow	38–45°
Cohesive	45–55°
Nonflowing (very cohesive)	$>55°$

The particle size and shape also influence the flow of powder in gravity discharge from hoppers, bins, and silos. Fine powders flowing out of a storage bin or hopper frequently form a rat hole through which powder flows and may extend from the outlet to the top of the surface (Fig. 1.29a). Many times, however, uncontrollable flow of powder through the system occurs due to the formation of arches, which are caused either by particle mechanically interlocking together or because of an increase in their cohesive strength (Fig. 1.29b).

Particle interlocking is often a result of their irregular shapes and sizes and cohesive arches form where particles bond together due to physical, chemical, or electrostatic forces. Hygroscopic materials in particular are very prone to arching because of the formation of liquid bridges between neighboring particles. This behavior may lead to no-flow, intermittent flow, or flooded flow due to sudden breakage of an arch or bridge, which would indeed affect downstream operations.

Problems

1.1 At what temperature does the Fahrenheit scale equal the Celsius scale? Show your calculations.

1.2 In an experimental measurement of freeze drying of beef, a vacuum of 752.5 mm Hg was maintained in the chamber. What would a pressure gauge attached to the chamber read in psig?

1.3 An "average" rod-shaped bacterium of 2 μm in length and 0.6 μm in diameter has a doubling time of 20 minutes. Estimate:

(a) Total surface area and volume of each bacterial cell
(b) Theoretical number of microorganisms after 12 hours of storage.
(c) Total mass of the bacterial colony in (b) if the specific gravity is assumed to be 1.05.

Fig. 1.29 Flow of powder in a hopper

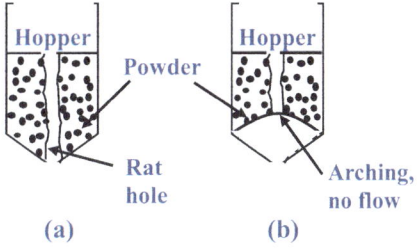

1.4 Your best friend is about to celebrate his/her birthday, and you planned to surprise him/her with his/her favorite triple chocolate cake. When you arrived at the cake shop, two sizes are offered: (a) a 7 in. × 10 in. rectangular cake and (b) a 10 in. round cake. Both have the same height of 4.5 in. and costs the same. Assuming that all other factors are identical, which cake is a better choice for your money?

1.5 A 25 cm diameter and 12.5 mm thick pizza pie is divided into eight equal slices, as shown below. Compute the following:
(a) Top surface area of each slice
(b) Volume of each slice
(c) Angle Θ
(d) Length of crust arc AB
(e) Mass of the pie if its specific gravity is 1.1

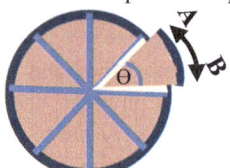

1.6 A cylindrical tank, 2.0 meters in diameter and 4.0 meters high, is filled with corn oil with a specific volume of 1.08 cm^3/g. Calculate (a) the mass of oil in kg and (b) the weight of oil in lb_f.

1.7 Estimate the pressure (in units of your choice) you will exert on the ice surface of the skating rink while standing still on skates. State all assumptions (e.g., size of skates, your weight, etc.).

1.8 By maintaining higher pressure and therefore higher internal temperature, a pressure cooker cooks food faster than an ordinary open pan. A metal piece, the petcock, is designed to sit on top of an opening in the middle of the cooker lid to safely maintain the desired pressure.
(a) Determine the mass of the petcock for the pressure cooker if the design pressure is 97.2 kPa above the atmospheric pressure and the cross-sectional area of the opening is 4 mm^2.
(b) Determine the inside temperature at the design pressure.

1.9 To install a new tank weighing 1200 kg, it needs to be lifted 4 meters and the power required to lift it is 300 W. How long will it take to lift the tank?

1.10 Show the relationship between moisture content on a dry basis and a wet basis by constructing a graph between the two moisture content measures.

1.11 A block of Havarti cheese has the following composition:

Component	Mass (g)
Water	38.6
Protein	21.2
Fat	29.1
Carbohydrate	1.9
Ash	4.0

Find the mass fraction of carbohydrate on (a) a dry basis and (b) a wet basis. If the cheese is reformulated to be fat-free, what will be the new mass fraction of carbohydrate on wet as well as dry basis?

1.12 A newly developed food product has the following composition: Water = 74%, Protein = 12%, Carbohydrate = 9%, Fat = 4%, and Ash = 1%
(a) What is the percentage of solids-not-fat (snf) on wet basis, on dry basis?
(b) If the product is allowed to lose 25% of its water during drying, what will be the new gross composition?

1.13 A slice subtending an angle of 42° is cut out from an 8-in diameter pizza pie, as shown on the right side. Estimate the following:
(a) Length of crust arc AB
(b) Top surface area of the slice
(c) If the wt.% composition of the pizza is: protein = 15; carb = 40; fat = 25; and moisture = 20, what is its composition on fat-free basis?

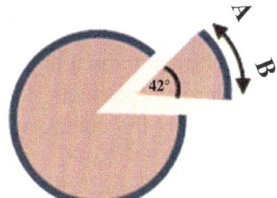

1.14 A 100-kg batch of dough contains 4 kg salt (NaCl). If the dough contains 30% moisture, what is the mass fraction of the salt in the dough?
(a) wet basis and (b) dry basis

1.15 The baker's percentage is yet another method used to express food formulations and recipes. In this method, flour is assigned the value of 100 percent, and the percentage of the rest of the ingredients is computed based on the flour. Convert the following baker's percent of bread formulation to weight percent and fill in the table.

Ingredient	Baker's %	wt.%
Flour	100	
Water	64	
Whey protein	7	
Sugar	6.4	
Monoglyceride	2.7	

1.16 The composition of a liquid stream is:
Water = 65.0 wt. %, Ethanol = 20.0 wt. %, and Acetic acid = 15.0 wt. %, compute the following

(a) mole fraction of each component

(b) wt. % of each component on a dry weight basis.

1.17 A wine is composed of 15.5% ethanol, 80% water, and 3% sugar by volume. The remaining volume consists of trace minerals and other compounds. If the density of ethanol is 789 kg/m^3, water 1000 kg/m^3, sugar 1590 kg/m^3, and the remaining minerals and compounds 1400 kg/m^3, what is the % ethanol by mass on a wet and dry basis?

1.18 The mass composition of milk is approximately as shown in the table below:

Component	Wt.%
Protein	3.5
Fat	3.7
Carbohydrate	4.8
Ash	0.7
Water	88.0

If the density of milk is approximately 1.03 g/cm^3, calculate (a) mass per unit volume concentration of fat and (b) protein mass fraction on a dry basis.

1.19 The table below shows the compositions of the two products. If 10 kg of product A is mixed with 5 kg of product B, what is the composition of the resulting product (mass percentages) on a wet basis?

Table: Mass fraction, dry basis, of products A and B

Component	Product A	Product B
Carbohydrate	0.40	0.60
Protein	0.35	0.40
Fat	0.25	0.00
Water	0.10	0.40

1.20 A conical silo is 10 m high and has a maximum radius of 2.5 m. The silo is filled with shelled corn, which has a bulk density of 0.72 g/cm^3. (a) What is the weight of corn? (b) What is the solid density of the corn if the porosity is 0.30?

1.21 Corn filled in a 1 m^3 hopper is milled into cornmeal, which reduces the porosity by a factor of 2. What is the bulk density of the milled corn, assuming the bulk density of the original corn was 0.75 g/cm^3 and the solid density of corn is 1.5 g/cm^3.

1.22 It is generally known that a human being loses consciousness when the average concentration of oxygen in the lungs drops below 11 vol.%. If a person enters a room that contains 100 vol.% CO_2, estimate how long it will take for that individual to become unconscious. A normal person will inhale approximately

30 L/min at 500 ml/inhalation, and the average concentration of oxygen in the lungs is 16 vol.%. (Assume a typical adult's lung volume is 5 L).

1.23 A gas sample contains 23% oxygen and 77% nitrogen, both by weight. Determine the volume of 1 kg of the sample at 0 °C and 1 atm (STP).

1.24 For the concentration of a liquid food under a vacuum of 744.5 mmHg and at 23 °C, water vapors are produced at a rate of $10,000.0$ ft^3 per hour. Compute the mass of water vapors that will have to be removed during an 8-hour shift of operation.

1.25 To produce plasticized fat like margarine, hydrogenation of vegetable oil like coconut or palm oil is carried out under controlled conditions and a typical reaction is shown below:

$$C_{57}H_{104}O_6(\text{liquid}) + 3H_2(\text{gas}) \xrightarrow{\text{Pd – Ni Catalyst}} C_{57}H_{110}O_6(\text{solid})$$

If a hydrogenation reactor of 1000 L capacity is charged with 200 °C hydrogen at 3 atm. and 20 kg of oil, compute the quantity of margarine produced.

1.26 A modified atmosphere package (MAP) of sliced apples contains 4% O_2, 4% CO_2, and 92% N_2 by volume at 1 atm and 4 °C. Calculate (a) the mole fraction of each gas component, (b) the partial pressure (kPa) of each gas component (assuming they behave as ideal gas), and (c) the density (g/L) of the gas mixture.

1.27 A modified oxygen environment is used to maintain quality in precut salad mixes during shipping. The atmosphere consists of O_2, N_2, and CO_2 with mole fractions of O_2 and N_2 of 0.07 and 0.50, respectively. What is the mass percentage of each gas if the bag contains a total of 100 moles of gas?

1.28 One mol of CO_2 is kept at 40 °C and 800 atm. Compute the volume the gas would occupy if (a) the gas behaves ideally and (b) the gas behaves nonideally. Additionally, determine the % error between the two volumes.

1.29 Carbon dioxide in a 12 cm diameter cylinder is compressed isothermally from 110 MPa to 200 MPa at 500 K. Compute: (a) the percent volume change through compression assuming the gas is nonideal and compressible and (b) the percent volume change if ideal gas behavior is assumed. (For CO_2, $T_c = 304.1$ K, $P_c = 7.38$ MPa.)

1.30 Compute the equilibrium solubility of CO_2 in water at 25 °C and 1 atm in moles/L and g/L under the following modified atmospheric conditions: 7.7% O_2, 79.2% N_2, and 13.1% CO_2.

1.31 A 2-L bottle of water was carbonated at 3 atm with pure CO_2 and stored at 5 °
C. The Henry's law constant under these conditions may be taken as
20 atm/(mol/L). After pouring out 250 mL of soda into each of 6 glasses, the
bottle was quickly capped and refrigerated at 5 °C. Assuming no CO_2 loss to
the environment, compute the concentration of CO_2 in the soda remaining in
the bottle. Will this soda be "flat" if the limit for "flatness" is set at 80% loss of
the initial CO_2 present?

1.32 A fly sitting at the liquid–air interface of a drink is struggling to get out. If the
surface area of the fly in contact with the drink is 5 mm^2 and the surface tension
of the drink is 0.03 N/m, estimate the amount of work it would need to do to
pull itself out.

1.33 Compute the temperature at which ice formation begins in an ice cream mix of
the following composition by weight:
Butter fat = 10%, solids-not-fat (snf) = 12%, sucrose = 15%, and stabilizer
= 0.2% (Hint: snf contains 54.5% lactose and the effect of small amounts of
salts may be ignored)

1.34 A cube (8 cm^3) of raspberry jelly on the plate is pushed laterally with 6 N force,
while a similar block of strawberry jelly on the same plate experiences a lateral
push of 8 N force, and their respective displacements are indicated in the
diagram. Calculate (a) the shear strain, shear stress, and shear modulus for
the raspberry and the strawberry jellies and (b) indicate which jelly is more
rigid and why. Assume the volume of the jelly remains constant.

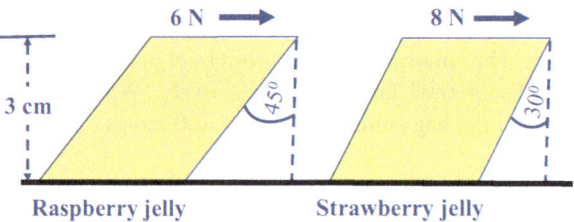

1.35 A marshmallow piece (height = 2 cm, diameter = 3 cm) has a Young's
modulus of 29 kPa and a Poisson's ratio of 0.48. Determine the dimensions
of the marshmallow after it is compressed with a force of 6 N.

1.36 Assume that a food product shaped like a rod (height = 10 cm, diameter =
1 cm) is subjected to an axial pull in the direction of its length (elongation) of
5 N. If the diameter of the product decreased by 0.3 mm and the length
increased by 8 mm, determine the values of the Young's modulus and
Poisson's ratio.

1.37 Assuming a density of 1000 kg/m^3 at the surface, estimate the density of water at the bottom of a 20-meter-deep lake. (Bulk modulus of water = 2.2×10^3 MPa.)

1.38 High-pressure processing (HPP) is performed by subjecting foods to isostatic pressure in the range of 400 to 600 MPa transmitted by compression of surrounding cold water (bulk modulus, 2.2×10^3 MPa). Estimate: (a) the percent change in the density of water at 500 MPa compared to its density at 4 °C and normal atmospheric pressure and (b) the final pressure required to change the density of water by 1%.

1.39 If 4.0 g of a seasoning (density 900 kg/m^3) is milled to reduce its average size from 5 micrometers to 1 micrometer, compute the percent change in the total surface area of the seasoning. (Assume spherical shape).

1.40 A cone 10 cm in diameter and 10 cm high is observed to form when 400 g of flour is poured from a funnel on the surface of a clean table.
   (a) Compute the volume and density of the flour.
   (b) Does the computed density represent true density or bulk density? Explain.
   (c) Is the flour free flowing? Explain why or why not?

Bibliography

1. Cheryan M (1998) Ultrafiltration microfiltartion handbook.2nd edn. CRC Press, New York
2. Carr RL (1965) Evaluating flow properties of solids. Chem Eng J 72:163–168
3. Ward IM, Sweeney J (2005) An introduction to the mechanical properties of solid Polymers.2nd edn. Wiley, West Sussex
4. Green DW, Southard MZ (2019) Perry's chemical engineers' handbook.9th edn. McGraw-Hill Education, New York

Systems, Processes, and Fluid Phase Equilibria

2

The job of the food engineer involves the design and/or operation of equipment needed to process food. Many different types of equipment are used to perform such processes as pasteurization, sterilization, evaporation, drying, freezing, and blending. If an engineer attempted to approach each piece of equipment as a totally new problem, the task would be overwhelming. Instead, he or she learns relatively few basic principles that are common to such equipment and applies them to solve the problems of design and operation.

The problems to be solved usually involve answering quantitative questions such as "How much energy will be lost to friction in this system of pipes?", "How large a pump do I need?", "What will be our energy costs, and how can I reduce them?", "What is the relative efficiency of two heating methods?", "Will the reduction in costs pay for a new piece of equipment?", etc. Answering these questions requires the use of mathematics, so this course will emphasize computing answers to engineering questions.

In this chapter, we review some of the basic principles from chemistry and physics that will help you in this endeavor.

2.1 Systems

The system is defined as any device or material chosen for evaluation, similar to an algebraic variable that can represent any number. We begin by defining a few terms.

2.1.1 Basic Terms

- **System**. The part of the universe that is chosen for analysis is called the system. It may consist of a complex set of processes, a part of that process, a simple object, or a quantity of material such as water. The selection depends on the problem that needs to be solved.

© The Author(s), under exclusive license to Springer Nature Switzerland AG 2024
S. S. H. Rizvi, *Food Engineering Principles and Practices*,
https://doi.org/10.1007/978-3-031-34123-6_2

- **Surroundings**. The entire universe that is not part of the system is called the surroundings. In most cases, however, only the immediate surroundings are of interest.
- **Boundary**. To analyze a system, one must clearly define the surface that separates it from the surroundings. This surface is called the boundary and may consist of real physical surfaces, imaginary surfaces, or both. "Clearly, defined" means that you can point to any place in the universe and say unambiguously whether it belongs to a particular system or to its surroundings.

Some examples of systems are as follows:

A. **A can of food**. A simple system might be defined as a sealed can and the food (e.g., the peas) within it. The boundary is the outside surface of the can. The surroundings consist of the entire universe outside the can.
B. *A pasteurizer*. A complex system might consist of all tanks, pipes, pumps, heat exchangers, etc. that make up a pasteurizer. The boundary consists of the walls of these tanks, pipes, etc. You can treat the pasteurizer as a single complex system or, if it better suits your purposes, you can break it down into many simpler subsystems. When you analyze a subsystem, the remaining parts become part of the surroundings.
C. **Water in a section of pipe**. A system might be defined as all the water in the 2-meter section of pipe between points A and B in Fig. 2.1. The boundary consists of the inside walls of the pipe and imaginary surfaces across the pipe at A and B. In this case, the rest of the water in the pipe is part of the surroundings.
D. **Gas in a model piston**. A cylinder that is open at one end has a piston slide into it, as shown in Fig. 2.2. The system can be defined as the gas that is trapped inside the cylinder by the piston. The boundary then consists of the inside wall of the cylinder and the inside surface of the piston. The surroundings include the walls of the cylinder and the air that surrounds it as well as the piston and anything the piston pushes against. This system is a useful model for studying many basic principles. We will return to it frequently.
E. **An Insulated tank**. A system might be defined as the contents of a tank that is completely enclosed in a very good insulator. The boundary of this system consists of the inside surface of the tank. A bomb calorimeter is such a system.

Fig. 2.1 A system of water in a section of pipe

Fig. 2.2 A cylinder and piston

2.1.2   Boundary Permeability

We are frequently interested in the exchange of matter and/or energy between a system and its surroundings. We can classify boundaries by what they allow to enter and leave the system:

A. **Permeable to matter.** If the boundary of a system allows matter to enter and/or leave, we say it is permeable to matter.

Example

1. *The pasteurizer, the section of pipe, and the cube of atmosphere are all systems that are permeable to matter. The can of the peas, the piston, and the insulated tank are impermeable to matter.* ◄

B. **Permeability to heat.** Heat is a form of energy that flows between two materials as a result of differences in temperature. If the boundary of a system allows this flow to take place, we say the system is permeable to heat

---

Example

2. *All of the examples above except the insulated tank are systems permeable to heat energy.* ◄

C. **Permeability to work.** Work is done when a force moves an object. A system can work on its surroundings by sliding or rotating a shaft or by simply expanding and thereby moving the surroundings. Similarly, the surroundings can do work on the system. If the boundary of the system is such that an exchange of work is possible, we say the system is permeable to work.

Example

3. *The model piston is permeable to work. If the piston is pushed in, the surroundings are doing work on the system. If the piston moves out, the system is doing work on the surroundings.* ◄

D. **Permeability to energy.** A system that is permeable to work and/or heat is said to be permeable to energy.

2.1.3 System Classification

One way of classifying systems is by their permeability:

A. **Isolated systems**. If a system such as the one in Fig. 2.3 is impermeable to both matter and energy (both heat and work), the system is said to be isolated.

Example

    4. *A sealed, rigid, insulated tank is an isolated system. None of the other examples above are isolated.* ◄

B. **Closed system**. A closed system is permeable to energy but not to matter as shown in Fig. 2.4. Energy may be transferred as either heat, work or both.

Example

    5. *The can of peas and the model piston are closed systems.* ◄

C. **Open System**. An open system is permeable to both matter and energy, as shown in Fig. 2.5.

Fig. 2.3 An isolated system

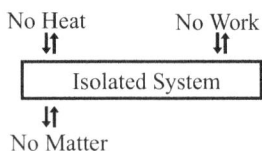

Fig. 2.4 A closed system

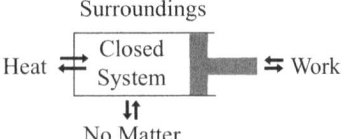

Fig. 2.5 An open system

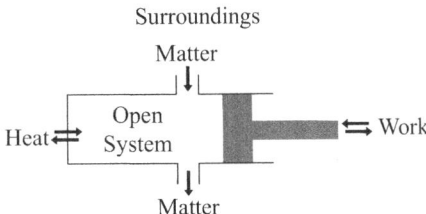

6. *Since some form of matter can enter and leave the pasteurizer, the section of pipe and the cube of atmosphere, they are all open systems.* ◄

D. **Adiabatic system**. Some closed systems, such as the one in Fig. 2.6, are permeable to work but not to heat or matter, and we call these adiabatic systems. An adiabatic process exchanges energy with the surroundings only as work.

7. *If the model piston was enclosed in a perfect insulator, heat could not flow between it and the surroundings. Nevertheless, if the piston moves, work will be exchanged between the piston and its surroundings, making it an adiabatic system.* ◄

2.2   System Properties

To understand a system, we must know its condition at any given moment. This is done by stating the values of such properties as the system's temperature, pressure, chemical composition, etc. Such properties are of two types:

Fig. 2.6 An adiabatic system

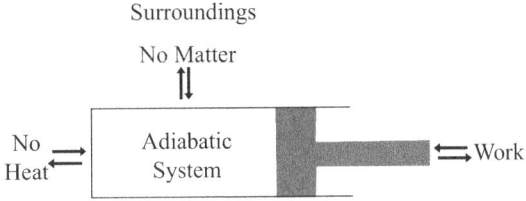

2.2.1   Extensive Properties

A system variable is an extensive property if it is dependent on the quantity or mass of the system. For example, the volume of a 1 kg butter is greater than that of a 1/4 kg bar. Volume, therefore, is an extensive property. Extensive properties include volume, mass, weight, energy, enthalpy, etc.

Figure 2.7 shows some extensive properties and indicates that they change in proportion to the change in mass of a system. Even in the same refrigerator, 10 kg of milk has greater mass, volume, weight, internal energy, etc. than 1 kg of the same milk.

Example

8. *Ten kilograms of milk in a refrigerator has the same density, protein concentration, temperature, etc., as 1 kg of the same milk in the same refrigerator, so all these are intensive properties.* ◄

2.2.2   Intensive Properties

A system variable is an intensive property if it is independent of the mass of the system. For example, the density of a bar of butter is the same whether that bar weighs 1/4 kg or 10 kg. Density, therefore, is an intensive property. Intensive properties include density, pressure, temperature, molar energy, chemical potential, viscosity, concentration, etc.

Figure 2.8 shows some intensive properties and indicates that they do not change just because the mass or size of a system changes.

Fig. 2.7 Extensive properties

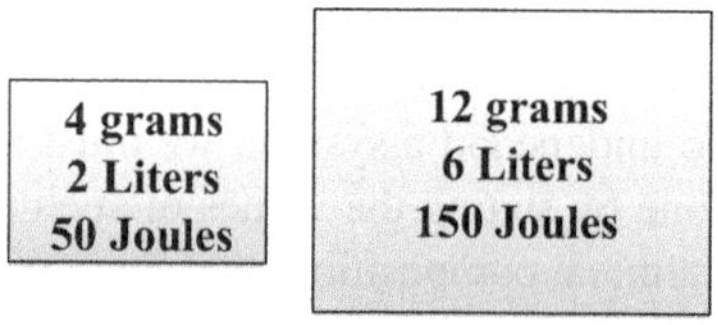

4 grams	12 grams
2 Liters	6 Liters
50 Joules	150 Joules

Fig. 2.8 Intensive properties

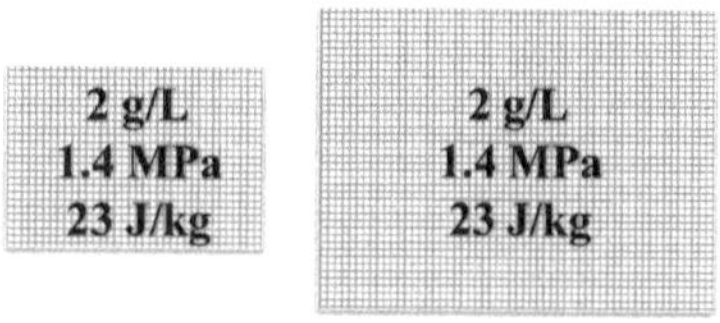

2 g/L	2 g/L
1.4 MPa	1.4 MPa
23 J/kg	23 J/kg

2.2.3 Making Extensive Properties Intensive

Extensive properties can be converted to intensive properties by dividing by the mass of the system. Thus, the volume of a system depends on mass and is extensive. On the other hand, volume per unit mass (or specific volume) does not depend on the mass of the system and so is intensive. Some extensive properties that have been made intensive include the following:

- Specific volume = volume per unit mass (reciprocal of density)
- Specific enthalpy = enthalpy per unit mass, etc.

Example

> 9. *Ten kilograms of milk may occupy a greater volume than 1 kg, but both have a specific volume of 1 L/kg.* ◄

2.3 System Processes

A process is an event that changes the state of a system by changing the value of one or more of its properties. When a process occurs, the system is in a new state, and there is a change in its temperature, pressure, mass, volume, chemical composition, or any other property. A process may involve internal changes in the system as well as exchanges of matter and energy between the system and its surroundings.

Example

> 10. *The systems described above might experience processes such as the following:*
>   - *Heating the can of food. Heating a can of peas from 30 °C to 115 °C is a process involving a change in system temperature and the transfer of heat energy from the surroundings to the system. If the can expands, some of the energy will be returned to the surroundings as work. There is no transfer of matter between the system and its surroundings.*
>   - *Filling a tank. In a liquid processing unit, it may be necessary to fill a tank within the system with milk, a process that changes the mass of the system by the transfer of matter from the surroundings. Furthermore, any energy contained in the milk, both thermal and chemical, will be transferred into the system.*
>   - *Reaction in a tank. Suppose hydrogen and oxygen are placed in a rigid, insulated tank and ignited, resulting in the combustion of hydrogen to form water. This process changes the temperature and composition of the system, but there is no transfer of energy between the system and its surroundings.* ◄

Check Your Understanding

1. Give an original example of each of the following types of systems and, for each, define the boundary and explain how your example illustrates the type of system.

   (a) Open System   (b) Closed System

   (c) Adiabatic System   (d) Isolated System.

2. For each of your examples, describe a process that might take place, indicating what changes take place in the system and what exchanges of matter and energy take place between the system and the surroundings. ◄

2.3.1   Types of Processes

A process usually involves a change in some but not all properties of a system. One way to classify processes is on the basis of the variable(s) that do not change. Since the prefix "Iso" means "constant value," it is used to name most types of processes.

A. **Isothermal process**. A process that runs at constant temperature is called isothermal.

Example

11. *Making yogurt in a constant temperature incubator is an isothermal process that changes the composition of the system.* ◄

B. **Isobaric process.** A process that runs at constant pressure is called isobaric (or isopiestic).

Example

12. *Heating soup in an open pan is an isobaric process since the system remains at atmospheric pressure throughout the heating process. This process changes the temperature and, to a very small extent, the volume of the system.* ◄

C. **Isochoric (isovolumic) process**. A process in which the volume does not change is called isometric or isovolumic.

13. *Cooking a meal in a pressure cooker is isochoric since the rigid cooker prevents changes in volume. Temperature, pressure, and composition are the variables that are changed.* ◄

D. **Adiabatic process**. A process in which there is no exchange of heat or matter between the system and its surroundings is called adiabatic. Quick expansion or contraction of a gas happens adiabatically. Putting topping from aerosol can on a desert may be considered *nearly* adiabatic. An adiabatic process is also an isenthalpic process.

14. *An adiabatic process takes place when an aerosol can is used to place topping on a desert. Within the can, the topping is at room temperature but under pressure. As it leaves the can, the pressure drops rapidly, allowing the topping to expand. Because the process is rapid and the topping is a poor thermal conductor, very little heat is exchanged between the system (topping) and the surroundings, making the process virtually adiabatic. The expanding material works against the surroundings, and this requires energy. Since no energy is absorbed during the process, the work must be at the expense of the internal energy of the topping, and this results in cooling. In the final state, the volume is increased while the pressure and temperature have decreased.* ◄

E. **Isentropic process**. Entropy is a measure of the disorder of a system. A process in which entropy does not change is called isentropic. To be isentropic, a process must not only be adiabatic but must also have no energy lost to friction.

15. *In refrigeration systems, a cold refrigeration fluid absorbs heat from the refrigerator. The fluid is then compressed isoentropically to raise its temperature without adding additional heat. In the new high-temperature state, the fluid can radiate the absorbed heat into the surroundings.* ◄

F. **Cyclic process.** A process that returns to its original state is called cyclic.

16. *Heating water in a sealed container and then cooling it to its original temperature is a cyclic process since in the final state, the system has the same temperature, pressure, mass, and composition as in the initial state. Heating and cooling in an open vessel would not be cyclic since some mass would be lost through evaporation. Cooking potatoes in a sealed container would not be cyclic since the composition of the potatoes would be altered.* ◀

2.4   Equilibrium State

If a system becomes isolated so that neither matter nor energy is exchanged with the surroundings and there is no net change in its macroscopic properties with time, the system is said to be in a thermodynamic equilibrium state. Once in equilibrium, the system will undergo no further changes until an exchange of matter and/or energy takes place with the surroundings. When disturbed, the system will attempt to return to the equilibrium state. At equilibrium, the potential energy of the system is always less than in any adjoining state.

17. *If a ball is placed in a U-shaped track such as the one in* Fig. 2.9, *it will roll back and forth, finally coming to rest at the bottom of the U. Here, the ball has its lowest potential energy, and this is its equilibrium state. If the ball is moved away from this state, it will tend to return to it.*

18. *If a mixture of water and ice is placed in a sealed insulated container (such as a Styrofoam bucket in* Fig. 2.10*), ice will melt or freeze, depending on the initial temperature of the system. Unless the system freezes or melts completely, this process will continue until the entire system is at the melting point temperature (0 °C at 1 atm. pressure). The system will remain*

Fig. 2.9 Rolling equilibrium

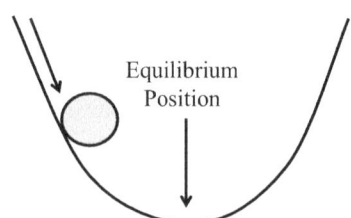

Equilibrium
Position

Fig. 2.10 Ice-water
equilibrium

> *as a mixture of ice and water in equilibrium, and no further changes take place in any system variable. If heat is added to the system, ice will melt until the system returns to 0° (unless the ice is used up). If heat is removed, water will freeze until the system returns to 0° (unless the water is used up).* ◀

2.5   Steady State

If matter and/or energy are entering and leaving a system, equilibrium will not be reached. However, when the system variables remain constant with time and the mass and energy input equals output, we say the system is in a steady state. On the other hand, if the mass and energy values of the system change with time, it will be an unsteady-state system. Many processes in the food industry operate under steady-state conditions.

Examples

19. *Consider the system in Fig. 2.11 consisting of the volume of milk occupying the section of heated pipe between points A and B. Milk is entering at point A at a constant flow rate of 5 gallons per minute. This makes milk out at point B at the same rate. Milk enters at 70 °F and is heated to 180 °F at point B. At any point in this system (in the shaded area), the volume, pressure, and temperature of the milk remain constant over time. This is a steady-state system.*

Fig. 2.11 A steady-state flow
and heating process

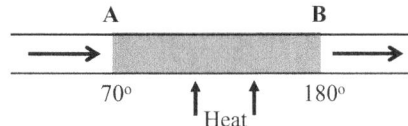

20. *It may be possible to adjust the faucet in the home water system so that the water flowing out exactly equals the inflow from the pump. When this is done, no change will take place in the mass in the tank, and the system is in a steady state. That is, if*

$$m_{\text{To Faucet}} = m_{\text{From pump}} \quad \text{and} \quad \Delta m_{\text{tank}} = m_{\text{From pump}} - m_{\text{To faucet}} = 0$$

21. *Food products such as spaghetti, Cheetos, etc. are produced by extrusion (see Fig. 2.12). In this process, ingredients are fed from a hopper at one end, and the food product comes out the other end. The temperature, pressure, and other physical properties of the material change as it moves down the extruder. However, in any short section of the extruder, the properties of the food moving through that section will always be the same, and hence, the system is in a steady state.* ◀

2.6 Other Types of Processes

Industrial manufacturing processes are classified according to how the raw material input and product output streams are carried out.

2.6.1 Batch Process

If a process begins at a fixed initial state and progresses in a specific order to a final state, it is called a batch process. Its advantages include lower initial and setup costs. Batch processes represent inherently an unsteady state.

Examples

22. *Bread dough is prepared by putting the ingredients in a tub and kneading them with a blender. This is a batch process.*
23. *In old pasteurizers, a volume of milk was placed in a tank or vat where it was heated and then cooled.* ◀

Fig. 2.12 Steady-state extrusion system

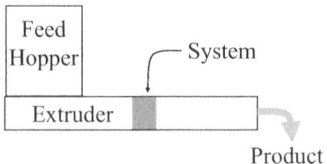

2.6.2   Continuous Process

If materials are continuously fed to a process via an input stream without any break in time sequence or extent and product is continuously removed through an output stream, the process is continuous. Advantages of continuous processes include higher efficiency and savings in time and energy. If such a process is operated so that at all points in the process, the temperature, pressure, and other system variables remain constant, it is also a steady-state process.

Frequently, a hybrid, **semi-batch** (or semicontinuous) process is also used. It is a combination of a batch and a continuous process with some elements that are batch and other elements that are continuous. In small bakeries, for example, dough kneading may be a batch operation while the steps of shaping and baking are done continuously.

Example

24. *In most modern pasteurizers, milk is fed continuously into a heat exchanger. As it flows through, it is first heated and then cooled. Done this way, pasteurization is both a continuous and a steady-state process.* ◄

2.7   State of a System

The state of a system is defined by the values of all its intensive variables. (In the next section you will find that only a limited number must be specified). The following intensive variables are particularly useful in defining the state of a system.

- **Temperature:** A measure of the average kinetic energy of the molecules in a system.
- **Pressure:** The force exerted per unit area of a surface.
- **Specific Volume:** The volume occupied by a unit of mass (kg, gm, or lb) of the system.
- **Concentration:** The amount of a particular component per unit mass or volume of the system. The amount can be expressed in grams, kilograms, gm moles, kg moles, pounds, or some similar unit.

Example

25. *The state of a particular system might be defined as follows:*

Composition = H_2O, Mass = 10 kg, Temperature = 45 °C, Pressure = 1 atm. Specific volume = 1 liter/kg ◄

2.7.1　Phase

You are familiar with the three phases of matter: solid, liquid, and gas. Liquids and gases are called fluids because they can flow. A phase is a homogeneous, physically distinct portion of the system that can be mechanically separated from the rest of the system. Mechanical separation can involve methods such as filtration, sedimentation, decantation, or even handpicking of crystals. It does not include absorption, evaporation, distillation, or extraction.

Example

26. *In a system consisting entirely of water, it is possible to have up to 3 phases: liquid, vapor (gas), and solid (except at extremely high pressures and temperatures).*
27. *Elemental sulfur can exist in 4 different phases, vapor, liquid, and two different solid forms, rhombic crystals and monoclinic crystals.*
28. *In a system containing two immiscible liquids, namely, oil and water, each liquid is a separate phase. Mayonnaise, for example, consists of an oil phase suspended in an aqueous phase.*
29. *All gases are miscible with each other, so in a mixture of gases, there can be only one phase.*
30. *In the molten form, bismuth and cadmium are miscible, and a mixture of the two forms a single liquid phase. Upon cooling, however, the two metals separate to form two separable solid phases.* ◄

2.7.2　Number of Components

A chemical constituent is a distinct chemical form, such as H_2O, NaCl, sucrose, H^+ ions, and OH^- ions. The number of components in a system is the smallest number of constituents whose concentration can be independently varied. We frequently classify systems according to the number of components.

Example

31. *A system consisting of water can exist in 3 different phases, but the composition of each is H_2O, so this is a one-component system. We could equally well describe the system by specifying the mass of oxygen, hydrogen, or hydroxyl ions, but once we specify the mass of one of these constituents, the rest are known.*
32. *A Na_2SO_4-water system is a two-component system, even though the various possible phases of the system can contain Na_2SO_4, $Na_2SO_4 \cdot 7H_2O$ and $Na_2SO_4 \cdot 10H_2O$, since specifying any two constituents is enough to allow us to compute the mass of the remaining constituents.*

33. *Pure water is a one-component system. Sugar–water is a two-component (binary) system. Oil, sugar, and water make up a three-component (tertiary) system.* ◄

2.7.3 Degrees of Freedom

The smallest number of independent intensive variables (pressure, temperature, concentrations, etc.) that must be specified to define the state of a system is called the degrees of freedom of the system. You can freely choose this number of state variables. The rest are then determined by state functions.

If you are told that a kg of nitrogen gas is at atmospheric pressure, you do not know its state because you do not know its temperature, specific volume, or other intensive properties. On the other hand, if, in addition to the pressure, you are also told that it is at 30 °C, its volume can be immediately computed using the ideal gas equation. We can also compute its other intensive properties. In other words, two (and only two) intensive properties were needed to define the state of the nitrogen gas, so it is a system with two degrees of freedom.

2.7.4 Gibb's Phase Rule

The state of a system can be defined by listing the properties of the system, including pressure, temperature, volume, mass, density, concentration of its components, etc. However, some of these variables are redundant. For example, if you specify that a system contains 1 kg mole of oxygen at 0.5 atmosphere pressure and 20 °C, we know from the ideal gas law that the volume is close to 48.1 cubic meters. In other words, because of various relationships between system properties, it is only necessary to specify a limited number to completely specify the state of the system. Gibb's phase rule provides a way to determine the minimum number of variables needed to specify a system.

When any system is in true equilibrium, Gibb's phase rule states that the number of intensive properties needed to unambiguously define its state is given by the equation:

$$df = 2 + C - P \tag{2.1}$$

where df = number of degrees of freedom (number of intensive properties),

C = number of independent components,
P = number of phases that exist in this state.

This rule does not specify which variables to use, only the minimum number of intensive properties needed to define a system. In general, there are many possibilities. Note that only intensive properties are considered. We consider 10 grams of water at 1 atmosphere and 100 grams at 1 atmosphere to be in the same state, even though their masses differ. It follows then that properties that depend on mass, i.e., extensive properties, do not determine the state of the system.

Example

34. *In a system consisting of pure liquid water, $C = 1$, $P = 1$, and the minimum number of intensive properties needed to define an equilibrium state is*

$$df = 2 + 1{-}1 = 2$$

Two intensive variables must be specified. Some possibilities include pressure and temperature or pressure and volume per mole.

35. *In a system consisting of water and ice, existing together in an equilibrium state,*

$$C = 1, P = 2 \text{ and } df = 2 + 1{-}2 = 1$$

In other words, only one intensive variable must be specified. For example, specifying a pressure of 1 atmosphere is adequate because at that pressure, a solid–liquid equilibrium can only exist at 0 °C. Similarly, temperature could be specified, and the pressure was determined from that.

36. *When the temperature of a pure water system is 0.0098 °C and the pressure is 4.58 mm Hg (0.006 atm), all three states can exist together in equilibrium. This is called the triple point, and in this state $C = 1$, $P = 3$ and*

$$df = 2 + 1{-}3 = 0$$

In other words, at the triple point, the system water has 0 degrees of freedom. This means that when the three phases exist together in equilibrium, it is unnecessary to specify any variables because such equilibrium can exist at only one state.

37. *A liquid consisting of a solution of sugar in water has $C = 2$, $P = 1$, and*

$$df = 2 + 2{-}1 = 3$$

To specify the state of such a system requires 3 intensive variables, which might be pressure, temperature, and sugar concentration. Equivalently, they could be temperature, volume per kg and water concentration or even temperature, pressure and refractive index since refractive index is a function of concentration. ◄

3. Suppose a tank contains steam and liquid water in equilibrium. How many degrees of freedom does this system have? Suggest variables that will define this system and give appropriate values for those variables.
4. A system consists of a solution of NaCl in water in equilibrium with ice. How many degrees of freedom does this system have and what variables could you use to define it? ◄

2.8 Phase Diagrams

In the last section, we pointed out that a system of pure water in a single phase has two degrees of freedom, and thus, any equilibrium state of water can be defined by two variables. These two variables can be represented by points on a two-dimensional Cartesian coordinate system (graph). It follows, therefore, that all possible states of water can be represented on such coordinates. This makes it possible to plot all states of water in a two-dimensional graph. Variable pairs that are commonly plotted include pressure versus temperature, pressure versus volume, enthalpy versus pressure, etc.

2.8.1 Pressure–Temperature Coordinates

Let us see what the states of water look like on pressure versus temperature coordinates. As shown in Fig. 2.13, the vertical axis represents pressure. We have drawn it here extending from 0.001 atmospheres absolute to 1000 atmospheres. Because of the wide range of pressures covered, this axis is logarithmically spaced.

The horizontal axis represents temperature. We have drawn it here extending from −100 to +400 °C. Any point on this diagram represents a particular state of a system. For example, point A represents a system at 0.1 atmospheres pressure and 200 °C.

Fig. 2.13 Coordinates for a phase diagram

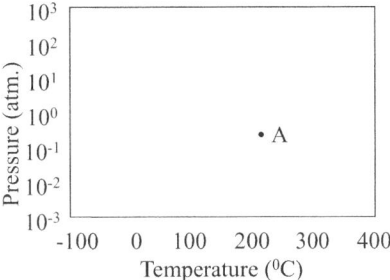

2.8.2   The Vapor Pressure Line

We started by measuring the boiling point of pure water at various pressures. At atmospheric pressure, it is 100 °C. If the pressure increases, the boiling point rises. If you decrease pressure, it falls. Plotting boiling point temperature versus pressure gives the curve in Fig. 2.14.

At any pressure–temperature combination to the right of this curve, water will be a gas at equilibrium. To the left of this line, it will be a liquid. If the pressure and temperature combination of a system falls exactly on the line, then liquid and vapor (gas) phases can exist together at equilibrium. When water is allowed to stand at any temperature, some evaporates and mixes with the air above it. This vapor creates a partial pressure that we call the vapor pressure of the water. In addition to telling us the boiling point at any temperature, the line we have drawn tells us the vapor pressure or the tendency to evaporate at any temperature.

The vapor pressure of liquid vapor in equilibrium at a desired temperature can be calculated using the following Clausius-Clapeyron equation, provided the vapor pressure is known at some other temperature, and if the enthalpy of vaporization is available.

$$\ln\left(\frac{P_1}{P_2}\right) = \frac{\overline{H}_{fg}}{R}\left(\frac{1}{T_2} - \frac{1}{T_1}\right) \tag{2.2}$$

where P_1 and P_2 are the vapor pressures at absolute temperatures T_1 and T_2, $\overline{H}_{fg}$ is the molar enthalpy of vaporization, and R is the gas constant.

2.8.3   Saturated Steam

Steam is used extensively as a heat source in the processing of foods. Much of the thermal energy contained in steam is released as the steam condenses into liquid water, which occurs at temperatures and pressures that fall on the vapor pressure line. For this reason, the properties of water along this line are of great interest and have been tabulated in Saturated Steam Tables as follows:

Fig. 2.14 The vapor pressure line

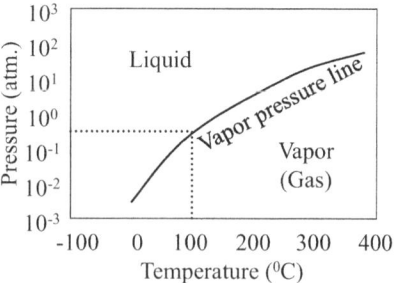

Table 2.1 Saturated steam table

		Specific volume (m³/kg)	
Temperature °C (°F)	Pressure kPa (atm)	Sat. liquid	Sat. vapor
0.01(32.02)	0.61 (0.006)	0.001000	206.1
20 (68)	2.34 (0.023)	0.001002	57.8
40 (104)	7.38 (0.073)	0.001008	19.5
60 (140)	19.94 (0.197)	0.001017	7.67
80 (176)	47.39 (0.467)	0.001029	3.41
100 (212)	101.32 (1.00)	0.001044	1.67
120 (248)	198.53 (1.96)	0.001060	0.892
140 (284)	361.3 (3.56)	0.001080	0.509

As you study the first two columns of this table, the following points should help clarify its meaning.

- The vapor pressure line in Fig. 2.14 is simply a plot of the pressures in column 2 versus the temperatures in column 1 of Table 2.1.
- If water is heated in an open vessel, the pressures listed in Table 2.1 are simply the partial pressures of water in the air above the surface of the liquid water.
- If a container is partially filled with water, then sealed and all air is removed, water will evaporate to fill the head space. In that case, the pressures listed in Table 2.1 represent the pressures developed in the head space.
- Note that at 100 °C, the vapor pressure of water is 101.32 kiloPascals. This is normal atmospheric pressure and is why water boils at this temperature.
- The temperatures listed in this table can be viewed as the boiling points of water at different pressures. As pressure increases, so does the boiling point.
- Canned foods are processed in high-pressure steam. From this table, we see that if the steam pressure is raised to 198 kPa (nearly 2 atmospheres), the temperature of the steam will be 120 °C. This temperature is frequently used in canning processes.

In addition to listing temperature and pressure, steam tables usually list other properties of water and steam that are functions of temperature (or pressure). Table 2.1 lists specific volumes, i.e., the volume occupied by 1 kg of water or 1 kg of steam. As you study these columns, note the following:

- Recall that the specific volume is the inverse of the density. A kg occupying a small volume must be denser than a kg occupying a large volume.
- The table shows that at 0.01 °C, 1 kg of water occupies 0.001 cubic meters or 1 liter. As the water is heated, it expands but very slowly so that at 100 °C, it occupies 1.044 liters, an increase of less than 5%. Even at 200 °C, a kg of water only occupies 1.157 liters. Remember, however, that for water to exist as a liquid at 200 °C, it must be kept at 15 atmosphere pressure (1553.8 kPa).

- At 0.01 °C, steam has a very low density and one kg occupies 206 cubic meters. As the temperature rises, the pressure increases, and the same kg becomes denser as it occupies progressively less volume. At 100 °C, a kg steam occupies 1.67 cubic meters (1670 liter).
- **Steam quality** is an important consideration in the use of steam. It refers to the relative proportion of vapor in the liquid and vapor mixture in steam, generally expressed as a weight percentage. For example, if 100 kg steam has 10 kg liquid water, then 90 kg is vapor, and the steam quality will be 90%.

2.8.4 Superheated Steam

Superheated steam is dry steam at a temperature higher than the saturation temperature. It is all vapor and contains no liquid water under thermodynamic equilibrium. The difference in temperature between a superheated steam and saturated steam at the same pressure is called the **degree of superheat**. Superheated steam will yield sensible heat first and must cool further to attain saturation conditions and then condense.

Saturated as well as superheated steam tables, covering a large range of temperature and pressure in both SI and Imperial units, are given in the Appendix A.4.

2.8.5 The Freezing Point Line

Now, the freezing point of water can also be determined at various pressures. At atmospheric pressure, it is 0 °C. At higher pressures, the freezing point declines slightly. Adding these freezing points to Fig. 2.14 yields the new line shown in Fig. 2.15. Locations to the right of this curve represent liquid phase conditions. Points to the left of this curve represent solid-phase conditions.

If the conditions of a system fall exactly on this line, both phases can exist together at equilibrium.

Fig. 2.15 The freezing point line

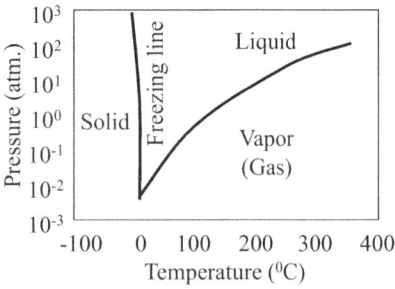

2.8.6 Sublimation Line

Note that the two lines meet near 0.06 atmospheres and 0 °C. Below this point, there is just a single line that separates the solid phase from the vapor (gas) phase. Adding the sublimation line to Fig. 2.15 yields Fig. 2.16. Under conditions along this line, the solid and gas phases can exist together at equilibrium.

Example

38. *On cold sunny winter days, one will sometimes observe snowbanks shrinking and leaving lacy crystals. These form when the vapor pressure of the air is low and the air temperature is below freezing. Under these conditions, ice sublimes into the gas phase.*
39. *If you place iodine crystals in the bottom of an enclosed container, you will eventually see iodine crystals higher up in the container. The iodine has gone from solid to gas, diffused through the container and converted back to the solid.* ◄

2.8.7 The Triple Point

The point where the three lines come together is called the triple point, as shown in Fig. 2.16. At this point only, all three phases can exist together at equilibrium.

Example

40. *The triple point for water is at a pressure of $T = 0.01°C$ and $P = 0.006$ atm.* ◄

Note that the triple point, where three phases can exist, has a pressure of 0.06 atm and a temperature of 0.0098 °C. When you know that all three phases can exist together you have no freedom to specify either a pressure or a temperature. This is exactly what the Gibb's phase rule predicts:

Fig. 2.16 The sublimation line

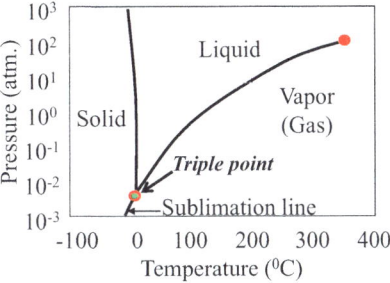

Table 2.2 Degrees of freedom in phase diagrams

No. phases at equilibrium			
1 Component	2 Components	A.Degrees of freedom	A.Geometry in diagram
3	4	0	Point
2	3	1	Line
1	2	2	Area
–	1	3	Volume

$$df = 2 + 1{-}3 = 0$$

Note also that there are 3 two-phase equilibria, solid–liquid, solid–gas, and liquid–gas, and that each of these falls on a line. If you specify either temperature or pressure, the other is immediately determined by the line. For example, if you have a liquid–gas equilibrium and you specify 1 atmosphere, the line tells you that the temperature must be 100 °C. Thus, only one variable can be freely chosen, again in agreement with the phase rule:

$$df = 2 + 1{-}2 = 1$$

Finally, note that each single phase is represented by an area. Within the limits of each area, you are free to select both the temperature and the pressure. In short, you have 2 degrees of freedom, as the phase rule suggests:

$$df = 2 + 1{-}1 = 2$$

These relationships are summarized in Table 2.2.

2.8.8   A Process on a Phase Diagram

To gain some insight into phase diagrams, let us follow a system of pure water in a process that heats it from temperature T_1 to temperature T_4 at a constant pressure of 1 atmosphere.

- We begin at point 1 in the diagram in Fig. 2.17. This is in the solid area, somewhere below 0 °C, so the system is in the form of ice. The vertical placement of this point corresponds to 1 atm of pressure.
- As we add heat to the ice, its temperature rises, and we move across the diagram to the right. Since we are heating at constant pressure, the line remains at the same vertical level.
- When the temperature reaches 0 °C, we are at the freezing point line (point 2) where the ice begins to melt and all the heat we add is consumed as latent heat of fusion. For this reason, the temperature does not rise as long as ice and water exist together as a two-phase equilibrium. From the point of view of the phase rule, we

Fig. 2.17 Heating under constant pressure

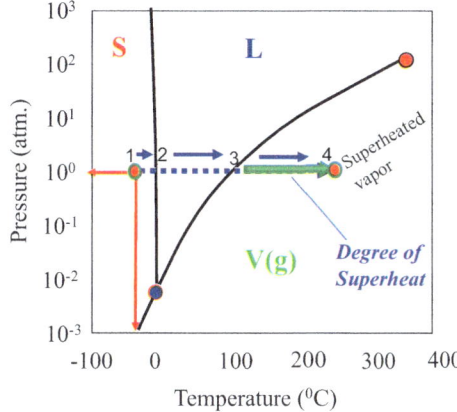

have one degree of freedom, and having selected 1 atm pressure, we are locked into a fixed temperature of 0 °C.

- When all the ice has melted, we return to a one-phase system with two degrees of freedom, and the temperature again begins to rise. This continues until we reach the boiling point line (point 3) at 100 °C.
- Now, water begins to change to gas (steam), and we again have a two-phase system with a single degree of freedom. The temperature now remains constant, and the heat is consumed as latent heat of vaporization.
- Once the water is completely converted to vapor, we again have a single-phase system with 2 degrees of freedom, and the temperature again rises until we stop the process at point 4. At this stage, the vapor (steam) is called **superheated steam,** and the degree to which it is away from the saturated conditions is known as **the degree of superheat.**

2.9   Critical Point and Supercritical Fluids

Now let us examine the phase diagram for water shown in Fig. 2.18. To illustrate several important properties, the pressure and temperature axes on the diagram are not drawn to a constant scale. Note that the vapor pressure line stops at a temperature of 374.4 °C and a pressure of 217.7 atmospheres. This is called the **critical point.** Any fluid above its critical pressure (Pc) and temperature (Tc) conditions is considered a supercritical (or dense) fluid. At the critical point, the meniscus between the vapor and liquid phases disappears, the two phases transition into one, and the surface tension effects disappear. In the supercritical region, 2-phase systems are not possible.

The following specific points with reference to the behavior of water, as shown in Fig. 2.18 above, should be understood:

Fig. 2.18 Pressure–
temperature (*P-T*) phase
diagram of water (*Not to
scale*) and illustration of some
of the unit operations in food
processing

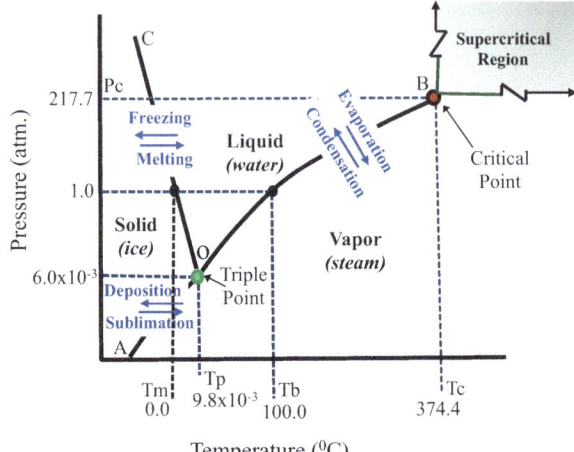

Temperature (^{0}C)

- Any point on the graph represents a state as defined by its pressure and temperature.
- Lines represent states in which phase changes take place and the two phases coexist at equilibrium.
- Various types of food processing unit operations that often involve transitions between two phases across the 2-phase equilibrium lines are also highlighted.
- Regions enclosed by lines represent states in which only a particular phase exists at equilibrium.
- At the **triple point** (0.0098 °C and 0.006 atm.) O, all three phases exist together at equilibrium.
- The O-B curve stops at the critical point, *B*, critical temperature (T_c) equal to 374.4 °C and critical pressure (P_c) equal to 217.70 atm. Beyond this point, the two-phase boundary disappears, and water exists as supercritical water.
- In the supercritical region, to the right and above critical point *B*, it is impossible to liquefy vapors by increasing pressure. A vapor above its critical temperature is referred to as gas. This is the distinction between a vapor and a gas, but often these terms are interchangeably used.
- A gradual transition between the properties of vapor and those liquids is observed as the critical point is approached.
- The highlighted region denotes space where supercritical fluids exhibit tunable densities and propensity to solubilize materials to various degrees.

The critical parameters of some of the more common supercritical fluids (SCFs) are shown in Table 2.3.

An order-of-magnitude comparison of the three physical properties of materials in various states is shown in Table 2.4. As may be observed, the supercritical state represents the more desirable properties between a gas and a liquid state. Their pressure tunability, ranging from gas-like to liquid-like, makes them very attractive

Table 2.3 Critical property data for selected supercritical fluids

Substance	Critical temperature (K)	Critical pressure (MPa)	Critical density (g/cm^3)
Ammonia	405.6	11.30	0.235
Acetone	508.1	4.70	0.278
Carbon dioxide	304.2	7.38	0.468
Ethane	305.4	4.88	0.203
Methanol	512.6	8.09	0.272
Propane	369.8	4.24	0.217
Water	647.3	22.00	0.322
Xenon	280.7	5.84	1.113
Nitrogen	−146.9	3.40	0.311
Oxygen	−118.6	5.05	0.432

Table 2.4 Typical physical properties associated with different fluid states

State of fluid	Density (g/cm^3)	Diffusivity (cm^2/s)	Viscosity (g/cm-s)
Gas			
$P = 1$ atm, $T = 15$–$30\,°C$	$(0.6$–$2) \times 10^{-3}$	0.1–0.4	$(1 - 3) \times 10^{-4}$
Liquid			
$P = 1$ atm, $T = 15$–$30\,°C$	0.6–1.6	$(0.2$–$2) \times 10^{-5}$	$(0.2$–$3) \times 10^{-2}$
Supercritical			
$P = P_c$, $T = T_c$	0.2–0.5	0.7×10^{-3}	$(1, 3, 6) \times 10^{-4}$
$P = 4P_c$, $T = T_c$	0.4–0.9	0.2×10^{-3}	$(3$–$9) \times 10^{-4}$

and "tunable solvents." Moreover, their one to two orders of magnitude greater than liquid diffusivity, low viscosity, and near-zero surface tension allow SCFs to have high mass transfer properties and to easily penetrate pores and crevices of solid matrices that are not easily accessible to liquids. This unique pressure- and temperature-dependent combination of density, diffusivity, viscosity, and surface tension make supercritical fluids ideally suited for a wide variety of industrial applications, including extraction, fractionation, sterilization, impregnation, chemical reaction, particle generation, cleaning and waste treatment, among many others.

Similar to gases, supercritical fluids are highly compressible, especially near the critical point where their densities are extremely sensitive to temperature and pressure. In Fig. 2.19, the variation in the molar density (ρ_m) with the pressure of an SCF is compared with its gaseous state. The density in the supercritical state is observed to vary from gas-like to liquid-like. Since the dissolving power of a solvent, as a first approximation, depends directly on its density, an SCF's solvent power is thus tunable, as represented by the isotherm in Fig. 2.19. This enables SCFs at high density to selectively extract solutes from solid or liquid matrices loaded in an extractor and deposit the extracted solutes from the SCFs by density reduction in a separator. Thus, the phase equilibrium data for the solute-SCF system are necessary to estimate the theoretical mass of SCF needed for extraction and separation of the

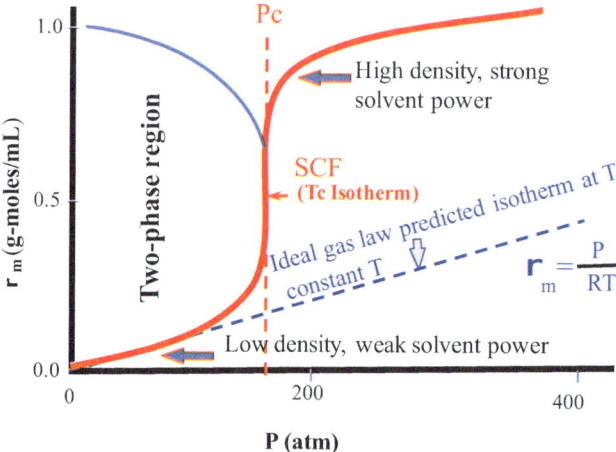

Fig. 2.19 Variation of molar density in gaseous and supercritical states

solute from the solid or liquid feedstock. Unfortunately, reliable thermodynamic equilibrium data are not as straightforward as they appear and must be determined from an appropriate equation of state. Thus, for the simplest case of a single solute, a widely used correlating equation for solubilities in supercritical fluids is the Chrastil equation:

$$\ln S = k \ln \rho + a/T + b \qquad (2.3)$$

where S is the solubility, ρ is the density of the supercritical fluid, T is the temperature, and k, a, and b are the solvent/solute pair specific constants.

The most common and important supercritical fluid of utility in food and bioprocessing is supercritical carbon dioxide (sc-CO_2), Fig. 2.20, since it is colorless, odorless, nontoxic, nonflammable, available in high purity, cost-effective, benign, and thermodynamically stable with low P_c (73 atm.) and near room temperature T_c (31.1 °C). It is used commercially for the supercritical fluid extraction (SFE) of caffeine from coffee and tea, various useful constituents from hops and spices and fat from different food matrices. Unlike methylene, hexane, and other organic solvents, it leaves no residual solvent in the products. Newer applications include the preparation of nutraceuticals, dietary supplements, controlled bioactive delivery systems, and other high-value products. Carbon dioxide is available in large quantities as a by-product of fermentation, combustion, and ammonia synthesis processes. Although carbon dioxide is recycled in large industrial operations, it is considered a "clean and green" solvent and is more environmentally friendly than most organic solvents typically used in extraction and separation processes. However, because of its low dielectric constant and zero molecular dipole moment, sc-CO_2 works well only for the extraction of nonpolar to moderately polar solutes

Fig. 2.20 Pressure–temperature phase diagram of carbon dioxide (*Not to scale*)

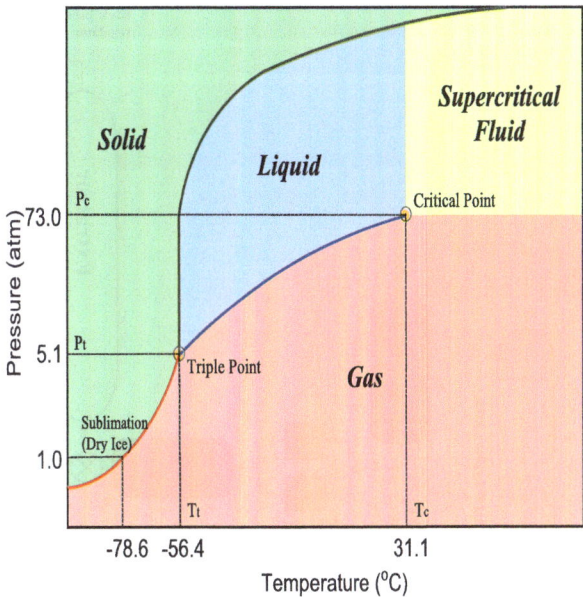

such as fats and oils. Polar solutes are better extracted using sc-CO_2 spiked with more polar cosolvents such as ethanol, propanol, water, etc.

A schematic diagram of a generic batch and a continuous extraction process using supercritical carbon dioxide as the solvent is shown in Fig. 2.21a, b, respectively. The extraction process occurs in two steps: (1) the extraction of solutes from the feedstock in the extractor/column and (2) the separation of the extracted components from the CO_2 stream. In a batch process, Fig. 2.21a, the feedstock is charged to the extractor and liquid, subcooled (to avoid cavitation), carbon dioxide is pumped through a heat exchanger to reach the desired supercritical state with high density. sc-CO_2 flows through the feedstock, extracting the soluble solutes. The solute-laden sc-CO_2 goes to the separator where separation of the extract is done either by lowering the solvent power (density) by changing the pressure or temperature or by adsorption on a bed of regenerable adsorbent (not shown).

In the case of multicomponent systems, depending on the extraction conditions used, several solutes with different solubilities may be simultaneously extracted because of the low selectivity of sc-CO_2. In such situations, extraction may be performed in successive steps at increasing pressures to obtain fractions containing similar components. Alternatively, fractional separation of the extracts may be carried out by adding several separators in series, each operating at a different but sequentially decreasing solvent density, as shown in Fig. 2.21b for a continuously operated extraction and fractionation setup using a packed column. For example, this protocol has been applied in the SFE and fractionation of milk fat into multiple fractions with different melting characteristics, Fig. 2.22, from the same extraction by modulating sc-CO_2 solvent power downward via alterations in pressure and

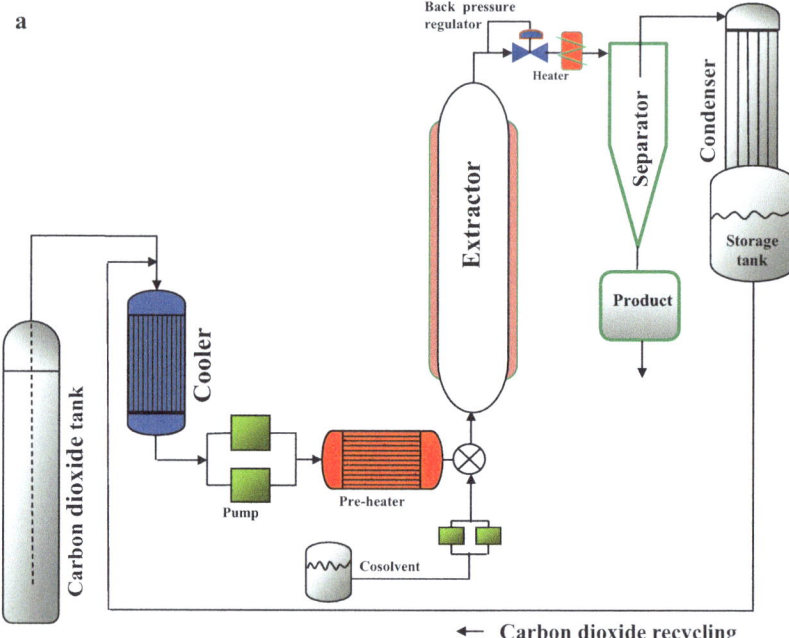

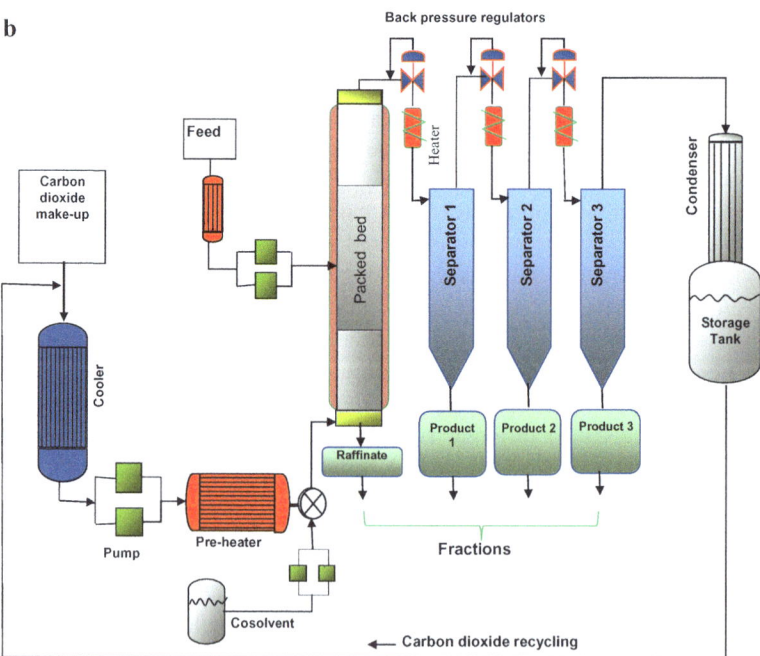

Fig. 2.21 Schematics of a batch extraction (**a**) and a continuous (**b**) supercritical fluid extraction and fractionation systems

Fig. 2.22 Fractions of milk fat obtained by supercritical carbon dioxide extraction

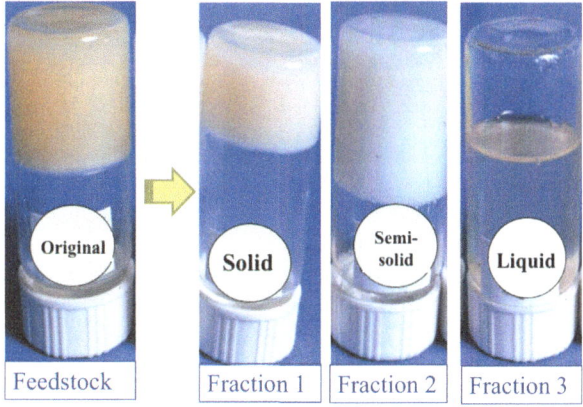

temperature conditions in each separator. In large commercial operations, carbon dioxide is recompressed and recycled to the extractor, and the process continues until the desired extraction is accomplished.

2.10   3-D Views of the Phase Diagram

Thus far, we have only examined pressure–temperature diagrams. Gibb's phase rule states that these two system properties are adequate to define the state of any system that, like water, contains 1 component and up to 3 phases. These are not, however, the only intensive properties that can be used to define the state of such a system. Other possibilities include the pressure and volume, temperature and enthalpy, or temperature and entropy. The second and third alternative must wait until we have introduced enthalpy and entropy. For now, let us examine the pressure–volume diagram of a single component system such as water. Figure 2.23 shows three diagrams for a system that behaves like carbon dioxide. The diagram on the left is the two-dimensional pressure versus temperature diagram we have been discussing. The solid region (S), the liquid region (L), and the gas or vapor region (G) can also be found. The lines separating these regions represent regions of two-phase equilibria. The triple point (TP) and critical point (CP) are shown.

The figure in the center is a three-dimensional plot of pressure versus temperature versus specific volume (volume per unit mass). This plot takes the form of a rather complex surface. Any point <u>on this surface</u> represents the temperature, pressure, and specific volume of a possible state of the system. Any point in this diagram that does not lie on the surface does not represent a possible equilibrium state. The pressure versus temperature plot we have been studying is a two-dimensional projection of this three-dimensional surface. The plot on the right shows a two-dimensional pressure versus volume projection of this surface.

Although the middle plot is in three dimensions, the fact that all possible states fall on a surface means that the system still has only 2 degrees of freedom. Plotting in

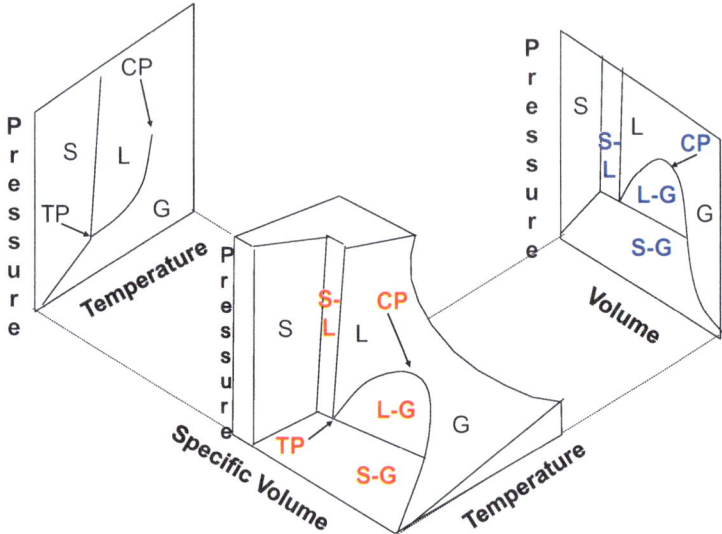

Fig. 2.23 Different views of the same phase diagram

three dimensions does not nullify the phase rule. Pick values for any two of these properties and the third property are determined by the surface.

Note that the lines that represent 2-phase states have become surfaces on the three-dimensional plot. The solid–liquid equilibria (S-L) lie on that narrow steep surface between the solid (S) and liquid (L) surfaces. The liquid–gas equilibria lie in a dome-shaped surface between the liquid (L) and gas (G) surfaces. The top of this dome is the critical point (CP), and the surface above this point represents the supercritical state. Below the liquid–gas surface is the solid–gas (S-G) surface representing sublimation states. Note that all of the two-phase surfaces (S-L, S-G, and L-G) lie parallel to the volume axis and at right angles to the temperature axis. This indicates that during these phase changes, the temperature remains constant while the volume increases. The width of these surfaces in the volume direction is an indication of the magnitude of the volume changes that take place during phase changes. Thus, the S-L surface is quite narrow, indicating a small volume change during melting. The L-G surface is much wider, particularly at lower pressures, indicating a larger volume change during evaporation. The S-G surface is very wide, indicating a very large volume change during sublimation.

2.10.1 A Constant Pressure Process

Figure 2.24 traces a process in which the system is taken from the solid phase to the gas phase by heating it at a constant pressure. Because the process is at constant pressure, the line is horizontal at a constant level on the pressure axis.

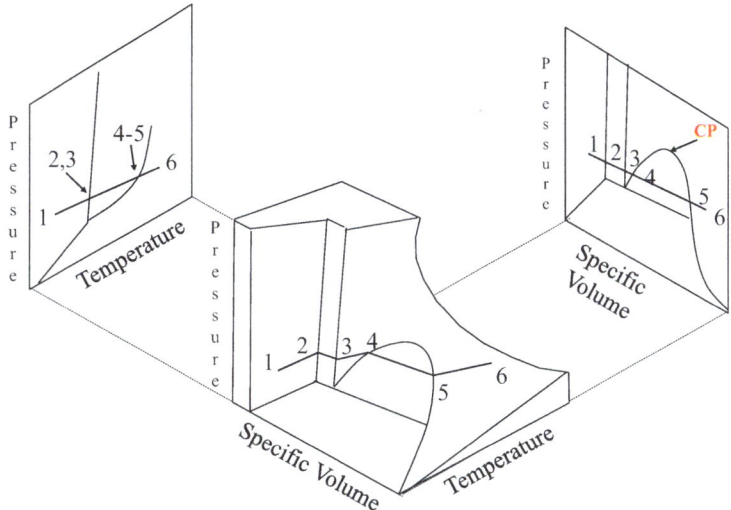

Fig. 2.24 A constant pressure process on phase diagram

- Point 1: Initially, the system is solid (S region).
- Point 2: As heat is added, the temperature increases with only a slight volume increase until the melting point is reached.
- Points 2 to 3: While the system melts, the temperature remains constant while the volume increases (S-L region). Because of this, the line between points 2 and 3 appears as a single point in the pressure–temperature diagram.
- Points 3 to 4: Once liquefied, the temperature begins to increase again with only a slight volume increase (L region).
- Point 4: At the boiling point, evaporation begins, and the temperature remains constant while the volume shows a large increase (L-G region). Again, the constant temperature makes this line appear as a single point on the pressure–temperature diagram.
- Points 5 to 6: Finally, in the gas phase, the system continues to increase in both temperature and volume (G region).

Check Your Understanding

5. Draw a line on the above diagram to represent a solid heated at a low enough constant pressure so that sublimation takes place. Describe the phases and pressure–temperature changes that occur during the process. ◄

2.10.2 A Constant Temperature Process

Figure 2.25 shows the same system being compressed to a smaller volume while holding the temperature constant. The line from 1 to 4 shows the changes that take place at "normal" temperatures. The line from 5 to 6 shows the changes that take place above the critical temperature.

- Point 1: The process begins with the system at a low pressure (G region). The pressure is low, and the volume is high.
- Point 2: As the volume is decreased, the pressure increases until the boiling (condensation) point is reached.
- Points 2 to 3: Now, as the system is compressed, the gas condenses into liquid with a much smaller volume. Because of this volume reduction, the pressure does not increase further until all the gas is converted to a liquid (L-G region).
- Points 3 to 4: Because liquid is relatively incompressible, any further reduction in volume is accompanied by a huge increase in pressure (L region). The changes in the supercritical region are somewhat simpler.
- Points 5 to 6: As the volume is reduced, the pressure of the gas phase simply increases as the gas changes gradually to a supercritical fluid. There is no distinct phase change in these conditions.

Check Your Understanding

6. Draw a line on the above diagram to represent a gas being compressed at a sufficiently low constant temperature so that deposition takes place. Describe the system and the pressure volume changes that occur during the process. ◄

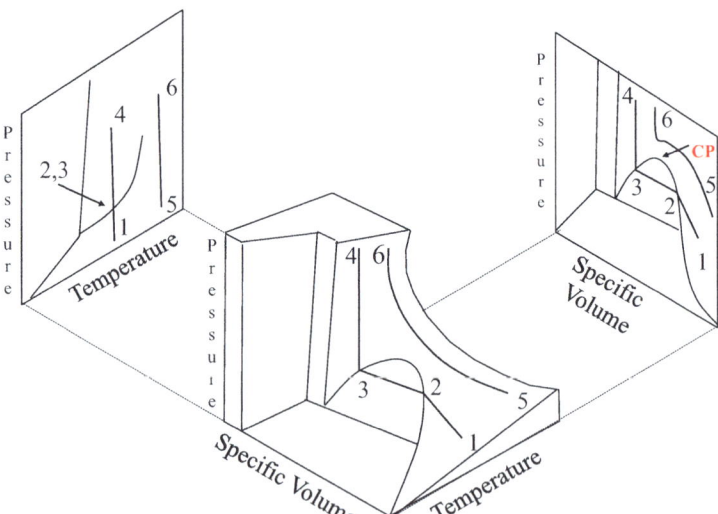

Fig. 2.25 A constant temperature process on phase diagram

2.11 Two-Component Systems: Vapor–Liquid Equilibria

In a two-component system such as a solution of alcohol in water, sugar in water or an iron-chromium alloy is called a **binary system**. In such a system, $C = 2$, and if there is only one phase, there can be as many as 3 degrees of freedom ($df = 2 + 2–1 = 3$), according to Gibb's phase rule. Describing all possible equilibrium states of such a system requires 3 axes. Therefore, we may select three intensive properties: temperature, pressure, and concentration (in percent or mole fraction), as shown in Fig. 2.26a. Since the concentrations of A and B must add to 100% (or 1.0 if in mole fractions), we can select either one. To avoid plotting in three dimensions, we fix the pressure and plot the other two variables, as shown in Fig. 2.26b. It is now a temperature-composition (T-xy) equilibrium phase diagram at a constant pressure.

In Fig. 2.27, three points *L, V,* and *Z* have been marked. Point *V* represents a system that is 25% *A* and 75% *B* at 60 °C. The diagram tells us that the entire system is in the vapor (gas) phase. Point *Z* represents a system that is 30% *A* and 70% *B* at 50 °C, and both *A* and *B* are present as liquids as well as vapors. In what follows, we will show how to determine their relative proportions in each phase. Point *L* represents a system containing 50% *A* and 50% *B* at 30 °C. It is entirely in the liquid phase.

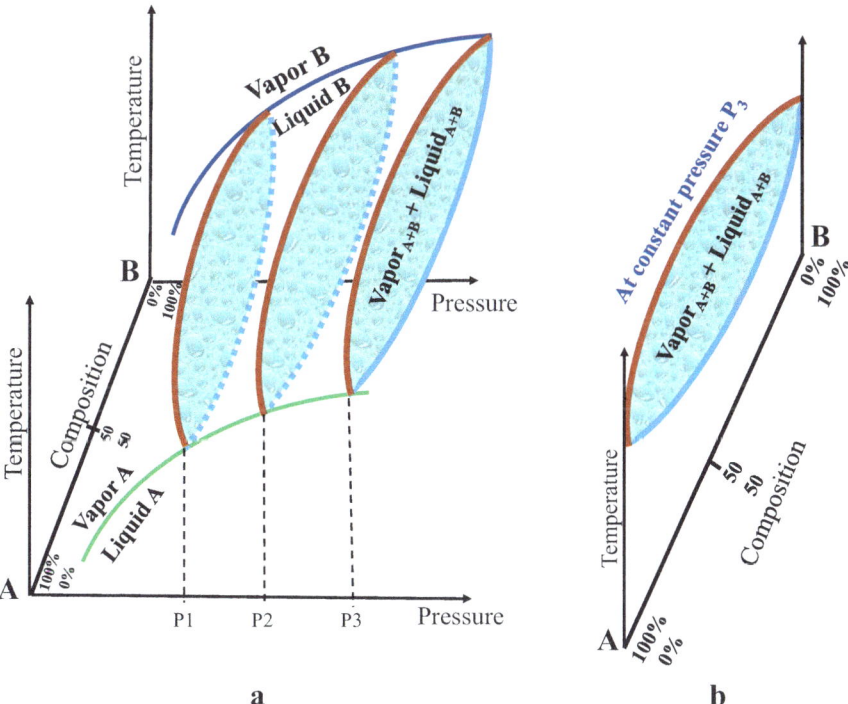

Fig. 2.26 3-D and 2-D plots of a binary system

Fig. 2.27 Temperature-composition (*T-xy*) diagram of a binary system at constant pressure

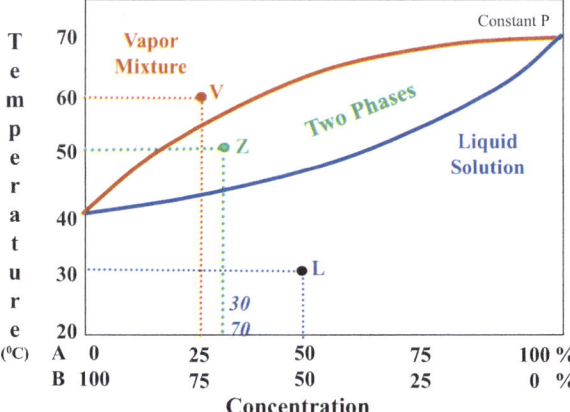

The diagram has been redrawn in Fig. 2.28. The upper line in this diagram indicates the temperatures at which a system that is being cooled just starts to condense and is called the saturated vapor line. Any point on this line is called the condensation or dew point (temperature) for a particular concentration of A and B. The lower line indicates the temperatures at which a system that is being warmed just starts to vaporize and is called the saturated liquid line. Since bubbles of vapor start to form at these temperatures, any point on this line is called the bubble point for a particular concentration. Figure 2.28 shows that pure *A* boils at 70 °C, while pure *B* boils at 40 °C. Now consider a mixture of *A* and *B* at the temperature indicated on the diagram by point *Z* (50 °C) in the two-phase region and it contains both liquid and vapor.

To know the compositions of the liquid and vapor phases at this temperature, an isothermal, horizontal line, called a **tie line**, is drawn through point *Z* on the phase diagram, as shown in the above diagram, intersecting the two saturated lines at points

Fig. 2.28 Descriptive terms for a two-component system

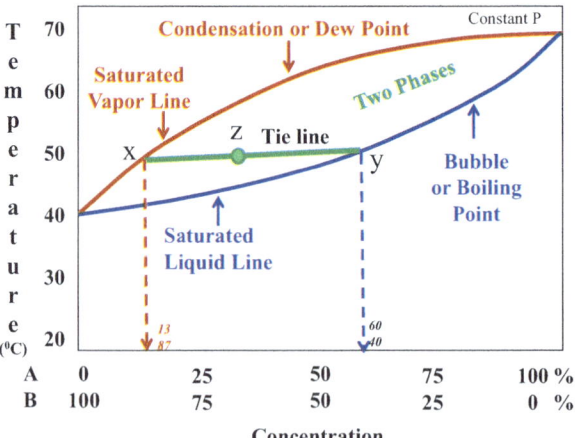

X and Y. The compositions of the two phases in equilibrium with each other at this temperature are obtained by dropping vertical lines from points X and Y to the composition axis and reading the values at the intersections. The diagram shows the composition of the vapor phase as 13% A, 87% B and the liquid phase as 60% A and 40% B. It is important to recognize that the overall composition of the mixture remains the same; it is only the relative proportions of the two phases and their compositions that change with temperature. The tie line information along with the lever rule and the principle of the conservation of mass may then be used to determine the ratio of the two phases.

To better understand the behavior of a mixture, consider a system that is 62% A that has been heated to 70 °C. Such a mixture falls at point E in Fig. 2.29, so it is a pure vapor. If you cool this mixture, the state of the complete system follows the vertical line from E to I. This tells us that the composition of the complete system does not change with cooling. However, the composition of the vapor phase follows the heavy line from E to U to H. The composition of the liquid phase follows the heavy line from V to G to I. When the temperature crosses the upper curve at point U, the condensation point, the mixture begins to condense into a liquid. However, at this temperature, the equilibrium composition of liquid is given by point V, so the

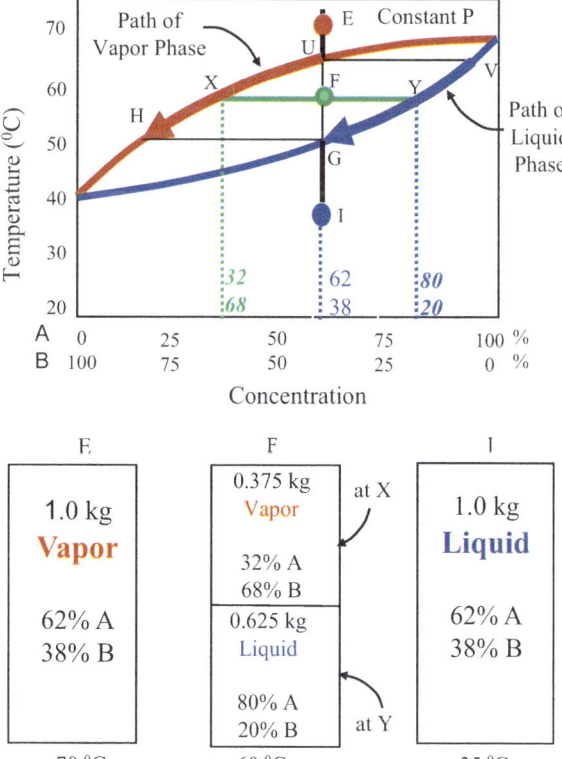

Fig. 2.29 Phase change in a two-component system

first tiny amount of condensate will have this composition, while the remaining vapor will have composition U. Notice that the liquid is initially rich in the component with the higher boiling point. As you continue cooling to 60 °C at point F, the equilibrium composition of the mixture is obtained by drawing a horizontal **tie line** through Y and connecting vapor and liquid lines at points X and Y. The liquid phase composition is **given** by point Y, while the equilibrium composition of the vapor phase is given by point X. From Fig. 2.29, we read that the liquid contains 80% A and 20% B while the vapor contains 32% A and 68% B.

With further cooling, the mass of liquid continues to increase, and the mass of the vapor continues to decrease until the system reaches point G. Here, the mixture will have become nearly all liquid with 62% A. An infinitesimal quantity of vapor will have the composition of point H. More cooling brings the system to point I at 35 °C with no further change in composition.

2.11.1 The Lever Rule

In the two-phase region, the relative masses of vapor and liquid can be obtained by performing a simple mass balance on the system. They can also be obtained by treating the horizontal line between X and Y as a lever with the fulcrum at point F, as shown in Fig. 2.30. The mass at point X must exactly balance the mass at point Y. This means that if

$A_x =$ Length from X to F, $M_x =$ Mass of vapor, $A_y =$ Length from F to Y and $M_y =$ Mass of liquid, then

$$A_x M_x = A_y M_y \qquad (2.4)$$

Example

41. *In the system shown in Fig. 2.29, the mixture at point I contains 62% component A. When heated to 60°C, we read from the graph that the vapor*

Fig. 2.30 The lever rule

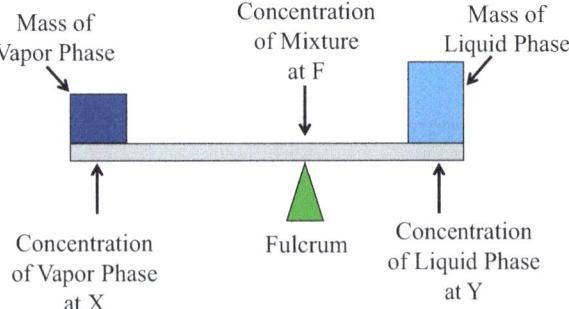

contains 32% component A, while the liquid contains 80% component A. What are the masses of the vapor and liquid at that temperature?

Solution (by mass balance): *The mass balance procedure is based on the fact that the mass entering a process will equal the mass leaving. Taking the mixture at E as the entering mass and the two phases at 60 °C as the leaving masses, we can write the following mass balance equations for 1 kg of mixture:*

$$1 \text{ kg} = V \text{ kg} + L \text{ kg (Overall mass balance)}$$

$$0.62 \text{ kg} = 0.32 \ V \text{ kg} + 0.80 \ L \text{ kg (Mass balance of component } A)$$

where V = the mass of the vapor phase and L = the mass of the liquid phase.

Solving these equations simultaneously yields:

$$V = 1 - L$$

$$0.62 = 0.32(1 - L) + 0.80 \ L$$

$$L = \frac{(0.62 - 0.32)}{(0.80 - 0.32)} = \frac{0.30}{0.48} = 0.625 \text{ (kg of liquid phase per kg of mixture.)}$$

$$V = 1 - 0.625 = 0.375 \text{ (kg of vapor phase per kg of mixture.)}$$

To verify, we enter these masses into the mass balance equations and show that both sides of those equations are equal:

$$1 = 0.375 + 0.625 \qquad\qquad\qquad \rightarrow \ = 1$$
$$0.62 = 0.32(0.375) + 0.80(0.625) \quad \rightarrow \ = 0.62$$

Solution (by lever rule): *The lengths of the lever arms from F to X and from F to Y are*

$$A_x = 62\text{–}32 = 30$$

$$A_y = 80 \ 62 = 18$$

For the lever to balance, the product of mass times length on each side must be the same so

$$30 \ V = 18 \ L$$

The masses V and L must sum to the mass of the mixture so that we start with 1 kg of mixture: $V + L = 1$

Solving these last two equations simultaneously yields

$$V = 1 - L$$

$$30(1-L) = 18\,L$$

$$L(30 + 18) = 30$$

$$L = \frac{30}{(30 + 18)} = 0.625\,(kg\ liquid\ phase\ per\ kg\ of\ mixture)$$

$$V = 1-0.625 = 0.375\,(kg\ of\ vapor\ phase\ per\ kg\ of\ mixture.)$$

To verify the results, we enter the masses into the lever equation and show that the two sides are equal:

$$30(0.375) = 18(0.625) = 11.25 \qquad \blacktriangleleft$$

Check Your Understanding

7. What is the percent A in each phase at 55 °C in Fig. 2.29? Verify that the mass of A in the liquid and vapor phases is 19.5 wt.% and 18.5 wt.%, respectively.
8. Verify that as the temperature drops, the percent of A in both liquid and vapor decreases. Additionally, verify that as this happens, the mass of liquid increases and the mass of vapor decreases. ◄

2.12   Vapor–Liquid Equilibria and Distillation

As discussed above, the vapor–liquid equilibrium information is essential for determining how a solution will separate between the vapor phase and a liquid phase. In addition, nowhere such information is more fundamental than in the **distillation** processes which are used extensively to separate liquid mixtures based on the relative volatilities (vapor pressures) of their components. It basically involves heating a solution to boiling to vaporize its components followed by condensing the vapor to recover it in liquid form, enriched in the more volatile fraction.

In the food industry, distillation is used to extract volatile flavor and aroma compounds from their natural matrices and in the production of alcoholic products such as gin, rum, and vodka from fermented wash or mash (the finished alcohol-water mixture following fermentation and ready to be distilled). Since the upper limit for yeast fermentation of grains is approximately 16% alcohol, distillation becomes necessary to produce hard liquors containing higher levels of alcohol. An interesting part of spirit distillation is that copper is the material of choice for the construction of stills since it reduces the unpleasantly pungent eggy smell due to sulfur compounds (e.g., hydrogen sulfide, sulfur dioxide, and other thiols) by sequestering sulfides from the vapor phase and forming copper salts that are left behind in the still after distillation.

To design any distillation process, we first need to understand the basics of the phase behavior of liquids in a mixture. For an ideal binary solution containing miscible components A and B at a constant temperature, according to **Dalton's law** of partial pressures discussed in Chap. 1, the total pressure is related to the partial pressures of the individual components as:

$$P_T = P_A + P_B \tag{2.5}$$

where P_T = total pressure, P_A = partial pressure of A, and P_B = partial pressure of B.

The partial pressure of each component in turn is related to its mole fraction as:

$$P_A = y_A P_T \text{ and } P_B = y_B P_T \tag{2.6}$$

where y_A and y_B are mole fractions of A and B in the vapor phase.

Assuming that the solution is nonreacting and shows ideal behavior obeying **Raoult's law,** we can also write:

$$P_A = x_A P_A^* \text{ and } P_B = x_B P_B^* \tag{2.7}$$

where x_A and x_B are mole fractions of A and B in the liquid phase and P_A^* and P_B^* are vapor pressures of pure A and pure B, respectively.

Now, applying Dalton and Raoult's laws to the vapor phase, we obtain:

$$y_A = \frac{x_A P_A^*}{P_T} \text{ or } \frac{y_A}{x_A} = \frac{P_A^*}{P_T} \tag{2.8}$$

The ratio of the mole fractions of a component in the vapor to liquid phases is known as **the distribution coefficient** (K) and thus:

$$K_A = \frac{y_A}{x_A} = \frac{P_A^*}{P_T} \quad \text{and similarly,} \quad K_B = \frac{y_B}{x_B} = \frac{P_B^*}{P_T} \tag{2.9}$$

As may be observed from the above expressions, if component B is more volatile than A, it would have a lower boiling point (i.e., $P_B^* > P_A^*$); thus, $y_B > y_A$, and the vapor phase will be enriched in the lower boiling (more volatile) component B. This phenomenon is the basis for the separation of a mixture of components by **distillation**.

To compare the volatilities of the two components, another term, called the **relative volatility (α),** is defined as:

$$\alpha_{BA} = \frac{y_B/x_B}{y_A/x_A} = \frac{P_B^*}{P_A^*} = \frac{P_B/x_B}{P_A/x_A} \tag{2.10}$$

As may be noticed from Eq. (2.10), a larger value of α will indicate a greater degree of separability. On the other hand, if the relative volatility between two components is unity, no separation will occur during distillation. For a mixture, the relative volatility varies with its composition. The above information can be used to determine the vapor–liquid equilibrium (VLE) plot at a fixed pressure.

The distillation operations in practice today may be categorized into two types: a) batch (simple, differential, or Rayleigh) distillation and b) fractional distillation. If the difference in boiling points of the components in a mixture is large, such as water and salt, a single-stage equilibrium-based simple distillation is used. If the boiling points are close, such as ethanol (78.40 °C) and water (100 °C), multiple-stage equilibrium or many redistillations may be needed, and the process is referred to as fractional distillation (or sometime as rectification).

2.12.1　Simple Batch (Differential) Distillation

As the name implies, it operates in a batchwise fashion. Consider a fixed (batch) quantity of the binary solution of A (less volatile) and B (more volatile), discussed earlier, that is loaded in a still (pot) fitted with a heating element that provides energy and connected to a water-cooled condenser that takes out energy to condense the vapor, Fig. 2.31. Heating the solution to boiling results in a vapor where the composition will be enriched in the more volatile component, B. The liquified vapor, called distillate or condensate, is collected in a receiver. As the distillation process progresses, the ratio of A and B in the pot keeps continually changing with time, becoming richer in less volatile component A. As a result, the boiling point of the solution rises, and consequently, the ratio of B and A in the vapor phase also changes, giving distillate with a changing ratio of B and A. The distillate may thus be collected as a composite product, or the receiver may be frequently replaced to collect separate **cuts** of the product. The composition changes can be analyzed using the phase diagram.

A batch process is inherently an unsteady state, the vapor is richer in the more volatile component B and the composition of the liquid and vapor is not constant. It

Fig. 2.31 Simple batch distillation

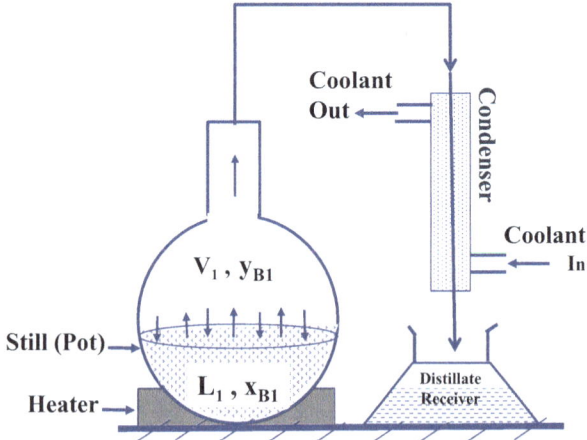

may be analyzed using time-dependent changes in the concentration in the pot using the Rayleigh approach, as discussed below.

Assuming L_1 and L_2 as the initial and final total moles with mole fractions of x_{B1} and x_{B2}, respectively, of the more volatile component B in the still, a differential mass balance at any time after the start of the process can be written. If the liquid and vapor leaving the still are in equilibrium and a small amount of the liquid, dL, is assumed to evaporate into the vapor phase with mole fraction of B as y_B, the mass balance on B can then be written as:

Initial moles of B in still $=$ Moles of B vaporized $+$ Moles of B left in the still, i.e.

$$x_B L = y_B dL + (x_B - dx_B)(L - dL) \tag{2.11}$$

$$\text{or,} \quad x_B L = y_B dL + x_B L - x_B dL - L\,dx_B + dL\,dx_B \tag{2.12}$$

The quantity $(dL\,dx_B)$ in the above expression is too small and insignificant and may be neglected. Eq. (2.12) can be rearranged and written as follows:

$$dL(y_B - x_B) = L\,dx_B \tag{2.13}$$

Separating the variables in the above equation and integrating it with limits from L_1 to L_2 and x_{B1} to x_{B2}, the following expression is obtained:

$$\int_{L_1}^{L_2} \frac{dL}{L} = \int_{x_{B1}}^{x_{B2}} \frac{dx_B}{(y_B - x_B)} \tag{2.14}$$

$$\text{Or,} \quad \ln\left(\frac{L_2}{L_1}\right) = \int_{x_{B1}}^{x_{B2}} \frac{dx_B}{(y_B - x_B)} \quad \text{or} \quad \ln\left(\frac{L_1}{L_2}\right) = \int_{x_{B2}}^{x_{B1}} \frac{dx_B}{(y_B - x_B)} \tag{2.15}$$

Equation (2.15) is called the **Rayleigh equation,** and differential distillation is thus also termed Rayleigh distillation. If the equilibrium data are available for the components of interest, the integration of the above equation can be done analytically or graphically by using equilibrium data to plot $1/(y - x)$ versus x and measuring the area under the curve. The vapor–liquid equilibrium data may be available either in graphical form or as an equation.

2.12.2 Fractional Batch Distillation

For separating solutions that are composed of components with boiling points that are relatively close together, the simple distillation discussed earlier will not be enough, and fractional distillation based on multiple-stage evaporation-condensation is used to achieve good separation. In a batch fractional distillation setup, Fig. 2.32, the solution mixture is heated to vaporize the volatile components in the mixture, and the vapors rise up through a **fractionating column**, where the separation occurs by a

Fig. 2.32 Fractional batch
distillation

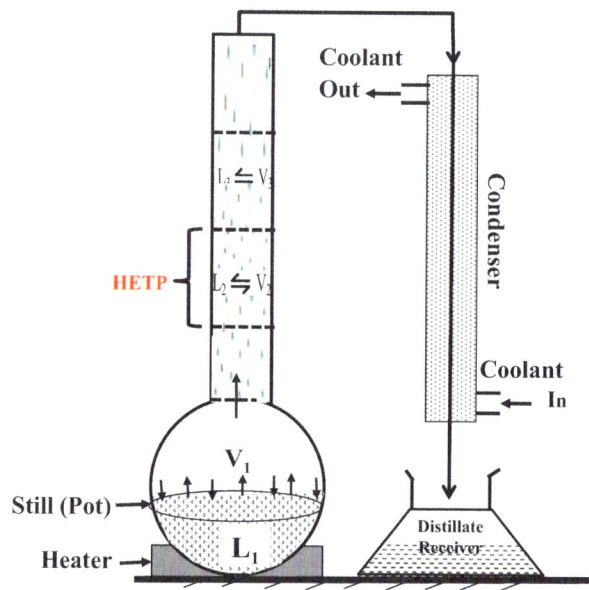

series of condensation and evaporation events, each mimicking a simple distillation
and enriching the vapor phase in the lower boiling component. The column as a
whole thus becomes a set of connected equilibrium stages, and each stage is called a
theoretical plate. The column is randomly filled with small pieces of packing
materials (such as glass beads, rings, short tubes, etc.) or structured metal sheets
that provide a large surface area for both condensation and evaporation to occur
more efficiently. The height of packing in the column that will give the same
separation of vapor and liquid as an equilibrium stage is called the **height equivalent
to a theoretical plate** (**HETP**). The size of the HETP depends on the packing
material's performance, and the lower the size is, the more efficient the separation
and shorter the distillation column. Increasing the interfacial area between the liquid
and vapor phases and making the liquid flow paths thin to speed up diffusion are
some of the ways that are used to reduce the HETP.

As the heating process continues, a temperature gradient develops along the
column, with temperature falling as vapors rise up the column. The vapors of higher
boiling components condense first and flow down toward the still. The vapors of
lower boiling liquid flow upward, condensing and evaporating sequentially. When
the top of the column reaches the boiling temperature of the lowest boiling component, vapors start to enter the condenser, where they are cooled, condensed back into
liquid, and collected in the distillate receiver.

The principle and mechanics of a fractional distillation unit are better understood
with the aid of a phase diagram. Consider a solution (L_1) of 90% *A* (boiling point 70 °
C) and 10% *B* (boiling point 40 °C) heated to 60 °C. Figure 2.33 shows its phase
diagram. The vapor (V_1) in equilibrium with the solution (L_1) at 60 °C enters and
flow upwards in the column where it experiences lower temperatures and condenses

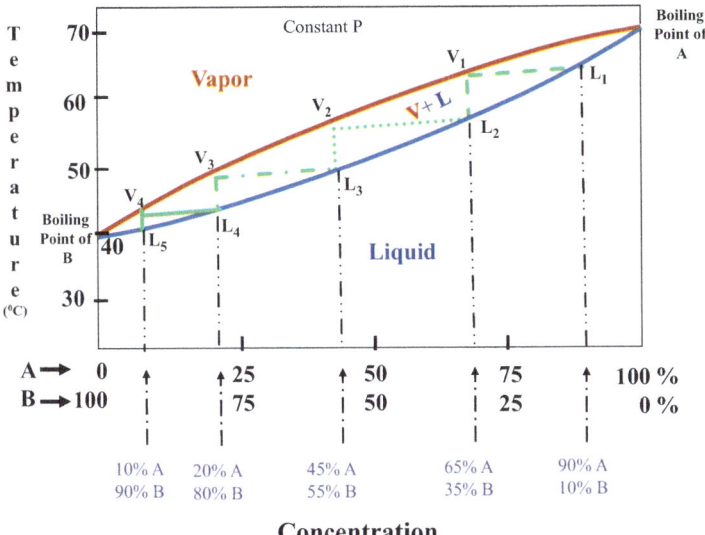

Fig. 2.33 Fractional distillation of a 2-component (*A* and *B*) solution with four theoretical plates

to a liquid L_2 (with the composition of V_1). This equilibrium vaporization-condensation event (L_1 to V_1 to L_2) represents one theoretical plate and would produce a liquid, L_2, containing 65% *A* and 35% *B*, which has a lower boiling point since it contains more of the lower boiling component *B*. Warmed up by the heat of condensation of vapors coming from below, a second vaporization-condensation event begins. Here, L_2 turns into vapor of composition V_2, which condenses further up the column to liquid L_3 of composition 45% *A* and 55% *B*. Depending on the designed column height and packing material efficiency, several successive vaporization-condensation steps (theoretical plates) can take place to produce a final distillate of desired composition. The process represented in Fig. 2.33 requires four theoretical plates to produce a distillate of 90% *B* from a solution initially containing 10% *B*.

A separation factor (SF) of *B* over *A* is defined as:

$$SF = \frac{(W_B/W_A)_D}{(W_B/W_A)_F} \tag{2.16}$$

where W_A and W_B are wt.% concentrations of *A* and *B* in the distillate (*D*) and feed (*F*), respectively. The higher the separation factor is, the more efficient the distillation process.

The process may be run with or without reflux (where a portion of the condensate is returned to the distillation system to cool and condense the up flowing vapors and make the distillation process more efficient). To minimize flavor loss from delicate mixtures, distillation is often performed without reflux. A batch operation without reflux is often called **differential** distillation.

In most real-life situations, deviations from the ideal solution behavior are observed. The above analysis, aided by more advanced modifications, is used to describe the behavior of such mixtures. It is also well known that distillation cannot produce a completely pure component from a mixture with close boiling mixtures since this would require the other component in the mixture to have no partial pressure. Several other separation techniques are required as adjunct processes to distillation to obtain very pure products. Fractional distillation is commonly used for separating, refining, and purifying different components from a mixture of liquid solutions. Ranging from crude oil refining to purification of reagents and gases to separation of ethanol from a water-ethanol mixture, distillation is extensively used for high value-added as well as commodity products. However, it has its limitations as well, as discussed below.

Figure 2.34 shows a typical but approximate equilibrium diagram for ethanol-water mixtures in terms of mole fractions $(x - y)$ and the corresponding temperature-composition (T-xy). It is well known that ethanol boils at 78.4 °C, water boils at 100 °

Fig. 2.34 Vapor–liquid equilibrium (T-xy and corresponding $x - y$) diagram of ethanol-water mixture

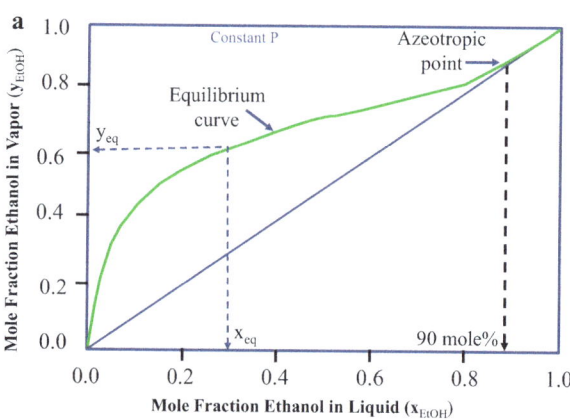

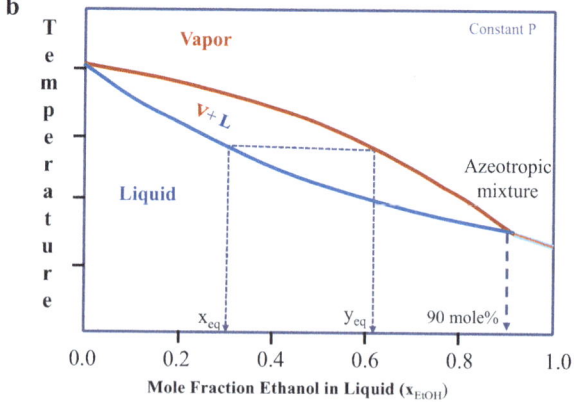

C, but their mixture boils at 78.2 °C, which is lower than either of its constituents. As may be noted from the figure, the equilibrium curve intersects the 45° line at approximately 90 mole% ethanol, where the composition in both the liquid and vapor phases become identical. The mixture cannot be further concentrated or purified by distillation. This is known as the azeotropic point, and the solution is referred to as an azeotropic (constant boiling in Greek) mixture or an azeotrope. By adding an entrainer that affects the volatility of one of the components, an azeotrope can be further distilled. Distillation under vacuum is also sometimes used since lower pressures are known to shift the azeotropic point.

2.12.3 Vacuum and Molecular Distillation

For heat-sensitive materials and for materials that normally boil at high temperatures, distillation is carried out at reduced, below atmospheric pressures or under vacuum. By reducing the pressure (50–15 Pa) in the system, the boiling point of the mixture is reduced, and distillation occurs at a lower temperature. Therefore, **vacuum distillation** is also known as low-temperature distillation. As may be expected, it increases the volume of vapor produced per unit mass of liquid distilled, and the size of the distillation setup thus increases.

When the pressure during vacuum distillation is so low (<2 Pa) that vapors are in their mean free path regime in the distillation unit, it is called **molecular distillation**. In the food industry, it is commercially used for the separation and purification of polyunsaturated fatty acids from oil.

2.12.4 Steam Distillation

For separating solutions containing immiscible and chemically nonreactive components that have a high boiling point and low volatility and are heat sensitive, steam distillation is alternatively used. In this method, steam is generated in the still to heat and evaporate the volatile components at temperatures below their normal boiling points.

In a mixture of immiscible components, the vapor pressure exerted by each component is independent of the vapor pressure of other components. In such cases, the total vapor pressure is equal to the sum of the individual pure vapor pressures:

$$P_T = \sum_i^n P_i^*$$ (2.16)

For example, if at 100 °C the vapor pressure of component A is 20 kPa and that of component B is 30 kPa, their total pressure of 50 kPa is well below the atmospheric pressure of 101 kPa for the two components to boil off. By adding steam at a partial pressure equal to or more than 51 kPa, the mixture can be made to boil and distill with steam from the solution. Examples of steam distillation include essential oil extraction from plant materials and fatty acids from soybean oils.

2.12.5 Fractional Continuous Distillation

While batch distillation is extensively used in many applications, most industrial operations opt for economically attractive and continuous fractional distillation systems, which in essence are many simple distillations in series combined into one unit. A schematic diagram of such a multistage fractional distillation column is shown in Fig. 2.35. The column is fitted with either a fixed number of sieve trays or filled with a packing material of high surface area containing the equivalent number of theoretical stages. The feed continuously enters the column somewhere at mid-stage, and the exact location depends on the feed stream composition and temperature. The section of the fractionating column above the feed point where the more volatile component is vaporized from a liquid feed is called the enriching (or rectifying) section, while the section below where the less volatile component is removed from a vapor feed is called the stripping section. Inside the column, the feed stream is heated and separates into vapor and liquid phases, staying in equilibrium with each other. The vapors move up the column while the liquid flows downward, passing through the sieve trays or over the packing materials. During this continuous, counter-current flow, the vapors condense and then re-evaporate and recondense under equilibrium conditions in each stage. During each re-evaporation, the new vapor phase becomes richer in the more volatile component. Thus, the greater the number of stages, the better the separation. A temperature gradient develops over the length of the fractionating column, and as the purity of the vapors increases, the temperature decreases and approaches the boiling (condensing) point of the more

Fig. 2.35 Continuous fractional distillation

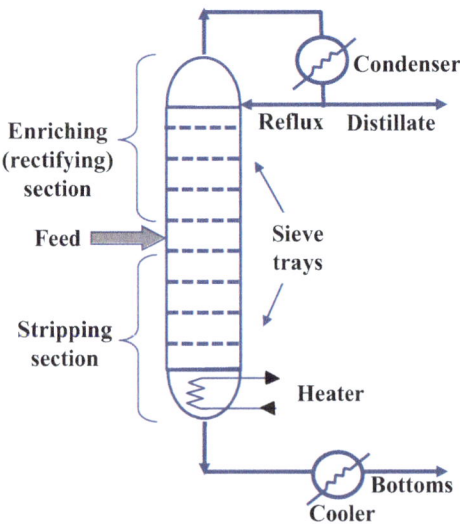

volatile component at the top of the column. Vapors of the more volatile (high vapor pressure) component finally move out of the column and condense in the condenser. To increase the distillation efficiency, a part of the condensate is recycled back to the column. This recycled portion of the liquid is called the reflux. During reflux, colder liquid back into the column helps maintain the column temperature gradient by bringing down the column top temperature and hence improves distillate purity. The distillate of highest purity is collected at the top as the product. A stream of liquid enriched in the less volatile component termed "bottoms" or residues exits from the bottom.

A very simple, graphical method for designing a continuous fractional distillation system is called the McCabe-Thiele method. This and other more advanced methods are available in conventional chemical engineering texts.

Problems

2.1 Are the sensory properties of foods like saltiness, viscosity, hardness, etc. intensive or extensive properties? Explain.

2.2 Using steam tables answer the following questions:
   (a) What is the saturation pressure of steam at 121°C, 267 °F, 373K and 815 °R?
   (b) At what temperature and pressure will the saturated water vapor have a density of 1.496 kg/m^3?
   (c) What is the total heat content (i.e., enthalpy) of 50 kg saturated steam at 150 °C?
   (d) Can saturated steam be produced at 100 °C and 70.14 kPa? Explain briefly.
   (e) A steam supply comes into a food processing plant at 700 kPa and 200 ° C. Is it saturated or superheated steam? If superheated, what is the degree of superheat?

2.3 Using a saturated steam table, answer the following:
   (a) What is the saturated pressure in psig of water at 120 °C, the temperature normally used in canning operations?
   (b) What is the difference in specific volume between saturated liquid and saturated vapor at 25 °C and 250 °C?

2.4 In an evaporator used for concentration of orange juice under a vacuum of 744.5 mmHg, saturated water vapor is produced at a rate of 12,500 ft^3 per hour. Compute (using saturated steam table):
   (a) Temperature of the water vapor
   (b) The mass of water vapor that will have to be removed during an 8-hour shift operation

2.5 For thermal processing of peas, cans were 90% filled with the product (peas + water) at 80 °C and hermetically sealed at atmospheric pressure of 765 mmHg. Following processing, the cans were allowed to cool to 20 °C. Assuming that there were no dissolved gases present in the cans, what should be the pressure inside the cans?

2.6 A 2.3 m^3 rigid tank contains saturated steam at 220 °C. If 1/3rd of the volume is in the liquid phase, the rest being vapor, find the (a) pressure of the steam, (b) quality of the saturated mixture, and (c) density of the saturated mixture.

2.7 Superheated steam was stored at 360 °C at 1500 kPa. Determine:
  (a) The specific volume assuming ideal gas behavior
  (b) The specific volume using steam tables
  (c) The percent deviation from ideal gas behavior

2.8 The phase diagram for an unknown compound is shown below.
  (a) At what temperature and pressure are the liquid and gas phases indistinguishable? What is this point called?
  (b) What is the triple point temperature and pressure?
  (c) If an isothermal process at 200 °C is carried out from 40 atm to 80 atm, what type of phase change will occur? Draw this process on the phase diagram.
  (d) What is the maximum pressure required for an isobaric process to undergo sublimation?
  (e) At what maximum temperature could an isobaric process undergo sublimation?

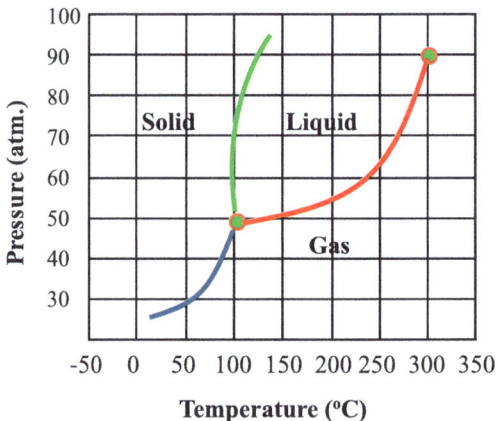

2.9 Vacuum cooling of leafy vegetables is a practical technology that utilizes the principle of evaporative cooling. It involved reducing the pressure within a hermetically sealed cooling chamber that results in evaporation of some water, and the heat of vaporization comes from the products, which lowers the product temperature. Using the 2-dimensional phase diagram shown on the right: (a) indicate the pathway of vacuum cooling from 25 °C and 1 atm to 5 °C and (b) determine the pressure or vacuum in the chamber needed to accomplish the job.

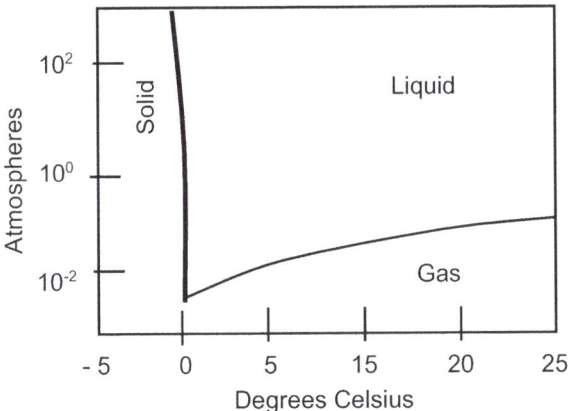

2.10 The P–T diagram on the right traces a freeze-drying process for coffee. In your own words, identify and describe each stage of the operation as it starts at point X and terminates at point Y.

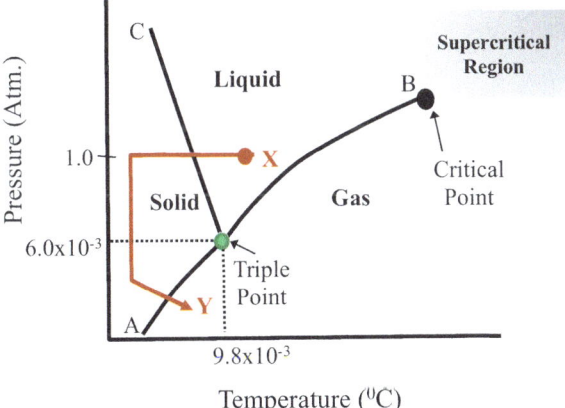

2.11 A 3D phase diagram for an unknown pure substance is shown below. Explain the processes following the paths indicated below and associated phase changes:
(a) 1 to 4;
(b) 5 to 11;
(c) If you drew these processes on a P–T diagram, which ones would appear as points instead of lines?

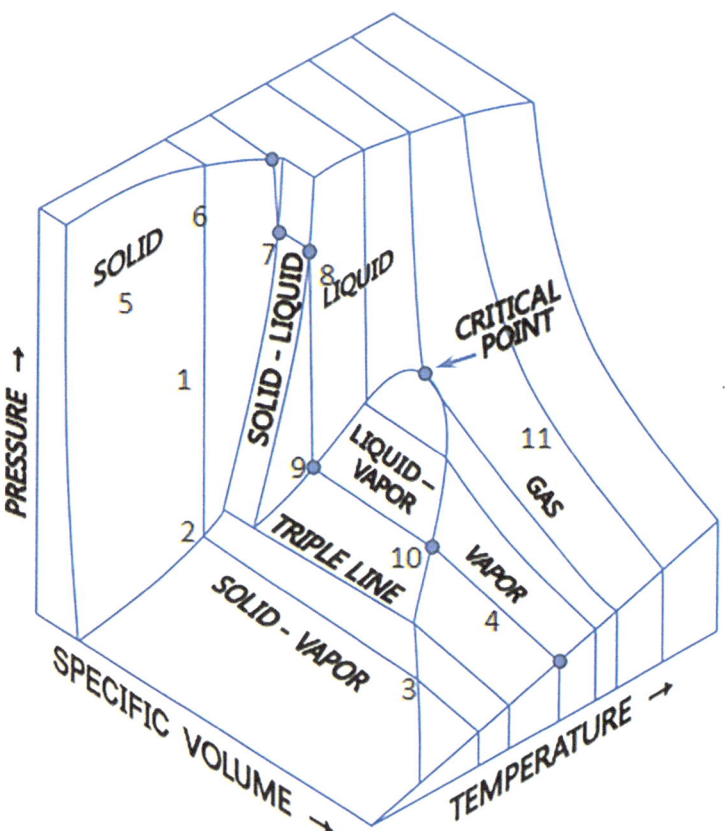

2.12 Shown is the phase diagram of a pure material in 3 dimensions and its two-dimensional projections. On each of these diagrams, trace the path of the following processes:

(a) A gas phase between its triple point and critical point is isothermally compressed into liquid and then undergoes isobaric cooling to change into a solid.

(b) A gas phase below its triple point is compressed isothermally into the solid phase.

(c) A supercritical fluid expands isothermally into the gaseous phase.

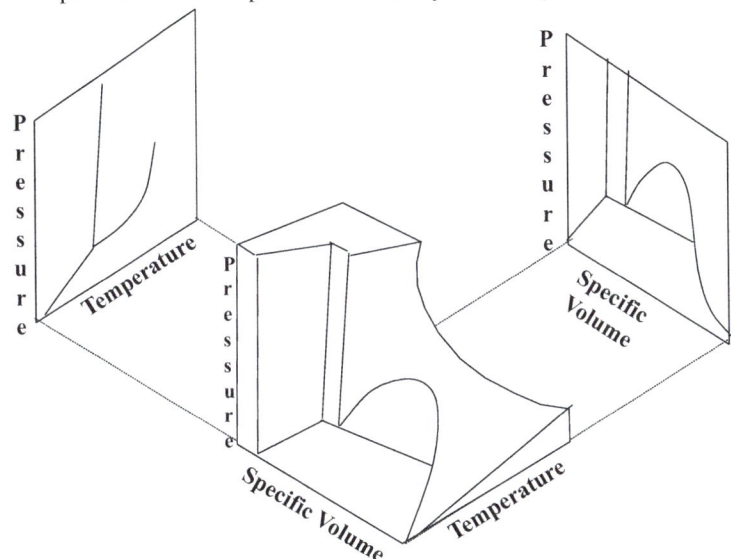

2.13 The phase diagram of a binary system is shown below. The system consists of two miscible components A and B.
   (a) What are the boiling points of A and B, respectively?
   (b) If a 50:50 mixture of A and B is heated to 120 °C, what would be the compositions and quantities of the vapor and liquid phases?

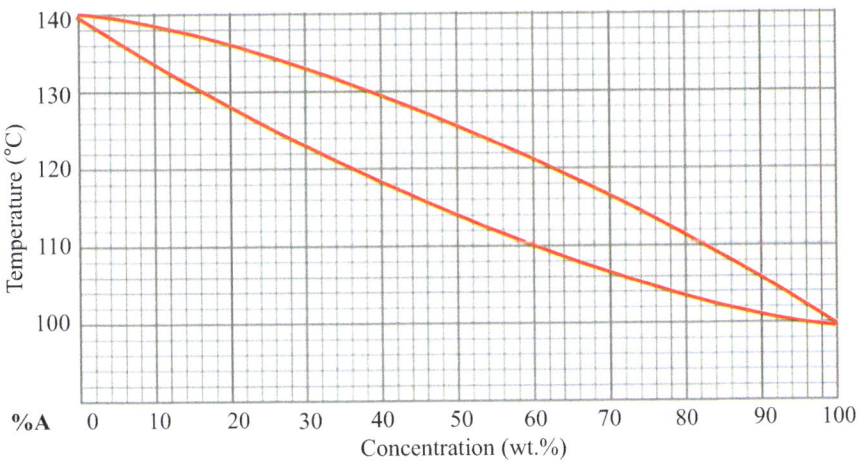

2.14 A fractional distillation method is used to separate a mixture of two miscible fluids, A and B. The initial mixture is 70% A.

(a) What is the bubble point of this mixture?
(b) How many theoretical plates would be needed to reach a concentration of B higher than 90%? Show this on the phase diagram.
(c) You are asked to design a fractional distillation process that will concentrate a mixture of these two fluids. Starting with a mixture that is 15% A, describe a process for achieving a concentration of at least 90% A.

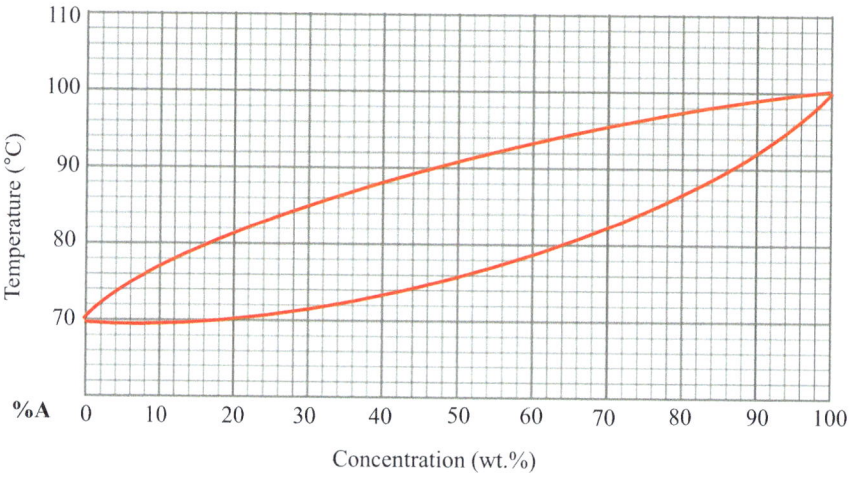

2.15 100 kg-moles mixture of A and B containing 60% of B is subjected to a differential distillation at atmospheric pressure till the composition of B in the still is 35%. Average relative volatility (α) may be assumed as 3.1. Compute the total moles of the mixture distilled. The equilibrium curve relationship is given by the below equation:

$$y_B = \alpha x_B / (1 + x_B(\alpha - 1)) = 3.1 x_B / (1 + 2.1 x_B)$$

Bibliography

1. Gorak A, Schoenmakers H (eds) (2014) Distillation operation and applications. Elsevier, New York
2. Geankoplis CJ, Hersel AA, Lepek DH (2018) Transport processes and separation process principles.5th edn. Prentice Hall, Upper Saddle River, NJ
3. McCabe WL, Smith JC, Harriott P (2005) Unit operations of chemical engineering.7th edn. McGraw Hill, New York
4. Treybal RE (1981) Mass transfer operations.3rd edn. McGraw Hill, New York

Thermodynamics: Basic Concepts

3

Thermodynamics studies the changes associated with the exchange of energy between a system and its surroundings. It establishes the conditions that determine the equilibrium states of systems. It is based on the following three laws, which describe the essential role of energy and predict and explain the direction of change in matter and processes with subtle and complex implications.

- **First Law**. It is about conservation of energy. Energy cannot be created or destroyed and only changes from one form to another. This tells us that if a system undergoes a process in which it gains or loses energy, the surroundings must lose or gain an equivalent amount of energy. Thus, in any process, it should be possible to account for all energy changes, and you will be doing just that very shortly. This law says nothing about the form of the energy (heat, electrical, light, mechanical, etc.) or whether it is in a useful condition, only that the total energy of the universe remains constant.
- **Second Law**. It is about the entropy or disorder of a system. Entropy is a measure of a system's energy that is not available for work. According to this law, the disorder in a system always increases, and the energy is transformed into a less usable form. The first law says nothing about the direction of a process. If you ski down a hill, you convert potential energy into kinetic energy and then into heat. If you were to suddenly convert that heat to kinetic energy and spontaneously ski up the hill, you would not violate the first law. You would, however, violate the second law, which says that heat will not spontaneously flow from an object at a lower temperature to one at a higher temperature or that a process cannot spontaneously increase the potential energy of a system. The second law, then, determines the direction of spontaneous processes. In an isolated system, natural processes are spontaneous when they lead to an increase in disorder or entropy. Processes that decrease the total entropy of the universe are not possible.
- **Third Law.** There is also a third law that states that a perfect crystalline solid has zero entropy at absolute zero. It should be noted that absolute zero is unattainable.

S. S. H. Rizvi, *Food Engineering Principles and Practices*,
https://doi.org/10.1007/978-3-031-34123-6_3

And then there is also the **zeroth law** of thermodynamics that says that if two systems are separately in thermal equilibrium to a third system, then they are also in thermal equilibrium with each other. Its importance was recognized after the first three laws were already established but was considered more fundamental and thus a lower number was assigned. The concept of temperature measurement is predicated on this law and permitted creation of thermometers.

3.1   Forms of Energy

Thermodynamics is the study of the movement of energy between a system and its surroundings. We begin by examining the forms that energy can take within a system and then look at the ways in which the exchange can take place.

3.1.1   Internal Energy

The internal energy (E_i) of a system is the energy associated with the molecules within the system and includes the following (Fig. 3.1):

A. **Nuclear energy** is associated with the binding of particles within atomic nuclei. This energy undergoes a change only when nuclear reactions take place.
B. **Chemical energy** is associated with the binding of atoms to make molecules. This energy changes when a chemical reaction occurs or the chemical makeup of the systems changes.
C. **Thermal energy** is associated with the random motion of molecules and electron shifts within the molecule. The motion includes translation, rotation, and vibration. This energy changes when the temperature of the system changes.
D. **Molecular energy** is associated with the attractions and repulsions between molecules. This energy changes primarily when the system undergoes a phase

Fig. 3.1 Forms of internal energy

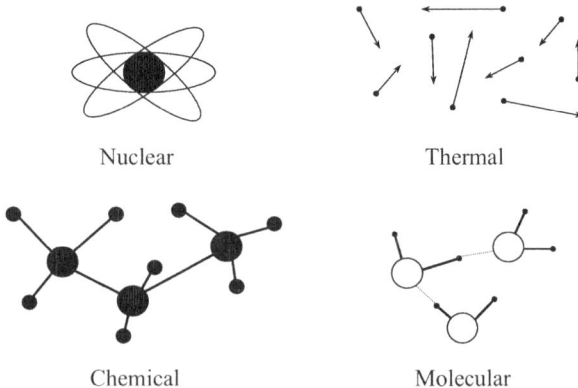

Nuclear   Thermal

Chemical   Molecular

<u>change</u>. For example, to convert water from a liquid to a vapor, it is necessary to add enough energy to overcome the attraction between water molecules.

Internal energy has several important properties:

• It depends on the <u>state</u> of the system.

Example

1. *A liter of water at 25 °C and 101.4 kPa will always have the same internal energy.* ◄

• It is not affected by the path followed in reaching that state.

Example

2. *One liter of water starts at 4 °C and 50 kPa pressure and is first heated to 25 °C and then raised to 101.4 kPa pressure. A second liter of water at 95 °C and 250 kPa is lowered in pressure to 101.4 kPa and then cooled to 25 °C. Since both liters finish in the same state, they will have the same internal energy, regardless of the difference in the paths followed to reach that state.* ◄

• It is an <u>extensive</u> property.

Example

3. *A liter of water at 25 °C and 101.4 kPa pressure has half the internal energy of two liters of water at the same temperature and pressure.* ◄

In general, it is not possible to determine the total internal energy of a system. This is not a problem, however, since we will only be interested in the change in internal energy(ΔE_i), and that can be measured. When the energy of the system increases, ΔE_i (or dE_i) will be positive.

3.1.2 Kinetic Energy

The kinetic energy (E_k) of a system is the energy associated with the motion of the system as a whole (as opposed to molecular motion), see Fig. 3.2.
The equation for kinetic energy then is given as:

Fig. 3.2 Kinetic energy

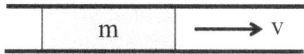

$$E_k = \frac{1}{2}mv^2 \tag{3.1a}$$

where $m =$ the mass of the system and $v =$ the velocity of the system. In the American Engineering system, this becomes:

$$E_k = \frac{1}{2}\frac{mv^2}{g_c} \tag{3.1b}$$

Example

4. *Consider a horizontal pipe of 1/2" inside diameter, with water flowing through it at an average velocity of 10 ft/s (Fig. 3.2). If we define our system as a portion of the water 5 inches long that is moving with this stream, compute its kinetic energy?*
   Solution: *The volume (V) of this system is*

   $$V = \pi r^2 x = (3.14)(0.25 \text{ in})^2 (5.0 \text{ in}) = 0.98 \text{ in}^3$$

   where $r =$ the radius of the pipe and $x =$ the length of the water making up our system.
   The mass of the system

   $$m = V\rho = 0.98 \text{ in}^3 \left(\frac{62.4 \text{ lb}_m/ft^3}{(12.0 \text{ in}/ft)^3} \right) = 0.0355 \text{ lb}_m$$

   The kinetic energy of the system is: $E_K = \frac{mv^2}{2g_c} = \frac{1}{2}\frac{(0.0355 \text{ lb}_m)(10 \text{ ft}/s)^2}{\left(32.17 \text{ ft·lb}_m/lb_f·s^2\right)} =$ $0.0552 \text{ ft} \cdot lb_f$ ◄

3.1.3 Potential Energy

The potential energy (E_p) of a system is the energy associated with the elevation of the system relative to some reference level. In other words, water at the top of a waterfall (h_1) has greater potential energy than water at the bottom (h_2), Fig. 3.3. The equation for potential energy at a given elevation is given as:

$$E_p = mgh \tag{3.2a}$$

where $m =$ the mass of the system, $g =$ the acceleration due to gravity, and $h =$ the elevation of the center of gravity of the system relative to a reference. Like, E_k, it may also be written as:

Fig. 3.3 Potential energy

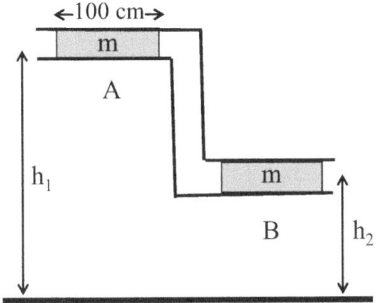

$$E_{\mathrm{p}} = \frac{mgh}{g_{\mathrm{c}}} \tag{3.2b}$$

where g_{c} is the gravitational constant.

Example

5. *Let a system consist of a 100 cm length of water in a pipe 3 cm in diameter at point A, 4 m above the floor, refer to Fig. 3.3. The pipe drops to point B, 2 m above the floor. What is the potential energy of the system relative to the floor? What is the change in potential energy of the system when it moves to point B in the pipe?*
   Solution: *The volume of the system is:*$V = \pi r^2 x = \pi(1.5 \times 10^{-2}m)^2 (1\ m) = 7.07 \times 10^{-4}\ m^3.$
      where $r =$ the radius of the pipe and $x =$ the length of the water composing the system.
   The mass of the system is:

$$m = V\rho = (7.07 \times 10^{-4}\ m^3)(1 \times 10^3 kg/m^3) = 7.07 \times 10^{-1}\ kg$$

   The potential energy relative to the floor is:

$$E_{\mathrm{P}} = mgh = (7.07 \times 10^{-1}\ kg)\left(9.81\frac{m}{s^2}\right)(4\ m) = 27.7\ J$$

   In other words, if the system dropped to the floor, 27.7 Joules of energy were converted to kinetic energy during the fall and to heat when it hit the floor. The change in potential energy in moving from the original position point A to point B is:

$$\Delta E_{\mathrm{P}} = mg(h_2 - h_1) = (7.07 \times 10^{-1}\ kg)\left(9.81\frac{m}{s^2}\right)(2 - 4)m = -13.9\ J$$

   Thus, a drop in elevation results in a negative change, indicating that potential energy is lost as the system loses elevation. ◀

3.1.4   Total System Energy

The total energy (E_T) of a system is the sum of its internal, kinetic, and potential energies.

$$E_T = E_i + E_K + E_P \qquad (3.3)$$

However, because the absolute value of E is unknown, the change in its value can be computed as shown below:

$$\Delta E_T = \Delta E + \Delta KE + \Delta PE \qquad (3.4)$$

For a stationary system, such as water sitting in a tank at datum, ΔE_K and ΔE_P are zero and drop out of the equation, making the change in total energy equal to the change in internal energy. If the system moves, say through a pipe or changes mass, say by adding an ingredient, ΔE_K reenters the equation. If the system changes elevation, ΔE_P reenters the equation.

3.2   Energy in Transition

3.2.1   Heat and Work

In thermodynamics, heat and work represent two forms of energy transfer. Work involves the transfer of **mechanical energy,** while heat entails the transfer of **thermal energy** that happens between a system and its surroundings or between two systems. Work and heat are not thermodynamic properties of a system, but heat can be transferred into or out of a system, and work can be done on or by a system. Additionally, mechanical energy can be 100% converted into thermal energy but not the other way around. Some heat is wasted when heat is converted into mechanical energy. It is also important to remember that kinetic energy and potential energy are the two main forms of mechanical energy, while thermal energy has only one form, heat.

Just approximately 200 years ago, heat was considered a fluid invisibly moving into and out of objects. Modern thermodynamics was born when it was at last recognized that heat is energy linked to the vibration of atoms. Because atoms and molecules are constantly moving, vibrating, and rotating, they contain kinetic energy. The molecules in a hot material have more kinetic energy than those in a cold material, as indicated in Fig. 3.4a. If materials at different temperatures are brought in contact, random collisions between molecules will take place, and energy will be transferred in the way that energy is transferred from one billiard ball to another. The result is a net flow of energy from hot to cold objects. Heat will stop flowing when both objects attain a uniform temperature, as shown in Fig. 3.4b. We use the symbol q to represent heat.

Heat differs from internal energy in some important ways:

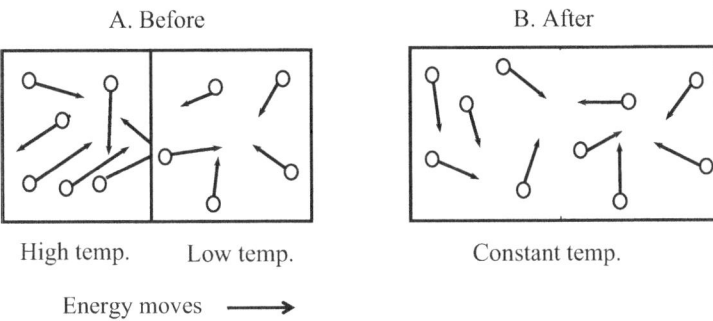

Fig. 3.4 Energy transfer as heat

- Internal energy is a function of the state of a system. It is in part a function of the temperature of a system. Heat, on the other hand, is not property of the state of a system but rather thermal energy that is entering or leaving a system during a change in its state.
- The quantity of heat that enters or leaves a system depends on the path followed.

Example

6. *If 1 kg of water at atmospheric pressure is cooled from 40 °C to 30 °C, it will lose approximately 10 kilocalories of heat to its surroundings. If 1 kg of water at the same pressure is heated from 25 °C to 30 °C, it will gain approximately 5 kilocalories from its surroundings, even though it ends up in the same state.* ◄

3.2.2 Point Versus Path Functions

A state diagram is a graph whose coordinates are intensive properties of a system. The phase diagrams in the last chapter are state diagrams. Since a state is defined by a limited number of intensive properties, it can be represented by the coordinates of a point on a state diagram. Any variable that is a function of the state of a system can be called a **point function**, such as internal energy, E. On the other hand, the values of some variables depend not on the state of a system but on the path followed in going from one state to another. These are called **path functions**. As you may have guessed it, work (w) and heat (q) are path functions since their values depend on the path followed, see Fig. 3.5. To distinguish between the two, different notations are used to represent small increments in their values: exact differential for change in point functions such as dE and dH. and inexact differentials such as δw and δq for change in path functions.

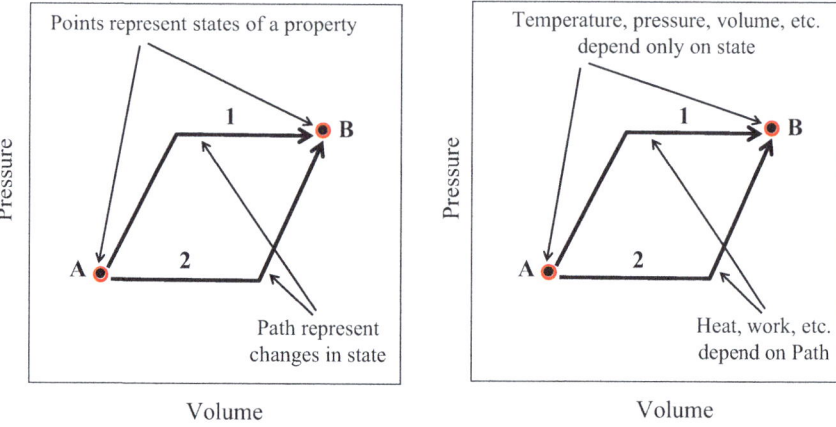

Fig. 3.5 Point versus path functions

Example

7. *Specifying the pressure and specific volume of a water system defines its state. They also define a point in* Fig. 3.5. *Pressure and volume are, therefore, point functions. When the pressure and volume of points A and B are each known, there are functions that allow us to compute the internal energy and temperature of the system, so these are also point functions. None of these variables are affected by the way the system reached these states.* ◄

Example

8. *Examples of path functions include heat and work. For example, in* Fig. 3.5, *the bent arrows represent two possible pressure–volume paths that take the system from state A to state B. Although the difference in internal energy between states A and B is independent of the path, along path 1, more of this energy was added in the form of work, while along path 2, more of the energy was added in the form of heat. Thus, the amount of heat or work added depends on the path, and these are path functions.* ◄

3.2.3   Types of Work

There are two major ways to accomplish work: pressure–volume and mechanical or shaft work.

A. **Pressure–Volume Work:** Work can also be computed as the product of pressure times change in volume, i.e.

$$w = -P\Delta V \qquad (3.5)$$

where w = work, P = pressure, and ΔV = the change in volume.

To see that this is equivalent to the previous definition, imagine a cylinder is fitted with a frictionless piston, trapping air inside, as shown in Fig. 3.6. This air is the system of interest. The air exerts a pressure that we will call P_i on the piston. The surroundings counter this with a pressure we will call P_e. The piston has area A, and since pressure = force/area, the forces (F) exerted on the piston equal

$$F_i = P_iA, \text{from the inside and } F_e = P_eA \text{ from the outside}$$

If the internal (system) force exceeds the external force, the piston will move outward and work on the surroundings. If it moves a distance Δx against a force F_e, it will do work equal to

$$w = F\Delta x = P_eA\Delta x. \qquad (3.6)$$

However, $A\Delta x$ has units of volume, specifically the volume swept out by the movement of the piston. If we replace $A\Delta x$ in this equation with ΔV (change in volume), we have

$$w = P_eA\Delta x = P_e\Delta V \qquad (3.7)$$

Thus, $P\Delta V$ has units of work and is referred to as pressure–volume work.

In summary, whenever a system increases in volume, it works against the surroundings. When a system decreases in volume, work is being done on it by the surroundings. Like heat, work is not a property of the state of a system but of the path taken in going from one state to another.

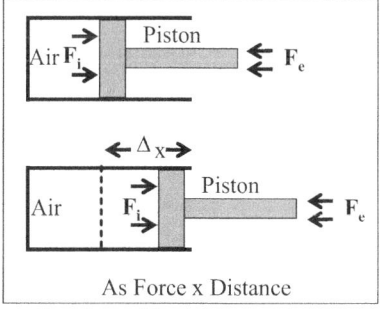

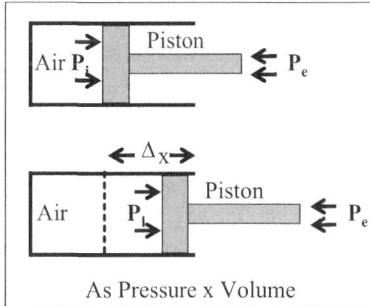

Fig. 3.6 Two views of the same work

Examples

9. *If a sealed can is heated, it will expand and work against the surroundings.*
10. *When gasoline is burned in a cylinder of a car engine, it moves the piston and works against the crankshaft, which turns the wheels and moves the car.*
11. *When a pot of water boils, the steam that is generated expands against the air around it, doing work by moving that air.* ◀

B. **Shaft Work:** Not all work on or by a system is done by volume changes. In many cases, a shaft enters a system and, by sliding or rotating, stirs, propels, or otherwise moves the system, Fig. 3.7. When this happens, the surroundings are doing work on the system. In other cases, movement within the system may cause the shaft to slide or rotate, allowing the system to work on the surroundings. When work is exchanged between a system and its surroundings in this manner, we call it shaft work. We use the symbol w_s to represent shaft work.

Example

12. *In Fig. 3.7, the system consists of a centrifugal pump together with the pipes between A and B. If the shaft is rotated, e.g., by a motor in the surroundings, the attached blades move water through the system. The surroundings are doing shaft work on the system. On the other hand, if water is already flowing through the system, it will rotate the shaft and do shaft work on the surroundings, perhaps driving a generator or other apparatus.* ◀

3.2.4   Sign Conventions

There are different conventions used in various books, but we use the following conventions:

Fig. 3.7 Shaft work

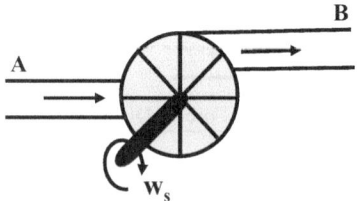

We symbolize heat with the letter q. When heat enters a system from the surroundings, q has a positive value. When heat leaves a system for the surroundings, q has a negative value, as shown in Fig. 3.8.

- We symbolize the volume change with ΔV. When the system expands, ΔV has a positive value. When the system contracts, ΔV has a negative value.
- We symbolize work with the letter w. When the surroundings work on the system, w has a positive value. When the system does work on the surroundings, w has a negative value. (The opposite convention is often used in some books.) To use these conventions, we rewrite the above equation as

$$w = -P_e \Delta V \tag{3.8}$$

3.3   First Law of Thermodynamics

The first law of thermodynamics is a statement of the "conservation of energy," which says that "energy cannot be created or destroyed." This can be expressed in terms of systems as follows:

$$dE_{\text{sytstem}} = \delta q + \delta w \tag{3.9a}$$

In this equation, a positive q indicates net heat entering the system, and a positive w indicates network being done on the system. This equation states that any heat (q) that enters a system must be added to the energy of the system or perform work on the surroundings ($-w$). No energy vanishes, and none is created. In principle, you should be able to account for all energy that enters and leaves the system, and you will do just this when you perform energy balances. Any energy that enters a system ($q + w$) must come from the surroundings ($-\Delta E_{\text{surroundings}}$), and any energy that leaves the system must go to the surroundings. Thus, we can write:

$$\Delta E_{\text{system}} = q + w = -\Delta E_{\text{surroundings}} \tag{3.9b}$$

The left-hand side of this equation represents gains in energy by the system. Therefore, we can write the first law equation as follows:

Fig. 3.8 The sign convention

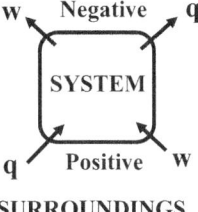

$$\Delta E_{\text{system}} + \Delta E_{\text{surroundings}} = 0 \qquad (3.10\text{a})$$

Since the surroundings plus the system add up to the entire universe, this equation says that the total energy of the universe does not change. We can write the first law as:

$$\Delta E_{\text{system}} = q + w \qquad (3.10\text{b})$$

which tells us that if the surroundings perform work on the system by compressing it, then this must either appear as an increase in the internal energy of the system or as heat leaving the system. This is the "work to heat" conversion of Joule's experiments.

3.3.1 Adiabatic Processes

Earlier, we defined an adiabatic process as one in which there is no exchange of heat between the system and its environment. Under those circumstances, $q = 0$, and the previous equation reduces to:

$$\Delta E_{\text{system}} = w \qquad (3.11)$$

which says that

- If you compress a system adiabatically, you must increase its total energy. If the system is stationary so that no kinetic or potential energy changes are involved, this becomes an increase in internal energy.
- If no nuclear or chemical reaction takes place and if there is no phase change, this increase in energy will all appear as thermal energy, and the temperature of the system will rise.

Similarly, if the system expands adiabatically, it will cool down. Adiabatic processes are important to operations such as refrigeration because they tell us that we can change the temperature of a system by doing work.

3.4   Second Law of Thermodynamics

The second law provides answers to such obvious questions as to why heat naturally flows from a hotter to a colder system and not the other way around. Although the latter will not violate the first law of thermodynamics since the energy will be conserved, we do not encounter such a system in nature. What determines the direction of a process? At its heart of this question is a useful thermodynamic property of systems called entropy (S), a measure of the degree of disorder or randomness in a system. Thus, a practical definition of the second law that is frequently used states that a natural process in any isolated system will proceed in the direction that causes the entropy of the system (ΔS_{sys}) plus the surroundings (ΔS_{sur}) to increase for an irreversible process and to remain unchanged for a reversible process. Stated mathematically

$$\Delta S_{sys} + \Delta S_{sur} \geq 0 \qquad (3.12)$$

The entropy change (ΔS) of a system is equal to the heat (ΔQ) reversibly entering the system divided by the absolute temperature (T) of the system:

$$\Delta S = \frac{\Delta q}{T} \qquad (3.13)$$

To determine the change in entropy, a perfect crystal at 0 K serves as a reference state, and its entropy is arbitrarily fixed to 0 J/K.

Entropy is consequently a measure of a system's energy that is unavailable for doing useful work, and since the entropy of spontaneous processes increases, no process operating between hot and cold systems can completely convert heat into work. In all energy transfer operations, the entropy of the universe increases, and the energy available to do work reduces (or remains the same in the extreme case).

As shown in Fig. 3.9, it is therefore not possible to take an amount of heat, q_h, from a hot reservoir and convert it all to work, w. Some amount of heat, q_c, representing unavailable energy, must be dumped into a cold reservoir. This is the basis for the operation of the heat engine and shows that no such engine can be 100% efficient. French physicist Sadi Carnot considered the father of thermodynamics, in 1824 published his treatise "Reflections on the Motive Power of Fire" and showed how to estimate the theoretical maximum efficiency of a steam engine from knowledge of the temperature of steam inside the engine cylinder (hot reservoir) and that of the air outside (cold reservoir).

Working in the reverse direction of a heat engine is a refrigerator in which heat is transferred from a cold reservoir to a hot sink, as also shown in Fig. 3.9. This is made possible only by doing work (w) on the system, equal to $q_h - q_c$. More on refrigeration system will be discussed in Chap. 12.

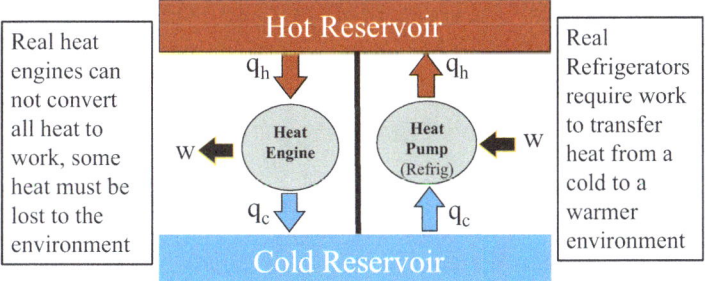

Fig. 3.9 Schematics of a heat engine and a heat pump (refrigerator)

3.5   Enthalpy

Additionally, since $E + PV$ appear in most processes undergoing energy exchange accompanied by changes in pressure and volume, the thermodynamic term enthalpy (H) of a system is defined as:

$$H = E + PV \tag{3.14}$$

where E is the internal energy of the system, P is the pressure at the boundary of the system and its environment, and V is the volume of the system.

While heat, q, is defined as thermal energy in transit, enthalpy represents the total energy content of a system.

We can represent a change in enthalpy (ΔH) with the equation:

$$\Delta H = \Delta E - P\Delta V \tag{3.15}$$

The following are important properties of enthalpy:

- **Enthalpy (H)** is an extensive property. That is, all else being equal, 2 kg of material has twice the enthalpy of 1 kg. Of course, expressed as a **specific enthalpy** (enthalpy per unit mass, $\overline{H}$), it becomes intensive.
- Enthalpy is a state or point variable. Similar to temperature, pressure, internal energy, and volume, its value depends on the state of the system and not on the path followed in reaching that state.
- Like internal energy, we cannot measure the total enthalpy content of a system. Only the change in enthalpy of a system can be measured. If we pick some standard conditions and arbitrarily assign it an enthalpy of zero, we can then report the enthalpy of any state relative to this reference state. In chemistry, we assign 0 enthalpy to the standard state of each element. This is taken to be 1 atm pressure, 25 °C, and the most stable form of the element. For gases, the stable form is the ideal gas. For liquids, the stable form is pure liquid. For solids, the stable form is an agreed upon crystalline state.
- For foods, the standard state is taken to be -40 °C and 1 atmosphere. Enthalpy is arbitrarily taken to be 0 at this state. For water, the standard state is frequently taken to be the triple point.

3.5.1   Enthalpy and Gibb's Free Energy

For a process occurring spontaneously at a constant temperature (T) and pressure (P), we learned that, according to the second law of thermodynamics, the system moves toward a state of higher disorder or entropy (S) and a low state of chemical energy or what constitutes its enthalpy (H). The latter represents the total energy available to do useful work, whereas the former represents lost energy that is not available to do work. Therefore, the energy that is available for doing useful work is the difference between these two quantities and is expressed as Gibbs free energy (G):

$$G = H - TS \tag{3.16}$$

In terms of changes for a system of interest, it is written as:

$$\Delta G_{sys} = \Delta H_{sys} - T \Delta S_{sys} \tag{3.17}$$

Now, by using the above expression, we can determine whether a system will move spontaneously in the forward direction, backward direction, or stay at equilibrium, as follows:

- When $\Delta G_{sys} < 0$, the process will proceed spontaneously in the forward direction
- When $\Delta G_{sys} > 0$, the process is spontaneous in the reverse direction
- When $\Delta G_{sys} = 0$, the process is in equilibrium

The Gibbs free energy is thus a useful tool to calculate the most reversible work that may be done by a system at a constant temperature and pressure and predict the direction of a process.

3.6 Saturated Steam Table

In the last chapter, we introduced the saturated steam table (Appendix A.4) as a tabulation of the properties of water and steam along the vapor pressure line on a phase diagram of water. Recall that this line represents all pressure–temperature combinations at which the liquid and vapor phases of water can exist together at equilibrium. The normal boiling point of water at 100 °C and 1 atm. pressure falls on this line.

At the time we introduced it, we showed how the steam table could be used to determine the temperatures, pressures, and specific volumes of liquid water and water vapor along the vapor pressure line. Any intensive property of water can be listed in such a table, and usually, steam tables also list specific internal energy and specific enthalpy. Table 3.1 is an abbreviated listing of these properties.

A few observations may help you make sense of the above table.

- As mentioned, the pressure and temperature combinations listed in this table all fall on the vapor pressure line of water. Similar to that line, the values in the table go from the triple point (0.01 °C at 0.61 kPa) to the critical point (374.14 °C at 22090 kPa).
- The standard condition is taken to be the triple point, and both internal energy and enthalpy are arbitrarily taken to be 0 at this state.
- The change in energy between any two states can be computed from this table. For example, suppose 2 kg of liquid water at 40 °C is heated to 100 °C. The table

Table 3.1 Saturated steam table sample

Temperature °C (°F)	Pressure KPa (atm)	Specific internal energy (kJ/kg)		Specific enthalpy (kJ/kg)	
		Sat. liquid	Sat. vapor	Sat. liquid	Sat. vapor
0.01 (32.02)	0.61 (0.006)	0.0	2375	0.0	2501
40 (104)	7.38 (0.073)	167.6	2430	167.5	2574
80 (176)	47.39 (0.467)	334.9	2482	334.9	2644
100 (212)	101.32 (1.00)	418.9	2507	419.0	2676
120 (248)	198.53 (1.96)	503.5	2529	503.7	2706
140 (284)	361.3 (3.56)	588.7	2550	589.1	2734
160 (320)	618 (6.09)	674.9	2568	675.6	2758

tells us that the internal energy at 40 °C is 167.6 kJ/kg, while at 100 °C, it is 418.9 kJ/kg. Therefore, this change in state requires that 418.9 − 167.6 = 251.3 kJ of energy be absorbed per kilogram of water. The 2 kg we heated would require twice that much energy or 502.6 kJ.

- Although the internal energies and enthalpies listed in this table are arbitrary, depending on the choice of standard conditions, the differences between them are not arbitrary. Thus, although the 418.9 and 167.6 values in the last example are arbitrary, the difference of 251.3 kJ/kg between them is not and would be the same regardless of the choice of standard conditions.
- The internal energy columns provide information about energy changes of constant volume processes, while the enthalpy column provides information about constant pressure processes. The example just computed used internal energy and, therefore, applies to water heated in a constrained space. If the same water was heated at constant pressure, e.g., in an open vessel, the enthalpy change would be 419.0 − 167.6 = 251.4 kJ/kg.
- Note that for liquid water at "normal" temperatures, the enthalpy change and internal energy changes are practically identical (251.4 versus 251.3 in the last example). This happens because liquid water is nearly incompressible, undergoes a very small volume change upon heating or cooling, and does very little work.
- On the other hand, water vapor is a gas and changes greatly in volume as its pressure or temperature changes. If water vapor is heated from 40 to 100 °C, the internal energy change per kilogram is: 2507 − 2430 = 77 kJ/kg, while the enthalpy change is 2676 − 2574 = 102 kJ/kg. The extra 25 kJ represents the work done by the water vapor as it expands against its surroundings.

3.7 Enthalpy of Phase Transformation (Latent Heat) vs. Sensible Heat

Usually, when heat is added to water, it increases its temperature. Since we can sense this temperature change, we refer to the heat as sensible heat. However, when water is at the freezing point, added heat does not change the temperature but instead promotes a phase change from solid to liquid. The added energy largely breaks the intermolecular bonds that maintain the crystal lattice of ice. Under these circumstances, the added heat is referred to as the **latent heat (enthalpy) of fusion or melting**. This is called fusion because when solid objects made from the same substance are melted, they can be mixed or fused together into a mass. Similarly, when water is at the boiling point, added heat causes a phase change from liquid to vapor and is called the **latent heat (or enthalpy) of vaporization or boiling**. The word latent means hidden since the temperature during phase change remains constant. This heat goes to both breaking the bonds that hold water molecules together as a liquid and to the work of expanding the vapor against the surroundings. We will use the following nomenclature to describe enthalpy transformations:

Enthalpy of vaporization (fluid to gas, specific), $\overline{H}_{fg} = \overline{H}_g - \overline{H}_f$
Enthalpy of condensation (gas to fluid, specific), $\overline{H}_{gf} = \overline{H}_f - \overline{H}_g$
Enthalpy of fusion (solid to fluid, specific), $\overline{H}_{sf} = \overline{H}_f - \overline{H}_s$
Enthalpy of solidification (fluid to solid, specific), $\overline{H}_{fs} = \overline{H}_s - \overline{H}_f$

$\overline{H}_g$, $\overline{H}_f$, and $\overline{H}_s$ represent the specific enthalpy in the gas, fluid (liquid), and solid phases, respectively.

Latent heat is important to food processing because it makes it possible to add or remove a large amount of energy without temperature changes. Ice, for example, will cool foods effectively because, as it melts, it absorbs a large amount of heat while remaining at zero degrees. In the same way, steam is used to heat foods because, as it condenses, it releases a large amount of heat while remaining at 100 °C. The latent heat of fusion at 0 °C and of vaporization at 100 °C in various units is given in Table 3.2. The latent heat of vaporization of water can be determined for other temperatures using the enthalpy data in the steam table.

Table 3.2 Latent heat of water at 1 atm

Latent heat of fusion at 0 °C ($\overline{H}_{fs}$)	Latent heat of vaporization at 100 °C ($\overline{H}_{fg}$)	Units
1.44	9.73	kcal/g-mol
79.72	538.6	kcal/kg
143.6	970.1	btu/lb
334.1	2257.1	kJ/kg

Example

13. *The steam table tells us that at 100 °C and 1 atmosphere, the enthalpy of liquid water is 419.0 J/kg. At the same temperature and pressure, steam has an enthalpy of 2676.1 J/kg. From this, we can calculate that as water boils at atmospheric pressure, it absorbs*

$$2676.1 - 419.0 = 2257.1 \ kJ$$

This is the latent heat of vaporization at atmospheric pressure.

14. *Suppose 3 kg of saturated steam at a pressure of 200 kPa is condensed and cooled to 40 °C. How much energy is released? From the steam table, we find that at 200 kPa (approximately 2 atm), the temperature of the steam is close to 120 °C, and its enthalpy is 2706.3 kJ/kg. Water at 40 °C has an enthalpy of 167.6 kJ/kg. Thus, each kilogram of steam releases*

$$2706.3 - 167.6 = 2538.7 \ kJ/kg$$

Our 3 kg will release 7616.1 kJ. An examination of the table shows that the latent heat of this change is

$$2706.3 - 503.5 = 2202.8 \ kJ/kg$$

while the sensible heat released is

$$503.5 - 167.6 = 335.9 \ kJ/kg$$

Clearly, most of the heat comes from the phase change. ◀

3.8   Heat Capacity and Specific Heat

Many food processing operations are temperature dependent. Canning, for example, destroys microbial spores by subjecting them to temperatures of approximately 121 ° C for a specified time. It is important, therefore, that we be able to attain specified temperatures. We cannot, however, add temperature to a system. Instead, we add energy, usually in the form of heat. Because of this, it becomes very important to know the relationship between heat and temperature. How much heat must we add to achieve a desired temperature? This is doubly important because energy, whether from coal, oil, wood, gas, or electricity, costs money.

The answer to this question turns out to be "it depends." It depends on the nature and mass of the material being heated. Every substance has its own characteristic response to energy transfer, meaning each takes a different amount of heat to raise its temperature by a certain degree.

The heat capacity (C) is simply the amount of heat energy needed to raise the temperature of an object by a unit degree and depends on the mass of the object. The specific heat (C_x), on the other hand, is the amount of energy required to raise the temperature per unit mass of an object by a unit degree.

3.8.1   Specific Heat of Water

Let us begin by investigating the heat content of water with the following conceptual experiment:

Place a known mass of water in an insulated container and immerse a heating element in the water, as shown in Fig. 3.10. Attach the element to a watt-hour meter so we can measure the electrical energy consumed. Measure the starting temperature, turn on the heater, wait for a few minutes, turn it off, and measure the final temperature. Suppose you repeat the experiment 3 times and obtain the following data (Table 3.3).

Since a Joule of energy equals 1 W-s, the energy consumption in experiment 1 is: (302 W-h)(3600 s/h) = 1,087,200 J.

We know that specific heat is proportional to mass involved and temperature rise, dividing the energy used by the mass of the water and the temperature change observed gives: (1,087,200 J)/((4000 g)(65 °C)) = 4.18 J/g-°C. If we repeat these calculations for the other experiments, we again obtain 4.18 J/g-°C or 4.18 kJ/kg-°C. We can conclude from these experiments that 4.18 J of heat energy will raise a gram of water by one degree Celsius. This quantity of energy is equivalent to 1 calorie. If we perform a similar experiment with another material, the heat capacity will be different.

We can summarize these findings with the equation:

Fig. 3.10 Measuring heat capacity

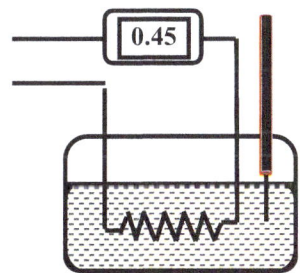

Table 3.3 Data on specific heat of water

Expt. no.	Mass of water	Elapsed time	Energy used	Temperature (°C)		
				Start	End	Change
1	4 kg	20 min	302 W-h	15.0	80.0	65.0
2	2 kg	10 min	151 W-h	15.0	80.0	65.0
3	4 kg	10 min	151 W-h	15.0	47.5	32.5

Table 3.4 Specific heat of some common materials at 1 atm. and 20 °C

Material	Specific heat (kJ/kg·K)	Material	Specific heat (kJ/kg·K)
Gold	0.13	Aluminum	0.90
Silver	0.24	Air	1.05
Copper	0.39	Wood	1.68
Iron	0.45	Steam (100 °C)	2.01
Sand	0.84	Water	4.18

$$C_x = \frac{q}{\Delta T \, m} \tag{3.18}$$

where ΔT is the change in temperature, q is the heat energy added to the system, and m is the mass of the system. Here, C_x represents the **specific heat** and provides a measure of the quantity of thermal energy required to raise the temperature of a unit mass by one degree. It is concerned only with sensible heat, and in SI units, it is expressed in kJ/(kg-K). The value of C_x depends on the nature of the material (Table 3.4).

Some values of specific heat for water to remember are 4.18 Joules per gram per degree C or $K = 1.00$ calories per gram per degree $C = 1.00$ Btu per lb per degree F or R. The specific heat of ice is 2.04 kilojoules per kilogram per degree C or K.

3.8.2 C_V and C_P

In the last section, we examined the relationship between heat energy (q) and two properties of a system, mass (m) and temperature change (ΔT), and observed that heat consumed is proportional to the mass of the system and the temperature rise. In this experiment, we paid no attention to the volume or pressure of the system. Since water changes little in volume when either its temperature or pressure changes, our results are affected very little by this oversight. When gas is heated, pressure and volume changes become important, and even with solids and liquids, they have some effect on heat capacity. Let us examine this effect.

When most materials are heated, there is an increase in either its volume or pressure or both. This is particularly true for a gas. There are two situations that require attention.

A. **Constant Volume (C_V)** is the specific heat of a system when it is heated in such a way that the volume does not change, and
B. **Constant pressure (C_P)** is its specific heat when the pressure does not change. We will see that the heat capacity is higher in the constant pressure case.

Let us define the specific heat for a unit mass of a material in terms of the following differential equation:

$$C_x = \frac{dq}{dT} \tag{3.19}$$

which says that specific heat (C_x) is the ratio of heat added (dq) to temperature change (dT) for a **unit mass**. For the special case where heating is performed at **constant volume**, we add the subscript V to this equation and thus,

$$C_V = \left(\frac{dq}{dT}\right)_V \tag{3.20}$$

In the case where heating is performed at **constant pressure**, we add the subscript P; thus,

$$C_P = \left(\frac{dq}{dT}\right)_P \tag{3.21}$$

Suffix P and V refer to constant pressure and constant volume conditions, respectively.

Now let us investigate the difference between these two. From the first law of thermodynamics, we observed that the heat entering a system must appear either as a change in internal energy (dE) or as work. If we limit work to pressure–volume work (PdV), we can write for a **unit mass** of a material:

$$dE = dq - PdV \tag{3.22}$$

Writing enthalpy Eq. (3.14) for a unit mass in differential form:

$$dH = dE + PdV + VdP \tag{3.23}$$

Substituting for $dE + PdV$ from Eq. (3.22), we obtain:

$$dH = dq + VdP \tag{3.24}$$

Now the two types of specific heats can be determined as follows:

- For a constant-pressure process, $dP = 0$, and we obtain:

$$dH = dq \tag{3.25}$$

$$\text{and thus } C_P = \left(\frac{dq}{dT}\right)_P = \left(\frac{dH}{dT}\right)_P$$

which yields $dH = C_p\,dT$ or $\mathbf{\Delta H = C_p \Delta T}$ (3.26)

- Similarly, for a constant-volume process, $dV = 0$, and Eq. (3.22) gives:

$$dE = dq \tag{3.27}$$

$$\text{and then } C_V = \left(\frac{dq}{dT}\right)_V = \left(\frac{dE}{dT}\right)_V$$

$$\text{or, } dE = C_v dT \text{ or } \Delta E = C_v \Delta T \tag{3.28}$$

The Relationship Between C_V and C_P: Using the definition of enthalpy ($\Delta H = \Delta E + P\Delta V$) and for a mole of gas at constant pressure ($\Delta V/\Delta T = R/\Delta P$), substitution into Eqs. (3.26) and (3.28), gives:

$$C_p = C_v + R \text{ or } C_p - C_v = R \tag{3.29}$$

Recall that internal energy consists of nuclear, chemical, molecular, and thermal energy and that temperature changes reflect changes in thermal energy. Thus, only changes in internal energy result in temperature changes. A comparison of C_p and C_v shows that under constant pressure conditions, more energy must be added to the system to bring about the same change in temperature, and C_P is, therefore, larger than C_v.

Another useful relationship between C_p and C_v, which is frequently used in the expansion and compression of gases, is the **specific heat ratio**, γ, which is defined as C_p/C_v and is dimensionless. For air, $\gamma = 1.4$ at STP.

3.9   Specific Heat of Food

Specific heat is an indicator of the heat storage capacity of a material. It is the relatively high specific heat of water (4.18 kJ/kg-°C) that helps maintain milder climates near the coastal areas via slow warming and cooling of the large mass of water. Both experimental and mathematical models have been used to estimate the specific heat of foods as a function of several parameters, such as temperature, water content, porosity, or other characteristics.

3.9.1   Specific Heat of Food Above Freezing

Siebel [8] made the first successful attempt to estimate the specific heat of foods rich in water, such as fruits, vegetables, purees, and concentrates. He proposed the following equations for estimating specific heat:

$$C_{pf} = 0.0335 \, (\%water) + 0.837 \, kJ/(kg\text{-}K) \tag{3.30a}$$

Table 3.5 Specific heat of major food components

Component	kJ/(kg K)	btu/(lb °F)
Water	4.187	1.000
Nonfat solids	1.256	0.300
Fat (solid)	1.675	0.400
Fat (liquid)	2.094	0.500
Protein	1.549	0.470
Carbohydrate	1.424	0.340
Ash	0.837	0.200

$$C_{pf} = 0.008 \; (\%water) + 0.2 \, btu/(lb\text{-}°F) \tag{3.30b}$$

Dickerson [5] proposed the following equation:

$$C_{pf} = 0.025 \; (\%water) + 1.675 \, kJ/(kg\text{-}K) \tag{3.31}$$

Charm [3] proposed the following equation. Since it considers the difference between fat and nonfat, it gives a better estimate.

$$C_{pf} = 4.187 \, x_w + 1.256 \, x_{snf} + 1.675 \, x_f \;\; kJ/(kg\text{-}K) \tag{3.32}$$

where x_w = the mass fraction of water in the food, x_{snf} = the mass fraction of solids-not-fat in the food, and x_f = the mass fraction of fat in the food.

Most foods are multicomponent systems, and their properties are composition dependent. Each component has its own thermal properties. A better estimate can be obtained using the method of mixtures described below.

Thermal Properties of Mixtures: In general, many thermal properties of mixtures can be computed by multiplying the properties of each component by its mass fraction in the mixture and summing the products.

$$P_m = \sum_{i=1}^{n} x_i M_{pi} \tag{3.33}$$

where P_m = a thermal property of the mixture, such as specific heat or specific enthalpy.

$x_i = m_i/m_T$ = the mass fraction of the ith component.

M_{pi} = corresponding thermal property of the ith component.

Table 3.5 gives the approximate specific heat for the major components of food.

Based on the constants in Table 3.5, the general equation for computing the specific heat of foods is:

$$C_{pf} = 4.187 \, x_w + 1.675 x_{fs} + 2.094 x_{fl} + 1.549 x_p + 1.424 x_c + 0.837 x_a \tag{3.34}$$

where x_w = mass fraction of water, x_{fs} = mass fraction of solid fat, x_{fl} = mass fraction of liquid fat, x_p = mass fraction of protein, x_c = mass fraction of carbohydrate, and x_a = mass fraction of ash.

The specific heats of materials often vary with temperature, but over a narrow range of temperatures, they may be treated as such. Many times, it is important to obtain a more precise value, and Choi and Okos [1] studied and reported the temperature dependence of the specific heat of major food components as follows:

$$C_{p,water} = 4081.7 - 5.3062\ T + 0.99516T^2 \quad (\text{for} - 40\ \text{to}\ 0\,^\circ C) \tag{3.35}$$

$$C_{p,water} = 4176.2 - 0.0909\ T + 5.4731 \times 10^{-3}T^2 \quad (\text{for}\ 0\ \text{to}\ 150\,^\circ C) \tag{3.36}$$

$$C_{p,carb} = 1548.8 + 1.9625\ T - 5.9399 \times 10^{-3}T^2 \quad (\text{for} - 40\ \text{to}\ 150\,^\circ C) \tag{3.37}$$

$$C_{p,protein} = 2008.2 + 1.2089\ T - 1.3129 \times 10^{-3}T^2 \quad (\text{for} - 40\ \text{to}\ 150\,^\circ C) \tag{3.38}$$

$$C_{p,fat} = 1984.2 + 1.4373\ T - 4.8008 \times 10^{-3}T_2 \quad (\text{for} - 40\ \text{to}\ 150\,^\circ C) \tag{3.39}$$

$$C_{p,ash} = 1092.6 + 1.8896\ T - 3.6817 \times 10^{-3}T^2 \quad (\text{for} - 40\ \text{to}\ 150\,^\circ C) \tag{3.40}$$

$$C_{p,ice} = 2062.3 + 6.0769\ T \tag{3.41}$$

where temperature (T) is in Celsius and specific heat values are in J/kg-°C.

Check Your Understanding

1. What is the specific heat of milk containing 3.4% liquid fat, 2.7% protein, 5.5% lactose, and 0.8% ash? ◄

3.9.2 Specific Heat of Food Below Freezing

Most liquid foods, such as milk or juice, start to freeze at −0.5 °C but do not become completely frozen until −90 °C. This happens because as the water freezes, the concentration of dissolved solutes in the unfrozen water increases, further depressing the freezing point. This process continues until all free water is frozen. Furthermore, a portion of the water in foods is energetically associated with other materials in the food, and this water (often called bound water) never freezes, even at very low temperatures. Before calculating the specific heat, it is thus necessary to determine the percentage of frozen water in the food.

Estimation of Frozen Water in Foods: Although this is best done experimentally, a good approximation can be obtained with one of the following formulas:

$$\text{Liquid foods} : x_{ice} = x_w \left(1 - \frac{T_f}{T}\right) \tag{3.42a}$$

As mentioned earlier, some water in foods remains unfrozen even at very low temperatures, depending on the composition of the food. It is more so in

Table 3.6 Initial freezing points and unfreezable water in selected food

Food	Initial freezing point (T_f), (°C)	Unfreezable or bound water (x_b)
Vegetables	−0.8 to −2.8	0.01 to 0.03
Fruits	−0.9 to −2.7	0.02 to 0.03
Meats	−1.7 to −2.2	0.10 to 0.12
Milk	−0.50	0.03
Liquid egg	−0.50	0.07
Fish	−0.6 to −2.0	0.06 to 0.08

concentrated foods and to account for this, the above equation is thus modified as follows:

$$\text{Concentrated foods}: x_{ice} = (x_w - x_b)\left(1 - \frac{T_f}{T}\right) \tag{3.42b}$$

where T_f = the initial freezing temperature, the temperature at which ice begins to form

x_{ice} = the mass fraction of ice at temperature T
x_w = the mass fraction of water at temperatures above T_f
T = the temperature of the frozen food ($T < T_f$)
x_b = the mass fraction of unfreezable or bound water

Table 3.6 lists the ranges of T_f and x_b for some common categories of foods. Once the percentage of frozen water is determined, the specific heat of food can be determined by considering it to be a mixture of three components: liquid water, solid water, and other solids.

The specific heat of foods (C_{pf}) below freezing can then be computed with the equation

$$C_{pf} = C_{ps}(1 - x_w) + C_{pw}\left(x_w\,\frac{T_f}{T}\right) + C_{pi}\left[x_w\left(1 - \frac{T_f}{T}\right)\right] \tag{3.43}$$

where C_{ps} = the specific heat of the solids in the food.

C_{pw} = the specific heat of the liquid water in the food = 4.18 kJ/(kg K).
C_{pi} = the specific heat of ice in food = 2.1 kJ/(kg K).
T_F = the initial freezing point of the food.
T = the temperature of the food ($T < T_F$).
x_w = the mass fraction of water in the food.

3.10   Enthalpy Calculations

The principal equations used to compute enthalpies during heating, cooling, and freezing are summarized below.

When there is no phase change and only sensible heating, or cooling occurs:

$$\Delta H = m \int C_p dT \tag{3.44a}$$

If the specific heat is independent of temperature, the above equation may be written as:

$$\Delta H = m\, C_p (T_2 - T_1) = m\, C_p \Delta T \tag{3.44b}$$

The changes in enthalpy of a substance between any two temperatures, T_1 and T_2, can also be calculated with the following equation:

$$\Delta H = m(\overline{H}_2 - \overline{H}_1) = m \Delta \overline{H} \tag{3.44c}$$

where ΔH is the change in total enthalpy, m is the mass of the material, and $\overline{H}_1$ and $\overline{H}_2$ are the specific enthalpies of the material at the two respective temperatures, T_1 and T_2.

Example

15. *Compute the enthalpy needed to heat a dry protein from 0 °C to 30 °C if its specific heat is given by the following relation:*

$$C_p = 4.10 – 5.30 \times 10^{-3}\, T + 9.95 \times 10^{-4}\, T^2 \ (kJ/kg\text{-}C)$$

   Solution: Using Eq. (3.44a) for specific enthalpy:

$$
\begin{aligned}
\Delta \overline{H} &= (4.10) \int_0^{30} dT - (5.30 \times 10^{-3}) \int_0^{30} T dT + (9.95 \times 10^{-4}) \int_0^{30} T^2 dT \\
&= (4.10)\, T - (5.30 \times 10^{-3})\, (1/2)\, T^2 + (9.95 \times 10^{-4})\, (1/3)\, T^3 \\
&= 4.10\,(30–0) – (5.30 \times 10^{-3})(30 - 0)^2/2 + (9.95 \times 10^{-4})(30 - 0)^3/3 \\
&= 132\ kJ/kg \quad\blacktriangleleft
\end{aligned}
$$

Example

16. *How much enthalpy is required to heat 20 kg of water from 40 °C to 80 °C?*

Solution: From the steam table (Appendix A.4), we find that the specific enthalpy of water is 167.6 kJ/kg at 40 °C and 334.9 kJ/kg at 80 °C. The change in enthalpy of 20 kg is

$$\Delta H = m(h_2 - h_1) = (20\,kg)(334.9\,kJ/kg - 167.6\,kJ/kg) = 3346\,kJ \qquad \blacktriangleleft$$

In situations where there is a phase change along with sensible heating or cooling, Eq. (3.44b) is modified to include the enthalpy of phase transformation:

$$\Delta H = m\,C_p\Delta T + m\lambda \qquad\qquad (3.45a)$$

where λ is the enthalpy of phase change. The above equation is frequently also written as:

For freezing:

$$\Delta H = m\,C_p\Delta T + m\overline{H}_{fs} \text{ or } m\,C_p\Delta T + m\left(\overline{H}_s - \overline{H}_f\right) \qquad (3.45b)$$

For melting:

$$\Delta H = m\,C_p\Delta T + m\,\overline{H}_{sf} \text{ or } m\,C_p\Delta T + m\left(\overline{H}_f - \overline{H}_s\right) \qquad (3.45c)$$

Similar expressions may be written for evaporation condensation, sublimation, and deposition with an appropriate understanding of the involved phase transformation.

Importantly, Eqs. (3.44b) and (3.45a) are the principal equations used to compute enthalpy changes during heat-transfer processes.

3.10.1 Isobaric Cooling with Phase Changes

Figure 3.11 shows a cooling process at constant pressure with phase changes. The same process is traced in Fig. 3.12 with the time course of cooling and change in

Fig. 3.11 An isobaric process on a phase diagram

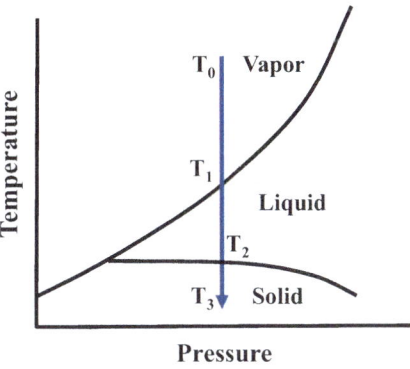

Fig. 3.12 Time course of an isobaric process

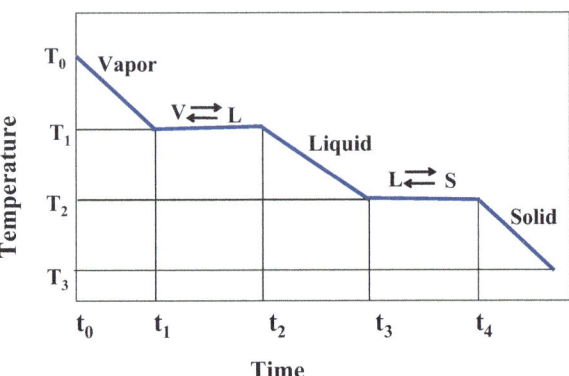

phase. The following observations may help visualize the energy changes involved during sensible heating/cooling and phase transformation processes:

- At time t_0, the material is in a vapor phase at temperature T_0.
- As heat is removed, the vapor cools until, at time t_1 and temperature T_1, the vapor begins to condense. The enthalpy being removed is the sensible heat, calculated as:

$$\Delta H_1 = mC_{pg}(T_1 - T_0) \tag{3.46a}$$

From time t_1 to time t_2, the vapor–liquid transition takes place. Since all heat removed comes from latent heat, the temperature remains constant at T_1 during this transition. For water at atmospheric pressure, $T_1 = 100\ °C$, and the enthalpy change is:

$$\Delta H_2 = m\overline{H}_{gf} = m\left(\overline{H}_f - \overline{H}_g\right) \tag{3.46b}$$

- From time t_2 to time t_3, the liquid cools for temperature T_2. Again, the heat being removed is sensible heat.

$$\Delta H_3 = mC_{pf}(T_2 - T_1) \tag{3.46c}$$

- Between times t_3 and t_4, the liquid–solid transition takes place. Since the heat removed again comes from latent heat, the temperature remains constant at T_2 during this transition. For water at atmospheric pressure, $T_2 = 0\ °C$.

$$\Delta H_4 = m\overline{H}_{fs} = m\left(\overline{H}_s - \overline{H}_f\right) \qquad (3.46d)$$

- Finally, after time t_4, the solid cools to T_3, giving up more sensible heat.

$$\Delta H_5 = mC_{ps}(T_3 - T_2) \qquad (3.46e)$$

Thus, the total energy to be removed becomes the sum of the energy needed at each step in the freezing process:

$$\Delta H_T = (-\Delta H_1) + (-\Delta H_2) + (-\Delta H_3) + (-\Delta H_4) + (-\Delta H_5) \qquad (3.46f)$$

Note the negative sign since the energy will be leaving the system. The same amount of energy will need to be added to the system to reverse the process, but then, according to our sign convention, it would be positive.

3.10.2 Enthalpy of Food

The amount of energy removed during freezing includes both sensible and latent heats, and both must be included when computing enthalpy changes.

Absolute enthalpy cannot be measured, so we focus on computing changes in enthalpy during a process. To do this, we select a reference temperature and compute the enthalpy relative to this reference. For foods that are not frozen, 0 °C or the triple point (0.01 °C) is frequently selected. However, to accommodate freezing operations, the enthalpy of foods is usually specified relative to −40 °C. This has the advantage of being below any temperature normally achieved in food processing so that we never have to deal with negative enthalpies. It also has the advantage that −40 °C is also −40 °F.

Enthalpy Calculations above the Freezing Point: The changes in enthalpy of any food between any two temperatures above its freezing point can be calculated with the aid of Eqs. (3.44a), (3.44b), and (3.44c).

Example

17. *Apples have a specific heat of approximately 4.0 kJ/kgK. How much enthalpy must we add to a 200 g apple to heat it from 16 °C to 54 °C?*
    Solution: Assuming the specific heat to be temperature independent and since we are concerned with differences in temperature, we can switch freely between Kelvin and Celsius and write:

$$\Delta H = mC_P \Delta T = (0.2\,\text{kg})\left(4.0\frac{\text{kJ}}{\text{kg}\,^\circ\text{C}}\right)(54\,^\circ\text{C} - 16\,^\circ\text{C}) = 30.4\,\text{kJ} \quad \blacktriangleleft$$

Check Your Understanding

2. Verify that 0.62 kg of ice must be melted to chill 5 kg of Atlantic salmon from 15 °C to 1 °C. The specific heat of Atlantic salmon is 0.71 kcal/kg°C. ◄

Enthalpy Calculations Below the Freezing Point: The quantity of heat removed during freezing then includes both sensible and latent heat, and both must be considered when computing enthalpy changes. Equations (3.44a) and (3.45b) are the principal equations used to compute these enthalpies.

Example

18. *Compute the enthalpy needed to transform 10 kg ice at –20 °C to 80 °C water in 1 atm. The values for the specific heat of water and ice are 4.18 and 2.12 kJ/kg-°C, respectively, and $h_{sf} = 334$ kJ/kg*
   Solution: Since both sensible and latent heats are needed, we may use Eq. (3.45b) as follows:

$$\Delta H = m\,C_{pi}\Delta T + m\overline{H}_{sf} + m\,C_{pw}\Delta T$$
$$= (10)\,\text{kg}\,[(2.12)(0 + 20) + (334.1) + (4.18)(80 - 0)]\,\text{kJ/kg} = 7109.0\,\text{kJ}$$

The change in enthalpy needed to freeze a food, however, can be determined by considering it to be a mixture of three components: liquid water, solid water (ice), and other solids. The equation to do this must consider the latent heat of fusion that must be removed to form the ice and the sensible heat that must be removed to cool each component to the desired temperature. The weight fraction of water that would be frozen into ice as a function of the temperature and its initial freezing point may be estimated using Eq. (3.42a or b), depending on the type of food. Once the percentage of frozen water is determined, Eqs. (3.44a) and (3.45b) may be employed to calculate the total enthalpy to be removed to accomplish freezing. When the freezing enthalpy is to be computed starting from the initial freezing point of the product, the equation becomes:

$$\Delta\overline{H} = [\text{Latent heat}] + [\text{Sensible heat}]$$

$$\Delta\overline{H} = \left(x_{ice}\overline{H}_{fs}\right) + (T - T_F)\left(x_s C_{ps} + x_{ice} C_{pi} + x_w C_{pw}\right) \tag{3.47}$$

where T_F = the initial freezing temperature of the food

T = any temperature below T_f

x_w = the mass fraction of liquid water in the food at T

x_{ice} = the mass fraction of ice in the food

x_S = the mass fraction of solids in the food

C_{pw} = the specific heat of liquid water (4.19 kJ/(kg C))

C_{pi} = the specific heat of ice (2.11 kJ/(kg C) at 0 °C)

C_{ps} = the specific heat of the other solids in the food

$\overline{H}_{fs}$ = the latent heat of solidification (opposite of fusion) of water to ice (−334.1 kJ/kg)

$\Delta \overline{H}$ = the specific enthalpy to convert unfrozen food at T_f to frozen food at T

Note: If there is sensible cooling before the initial freezing is reached, the sensible cooling enthalpy must be added to Eq. (3.47) ◄

Example

19. *A particular food is 70% free water, 10% of which is bound water. The remaining solids have a specific heat of 2.3 kJ/kg °C. The initial freezing point of the food is −0.9 °C. Using the above procedure, how much enthalpy must be removed from this food to lower its temperature to −5 ° C? To -40 °C?*

   Solution: *Use Eq. (3.42b) to find the mass fraction of water at each temperature:*

$$At -5\,°C : x_{ice} = (x_w - x_b)\left(1 - \frac{T_f}{T}\right) = (0.70 - 0.1(0.7))\left(1 - \frac{-0.9}{-5}\right) = 0.52$$

$$At -40\,°C : x_{ice} = (x_w - x_b)\left(1 - \frac{T_f}{T}\right) = (0.70 - 0.1(0.7))\left(1 - \frac{-0.9}{-40}\right) = 0.62.$$

   Use Eq. (3.47) to find the enthalpy change needed to reach each temperature:

$$At -5\,°C : \Delta\overline{H} = 0.52(-334.1) + (-5.0 + 0.9)[(1.0 - 0.7)2.3 + 0.52(2.11)$$
$$+ (0.7 - 0.52)4.19] = 184.1\ \text{kJ/kg}$$

$$At -40\,°C : \Delta\overline{H} = -207.14 - 39.14(0.69 + 1.31 + 0.34) = -298.6\ \text{kJ/kg} \ ◄$$

20. *Taking −40 °C as a reference, what is the enthalpy of 6 kg of the food from the previous example at −5 °C?*
    Solution:
        The specific enthalpy, relative to −40 °C, is: $\overline{H} = 298.6 - 184.1 = 114.5\ kJ/kg$
        The total enthalpy, relative to −40 °C, is: $H = (6\ kg)114.5\frac{kJ}{kg} = 687.0\ kJ$ ◄

3.11 General Observations on Specific Heat and Enthalpy of Foods

Heat transfer is one of the most important unit operations in food manufacturing, and thermal property data are needed for the engineering and design of these processes. Most foods are either heated or cooled to accomplish preservation and to make them more palatable. Prediction of the time necessary for many processes depends on good knowledge of the thermal properties of the food, which are often considered constant values. This may not always be true, and variations due to temperature, composition, and process variables become important factors in many applications that influence food properties. Given the diversity of foods and their compositions, it is indeed daunting to experimentally determine and tabulate all the thermal properties of utility to food processing. Since the composition data for most foods are available in the literature, many mathematical models to predict thermal properties based on composition and temperature as discussed above are in use, and more have become available in recent years.

The general behavior of the two properties of significant interest is illustrated in Fig. 3.13 and includes the following:

- **Specific Heat (C_p):** This measures the quantity of thermal energy required to raise the temperature of a unit mass by one degree. It is concerned only with sensible heating and cooling.
- **Specific enthalpy ($\overline{H}$):** This measures the quantity of thermal energy required to heat a unit mass from a reference temperature to the current temperature. It includes both the sensible heat needed to change the temperature and the latent heat needed for any phase changes that take place.

Some foods are stored at fairly low temperatures, while others are processed at fairly high temperatures, so the values of these parameters of interest to food processors range from −50 °C to 150 °C. Figure 3.13 compares these values over a range from 40° below freezing to 40° above freezing. Since most foods have high

A. Specific Heat (kJ/kg-K)

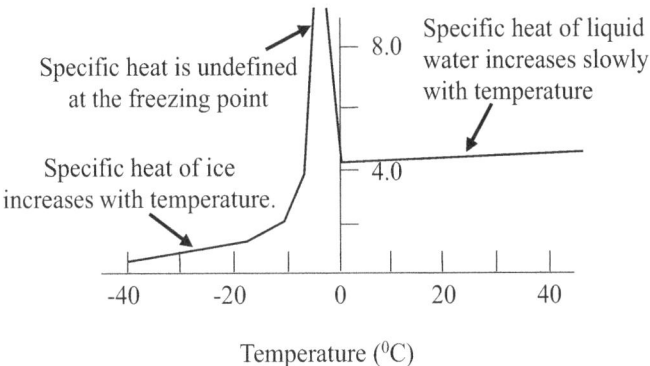

Temperature (^{0}C)

B. Specific Enthalpy (kJ/kg)

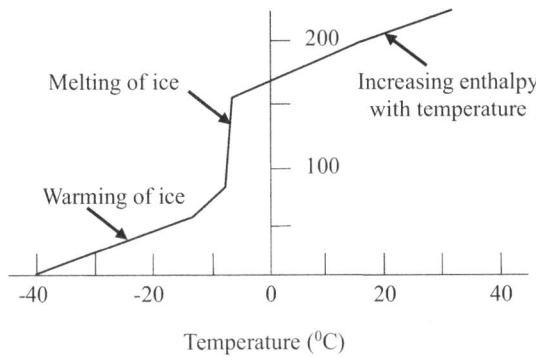

Temperature (^{0}C)

Fig. 3.13   General trends in the behavior of thermal properties of foods

water contents, their thermal properties are largely determined by water. Note the following:

- The specific heat of ice is lower than that of liquid water but increases with increasing temperature. Because adding energy at the freezing point produces no temperature rise, the specific heat at this temperature becomes nearly infinite and undefined. This explains the discontinuous peak in A.
  Both curves in Fig. 3.13 show a break a few degrees below 0. This is the freezing point of the foods that has been depressed by dissolved solutes. Since enthalpy must be added to raise the temperature, curve B trends continuously upward. At the freezing point, the enthalpy is used for the phase transition, so at this point, there is a sharp increase in enthalpy with little temperature change.

Problems

3.1 Energy density is defined as the amount of energy per unit volume. Calculate the "kinetic energy density" of water flowing through a horizontal pipe with inner area of 5 m^2, at a mass flow rate of 2 kg/s. Density of water is 1000 kg/m^3. Report your result in J/m^3 unit.

3.2 A 50 kg crate of fruits is raised to a platform 10 m above the ground in 5 s. Compute the potential energy and the power requirement for the job.

3.3 A 5.0 kg block of ice at 0 °C is left in a room at 25 °C. Determine the amount of enthalpy the ice (and then water) will absorb for melting and then warming up to the room.

3.4 Determine the amount of enthalpy that would be released if 100 lb$_m$ of saturated steam at a pressure of 20.72 psig is condensed and then cooled to 80 °F. What percent of the total released enthalpy will come from cooling?

3.5 Determine the amount of enthalpy needed to (a) heat 18 kg of water from 25 ° C to 100 °C and (b) convert the water from liquid to steam (at 1 atm).

3.6 An insulated kettle containing 120 kg of water at 20 °C is fitted with three electrical heating elements of 7500 W each. How long will it take (in minutes) for the water to reach 90 °C?

3.7 A food has the composition given in the table below. Estimate the specific heat of the product above the freezing temperature. If 8 kg of the food is homogeneously mixed with 3 kg of a sucrose-water solution of 25 °Brix, calculate the new product's specific heat.

Component	Mass fraction, dry basis
Water	0.15
Protein	0.32
Carbohydrate	0.53
Solid fat	0.15

3.8 The specific heat of a food product is given as:

$$C_p \ (kJ/kg\!-\!K) = 9.86 + 0.128T + 0.16T^2 \text{ where } T \text{ is in } (°C)$$

Calculate the heat taken out per unit mass if the product is cooled from 80 °C to 5 °C.

3.9 A jacketed kettle is filled with 15 kg of a solution of sucrose in water (65 ° Brix) at 25 °C. Determine the mass of saturated vapor at 120 °C that would be needed to heat the product to 75 °C. Assume that the specific heat of the product is constant, the kettle is perfectly insulated, and the condensate leaves the kettle at 120 °C.

3.10 A refrigerator is designed to remove 1500 kW heat from a food product. If the food product is 85% water, of which 10% is bound, how long would it take to reduce the temperature of 500 kg of this product from −5 °C to −40 °C? The initial freezing point of the food is -2 °C. (Cwater $= 4.18$ kJ/kgK, Cice $= 2.11$ kJ/kgK, Csolids $= 2.3$ kJ/kgK, $h$fs,water $= 334.1$ kJ/kg)

3.11 A batch evaporator operating at a vacuum of 610.3 mmHg is used to concentrate 500 kg of fresh tomato juice from 20% solids to 40% solids. Compute the following.

(a) The amount of water evaporated
(b) The latent heat of evaporation at the above pressure
(c) The total energy required to carry out the above operation. Assume that the initial temperature of the juice is 20 °C and the average specific heat (Cp) may be taken as 3.85 kJ/kg·K.

3.12 Mango pulp has the composition given in the table below. The initial freezing point of the product is -0.9 °C. For a batch of 50 kg of this product at 5 °C, compute the enthalpy that needs to be removed to lower its temperature to -20 °C. To determine the amount of bound water, the following equation was used:

$x_b = 0.4 \times x_p$, where x_b is the mass fraction of bound water and x_p is the mass fraction of protein in the food item.

Component	Mass fraction, dry basis
Water	4.47
Protein	0.03
Fat (liquid)	0.02
Carbohydrate	0.92
Ash	0.03

3.13 A well-insulated vacuum flask (mass 0.05 kg, specific heat 0.85 KJ·.kg^{-1}. K^{-1}) contains 0.10 kg water at 20 °C. Next, 0.05 kg of milk at 90 °C is mixed with the water, and the whole system (flask, water, and milk) is allowed to attain thermal equilibrium without any loss of energy to the surroundings. If the system temperature after equilibrium is 44 °C, estimate the specific heat of the milk.

3.14 Calculate the energy required to raise the temperature of an infant formula from 4 °C to 65 °C. The formula has the following composition.

Component	Mass (g)
Water	72
Nonfat solids	8.3
Fat (liquid)	14.4
Protein	5.1
Carbohydrate	15.8
Ash	9.4

3.15 An ice cream mix used in a manufacturing plant has the following composition:

14% fat (liquid); 12% protein; 16% sucrose; 0.5% ash

(a) Calculate the specific heat (C_p) of the mix

(b) Calculate the change in enthalpy of the mix if its temperature decreases from 40 °C to −15 °C during freezing, given the following information:

   (i)   The initial freezing point of ice cream is −2.8 °C.
   (ii) 10% of the total water content is bound water.
   (iii) The specific heat of ice at 0 °C is 2.11 kJ/kg-°C.
   (iv) The specific heat of other solids in the mix is 0.711 kJ/kg-°C

Bibliography

1. Atkins P (2007) Four laws that drive the universe. Oxford University Press, London
2. Cengel YA, Boles MA (2010) Thermodynamics, an engineering approach. McGraw Hill, Boston, MA
3. Charm SE (1978) The fundamentals of food engineering, 3rd edn, AVI Publishing Company, Westport, CT
4. Choi Y, Okos MR (1986) Effects of temperature and composition on the thermal properties of foods. In: Maguer LM, Jelen P (eds) Food engineering and process applications. Elsevier, New York
5. Dickerson RW (1964) Thermal properties of foods. In: Tressler DK , Van Arsdel WB, Copley MJ (eds) The freezing preservation of foods, 4th edn, AVI Publishing Company, Westport, CT
6. Mohsenine NN (1980) Thermal properties of food and agricultural materials. Gordon & Breach, London
7. Rahman MS (1995) Food properties handbook. CRC Press, Boca Raton, FL
8. Siebel JE (1892) Specific heat of various products. Ice Refrigeration 2:256

Mass and Energy Balances

4

One of the most important skills you can develop in Food Engineering is that of performing mass and energy balances. They are used for the design, operation, control, and analysis of processing operations as well as in product development and waste reduction. Mass and energy balances are based on the laws of the conservation of matter and energy, which state that neither matter nor energy can be created or destroyed. This being so, you should, with careful measurement, be able to account for all the matter and energy involved in a process. Mass and energy balances provide a framework for accounting of material and energy flows in and out of a process of interest. The same thing is true of your bank account. You should be able to account for every penny. When you balance your checkbook, you list the following information:

- **Deposits (Input).** You make a list of the deposits into the account. The engineer calls these inputs.
- **Interest (Generation).** You list any interest that is paid into your account. The engineer calls this generation.
- **Change in Balance (Accumulation).** You make a note of the old and new balances in your account and compute the change. The engineer calls this change accumulation.
- **Withdrawals (Output).** You make a list of the checks you have written. The engineer calls these outputs.

You can perform some arithmetic to verify that deposits plus interest equal withdrawals plus the change in your account balance. This process can be summarized in the following equation:

$$\{Deposits\} + \{Interest\} = \{Change\ in\ Balance\} + \{Withdrawals\} \qquad (4.1)$$

One of the three things will happen during the month:

© The Author(s), under exclusive license to Springer Nature Switzerland AG 2024
S. S. H. Rizvi, *Food Engineering Principles and Practices*,
https://doi.org/10.1007/978-3-031-34123-6_4

- If your deposits and interest add to more than the checks you have written, your balance will increase. We say you show a positive accumulation.
- If your deposits and interest add to less than the checks, your balance will decrease. We say you show a negative accumulation.
- If your deposits exactly equal your checks and, if there is no interest, your account balance will remain unchanged. We say that your account is in a steady state.

When you perform a mass or energy balance, you will do the same thing, writing equations of the form:

$$\{Input\} + \{Generation\} = \{Accumulation\} + \{Output\} \qquad (4.2)$$

For example, in analyzing the energy in a hot water heating system, "input" could refer to the heat contained in the water as it enters the furnace, "generation" to the heat added by the furnace, "accumulation" to the increase in heat of the water that stays in the furnace and "output" to the heat contained in the water as it leaves the furnace.

Not all balance equations require these 4 parts. Frequently, nothing is generated in an operation, and that term can be dropped from the balance equation, leaving

$$\{Input\} = \{Accumulation\} + \{Output\} \qquad (4.3)$$

For example, in a process that uses filters to separate protein solids from milk, "input" would be the mass of milk entering. "Output" would be the combined masses of the milk and the protein leaving, and "accumulation" would be the material that builds up on the filter and eventually clogs it. In the case of your bank account, this would be equivalent to an account that paid no interest.

In many processes, all the quantities of mass or energy entering a process will exactly equal the quantity leaving. Under these circumstances, the equation reduces to:

$$\{Input\} = \{Output\} \qquad (4.4)$$

and we say that the process is in a steady state. In the mass balance of a bottling plant, for example, the "inputs" would be the syrup, water, and carbon dioxide that enter, and the "outputs" would be the bottled soda. No material accumulates, and none is produced.

The remainder of this chapter contains techniques for dealing with equations of this form. In your study of Food Engineering, you will have occasion to perform balances on such things as:

- Mass in a manufacturing process
- Ingredients in a recipe
- Thermal energy

- Energy in moving fluids
- Momentum, etc.

In each case, you will set up one or more equations in the pattern of Eqs. (4.2), (4.3) or (4.4) and, if necessary, solve them for any unknown values. This will be similar to the situation where you forgot to write in the amount for one of your outstanding checks. If it is the only unknown transaction, you can easily calculate its amount.

Of course, as engineers, you will find yourself balancing several quantities simultaneously. For example, in a problem with a cream separator, one might balance the fat, solids-not-fat, and the water. Each of these would yield a separate equation. If there is more than one unknown, you will have to solve these balances simultaneously using simple algebra.

4.1   A Systematic Approach to Solving Mass and Energy Balance Problems

(a) **Select a system**
  - Draw a process flow diagram (pfd), showing all the input and output streams with known details.
  - Select the system's process segment of interest and draw a **boundary** around it to indicate streams of interest. Process streams that do not cross the boundary, e.g., recycled flows, are not included in the mass balance calculations.

(b) **Select a basis**
  - Select a basis for the mass balance calculations either in terms of mass of some material or component entering or leaving the process segment of interest or in terms of time.
  - Generally, choose as a basis a component that enters the process in either
    (i) only one stream and leaves the process in only one stream or
    (ii) the least number of streams

(c) **State assumptions**
  - Clearly, indicate any assumptions made to make up for missing information or to obtain estimates to facilitate solutions.

(d) **Set and solve equations**
  - Develop an independent equation involving each unknown quantity. This is achieved by writing the mass balance equation for the total mass as well as for each component entering and leaving the process. Often, many components are interrelated, which helps reduce the number of independent equations.

4.2   Mass Balance for One Operation

Let us start with a simple situation, like your bank account. For mass balance, we set up an equation to account for the **total mass** entering and leaving an operation. In the next section, we will look at several components at a time.

4.2.1   Definitions

We will start by defining a few terms:

- **Process.** We will use the term process to refer to the entire manufacturing process that you are analyzing. Some examples include a pasteurizer or a bread-making process.
- **Unit Operation.** A process can be divided into one or more unit operations, such as a mixer, a heater, a crystallizer, an oven, etc.
- **Input**. Various materials will enter each unit operation. These are the inputs. For example, in a bread mixing operation, inputs would be flour, shortening, water, etc.
- **Feed.** Where there is one predominant input to an operation, we frequently call it the "feed." For example, in a bean canning operation, the inputs would be beans, water, and salt, but beans would be considered to be the "feed."
- **Output.** Various materials will leave each operation. These are the outputs. In a baking operation, the outputs would be the finished bread plus moisture that escapes during baking and crumbs left in the baking pan.
- **Product.** Where there is one predominant output, we call it the product. Bread, for example, is the product of the above baking operation.
- **Stream.** The term stream will be used to refer to either an input or an output.
- **Path.** A sequence of one or more streams and operations that connect two parts of a process.
- **Accumulation.** Just as your bank balance can grow or shrink if your inputs do not equal outputs, mass can accumulate in an operation if the input exceeds the output. If the input is less than the output, there will be a negative accumulation.
- **Steady State.** If the mass (or energy) entering a system exactly equals the mass (or energy) leaving, there will be no accumulation, and we say the operation is in a steady state. Notice that, unlike equilibrium, a steady state is not a natural unchanging state. A steady state occurs only because we are adding and removing material at the same rate.

4.2.2   Overall Mass Balance

An overall mass balance for a single unit operation simply tracks the total mass entering and leaving the operation.

A. **Overview:** To perform a mass balance correctly, the following six steps are needed:
 1. Draw a **process flow diagram** of the operation and label it with total masses.
 2. State **assumptions** you will make.
 3. From the flow diagram, we write an **equation** for the total masses entering and leaving.
 4. If possible, **solve** the equation for any unknown masses.
 5. **Verify** that the equation is in balance.
 6. Use the solutions to compute **answers** to questions.
B. **Process Flow Diagram:** The first step in performing most mass balances involves drawing a process flow diagram of the operation. This diagram consists of three parts:
 1. **Operation**. Draw **a rectangular box** to represent the unit operation that is being balanced.
 2. **Inputs.** Draw an **arrow pointing into the box** for each mass input. Label each arrow with the name of the material it represents and with the quantity of that material. If a quantity is unknown, invent a variable to represent it.
 3. **Outputs.** Draw an **arrow pointing out of the box** for each mass output. As with the inputs, label each arrow with the name of the material it represents and with the quantity of that material. Again, invent variables to represent unknown quantities.

   Example 1 illustrates the above principles and practice of doing an overall mass balance.

Example

1. *In the process of making French fried potatoes, 5.1 kg of potatoes are lowered into 20.2 kgs of hot oil. At the end of the process, 3.6 kg of fried potatoes are removed, leaving behind 19.8 kg of oil. During frying, an unknown amount of volatile material was driven off. What is the mass of volatiles lost?*
   Solution: The operation is the fryer. The input streams are the raw potatoes (the feed) and the oil. The output streams are the fried potatoes (the product), the remaining oil, and the volatiles The path followed by the potatoes consists of the sequence Feed-Fryer-Product. This is diagrammed in Fig. 4.1:

Fig. 4.1 A process flow
diagram for mass balance

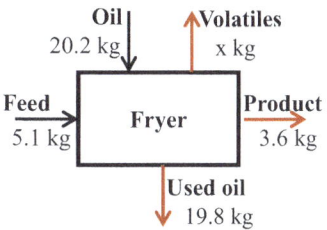

Assumptions: There will usually be missing information, and it may be necessary to make certain assumptions to make up for this. Furthermore, if very precise results are not needed, assumptions will allow you to make a good estimate without unneeded complications.

For the process in example 1, we make the following assumption:

- *The process is in a steady state.*
- *No matter is being generated, none accumulates.*
- *The balance equation, therefore, contains only input and output terms.*

Balance Equation: Once a flow diagram is constructed, write an equation to represent the movement of materials through the diagram. The equation is written according to the following rules:

- *Write **one term** in the equation for each <u>input</u> and <u>output</u> stream. The value of each term equals the mass of that stream.*
- *Place terms for the **input** streams to the <u>left</u> of the equal sign.*
- *Place terms for the **output** streams to the <u>right</u> of the equal sign.*
- *If the operation is not in steady state (or if you do not know), add a term for **generation** to the **left** of the equal sign and for **accumulation** to the **right** of the equal sign.*

For example, 1: The flow diagram in Fig. 4.1 yields the following overall mass equation.

$$Feed + Oil = Product + Used\ Oil + Volatiles \qquad (4.5)$$

$$5.1\ kg + 20.2\ kg = 3.6\ kg + 19.8\ kg + x\ kg \qquad (4.6)$$

Solving the equation: If there is only one unknown, you can solve for this quantity. Later, we will show how to deal with several unknowns.

Solving the equation above, we have

$$x = 5.1\ kg + 20.2\ kg - 3.6\ kg - 19.8\ kg = 1.9\ kg \qquad (4.7)$$

This tells us that 1.9 kg of volatiles were lost during this particular frying operation. This conclusion is valid only if the flow diagram and equation account for all inputs, outputs, and accumulations. If some of the material lost from the potatoes remained in the oil, the solution above would not be accurate. Depending on your needs, it might or might not be a useful approximation.

Verifying the Equation: All balances should be verified. To do this for an overall mass balance:

- *Substitute any solutions into the original equation.*
- *Separately add the left and right sides to the equation.*
- *Verify that the two <u>sums are equal</u>.*

To verify the solutions for Example 1,

$$5.1 \ kg + 20.2 \ kg = 3.6 \ kg + 19.8 \ kg + 1.9 \ kg$$
$$25.3 \ kg = 25.3 \ kg \rightarrow \textit{The equation balances.}$$

In this example, verification is simply a check on the solution. If there were no unknowns, a failure to balance could mean:

- *A measurement error.*
- *A math error.*
- *A stream or accumulation that was overlooked in drawing the flow diagram and setting up the equation. Perhaps there are some losses that need to be located.*
- *The existence of generation or accumulation in the operation.*   ◀

4.2.3   Component Balance

An overall balance, such as the one done in the last section, is not by itself very interesting. Usually, the material entering and leaving an operation consists of several components, such as protein, fiber, fat, and water. You may wish to track each of these components through an operation.

A. **Overview:** To do a component balance on a unit operation, the following six steps are performed:
  1. Draw a **process flow diagram** of the operation labeled with total and component masses.
  2. **State assumptions**.
  3. Write **equations** for the **total masses** and the masses of **each component** entering and leaving.
  4. If possible, **solve** the equations simultaneously for any unknown masses.
  5. **Verify** that each equation is in balance.
  6. Use the solutions to compute **answers** to questions.
B. **Process Flow Diagram:** The flow diagram is drawn for the overall balance with the following additional steps:
  1. Choose two or more **components** for the materials entering and leaving the operation.
  2. Make sure the selected components account for **all the materials** in the stream. If they do not, create an "other" component, such as "solids-not-fat" to account for the rest.
  3. **Label** each stream with the **masses** or **percentages** of the components in that stream. Invent variables where values are unknown.

Example 2 below illustrates the practice of writing a component balance in addition to the overall balance.

Example

2. *If the raw potatoes in Example 1 are 18% solids, determine the solids, water, and oil content of the fried potatoes.*

Solution: *For this problem, we view the materials as being composed of just water, solids, and oil.*

This selection of components accounts for all masses. For example, raw potatoes consist of just solids and water, while fried potatoes consist of just solids, oil, and water. In the flow diagram in Fig. 4.2, each stream is labeled with the percent of each component. The composition of the product is unknown, so we represent these percentages as variables.

Assumptions: *For this example, we make the following simplifying assumptions:*

- *That the process is in a steady state, i.e., the sum of inputs equals the sum of outputs.*
- *That the volatiles contain nothing but water.*
- *That no material from the potatoes is lost in the oil, allowing us to treat the used oil as pure oil.*

Balance Equations: *We write equations as follows:*

- *Write an **overall mass balance equation** in which each term represents the total mass in one stream.*
- *For a process involving n components, select **any** n − 1 components.*
- *Write a separate balance **equation** for **each selected component**.*

Fig. 4.2 A process flow diagram for mass and component balances

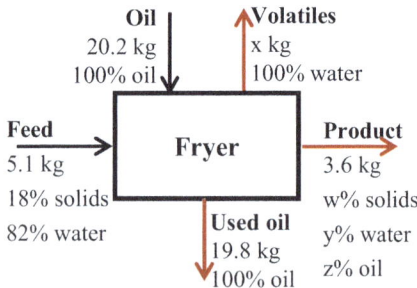

- **Each term** of a component equation represents the **mass of that component** for a single stream.
- **Each term** of the component equations is usually written as the **product** of the **fraction** of the component (percent/100) and the **total mass** of the stream.
- When a stream contains only one component, the **fraction is 1** for that component. When it contains none of a component, **the fraction is 0 for that component.**

Example 2 has 3 components, so we need an overall balance equation and $3 - 1 = 2$ component balance equations. The equations, taken from Fig. 4.2, are as follows:

$$Feed + Oil \; = Product + Used\; Oil + Volatiles \tag{4.8}$$

$$Overall: \;\; 5.1\,kg + 20.2\,kg \; = 3.6\,kg \;\; + 19.8\,kg + x\,kg \tag{4.9}$$

$$Components: \tag{4.10}$$
$$Solids: \;\; (0.18)5.1\,kg + (0)20.2\,kg = (w)3.6\,kg + (0)19.8\,kg + (0)x\,kg$$

$$Water: \;\; (0.82)5.1 \;\; kg + (0)20.2 \;\; kg$$
$$= (y)3.6 \;\; kg + (0)19.8 \;\; kg + (1)x \;\; kg \tag{4.11}$$

$$Oil: \;\; (0)5.1\,kg + (1)20.2\,kg = (z)3.6\,kg + (1)19.8\,kg + (0)x\,kg \tag{4.12}$$

The variables w, y, and z represent unknown fractions between 0 and 1. The variable x represents a mass in lbs. Notice where the fractions 0 and 1 are used. Since we only need 2 component equations, we will arbitrarily discard the solids equation.

Solve the Equations: For an n component system, there will be n equations. If there are no more than n unknowns, the equations can be solved simultaneously for the unknowns. This can be done using pencil and paper or the equations can be set up in matrix form and solved using a spreadsheet or other computer programs capable of handling matrices. Here is the pencil and paper solution. The matrix solution will be shown later.

We will solve the overall water and oil equations from Sect. 4.2.3 above. They are:

$$Overall: \;\; 5.1 \;\; kg + 20.2 \;kg = 3.6 \;\; kg + 19.8 \;\; kg + x \;\; kg \tag{4.13}$$

$$Water: \;\; (0.82)5.1 \;\; kg + (0)20.2 \;\; kg$$
$$= (y)3.6 \;\; kg + (0)19.8 \;\; kg + (1)x \;\; kg \tag{4.14}$$

$$Oil: \;\; (0)5.1 \;\; kg + (1)20.2 \;\; kg = (z)3.6 \;\; kg + (1)19.8 \;\; kg + (0)x \;\; kg \tag{4.15}$$

Solution: Equation (4.13) *above has only one variable, x, so it is a good place to start. Solving for x, we obtain*

$$x = 5.1 \ kg + 20.2 \ kg - 3.6 \ kg - 19.8 \ kg = 1.9 \ kg \ in \ the \ steam \ stream.$$

Substitute this solution into the above Eqs. (4.14) and (4.15) and compute the masses of water and oil in each stream.

$$Mass \ water : \quad 4.2 \ kg + 0 \ kg = 3.6y \ kg + 0 \ kg + 1.9 \ kg$$
$$Mass \ oil : \quad 0 \ kg + 20.2 \ kg = 3.6z \ kg + 19.8 \ kg + 0 \ kg$$

Solve for y and z, the fractions of water and oil in the product stream.

$$y = \frac{4.2 + 0 - 0 - 1.9}{3.6} = 0.64 \ (64\% water \ in \ the \ fried \ potatoes)$$

$$z = \frac{0 + 20.2 - 19.8 - 0}{3.6} = 0.11 \quad (11\% oil \ in \ the \ fried \ potatoes)$$

Since water, solids, and oil account for the entire mass of the fried potatoes, the percent solids is easily obtained by difference.

$$w = 1.00 - 0.64 - 0.11 = 0.25 \ (25\% solids \ in \ fried \ potatoes)$$

Verify: *To verify a set of balance equations, we should now substitute the solutions in the original equations and show the following:*
- *For each equation, the sum of the left side equals the sum of the right side.*
- *All fractions in the solution are numbers between 0 and 1.*
- *For each stream, the sum of the component fractions is 1.00.*
- *For each stream, the sum of the component masses equals the total mass for that stream in the overall balance equation.*

1. *For the fried potatoes, we first substitute the solutions found for x, y, and z.*

$$Feed + Oil \ = Product + Oil + Volatiles \qquad (4.16)$$

$$Overall : \quad 5.1 \ kg + 20.2 \ kg = \quad 3.6 \ kg + 19.8 \ kg + 1.9 \ kg \qquad (4.17)$$

Components :

$$Solids : \ (0.18)5.1 \ kg + (0)20.2 \ kg = (0.25)3.6 \ kg + (0)19.8 \ kg + (0)1.9 \ kg$$
$$(4.18)$$

$$Water : \ (0.82)5.1 \ kg + (0)20.2 \ kg$$
$$= (0.64)3.6 \ kg + (0)19.8 \ kg + (1)1.9 \ kg \qquad (4.19)$$

$$Oil: \ (0)5.1 \ kg + (1)20.2 \ kg$$
$$= (0.11)3.6 \ kg + (1)19.2 \ kg + (0)1.9 \ kg \tag{4.20}$$

Compute the masses for each stream:

Overall :   5.1 *kg*   +   20.2 *kg*   =   3.6 *kg*   +   19.8 *kg*   +   1.9 *kg*

Components :

Solids :   0.92 *kg*   +   0 *kg*   =   0.90 *kg*   +   0 *kg*   +   0 *kg*

Water :   4.18 *kg*   +   0 *kg*   =   2.30 *kg*   +   0 *kg*   +   1.9 *kg*

Oil :   0 *kg*   +   20.2 *kg*   =   0.40 *kg*   +   19.2 *kg*   +   0 *kg*

Balance each equation by summing each side:

Overall :   25.30 *kg*   =   25.30 *kg*

Solids :   0.92 *kg*   =   0.90 *kg*

Water :   4.18 *kg*   =   4.20 *kg*

Oil :   20.20 *kg*   =   20.20 *kg*

Except for rounding error, the equations balance.

2. *The fractions in the solution are 0.64, 0.11, and 0.25, all between 0 and 1.*
3. *Add the fractions for each stream across all component equations and verify that they add to 1.00. For example,*

(a) *Feed Stream:* 0.18 + 0.82 + 0.00 = 1.00
(b) *Product Stream:* 0.25 + 0.64 + 0.11 = 1.00.

4. *Add the component masses for each stream across all component equations and verify that they add to the mass in the corresponding overall equation. For example,*

Feed stream :   0.92 *kg solids*   +   4.18 *kg water*   +   0.00 *kg oil*   =   5.1 *kg*

Product stream : 0.90 *kg solids*   +   2.30 *kg water*   +   0.40 *kg oil*   =   3.6 *kg*

◀

Check Your Understanding

1. How many kilograms of 28% cream and 3% milk will be required to make 1000 kg of 4% fat milk? ◀

4.3   Mass Balance for More than One Operation

When there is more than one unit operation,

- Draw a **flow diagram** with a **separate box** for each operation.
- State your **assumptions**.
- Draw arrows to represent input and output **streams**, noting that some outputs of one operation will become inputs for the next.
- Write a separate set of **equations** for each box (operation), following the same rules as for a single operation. If your problem involves n components and m operations, you will write m equations n times.
- **Solve** these n times m equations for any unknowns.
- **Verify** the balances as before.
- Use the solutions to compute **answers** to questions.

Example

3. *In the making of a cake from mix, 1.34 lbs of mix with a moisture content of 6.2% are combined with 0.75 lb of water. After transferring the batter to a baking pan, it is found that 0.14 lb of batter remained in the mixing pan. After baking, the cake weighs 1.72 lbs. What is the percent yield of baking operation and of the overall process? What is the percent moisture in the finished cake?*

   Solution: *This process involves a two-unit operation, mixing and baking. The flow diagram is shown in Fig. 4.3. While labeling this diagram, it was necessary to invent variables for the masses of steam and batter and for the fraction moisture in the batter, waste, and cake. Since it seems reasonable that the batter and waste are identical in composition, the same variable was used for water fraction in each stream.*

   Assumptions: *For this example, we will assume that*

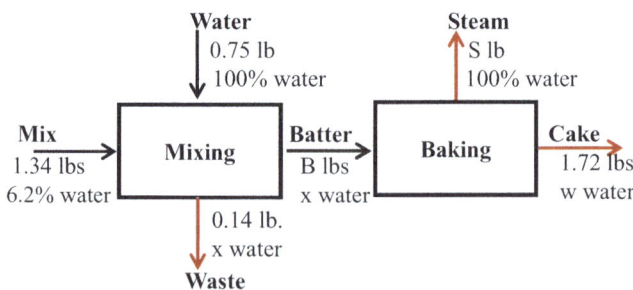

Fig. 4.3   A process flow diagram with two-unit operations

- *The process is in a steady state.*
- *That the batter and waste have the same composition.*
- *That there is negligible waste left in the baking process.*
- *That the weight loss during baking is due entirely to lost steam.*

Equation: When there is more than one operation, the equations are set up as follows:

- *Set up a **separate set** of equations for each operation.*
- *For each operation, **one overall** balance and **n − 1 component** balances were set up.*
- *When a stream **connects** two operations, the terms for that stream will appear in both sets of equations.*

For the problem in Example 3, we can view all materials as being composed of water and solids in various proportions. With 2 components and 2 operations, we write 4 equations.

For the first operation, the equations are:

$$Mix \quad + \quad Water \quad = \quad \underline{Batter} \quad + \quad Waste$$

$$1. \, Overall: \quad 1.34 \quad + \quad 0.75 \quad = \quad \underline{B} \quad + \quad 0.14 \qquad (4.21)$$

$$2. \, Water: \quad (0.062)1.34 + (1)0.75 \; = (x)B + (x)0.14 \qquad (4.22)$$

For the second operation, the equations are:

$$\underline{Batter} \quad = \quad Cake \quad + \quad Steam$$

$$3. \, Overall: \quad \underline{B} \quad = \quad 1.72 \quad + \quad S \qquad (4.23)$$

$$4. \, Water: \quad (x)B \; = (w)1.72 + \; (1)S \qquad (4.24)$$

Note that the batter terms (underlined) appear in both sets of equations, in the first set on the output side and in the second set on the input side.

Solution: *All equations for the process can be solved together as a set of simultaneous equations.*

The equations can be solved quite simply. Equation (4.1) above has only one unknown so it can first be solved for B:

$$B = 1.34 + 0.75 - 0.14 = 1.95 \; lbs.of \; batter \; in \; the \; cake.$$

Substitute this value for B in the above Eqs. (4.2) and (4.3) and solve for x and S:

$$x = \frac{(0.062)1.34 + (1)0.75}{1.95 + 0.14} = 0.40 \,(The\ batter\ is\ 40\%moisture)$$

$$S = 1.95\text{–}1.72 = 0.23\ lb.(Steam\ lost)$$

Substitute B, x, and S into Eq. (4.4) and solve for w:

$$w = \frac{(0.40)1.95 - (1)0.23}{1.72} = 0.32 \,(The\ cake\ is\ 32\%moisture)$$

The percent yield in the baking process is:

$$\%yield = 100\left(\frac{Mass\ cake}{Mass\ batter}\right) = 100\left(\frac{1.72}{1.95}\right) = 88\%$$

The percent yield for the overall process is:

$$\%yield = 100\left(\frac{Mass\ cake}{Mass\ mix + Mass\ water}\right) = 100\left(\frac{1.72}{1.34 + 0.75}\right) = 82\% \quad \blacktriangleleft$$

4.3.1   Multiple Paths

In the process shown above, there is only one path, the batter stream, connecting the operations. Sometimes there may be more than one path between operations. This makes the flow diagram slightly more complicated, but the procedure is the same.

- Draw and label the flow diagram.
- State assumptions.
- For each operation, write an overall balance equation and $n - 1$ component equations.
- Solve the equations.

Example

4. *In a 2-stage filtration process to concentrate proteins, a suspension containing 16% protein in water is mixed with water that is recycled from stage 2 at a rate of 500 kg/h of suspension to 2000 kg/h of water. The diluted suspension is passed through the stage 1 filter that holds back a 50% protein suspension (Retentate 1). (The diagonal lines in the boxes represent the filter.) The suspension that passes through the first filter (Permeate 1) contains 1.25% protein. This was passed through a stage 2 filter that held back 50 kg/h of suspension (Retentate 2) containing all remaining protein. Of the water that passes through the second filter, 2000 kg per hour is recycled for use in the first filter. The rest is waste. What is the mass flow rate of*

Retentate 1? What is the protein concentration of Retentate 2? What is the mass flow rate of the waste?

Solution: *Fig. 4.4 is a flow diagram of this process. Since there are two components, we must set up two equations for each operation, for a total of 4 equations. These are:*

$$\text{Operation 1}: \quad Feed + \underline{Recycle} \;\; = Retentate\ 1 + \underline{Permeate\ 1} \qquad (4.25)$$

$$\text{Overall}: \quad 500 + 2000 = X + Y \qquad (4.26)$$

$$\text{Protein}: (0.16)500 + (0)2000 = (0.50)X + (0.0125)Y \qquad (4.27)$$

$$\text{Operation 2}: \quad \underline{Permeate\ 1} = \underline{Recycle} + Retentate\ 2 + Waste \qquad (4.28)$$

$$\text{Overall}: \quad Y = 2000 + 50 + Z \qquad (4.29)$$

$$\text{Protein}: \quad (0.0125)Y = (0)2000 + (w)50 + (0)Z \qquad (4.30)$$

Because the Permeate 1 stream and the Recycle stream are both paths between the two operations, terms for those streams appear in both sets of equations.

There is no equation with only one unknown, so we must use a new technique to solve these equations. We will start by eliminating X from the above Eqs. (4.26) and (4.27). To do this, multiply the above Eq. (4.27) by 2 and subtract it term by term from Eq. (4.26).

1. *Overall:* $500 + 2000 = X + Y$

2. *Protein:(x2)* $\underline{160 + 0 = X + (0.025)\ Y}$

Difference: $340 + 2000 = (0.975)\ Y$

Solve the new equation for Y:

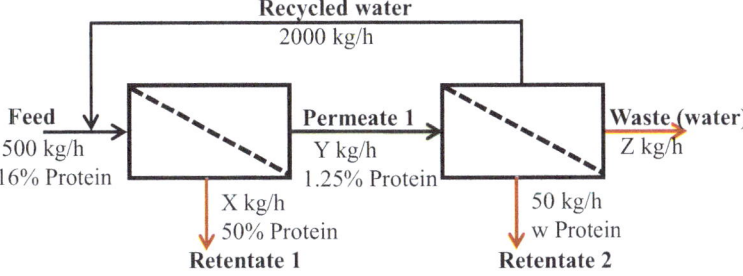

Fig. 4.4 A process flow diagram with recycling

$$Y = \frac{340 + 2000}{0.975} = 2400 \text{ kg/h Permeate 1}$$

Substituting Y into Eqs. (4.26) and (4.29), we obtain

$$X = 500 + 2000 - 2400 = 100 \text{ kg/h Retentate 1}$$

$$Z = 2400 - 2000 - 50 = 350 \text{ kg/h of waste}$$

Finally, substitute Y and Z in Eq. (4.30) and solve for w.

$$w = \frac{(0.0125)2400}{50} = 0.60 \text{ (Retentate 2 is 60\%protein)}$$

We leave it to the reader to verify these balances. ◀

4.4   Energy Balance

An energy balance is similar to a mass balance except that it deals with enthalpies (heat contents) rather than masses. Energy balances are usually done together with mass balances because the energy content of the various streams is dependent on their masses. The steps in such a balance are as follows:

- Draw a **process flow diagram** and label it with masses and enthalpies.
- State **assumptions**.
- Write **mass balance equations** for this diagram.
- Write an **energy balance equation** for this diagram.
- **Compute the enthalpy** for each stream. If information is incomplete, invent variables and write the expression needed to compute the enthalpy for the stream. Insert these in the energy equation.
- **Solve** the mass and energy balance equations together.
- **Verify** both the mass and energy balances.
- Use the solutions to **answer** questions.

4.4.1   Process Flow Diagram

The process flow diagram is drawn as follows:

- Draw a flow diagram for a mass balance.
- Label the streams with total masses and, if they are involved, component masses.
- Label each stream with a symbol representing the enthalpy contained in that stream, using the letter "H" with a subscript to represent each enthalpy.
- Label the streams with their pressure and/or temperature.

Example

5. *A total of 1520 kg per hour of a juice enters an evaporator at 25°C, where it is heated at a pressure of 0.75 atmospheres, causing the water to evaporate from the juice. The concentrated juice emerges at a rate of 450 kg per hour. Heat is supplied by the condensation of saturated steam at 1.2 atmospheres pressure. The steam is separated physically from the juice so that heat, but no mass is transferred between them. How many kilograms of steam must be supplied each hour?*

   Solution: *In drawing the flow diagram for this problem (Fig. 4.5), we show the operation as a box with a dividing line. This line indicates that steam and juice are in separate chambers and that energy but not matter flows from one chamber to the other.*

   The enthalpies of the streams are represented by the symbols H_F, H_v, H_P, H_S, and H_C. Note that the steam becomes the condensate, so their masses are the same and have been assigned the same variable.

   In the SI system, the pressures are:

   $$0.75(101.4) = 76.1\,kPa\,and\,1.2(101.4) = 121.7\,kPa.$$

Assumptions: In this example, we make the following assumptions.
* *The process is in a steady state.*
* *The steam is saturated (not superheated) at the given pressure.*
* *The steam is 100% quality, i.e., it is pure vapor with no liquid water.*
* *The condensate is saturated liquid at the same temperature as the saturated steam. This implies that all heat is obtained from latent heat, and none is obtained from cooling the condensate.*
* *The condensation of the steam is complete within the heat exchanger so that the water emerging is completely liquid.*
* *The masses of the steam and condensate are equal.*

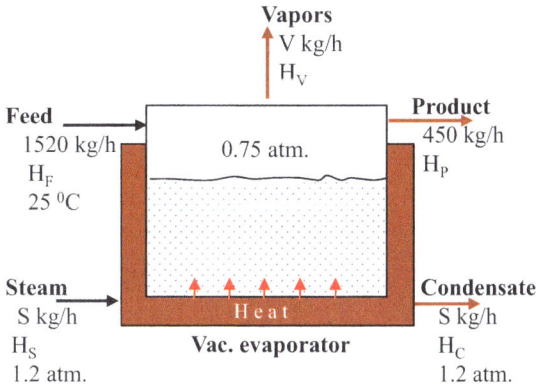

Fig. 4.5 A process flow diagram for mass and energy balance

- *The vapor and product can both be treated as saturated at the given pressure, and the product emerges at the temperature of saturated vapor.*
- *That both the feed and product are high enough in water content so that their heat capacities are those of pure water.*
  The Balance Equations: *Balance equations are written as follows:*
- *The mass balance equations are written as previously described.*
- *Each term in the energy balance equation consists of the total enthalpy content of one stream.*

In this example, since no mass can be exchanged between the juice and steam sides of the heat exchanger, we can write separate balance equations for each.

$$Feed = Product + Vapor \tag{4.31}$$

$$1. \; Juice : 1520 \; kg/h = 450 \; kg/h \; + V \; kg/h \tag{4.32}$$

Since heat is exchanged between chambers, we must write a single energy balance equation for the entire operation.

$$Feed + Steam \; = Product + \; Vapor + \; Condensate \tag{4.33}$$

$$2. \; Energy : \; H_F + H_S = H_P + H_V + H_C \tag{4.34}$$

Compute Enthalpies from Steam Tables: *For streams that contain saturated water or vapor, enthalpies can be computed from a saturated steam table. For superheated steam, values can be computed from superheated steam tables. The following formula is used:*

$$H = m \, \overline{H} \tag{4.35}$$

where $H =$ the enthalpy of the stream, $m =$ the mass of the stream, and $\overline{H} =$ the specific enthalpy obtained from the steam table.

There are 4 streams in Example 5 that contain saturated vapor or liquid water. The steam and condensate streams are both saturated at 1.2 atmospheres (121.7 kPa). The vapor and product streams are saturated at 0.75 atm (76.1 kPa). Turning to a steam table, we obtain the following information:

Pressure (kPa)	Temperature (°C)	Specific enthalpies (kJ/kg)	
		Saturated liquid	Saturated vapor
76.1	92	380	2663
121.7	105	505	2708

From this information and the mass of the streams, we write the following expressions for the enthalpy of the streams:

$$H_s = (2708 \, kJ/kg) \, (S \, kg/h) = 2708 \, S \, kJ/h$$
$$H_C = (505 \, kJ/kg) \, (S \, kg/h) = 505 \, S \, kJ/h$$
$$H_V = (2663 \, kJ/kg) \, (V \, kg/h) = 2663 \, V \, kJ/h$$
$$H_P = (380 \, kJ/kg) \, (450 \, kg/h) = 171,000 \, kJ/h$$

Compute Enthalpies of the Other Streams: When a stream does not contain saturated liquid or vapor or superheated steam, the steam table is not appropriate. In that case, the enthalpy of a stream can be computed from its temperature, mass, and heat capacity (C_P). If the material is predominantly water, the C_p for water can be used. Otherwise, use C_p values for the specific material. The equation is

$$H = mC_p\Delta T \qquad (4.36)$$

where H = the enthalpy of the stream.
m = the mass of the stream.
C_P = specific heat of the stream,
ΔT = the temperature of the stream relative to a reference temperature. Since the steam table uses $0 \,°C$ as a reference, it is best to use $0 \,°C$ as a reference for all calculations.

The enthalpy of the feed stream can be approximated using the heat capacity of water. The temperature of the feed is given as $25 \,°C$. The enthalpy of this stream is:

$$H_F = mC_P\Delta T = (1520 \, kg/h) \, (4.190 \, kJ/kg \,° C) \, (25 - 0 \,° C) = 159,220 \, kJ/h$$

Solution: The enthalpy expressions are substituted into the energy balance equation, and the mass and energy balance equations are solved together.

Substituting the enthalpies into the above Eq. (4.2), we now have 2 unknowns. Then the above Eqs. (4.1) and (4.2) can be solved together.

$$Feed + Steam = Product + \quad Vapor + Condensate \qquad (4.37)$$

$$1. \, Juice \quad 1520 \, kg/h = 450 \, kg/h \ + \ V \, kg/h \qquad (4.38)$$

$$2. \, Energy \quad 159220 + 2708 \, S = 171000 + 2663 \, V + 505 \, S \qquad (4.39)$$

These can be solved as follows:

$$Equation \, (4.38) : V = 1520 - 450 = 1070 \, kg/h \, of \, vapor$$

$$Equation \, (4.39) : S = \frac{2663(1070) + 170198 - 159220}{2708 - 505} = 1.298 \times 10^3 \, kg/h \, steam$$

Verify: To verify the solutions, we first check the mass balance:

$$1520 \, kg/h = 450 \, kg/h + 1070 \, kg/h$$
$$1520 \, kg/h = 1520 \, kg/h$$

Then, the energy balance gives:

$$159220 \, kJ/h + 2708 \, kJ/kg(1298 \, kg/h)$$
$$= 171000 \, kJ/h + 2663 \, kJ/kg(1070 \, kg/h)$$
$$+ 505 \, kJ/kg(1298 \, kg/h)3.674 \times 10^6 \, kJ/h = 3.675 \times 10^6 \, kJ/h$$

◀

Check Your Understanding

2. Electrical power is frequently used in food processing. How is the electric power converted into energy? ◀

4.5   Solving Problems Using Matrices

As balance equations become larger and more complicated, the pencil and paper solution becomes less desirable. An alternate approach involves converting the balance equations to matrix form and solving them with a computer.

4.5.1   Basic Terms

What is a Matrix? A matrix is a rectangular array of numbers surrounded by brackets.

Example

6. *Examples of a matrices*

$$\begin{pmatrix} 2 & 3 & -7 & 15 \\ 4 & 18 & -5 & 3 \\ 2 & -0.5 & 0 & 1 \end{pmatrix} , \quad (5 \quad 2 \quad -0.25 \quad 0) , \quad \begin{pmatrix} 5 \\ -2 \\ 0 \end{pmatrix} , \quad (4)$$

◀

Parts of a Matrix. The individual numbers in a matrix are called elements. These are arranged in rows and columns. The first matrix in Example 6 has 12 elements. These are 2, 3, −7, etc. The first row of that matrix contains elements 2, 3, −7, and 15. The first column of that matrix contains the elements 2, 4, and 2.

Dimensions of a Matrix. The dimensions of a matrix are given as the number of rows times the number of columns. The examples given above have the following dimensions:

$$3 \times 4, 1 \times 4, 3 \times 1 \text{ and } 1 \times 1$$

What is a Vector? A matrix containing a single column is called a **column vector**. A matrix containing a single row is called a **row vector**. (If you are used to thinking of vectors as arrows, think of each element of a row or column vector as a coordinate of the point of such an arrow). In Example 6, the second matrix is a row vector, and the third is a column vector.

4.5.2   Rearrange the Equations

A set of simultaneous equations can be expressed in matrix notation. This provides both a convenient and compact way of writing a set of equations and a simple method for solving them. Before putting these equations in matrix form,

• Stack the equations one above the other.
• Place the terms involving variables to the left of the equal sign.
• Place the constant term to the right of the equal sign. If there is no constant term, use a zero. If there is more than one constant term, combine them into one.
• Arrange the terms on the left so the variables are in the same order in each equation.
• If a variable has no coefficient, give it a coefficient of 1.
• If a variable is missing from an equation, insert a term for it with a 0 coefficient.
• Check signs carefully, they change when a term is moved from one side of the equation to the other. (To reduce the number of minus signs, you can change the sign of all terms of any equation by multiplying all by -1.)

Example

7. *Let us solve Example 4 (Fig. 4.4) with matrices. That example produced the following equations:*

Operation 1   *Feed + Recycled = Permeate 1 + Retentate 1*
1. Overall: *500 + 2000 = X + Y*
2. Protein: *(0.16)500 + (0)2000 = (0.50)X + (0.0125) Y*

Operation 2   *Retentate 1 = Recycle + Retentate 2 + Waste*
3. Overall: *Y = 2000 + 50 + Z*
4. Protein: *(0.0125)Y = (0)2000 + (w)50 + (0)Z*
        If these equations are rearranged according to the rules above, they look like this:

$$\underline{aX + bY + cZ + dw = Constant}$$
$$(-1) X + (-1) Y + 0 Z + 0 w = -2500$$
$$(-0.5) X + (-0.0125) Y + 0 Z + 0 w = -80$$
$$0 X + 1 Y + (-1) Z + 0 w = 2050$$
$$0 X + 0.0125 Y + 0 Z + (-50) w = 0 \blacktriangleleft$$

4.5.3 Coefficient Matrix

Translate the set of equations into matrix form. First, create a matrix of the coefficients in the order they are arranged in the stack of equations.

- Each row of the matrix should contain coefficients from one equation.
- Each column should contain coefficients for one variable. The coefficient matrix for example 7 is:

$$\begin{pmatrix} -1 & -1 & 0 & 0 \\ -0.5 & -0.0125 & 0 & 0 \\ 0 & 1 & -1 & 0 \\ 0 & 0.0125 & 0 & -50 \end{pmatrix}$$

4.5.4 Variable Vector

Following this matrix, place a column vector of the variables in the order they appear in the rearranged equations. In the equations in Example 7, the variables are in the order X, Y, Z, and w from left to right. We place them in the same order but from top to bottom in the variable vector:

$$\begin{pmatrix} X \\ Y \\ Z \\ w \end{pmatrix}$$

4.5.5 Constant Vector

Follow this with an equal sign and a column vector of the constants, ordered as they are in the stacked equations. Adding the constant terms for the 4 equations, we have the final matrix equation:

$$\begin{pmatrix} -1 & -1 & 0 & 0 \\ -0.5 & -0.0125 & 0 & 0 \\ 0 & 1 & -1 & 0 \\ 0 & 0.0125 & 0 & -50 \end{pmatrix} \begin{pmatrix} X \\ Y \\ Z \\ w \end{pmatrix} = \begin{pmatrix} -2500 \\ -80 \\ 2050 \\ 0 \end{pmatrix}$$

If we let $\mathbf{A}$ = the coefficient matrix, $\mathbf{x}$ = the variable vector and $\mathbf{c}$ = the constant vector, this equation can be written in the very compact form:

$$\mathbf{A}\mathbf{x} = \mathbf{c}$$

4.5.6 Solving Matrix Equations

In regular algebra, an equation of the form

$$ax = c$$

can be solved by multiplying both sides by the inverse of a, thus:

$$\frac{1}{a}ax = \frac{1}{a}c$$

This can also be written as follows:

$$a^{-1}ax = a^{-1}c$$

Since $a^{-1}a = 1$, this equation reduces to the solution:

$$x = a^{-1}c$$

For example, if $a = 6$ and $c = 18$, we have

$$6x = 18$$
$$(6^{-1})6x = (6^{-1})18, x = 3$$

Solving a matrix equation involves a similar process.

- Starting with the matrix equation: $\mathbf{A}\mathbf{x} = \mathbf{c}$
  Find a new matrix that is the "inverse" of the coefficient matrix. We will represent this matrix as $\mathbf{A}^{-1}$
- "Multiply" the constant vector by this inverse matrix to yield the solution. (We will not define matrix multiplication here, but there are computer programs that know how to do it.) In matrix notation, this solution is expressed as follows:

$$\mathbf{x} = \mathbf{A}^{-1}\mathbf{c}$$

Using the matrix equation developed in Sect. 4.5.5, the spreadsheet will look something like this:

	A	B	C	D	E	F
1	-1	-1	0	0		-2500
2	-0.5	-0.0125	0	0		-80
3	0	1	-1	0		2050
4	0	0.0125	0	-50		0

Inverting the coefficient matrix yields:

	A	B	C	D
5				
6	0.02564	-2.05128	0	0
7	-0.02564	2.05128	0	0
8	-0.02564	2.05128	-1	0
9	-0.00026	0.00051	0	0.02

Now multiply the constant matrix by the inverse matrix, the solution becomes:

E	F
5	
6	100
7	2400
8	350
9	0.6

which agrees with the solution in Example 4 that:

$X = 100$ kg/h retentate 1,
$Y = 2400$ kg/h of permeate 1,
$Z = 350$ kg/h of waste, and
$w = 0.60$ protein in permeate 2

Problems

4.1 Calculate the quantity of dry salt that needs to be added in 125 kg of aqueous salt solution to increase its concentration from 15% to 55%?

4.2 A new product X with 20 wt.% fat is to be formulated by blending ingredient Y which contains 30 wt.% fat and ingredient Z which has 5 wt.% fat. Calculate the proportions of Y and Z that should be mixed to get product X with the required fat content.

4.3 Fresh orange juice (15000 kg/h) is concentrated in an evaporator from 14% to 66% soluble solids. To improve the quality of the final product, the concentrated juice is then mixed with fresh juice (cut back) to change the concentration of the final mixture to 45% soluble solids. Assuming a steady-state operation, calculate the following:
(a) Rate of water evaporation in the evaporator,
(b) Rate of fresh juice addition following evaporation and
(c) Rate of final product production.

4.4 One hundred kilograms of oat grains are precooked with steam to make instant oatmeal. The steam is supplied at a rate of 8 kg per 100 kg of oat grains. The waste generated is 2 kg per 100 kg of oat.
(a) Draw and label a flow diagram of the process.
(b) State the assumptions required for this problem.
(c) Set up and solve the mass and components balance equations
(d) Verify that the equations are in balance.
(e) Compute the final water content of the steamed oat if the initial water content of oatmeal is 4%.

4.5 A dairy plant produces cream from milk containing 5% milk fat. The cream has 60% milk fat, while skim milk contains 1% milk fat. If 100 tons of milk is supplied to the creamery each day, what is the daily production of the cream and skim milk?
(a) Draw and label a flow diagram of the process.
(b) State the assumptions required for this problem.
(c) Set up and solve the mass and component balance equations.
(d) Verify that the equations are in balance.
(e) Determine the daily production of cream and skim milk.

4.6 After HTST pasteurization of milk at 72 °C for 16 s, the milk is to be chilled in a counter-current heat exchanger. The cooling medium is cold raw milk entering the heat exchanger at 4 °C and leaving at 50 °C. The flow rates of the HTST milk and raw milk are 600 and 800 kg per hour, respectively. The specific heat of milk is 3.9 kJ/kg-°C.
(a) Draw and label a flow diagram of the process.
(b) State the assumptions required for this problem.
(c) Set up and solve the energy balance equation
(d) Verify that the equations are in balance.
(e) Determine the final outlet temperature of the chilled HTST milk.

4.7 One hundred kilograms of a dry meal replacement is to be formulated with the following composition.
Protein=30%, fat=10%, carbohydrate=30%, lecithin= 5%, and the other 25% is a filler.
   Ingredients available are soy protein isolate, 90% protein; whey protein concentrate, 40% protein; corn flour, 10% fat and 80% carbohydrate; soybean oil, and lecithin.
(a) Draw and label a flow diagram of the process.
(b) State the assumptions required for this problem.

(c) Set up and solve the mass and component balance equations (d) Verify that the equations are in balance.

(d) Determine the quantity of each ingredient needed for this product.

4.8 During making of potato chips, 20% of the initial mass of 100 kg of potatoes were lost to peeling and cutting. After frying in oil, the final mass of potato chips was found to be 80 kg with absorbed oil accounting for 10% of the final mass.

(a) Draw and label a flow diagram of the process.

(b) State the assumptions required for this problem.

(c) Set up and solve the mass balance equations

(d) Verify that the equations are in balance.

(e) Determining the yield (%) of potatoes after peeling and cutting.

(f) Determine the amount of water lost during frying.

4.9 One hundred kilograms per hour of tomato puree is to be cooked in a continuous cooker with direct steam injection. The initial solid content of tomato puree is 15%. The steam is provided at 120 °C and 198 kPa and mixed with the puree to heat it from 20 to 95 °C. The specific heat of the solids in puree is 2.2 kJ/kg-°C. The specific heat of water is 4.19 kJ/kg-°C. The specific heats of steam and condensate at 120°C are 2706 kJ/kg and 504 kJ/kg, respectively.

(a) Draw and label a flow diagram of the process.

(b) State the assumptions required for this problem.

(c) Set up and solve the mass and energy balance equations.

(d) Verify that the equations are in balance.

(e) Determine the flow rates of the steam required for the heating process.

(f) Determine the solid content (%) of the final product.

4.10 Milk (specific heat =3.79 kJ/kg-°C) flowing at the rate of 1000 kg/h is to be heated in a plate heat exchanger from 40 °C to 70 °C using water as the heating medium. If water enters the heat exchanger at 93 °C and leaves at 78 °C and the heat losses to the surroundings are assumed to be negligible, calculate the mass flow rate of water required for the process.

4.11 Pectin enzymes of tomato pulp need to be deactivated by direct injection of steam into the pulp with a total solids concentration of 7.5. wt.%. Steam at 110 °C is mixed with the pulp to heat it from 30 °C to 90 °C. The specific heat of the solids is 3.3 kJ/(kg K) and that of water is 4.19 kJ/(kg K). Determine the concentration of solids in the mixture.

4.12 A new product is to be formulated using fruit pulp (85% moisture content) and sugar (95% total solids). The formulation requires the use of equal amounts of sugar and fruit pulp along with 0.005% of an additive X in the final product. The process involves mixing the ingredients followed by removal of water until a total solids concentration of 65% is achieved. To produce 150 kg of the new product, determine:

(a) The mass of sugar and fruit pulp required

(b) The mass of water to be removed

4.13 Milk at 5 °C flows at 125 kg/h through a preheater where the heating medium is milk previously heated in a heater at 71.7 °C. In the heater, water is used as the heating medium, where it enters at 96 °C and exits at 78 °C. After the milk exits the preheater (as a heating medium), its temperature is 50 °C. Determine the mass flow rate of water and the temperature of the milk as it exits the preheater. The specific heat of water and milk are 4.18 and 3.9 kJ/kg °C, respectively.

4.14 Laboratory tests show that raw food coming to a drying plant contains 75% water. Following drying of the product, 60% of the original water was found to have been removed.
(a) Draw and completely label a flow diagram of the process
(b) Set up and solve the mass balance equations from this diagram.
(c) Determine the water content of the dried food.
(d) Determine the mass of water removed per unit mass of wet weight.
(e) Verify the balances.

4.15 A microwave rated at 100 W connected to the 220-volt source is used for heating foods for 10 min. What will be the current in the microwave and the total energy used in the process?

4.16 During the formulation of a new food product, 20 kg of component A containing 40% solids is mixed with an unknown quantity of component B containing 70% solids. A total of 130 kg of the product is desired.
(a) Draw and completely label a flow diagram of the process
(b) Set up and solve the mass balance equations from this diagram.
(c) Determine the amount of B needed.
(d) Determine the % solids of the product.
(e) Verify the balances.

4.17 In the past, a major fault of concentrated orange juice was flat flavor resulting from the loss of volatile constituents during evaporation. A process overcomes this difficulty by concentrating the extracted juice to 60% solids in vacuum evaporators and blending the concentrate with unconcentrated juice so that the blend contains 40% solids. This process starts with the fresh extract containing 15% solids fed to a juice finisher from which a pulpy juice (20% of feed) and a strained juice (80% of feed) streams are obtained. No losses occur in the finisher. The strained juice stream from the finisher then goes to vacuum evaporators. The 20% of the feed not passing to the evaporators, i.e., the pulpy juice, is then used to dilute the juice concentrated (60% solids) in the mixer to the desired final strength of 40% solids.
(a) Draw and label a flow diagram of the process.
(b) Calculate the weight of water evaporated per 100 kg of juice fed to the system.
(c) Calculate the concentration of the solids in each stream leaving the finisher.
(d) Calculate the weight ratio of concentrated to unconcentrated juice in the final product.

4.18 A peach puree was concentrated in a continuous vacuum evaporator at a rate
      of 100 kg/h. The feed temperature was 20 °C and it had a total solids content
      of 12%. A product of 40% total solids was withdrawn at a temperature of 40 °
      C. To perform heating, saturated steam to the evaporator entered at 140 °C
      and exited as a condensate at 110 °C. The specific heat of the puree was given
      as 2.10 kJ/(kg K). Water has a specific heat of 4.19 kJ/(kg-K).
      (a) Set up, solve, and verify the mass and energy balance equations for the
          evaporator.
      (b) Determine the flow rate of the product and the condensate streams.
      (c) Determine the rate of steam consumption.

Bibliography

1. Ghasem N, Henda R (2015) Principles of chemical engineering processes: material and energy
   balances.2nd edn. CRC Press, Boca Raton, FL
2. Oloman C (2005) Material and energy balances for engineers and environmentalists. Imperial
   College Press, Hackensack, NJ
3. Reklaitis GV (1983) Introduction to material and energy balances. Wiley, Hoboken, NJ

Fluid Mechanics: Basic Concepts

5

The various unit operations used in the processing and manufacturing of materials have been grouped, based on the commonality of some underlying principles, together into three fundamental, molecular-level transfer (or transport) processes: mass transport, heat transport, and momentum transport. As we will see later, there exists mathematical analogy among these three processes, and although their physical mechanisms are totally different, in the elementary sense they can be described by the same general equation. Fluid mechanics addresses the study of momentum transport in a fluid and much of food engineering is concerned with the handling of fluids. Many food products are fluids or are in a fluid stage at some point in their processing. In addition, fluids such as water, steam, refrigerants, and air are used to heat, cool, and otherwise process foods. This chapter will introduce the engineering principles of fluid mechanics that underlie the study of the forces on fluids and their utility in the use and handling of fluids. Fluid mechanics includes both fluid statics and fluid dynamics (Fig. 5.1). It is based on the principle of conservation of mass, energy, and momentum (following Newton's second and third laws). Additionally, fluids are assumed to represent a continuum rather than discrete parts, and the fact that a fluid is composed of discrete molecules is ignored.

First, we will examine fluid statics and the behavior of fluids at rest and then start to examine fluid dynamics and look at the additional problems introduced when fluid moves. We will review and introduce several basic concepts but first some definitions:

5.1 Fluid Statics

In this part of fluid mechanics, we will focus on the pressure exerted by a fluid at equilibrium and at rest. When dealing with incompressible fluids such as liquids, it is also referred to as hydrostatics.

© The Author(s), under exclusive license to Springer Nature Switzerland AG 2024
S. S. H. Rizvi, *Food Engineering Principles and Practices*,
https://doi.org/10.1007/978-3-031-34123-6_5

Fig. 5.1 The study of fluids

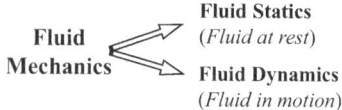

Fluid
Mechanics

Fluid Statics
(*Fluid at rest*)

Fluid Dynamics
(*Fluid in motion*)

5.1.1   Hydrostatic Pressure

Hydrostatic pressure is the pressure exerted by a static (resting) fluid above a measurement point due to the force of gravity. In a column of fluid, it progressively increases from the surface because of the increasing mass of fluid exerting downward force from above. The denser the fluid is, the more pressure it exerts. Thus, the three variables controlling hydrostatic pressure in a fluid are the height or depth of the fluid, the density of the fluid, and the local acceleration due to gravity. If fluid is in a confined space, there will be some pressure on the wall of the container. Hydrostatic pressure is fundamental to understanding numerous processes that involve storing and using fluids to do useful things.

A very important principle regarding the transmission of fluid pressure is expressed by **Pascal's law**, which states that pressure applied to an incompressible confined liquid is instantaneously transmitted undiminished to every part of the liquid, including the walls of the container. This does not imply that the pressures are the same throughout the confined liquid but that the pressure is increased everywhere by the same amount as the applied external pressure. The exerted pressure is independent of the container size and shape. Applications of this include high-pressure processing of foods. High-pressure processing (**HPP**), also called **Pascalization or cold pasteurization**, has emerged as a novel food preservation method. Figure 5.2 shows HPP in comparison with other high-pressure operations in food processing. HPP utilizes cold water and extremely high hydrostatic pressure of up to 6000 bar (87,000 psi), which is about six times the pressure at the deepest ocean trench (Mariana Trench, ~11,034 meters, 1086 bar.) on Earth, for a few minutes to inactivate foodborne pathogens and spoilage organisms while minimizing collateral damage to food quality compared to conventional thermal processing. The microbial inactivation follows Le Chatelier's principle which says that any system in dynamic equilibrium when disturbed by constraints like pressure, temperature, and concentration will shift to a new state to minimize the effect of the constraint. Consequently, at high pressures, such as in HPP, processes that favor a decrease in volume are preferentially promoted and result in microbial destruction.

5.1.2   The Static Pressure Equation

Figure 5.3a shows a column of fluid of density ρ, with cross-section area A and height h. We imagine a thin horizontal disk of the fluid at some arbitrary level in the column. Since this disk extends across the entire column, it also has area A but with an infinitesimal height dh. Figure 5.3b is an "edge on" or side view of this same disk showing the forces acting on it.

The weight of the fluid above the disk bears down on it with force F_1. Add to this the weight of the disk itself, dF, and a total force of $F_1 + dF$ is pushing against the

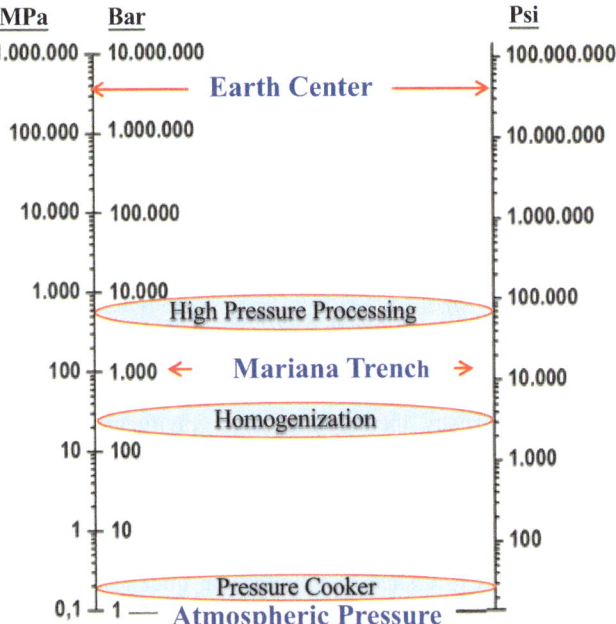

Fig. 5.2 Hydrostatic pressure scales

Fig. 5.3 Static fluid pressure

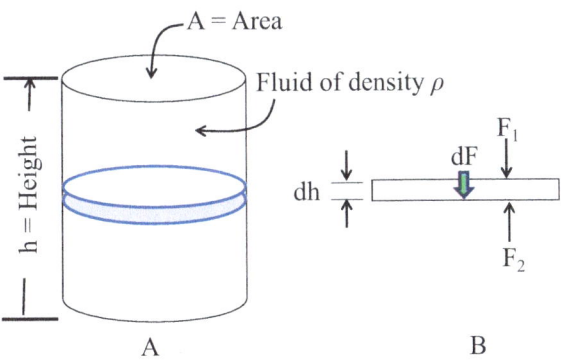

fluid beneath the disk. However, the fluid is at rest, so the net force at all locations must be 0. Therefore, the fluid below the disk must be exerting an upward force F_2 that equals the downward force. Thus:

$$F_1 + dF = F_2 \text{ or } dF = F_2 - F_1 \tag{5.1}$$

The weight of the disk (dF) is equal to its infinitesimal mass (dm) times the acceleration of gravity (g); thus:

$$dF = gdm \tag{5.2}$$

However, the mass is equal to the density (ρ) times the volume (V), which in turn is equal to the area of the disk times its height (Adh). Substituting into Eq. (5.2) gives:

$$dF = \rho gdV = -\rho gAdh \tag{5.3}$$

The minus sign indicates that g acts in the negative direction along the height scale. Dividing the above equation by the area, we obtain:

$$\frac{dF}{A} = -\rho gdh \text{ or } dP = -\rho gdh \tag{5.4}$$

It states that the difference in pressure from one side of the disk to the other equals the product of the fluid density, the acceleration of gravity, and the thickness of the disk. Notice that the area of the column canceled out. Pressure, therefore, depends only on depth and not on the cross-section area of the column of fluid. Equation (5.4) indicates that pressure decreases with elevation and expresses the most fundamental relationship of utility in fluid statics.

Equation (5.4) applies to a slice of infinitesimal thickness (dh). To obtain the equation for the pressure change over a finite height (h), this equation is integrated between two depths h_1 and h_2, as shown in Fig. 5.4. The inverted triangle on the top surface symbolizes atmospheric pressure.

$$\int_{P_1}^{P_2} dP = -\int_{h_1}^{h_2} \rho gdh \tag{5.5}$$

For an incompressible liquid, ρ is constant, and the integration of Eq. (5.5) gives:

$$P_2 - P_1 = -\rho g \int_{h_1}^{h_2} dh = -\rho g(h_2 - h_1) \tag{5.6}$$

$$\text{Or, } P_2 = P_1 - \rho g(h_2 - h_1) \text{ and } P_1 = P_2 + \rho g(h_2 - h_1) \tag{5.7}$$

This may be written more simply as follows:

Fig. 5.4 Pressure distribution in a fluid column

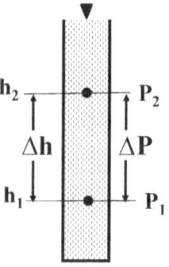

$$\Delta P = \rho g \Delta h \tag{5.8}$$

In the American Engineering set of units, this equation becomes:

$$\Delta P = \rho \frac{g}{g_c} \Delta h \tag{5.9}$$

5.1.3   Container Shape and Pressure

Since the pressure integral is independent of the cross-section area and depends on height only, the pressure must be independent of the shape of the container. Figure 5.5 shows a variety of containers of different shapes and sizes. If filled with fluids of the same density, at any given depth (h), the pressure will be the same in each container. The pressure will indeed increase as the height of the fluid increases as depicted by the last container.

5.1.4   Pressure Calculations

When a liquid is open to the atmosphere, the top of that liquid will be at atmospheric pressure and it must be added to the computed pressure to obtain absolute pressure. If several layers of immiscible liquids are stacked, the pressure change in each is simply computed and then summed.

Example

1. As Fig. 5.6 shows, the container is filled to a depth of 0.3 m with mercury ($\rho = 13,600$ kg/m³). Above this is 1.2 m of H_2O ($\rho = 1,000$ kg/m³). What is the absolute pressure at a point 0.1 m above the bottom of the tank?

Fig. 5.5 Pressure in containers of different shapes

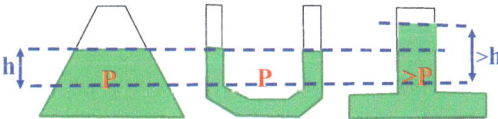

Fig. 5.6 Pressure changes with height

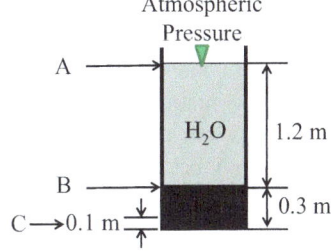

Solution: *The top of the liquid (A) is open to the atmosphere, so point A is at atmospheric pressure.*

$$P_A = 101.4 \ kPa$$

The change in pressure between points A and B in the water is

$$\Delta P_{AB} = \rho g \Delta h = \left(1000 \frac{kg}{m^3} \right) \left(9.81 \frac{m}{s^2} \right) (1.2 \ m) = 11.8 \ kPa$$

The change in pressure between points B and C in the mercury is

$$\Delta P_{BC} = \rho g \Delta h = \left(1.36 \times 10^4 \frac{kg}{m^3} \right) \left(9.807 \frac{m}{s^2} \right) (0.3 - 0.1 \ m) = 26.7 \ kPa$$

The absolute pressure at point C, 0.1 m above the bottom, is the sum of these three pressures:

$$P_C^{abs} = P_A + P_{AB} + P_{BC} = 101.4 \ kPa + 11.8 \ kPa + 26.7 \ kPa = 139.9 \ kPa \quad \blacktriangleleft$$

Check Your Understanding

1. Corn oil has a density of 921 kg/m^3. A tank that is open to the air is filled to a depth of 3.20 m with water. A layer of corn oil 0.50 meters deep is floated on top of this. What is the absolute pressure at the oil/water interface? Verify that the absolute pressure at the bottom of the tank is 137 kPa. ◄

5.1.5   Pressure Measurement Devices

Several devices are available for measuring pressure. These include:

(A) Manometers

The manometer consists of a simple U-shaped tube, as shown in Fig. 5.7a, partially filled with a manometric liquid M, typically mercury. Above this fluid, the tube is filled with fluid A in the left arm and fluid B in the right arm. One arm connects to the container whose pressure is to be measured, and that arm will normally contain the fluid (A) in that container. The other arm may be open to the atmosphere or connected to another container and will also be filled with a fluid (B) that may be the same as A or different. Both A and B must be lighter than the manometric fluid (M) and not miscible in it.

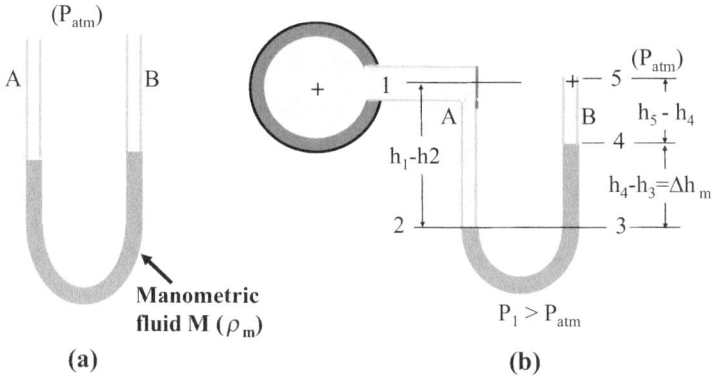

Fig. 5.7 A U-tube manometer

For example, if you are measuring the pressure in a water tank, fluid A would be water. This water should completely fill the tube connecting the tank to the manometer. If the pressure in the tank is simply measured, the second arm will be open to the air, and fluid B will be air. On the other hand, if the difference in pressure between two locations on a pipe filled with water is measured, both fluids A and B will be water. Mercury makes a good manometric fluid because it is heavier than virtually any other fluid and immiscible in almost anything else.

We use a manometer to measure the difference in pressure between point 1 and point 5. Note that in Fig. 5.7b, the level of the manometric fluid is higher in the right-hand arm (point 4) than in the left arm (point 2). The pressure at point 2 must be greater than that at point 4 to support this difference in level. However, in any continuous fluid, all points at the same depth must be at the same pressure. Therefore, point 3 must be at the same pressure as point 2. With this information, let us compute the difference in pressure between points 1 and 5.

The pressure at point 2 equals the pressure at point 1 plus the change in pressure between points 1 and 2. Using Eq. (5.8), we write:

$$P_2 = P_1 + \rho_A g(h_1 - h_2) \tag{5.10}$$

The pressure at 3 must equal the pressure at point 5 plus the change in pressure between points 5 and 4 plus the change in pressure between points 4 and 3. Again, using Eq. (5.8), we write

$$P_3 = P_5 + \rho_B g(h_5 - h_4) + \rho_M g(h_4 - h_3) \tag{5.11}$$

Since points 2 and 3 are at the same pressure, Eqs. (5.10) and (5.11) must have the same value, so we can write:

$$P_1 + \rho_A g(h_1 - h_2) = P_5 + \rho_B g(h_5 - h_4) + \rho_M g(h_4 - h_3) \tag{5.12}$$

Solving this for the pressure difference between points 1 and 5 and letting $\Delta h_m = h_4 - h_3$, the differential height of manometric fluid, we obtain the general equation for a manometer

$$P_1 - P_5 = g[\rho_B(h_5 - h_4) + \rho_m \Delta h_m - \rho_A(h_1 - h_2)] \tag{5.13}$$

The first term on the right describes fluid B, the second term describes the manometric fluid, and the third term describes fluid A. Notice that only the fluid above the common level is important since, below that level, the fluid on the left balances the fluid on the right.

Under certain conditions, Eq. (5.13) can be simplified. For example, if fluids A and B are identical, the densities of fluids A and B will be the same, and the first and third terms can be combined. The equation then becomes:

$$P_1 - P_5 = g[\rho_A\{(h_5 - h_4) - (h_1 - h_2)\} + \rho_m \Delta h_m] \tag{5.14}$$

Observe in Fig. 5.7 that $(h_5 - h_4) - (h_1 - h_2)$ is equal to $-\Delta h_m$, so the equation becomes:

$$P_1 - P_5 = (\rho_m - \rho_A)g\Delta h_m \tag{5.15}$$

In other words, the pressure difference is equal to the difference in density between the manometric fluid and fluid A times the difference in the level of the manometric fluid times the acceleration of gravity. As another example, if fluids A and B are gases so that $\rho_A = \rho_B << \rho_M$, then the density of the upper fluids can be ignored, and Eq. (5.15) reduces to:

$$P_1 - P_5 = \rho_m\, g\, \Delta h_m \tag{5.16}$$

The pressure difference is thus equal to the density of the manometric fluid times the difference in height multiplied by the acceleration due to gravity. In cases where the pressure difference is small, the manometer column height may be hard to read accurately. In such situations, accuracy is enhanced by employing an **inclined-tube manometer.** With one arm inclined 20 to 30 degrees to the horizontal, a small change in pressure causes greater displacement of the manometric liquid in the column. The slanted reading (h) is converted into vertical distance (Δh_m) using the angle (θ) of inclination,$(\Delta h_m = h \sin \theta)$.

Example

2. *The left arm of a mercury manometer is connected to a vacuum tank. The other arm is open to the air. The surface of the mercury in the left arm is 3.5 cm higher than that in the right arm. What is the absolute pressure in the tank?*
   Solution: Since fluids A and B are both air, we will use Eq. (5.15). Mercury has a density of 13,600 kg/m³. Since the level in the arm connected to the tank

is higher than that in the other arm, $\Delta h = -0.035$ m. *The difference in pressure between points 1 (the tank) and 5 (the atmosphere) is, therefore,*

$$P_1 - P_5 = \left(13,600 \frac{kg}{m^3}\right)\left(9.807 \frac{m}{s^2}\right)(-0.035\ m) = -4.67 \times 10^3 Pa = -4.67\ kPa$$

Thus, the vacuum tank is 4.67 kPa below atmospheric pressure. Since this pressure is relative to atmospheric pressure, it is gauge pressure. The absolute pressure in the tank is obtained by adding gauge pressure to atmospheric pressure:

$$P_a = P_g + P_{atm} = -4.67\ kPa + 101.32\ kPa = 96.65\ kPa \qquad \blacktriangleleft$$

Check Your Understanding

2. As shown in Fig. 5.8, the left arm of a mercury manometer is connected to the bottom of tank A and the right arm is connected to the bottom of tank B. Both tanks contain water, and this water extends into and fills both arms above the manometric fluid. The mercury in the right arm is 16.2 cm above the level in the left. A pressure gauge at the bottom of tank A reads 59.0 kPa (gauge). Verify that the absolute pressure at the bottom of tank B is 39 kPa. ◄

(B) **Pressure Gauges: Bourdon tube-based**

A mechanical pressure gauge is based on the Bourdon tube, which is named after Eugéne Bourdon, a French watchmaker and engineer who invented the Bourdon gauge in 1849. It works on the simple principle that a flattened tube tends to straighten when pressurized to regain its circular form. The most common Bourdon pressure gauges have a slightly elliptical cross-section, and the tube is generally bent into a C-shape, as shown in Fig. 5.9. The tube is connected to the system, whose pressure is to be measured. As pressure increases inside the Bourdon tube, it tends to unwind and straighten out. The displacement of the tube's free end provides a measure of the pressure. The tube is linked to a pointer through a set of levers so

Fig. 5.8 Differential pressure in a system

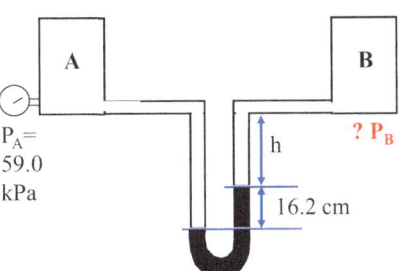

P$_A$=
59.0
kPa

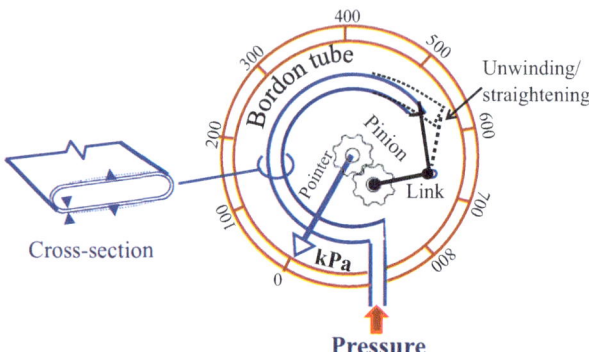

Fig. 5.9 Bourdon tube-based pressure gauge

that, as the tube moves, the pointer moves proportionately and displays the pressure on the scale.

(C) **Strain gauges**

Strain, as we learned earlier, is the amount of deformation a material experiences due to an applied force. The most common method of measuring it is with a strain gauge. Lord Kelvin first reported in 1856 that metallic conductors subjected to mechanical strain exhibit a change in their electrical resistance. If a wire is strained, it becomes slightly longer, and its cross-sectional area is reduced and changes its electrical resistance (R) in proportion to the amount of strain in the wire. This small change in resistance is reflected in a change in the measured voltage (V) or current (I) according to Ohm's Law:

$$R = \frac{V}{I} \qquad (5.17)$$

Using this phenomenon, the metallic strain gauge is designed, and it consists of a very fine wire or metallic foil arranged in a grid or zig-zag pattern to increase its length. It is then attached to the test specimen.

To measure pressure, a thin metal disk is placed across a tube, and a strain gauge is attached to the disk, as shown in Fig. 5.10. As the pressure changes on one side of the disk, the disk flexes, and the strain gauge converts this change to an electrical current that can be measured and transformed into a numeric value by a micropro- cessor. Figure 5.10 shows a single strain gauge. In actual practice, several strain gauges are attached and connected in a circuit like a Wheatstone bridge. This configuration produces a more sensitive measurement than would be obtained with a single strain gauge.

(D) **Pressure Head**

When using a manometer, pressure (in units of force per unit area) is computed from the height of manometric fluid that the pressure supports (in units of length). As

Fig. 5.10 Pressure measurement with a strain gauge

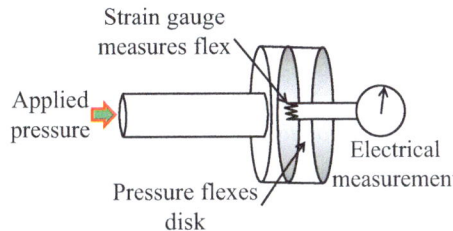

an alternative, we can report pressure by giving the height supported. When pressure is given this way, we call it the **pressure head**. It is the equivalent height of fluid that would produce the same pressure, or elevation equivalent of pressure, and is expressed in meters (or feet).

Example

3. *The top of the vertical pipe shown in Fig. 5.11 is open to the atmosphere. The bottom is connected to a pressurized water system. What is the pressure in the water system if water in the pipe rises to a height of (a) 3 m expressed in kPag and kPaa, and (b) 12 ft, expressed in psig and psia?*
   Solution:
(a) *The pressure expressed as the head of water is 3 m. Since water has a density of 1000 kg/m³, the pressure expressed in kPag is*

$$P_g = \rho g \Delta h = \left(1000 \frac{kg}{m^3}\right)\left(9.81 \frac{m}{s^2}\right)(3m) \; or \; P_g = 29.4 \; kPag$$

Since atmospheric pressure is 101.3 kPa, the absolute pressure of the water system is

$$P_a = P_g + P_{atm} = 29.4 + 101.3 = 130.7 \; kPa$$

(b) *The pressure expressed as the head of water is 12 ft. Since water has a density of 62.4 lb$_m$/ft³, the pressure expressed in psig is*

Fig. 5.11 Pressure measurement as head of a fluid

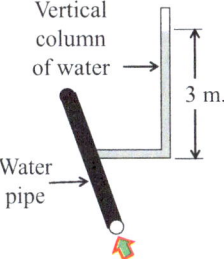

$$P_g = \rho \frac{g}{g_c} \Delta h = \left(62.4 \frac{lb_m}{ft^3}\right) \frac{\left(32.17 \frac{ft}{s^2}\right)}{\left(32.17 \frac{lb_m ft}{lb_f s^2}\right)} (12 ft) \left(\frac{1 ft}{12 in}\right)^2 \ or \ Pg = 5.2 psig$$

Since atmospheric pressure is 14.7 psig, the absolute pressure of the water system is

$$P_a = P_g + P_{atm} = 5.2 + 14.7 = 19.9 \ psia \qquad \blacktriangleleft$$

Check Your Understanding

3. A water column shows a pressure of 29.8 psia at the bottom. Verify that this is equivalent to a pressure head of 34.8 ft of water. ◄

5.2   Fluid Dynamics

Above, we examined the behavior of fluids at rest. In this section, you will see what happens when fluids move. This information is needed to answer such questions as why do some drinks leave a slight coating on the tongue and change the mouthfeel? Or how rapidly will suspended particles separate from the fluid in a separator/ centrifuge? Or how long must a holding tube of a pasteurizer be to kill the pathogens, among many others?

5.2.1   Measurements of Flow

The food industry has a frequent need to move fluids through pipes, and it is often necessary to know the speed at which the fluid is moving. This speed can be expressed in many ways, and each is important in different situations. We will start by describing some of these.

(A) Gradients

A gradient is defined as the rate at which a property such as pressure, temperature, or concentration changes per unit distance. As Fig. 5.12 shows, a gradient can be viewed as the slope of a function on a graph. For example, if a hallway is 20 meter long and the temperature in the hallway varies at a constant rate from 20 °C at one end to 25 °C at the other, we say there is a temperature gradient of 0.25 degrees per meter:

Fig. 5.12 Gradient

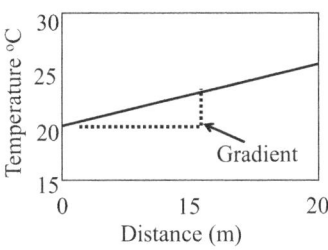

$$\text{Temperature gradient} = \frac{\Delta T}{\Delta x} = \frac{(25-20)\,^\circ\text{C}}{20\text{ m}} = 0.25\,^\circ\text{C/m}$$

Gradient is introduced here because it is the gradient that makes transport processes happen and should be well understood.

Check Your Understanding

4. Sugar syrup was poured into a beaker. A layer of distilled water was then carefully placed on top. After many hours, the water 0.2 cm above the sugar solution was found to contain 0.12 g of sugar per $cm^{3,}$ while the water 1.4 cm above the solution contained 0.04 g of sugar per cm^3. Verify that the concentration gradient going from the lower location to the upper is − 0.067 g/cm^3 per cm. ◄

(B) Mass Flow Rate

Mass flow rate measures the mass of fluid that passes a given point per unit of time. It is a measure of how much fluid is being moved, not its speed. To measure the mass flow rate in a pipe, discharge the fluid into a container for a measured interval of time. Weigh the fluid and divide the mass by the time. We will represent this rate with an m (for mass) with a dot over it. In general, we will use a dot over a quantity to mean "per unit time," so ("m dot") means mass per unit time (Fig. 5.13).

Fig. 5.13 Measuring mass flow rate

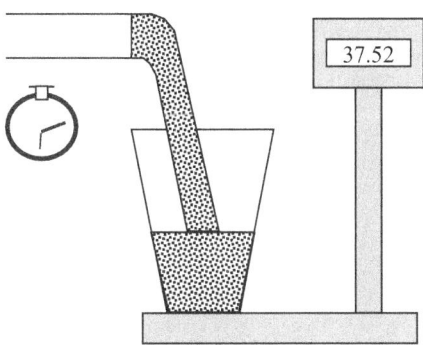

$$Mass\ flow\ rate = \dot{m} = \frac{m}{t} \tag{5.18}$$

---

4. *A bucket when empty weighs 6.8 kg. After a pipe discharges water into the bucket for 32 seconds, the bucket weighs 15.3 kg. What is the mass flow rate in the pipe?*
   Solution: *The mass of the water is $m = 15.3\ kg - 6.8\ kg = 8.5\ kg$*
      The mass flow rate is $\dot{m} = \frac{8.5\ kg}{32\ s} = 0.27\frac{kg}{s}$. ◀

(C) Mass Flux

A <u>flux</u> is defined as the rate at which a property moves through a unit area per unit time. If the quantity is a mass, then we call it mass flux (Fig. 5.14).

$$Mass\ flux = \frac{\dot{m}}{A} = \frac{m}{tA} \tag{5.19}$$

---

5. *If oxygen diffuses through a membrane at a rate of 0.5 g each hour and the membrane has an area of 2 square meters, find the oxygen flux.*
   Solution: Oxygen flux $= \frac{m}{tA} = \frac{(0.5\ g)}{(1\ h)(2\ m^2)} = 0.25\frac{g}{h-m^2}$ ◀

---

5. Your screen door has an area of 0.7 m². Because the mesh is too large, a no-see-um (a small insect) enters your house through the screen every 25 s. Verify that the no-see-um flux of your screen door is 206 insects per hour per square meter of screen. Suppose you replaced your door with a one that has

---

Fig. 5.14 Flux definition

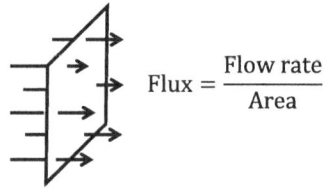

0.9 m² of screen. Verify that if the no-see-um flux remains constant, 185 insects will enter each hour. ◄

(D) **Volumetric Flow Rate**

Like the mass flow rate, the volumetric flow rate measures the amount of fluid that is moving, but in volume units. The volumetric flow rate can be measured by discharging fluid into a graduated cylinder, as shown in Fig. 5.15a, or by dividing the mass flow rate by the fluid density. We will use $\dot{V}$ for volumetric flow rate.

$$\text{Volumetric flow rate } (\dot{V}) = \frac{V}{t} = \frac{m}{\rho\, t} \qquad (5.20a)$$

The volumetric flow rate can also be computed from knowledge of the pipe area and velocity of the fluid, as shown in Fig. 5.15b, as follows:

$$\text{Volumetric flow rate } (\dot{V}) = \frac{V}{t} = \frac{Ax}{t} = A \cdot v \qquad (5.20b)$$

Example

6. *What is the volumetric flow rate in Example 4 in cubic meters per hour?*

   Solution: *The volumetric flow rate is:* $\dot{V} = \dfrac{\left(0.27\frac{kg}{s}\right)}{\left(1000\frac{kg}{m^3}\right)}\left(3600\frac{s}{h}\right) = 0.972\,\frac{m^3}{h}$ ◄

(E) **Volumetric Flux**

Volumetric flux is simply the volume moved per unit time per unit area. In other words, it is the volumetric flow rate divided by the cross-section area of the stream of fluid. Frequently, this is the cross-section of a pipe.

Fig. 5.15 (**a**) Measuring volumetric flow rate. (**b**) Estimating volumetric flow rate from area (A) and velocity (v)

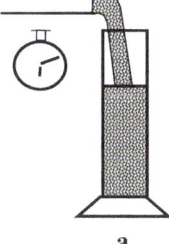

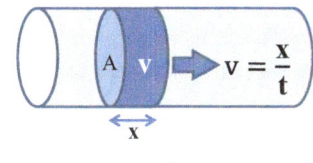

a

b

$$Volumetric\ flux = \frac{\dot{V}}{A} = \frac{V}{A\,t} \tag{5.21a}$$

The SI units of volumetric flux are:

$$\frac{m^3}{s \cdot m^2} = \frac{m}{s} \tag{5.21b}$$

(F) Local Velocity

A fluid does not flow at the same velocity in all portions of a conduit or pipe. The layer of molecules in immediate contact with the pipe wall will have **zero velocity**. On the other hand, the fluid in the center of the pipe will travel at its **maximum velocity** (v_{max}). For example, when water is flowing at low velocities (we will expand on this in a later section), the water in the center of the pipe will travel at twice the average velocity. Figure 5.16 shows two typical local velocity profiles, laminar and turbulent, in a pipe.

(G) Average Velocity

As mentioned earlier, invariably there is a distribution of velocity during fluid flow in a conduit, sometimes we are interested in just the **average velocity** ($<v>$) with which streams move (Fig. 5.17). This is a measure of the velocity of the fluid and not the quantity of fluid being moved and can be computed by using either the volumetric or mass flow rate and the cross-section area of the pipe, as shown below:

$$Average\ velocity,\ <v> = \frac{\dot{V}}{A} = \frac{V}{At} = \frac{m}{\rho At} = \frac{\dot{m}}{\rho A} \tag{5.22}$$

Fig. 5.16 Fluid-velocity profiles of (**a**) laminar and (**b**) turbulent flows in conduits

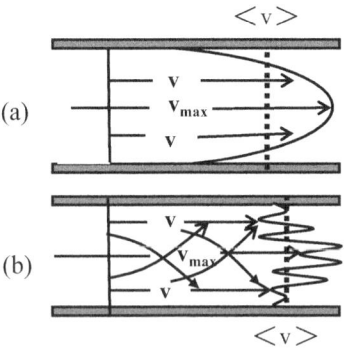

Fig. 5.17 Average velocity

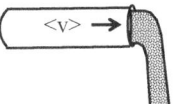

Notice that this is identical to the formula for volumetric flux. In fact, volumetric flux and average flow rate are two ways of looking at the same thing. When we speak of flux, we are focusing on a fixed location and measuring the volume of a fluid that passes through a unit area in unit time at that location. When we speak of velocity, we are focusing on a moving portion of a fluid and measuring its speed. Mathematically, however, they have the same value and the same units.

Example

7. *If the fluid in Example 6 is moving through a pipe of 12.5 cm inside diameter, what is the average velocity of the fluid in meters per second?*
   Solution: The cross-section area is

$$A = \pi \left(\frac{d}{2}\right)^2 = 3.1416\left(\frac{12.5}{2}\right)^2 \left(\frac{1 \text{ m}}{100 \text{ cm}}\right)^2 = 0.0123\,\text{m}^2.$$

   The average velocity: $<v> = \left(\frac{0.972\frac{\text{m}^3}{\text{h}}}{0.0123 \text{ m}^2}\right)\left(\frac{1 \text{ h}}{3600 \text{ s}}\right) = 0.022\,\frac{\text{m}}{\text{s}}.$

   We can view this as water moving an average of 0.022 m every second or as 0.022 m³ of water passing through a square meter opening every second. ◀

Example

8. *Assuming that the overall velocity is low enough, what is the average velocity of water in Example 7 at the wall and in the center of the pipe?*
   Solution: At the wall, the velocity is 0 m/s. In the center, it is: $2\left(\frac{0.022m}{s}\right) = 0.044\ m/s$ ◀

Check Your Understanding

6. A pipe with an inside diameter of 9.2 mm fills a graduated cylinder with water to a volume of 65 ml in 3.52 s. Verify that the volumetric flow rate is 0.0665 m³/h and that the mass flow rate is 1.11 kg/min. and that the average velocity is 0.277 m/s. Without further calculations, what is the volumetric flux? Assuming that this flow rate is sufficiently slow, verify that the local velocity in the center of the pipe is 0.55 m/s. Assuming the pipe is 20 m long, verify that a particle traveling down the middle of the pipe will take 35 seconds to reach the other end. ◀

5.2.2   Fluid Flow Characteristics

Some fluids flow more readily than others. For example, water flows more easily than many oils and concentrates. The ease of flow is determined by the organization of molecules within the fluid. A good understanding of the properties of flowable foods is needed for product development, quality control, and process design and evaluation. To treat this phenomenon quantitatively, we first review and introduce several concepts.

(A) Shear

Like solids, discussed in Chap. 1, liquids also undergo shear but with a different twist. If you place a deck of cards on a table and push the top card sideways, the cards in the deck will slide over each other, as shown in Fig. 5.18. Each card moves slightly faster than the one below it. Similarly, when a fluid moves, some parts of the fluid will move faster than the rest. When this happens, you can think of layers of fluid (lamella) sliding over each other like cards in a deck. Fluids that flow in this way are said to have **laminar** (streamline) flow. Such shearing occurs when fluid flows at low velocity through a slit or pipe, Fig. 5.18b, c.

(B) Shear Rate

When two adjacent layers slide over each other, it is useful to know the difference in velocity between layers. As depicted in Fig. 5.18b, c, a fluid layer at y travels at a velocity v_x in the x direction due to a force F applied over area A. An adjacent layer at $y + dy$ moves with a velocity $v_x + dv_x$. The shear strain (γ), as discussed in Chap. 1, is dx/dy. We define **shear rate**, symbolized by the Greek letter gamma with a dot $(\dot{\gamma})$, as the time rate of change of shear strain, then:

$$Shear\ Rate\ (\dot{\gamma}) = \frac{d}{dt}(\gamma) = \frac{d}{dt}\left(\frac{dx}{dy}\right) = \frac{d}{dy}\left(\frac{dx}{dt}\right) = \frac{dv_x}{dy} \qquad (5.23a)$$

where v_x = the local velocity in the x direction at any point in a fluid
dv_x = the change in this velocity as you move a distance dy in the y direction

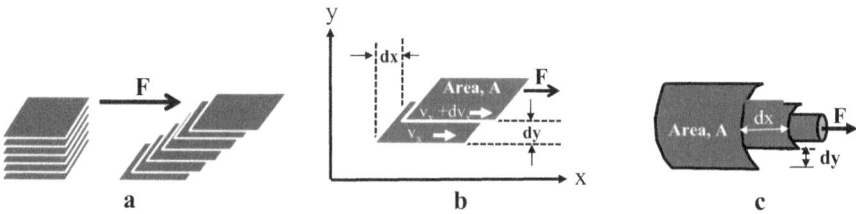

Fig. 5.18 (a) Shear of a deck of cards. (b) Laminar flow in a slit. (c) Laminar flow in a tube

The shear rate is thus the ratio of the difference in velocity between two layers of fluid, divided by the distance between them (Fig. 5.19). This distance must be measured at right angles to the direction of flow. Thus, the shear rate is also called the **velocity gradient**.

The shear rate has units of reciprocal time in any system of units.

$$\dot{\gamma} \, [=] \frac{m/s}{m} = s^{-1} \tag{5.23b}$$

A high shear rate indicates a rapid change in velocity with distance. High shear rates are encountered when a fluid passes rapidly through a small orifice. They also occur close to the surface of a spacecraft during reentry into the atmosphere or when a liquid is stirred rapidly. It is frequently used in determining the processability of materials in many industries. Typical ranges of shear rates for various food processing operations are listed in Table 5.1

Example

9. Figure 5.19 *shows a flowing fluid with a local velocity of 3.1 m/s at one location and a local velocity of 2.4 m/s at another. Find the average shear rate.*
   Solution: The two locations are separated by 0.02 m. The average shear rate is:

$$\dot{\gamma} = \frac{(3.1 - 2.4) \, m/s}{0.02 \, m} = 35 \, s^{-1} \qquad \blacktriangleleft$$

(C) Shear Stress

When cards of a playing deck slide over each other, there is friction between them. If this were not so, sliding the top card would not move the cards underneath. This friction has two complementary effects:

Fig. 5.19 Shear rate

		3.1 m/s
0.02 m		
		2.4 m/s

Table 5.1 Ranges of shear rate in selected operations

Process	Shear rate (s^{-1})
Pumping	$10^{\circ}-10^{3}$
Chewing	$10-10^{2}$
Extrusion, mixing, agitation	$10-10^{3}$
Homogenization	$10^{2}-4 \times 10^{2}$
Nozzle spraying	$10^{3}-10^{6}$

- It allows fast moving layers to move the slower ones and
- This makes the slow layers to retard the movement of the faster layers.

Thus, in a fluid with high internal friction, motion will be transmitted from layer to layer more readily, but greater force will be needed to create the shear. For example, if you stack 52 sheets of sandpaper, sliding the top one will move those below, but it will take more work than sliding playing cards because of the greater friction. Additionally, large pieces of sandpaper will require more effort to move than smaller pieces. Figure 5.20 shows the opposing forces needed to produce shear. Note that they are parallel to the direction of shear but not in line with each other (if they were in line, they would produce stretch rather than shear.). The applied force is transmitted from layer to layer throughout area A. As in Chap. 1, we define shear stress, symbolized by the Greek letter tau (τ), as the magnitude of the force per unit area.

$$Shear\ stress, \tau = \frac{F}{A} \tag{5.24}$$

Shear stress has units of force per area, the same unit as pressure. The difference lies in the fact that the force in shear stress is parallel to the area, while with pressure, the force is perpendicular to the area.

5.2.3   Newtonian Fluids and Viscosity

We all know that oil is thicker than water. To achieve the same shear rate with oil and water, one must apply a greater shear stress to the oil. To look at this in a different way, if the same shear stress is exerted to both, water will shear faster. This suggests that a good measure of the difference in flow ability or resistance to flow between fluids is the ratio of the shear force per unit area to the shear rate. This concept is the basis of Newton's law of **viscosity**, which states that the rate of shear in a fluid is proportional to the applied shear stress. This proportionality constant or their ratio is called **viscosity, absolute viscosity, or often dynamic viscosity** and is symbolized by Greek letter eta (η) and sometimes by mu (μ). A fluid that follows the Newton law of viscosity is called a **Newtonian fluid**. The relationship can be mathematically expressed as:

$$Viscosity\ (\eta) = \frac{\tau}{\dot{\gamma}} = \frac{Shear\ stress}{Shear\ rate} \tag{5.25}$$

The units of viscosity in different systems are:

Fig. 5.20 Shear stress

- In SI units: $\eta[=]\frac{N \cdot s}{m^2} = Pa \cdot s$ Note that viscosity has the units of pressure times time.
- It may also be expressed as: $\eta[=]\frac{N \cdot s}{m^2} = \frac{kg\frac{m}{s^2}s}{m^2} = \frac{kg}{m \cdot s}$
- In CGS units: $\eta[=]\frac{g}{cm \cdot s} = $ Poise (Centipoise (cP) $= 10^{-2}$ Poise $= 1$ mPa·s)
- In American Engineering units: $\eta[=]\frac{lb_m}{ft \cdot s}$

Rewriting Eq. (5.25) in terms of τ and substituting Eqs. (5.23a) and (5.24), we obtain:

$$\frac{F}{A} = \eta\frac{dv_x}{dy} \tag{5.26a}$$

$$\text{Or}, \tau = \eta\dot{\gamma} \tag{5.26b}$$

Equations (5.26a and 5.26b) are two ways of expressing Newton's law of viscosity, and their plot will be a straight line through the origin with slope η. Figure 5.21 is such a plot for water and for thick oil, both of which obey Newton's law. Note that the more viscous fluid has a steeper slope, indicating higher viscosity. Because the viscosity of a Newtonian fluid does not change with increasing shear rate, the plot of τ vs $\dot{\gamma}$ is a straight line, as shown in Fig. 5.21. Most fluids that have 80–90 percent water, such as milk, fruit juices, sugar and salt solutions, and many oils are Newtonian.

Viscosity can thus be considered a quantitative measure of a fluid's resistance to flow, recognizable in the difference between easily flowing water and slow-flowing honey. It represents the internal friction between adjacent layers of a moving fluid and provides a measure of the ability of a moving layer of fluid to transfer momentum to an adjacent, slower-moving layer. Thus, a fluid of high viscosity will transfer momentum efficiently. Viscosity has been observed to vary over many orders of magnitude, from very low values for helium gas to extremely high values for materials near the glass transition. Table 5.2 below provides a useful order of magnitude viscosity estimate for some of the common products.

Fig. 5.21 Viscosity of Newtonian liquids

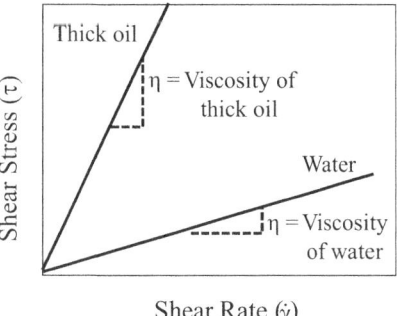

Table 5.2 Viscosity of some Newtonian fluids at room temperature

Fluid	Viscosity (mPa·s)
Corn syrup	10^5
Honey	10^4
Glycerol	10^3
Cooking oils	10^2
Water	1
Air	10^{-2}

When liquids flow under the force of gravity alone, density, in addition to viscosity, becomes very important. For such situations, viscosity is expressed as **kinematic viscosity (υ, Nu),** defined as $\upsilon = \eta/\rho$ with units of m^2/s. In the CGS system, the unit of kinematic viscosity is **Stokes (S),** named after physicist Sir George Stokes and has units of cm^2/s. One centistokes (cS) equals 1 mm^2/s.

5.2.4   Fluid Flow and Momentum Transport

To understand the physics behind Newton's law discussed above, imagine two infinitely large parallel plates separated by a space filled with a fluid, as shown in Fig. 5.22. While the bottom plate remains stationary, due to force F, the top plate slides sideways at a constant velocity, v_m. Assume that the following is true about this system:

- The temperature remains constant.
- Force F is applied to move the top plate only.
- Movement is in the x direction only.
- The fluid between the plates undergoes laminar flow.
- This process continues for a long time so that the system reaches a steady state.

When steady state is reached, we examine the local velocities between the plates. In Fig. 5.23, we have plotted 4 of those velocities. Under the **"no-slip condition,"** meaning no relative movement between the contact surface and the fluid layer, the velocity varies from zero next to the stationary plate to v_m next to the moving plate. The vectors shown in Fig. 5.23 represent the velocities at distances y_1, y_2, y_3, and y_4 from the stationary plate. The sloping line that connects the heads of the velocity vectors is called a **velocity profile**. The ratio of the change in velocity to the change

Fig. 5.22 A fluid between infinite parallel plates

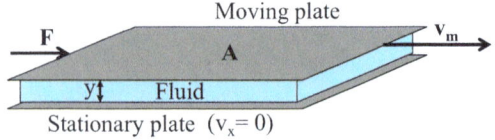

Moving plate

Stationary plate $(v_x = 0)$

Fig. 5.23 Steady-state velocity profile

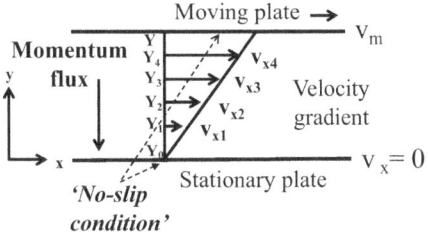

in distance is a velocity gradient. In this case, the gradient is uniform across the distance between the plates.

Consider what is happening here. The fluid layer next to the moving plate has acquired a certain momentum in the x direction. Because of friction within the fluid, some of this momentum is transferred to the adjoining layer. This layer in turn transfers some of its momentum to the next layer. In this manner, momentum in the x direction is transferred from layer to layer in the y direction. The result is a gradual change in momentum from top to bottom along a momentum gradient. We call this process **momentum transport**.

According to Newton's law (Eq. 5.26b), the shear stress between any two vectors may be written as:

$$\tau = \eta \frac{v_{x1} - v_x}{y_1 - y_0} = \eta \frac{v_{x2} - v_{x1}}{y_2 - y_1} = \eta \frac{v_{x3} - v_{x2}}{y_3 - y_2} \tag{5.27}$$

Or, for the overall flow:

$$\tau = \eta \frac{v_m - v_x}{y - y_0} = \eta \frac{\Delta v}{\Delta y} \tag{5.28}$$

where τ is the shear stress and indicates momentum transfer in the y direction due to movement in the x direction, written as τ_{yx} and shown below:

$$\tau = \frac{\text{Force}}{\text{Area}} = \frac{\text{Mass} \times \text{Acceleration}}{\text{Area}} = \frac{\text{Mass} \times \text{Velocity}}{\text{Area} \times \text{Time}} = \frac{\text{Momentum}}{\text{Area} \times \text{Time}}$$
$$= \text{Momentum flux}, \tau_{yx}$$

In other words, τ represents the transfer of momentum in the y direction due to velocity in the x direction per unit time per unit area or the rate of movement of momentum per unit area. By the definition of flux, this is the **momentum flux** of the system, written explicitly as τ_{yx}, where the first subscript indicates the direction of momentum transfer, and the second subscript shows the velocity direction. Therefore, Eq. (5.28) tells us that:

$$(\text{Momentum flux}) = (\text{viscosity}) * (\text{velocity gradient}) \qquad (5.29)$$

Check Your Understanding

7. Imagine that in the system described in Fig. 5.23, the upper plate is 1.25 cm above the lower plate and that the upper plate is moving to the right at a rate of 2.30 cm/s. If all the assumptions are met, verify that at steady state, the velocity gradient will be 1.84 meters per second per meter or $1.84~\text{s}^{-1}$.
8. Water at 20 °C has a viscosity of 1.00 Pa·s. Verify that momentum is being transferred in the previous problem at a rate of $1.84~\text{N/m}^2$. ◄

5.2.5 Non-Newtonian Fluids

Not all fluids obey Newton's law. Most fluids are non-Newtonian; they deviate from Newton's law, and their viscosity will change due to agitation (shear) or pressure (stress). There are two major classes of non-Newtonian fluids: time independent and time dependent.

(A) **Time-Independent Fluids**

The viscous behavior of many fluids is independent of time, and they have no memory of their history. This simply means that if you apply a certain stress, you will obtain a predictable shear rate, and this shear rate will remain constant as long as the stress does not change. In other words, the shear stress at any location in the fluid is a function of the instantaneous shear rate. Figure 5.24 shows the general flow curves (rheograms) of time-independent Newtonian and non-Newtonian fluids described below. The relative position of each curve depends on material's viscous behavior.

• **Pseudoplastic Fluid:** For many fluids, as the rate of shear increases, the friction between the layers appears to decrease. These fluids appear to become thinner as the shear increases and are called "**shear thinning**" or pseudoplastic. Figure 5.24 shows that a τ vs $\dot{\gamma}$ plot of a pseudoplastic fluid decreases in slope with increasing shear rate. In such fluids, shearing reversibly breaks aggregates and their secondary structures, such as hydrogen and other bonds, which renders them more flowable. They revert back to their original state when the shear force is removed. Salad dressings, concentrated fruit juices, dairy cream, and fruit and vegetable purees are pseudoplastic materials.
• **Dilatant Fluid:** A few fluids show the reverse pattern. As the shear rate increases, the internal friction and, hence, the viscosity increase. Such fluids are suspensions of granular materials and behave almost like a Newtonian fluid at low shear rates. At high shear rates, however, they expand (dilate) and increase the void space. Consequently, the liquid becomes insufficient to fill the increase in the volume of

the voids, and solid–solid friction increases, causing the shear stress to shoot up. In a τ vs $\dot{\gamma}$ plot, the slope increases dramatically with increasing shear. These fluids are called "**shear thickening**" or dilatants. Dilatant fluids are rare. Concentrated (60%) suspensions of solid spheres, such as corn starch granule, quicksand, and synovial (joint), are examples of fluids that show this behavior. The synovial fluid is interesting in that it usually has low viscosity and provides lubrication for the joint in the knees and elbows to move freely. When bumped against a hard surface, the synovial fluid quickly turns into a shear-thickening fluid to cushion and protect the joints.

- **Bingham Plastic:** Another group of materials fail to shear until a threshold shear stress, called yield stress (τ_0), is exceeded. Below this threshold, the material behaves as a solid. Above this threshold, shearing occurs, and the material behaves like a fluid. If, above the threshold, the shear rate is proportional to the increase in stress, as shown in Fig. 5.24, the material is called a Bingham plastic. Molten chocolate, mayonnaise, sludge, blood, and toothpaste are examples of Bingham plastics.
- **Non-Bingham Plastic (or Viscoplastic):** The lower curve in Fig. 5.24 shows a non-Bingham plastic. Similar to a Bingham plastic, it also requires a threshold stress, but the behavior after that is nonlinear. In this example, like ketchup, the material is shear thinning.

    Several models like Herschel-Bulkley, oswald, Ellis and Erying, Bingham, etc. are used to mathematically model non-Newtonian fluid flow behavior.

(B) **Time-Dependent Behavior**

When some fluids are subjected to a constant shear stress or shear rate, they become thinner (or occasionally thicker) with time, even though the stress does not change. Their viscosities depend not only on the applied shear rate or shear stress but also on the duration and history of shearing. We call this "time-dependent" behavior. Time-dependent fluids fall into two categories.

- **Rheopectic Fluids**. A fluid that, when subjected to shearing, becomes thicker with time is called rheopectic (or **rheoplastic**). It may be thought of as a time-dependent dilatant. The upper curve in Fig. 5.25 shows a rheopectic

Fig. 5.24 Time-independent Newtonian and non-Newtonian fluids

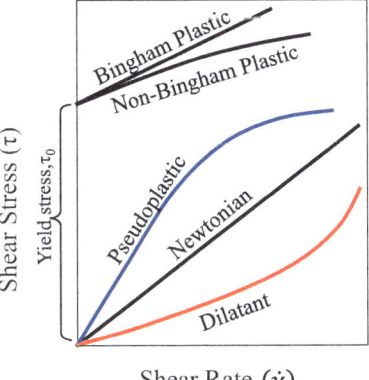

Shear Rate ($\dot{\gamma}$)

fluid. This type of fluid thickens or solidifies when shaken and is rare. There are no good food examples, and starch-based suspensions approximate such behavior. Nonfood examples include gypsum, printer ink, and bentonite clay (silica-based, derived from volcanic ash). Sodium bentonite clay is available as food grade.

- **Thixotropic Fluids**. A fluid that, when subjected to a constant shear rate, becomes thinner with time is called thixotropic, shown as the lower curve in Fig. 5.25. The example below describes a thixotropic fluid. Condensed milk, yogurt, egg white, honey, xanthan gum, peanut butter, ice cream and shortening are examples of this type of fluid.

Example

10. *The upper curve in* Fig. 5.26 *represents the behavior of a pseudoplastic fluid when first subjected to shear rate. The middle curve represents the behavior of the fluid if the shear rate is held constant for t_1 seconds. The lower curve represents the behavior after t_2 seconds, still at the same shear rate. The slopes of the straight lines (η_0, η_1 and η_2) represent the apparent viscosities at these times. We see that the fluid becomes less viscous with time and that the stress needed to maintain the shear rate decreases. In* Fig. 5.25, *the apparent viscosities from* Fig. 5.26 *are plotted versus time in the lower curve. Again, we see that this fluid becomes thinner with time.* ◄

(C) Apparent Viscosity

With a Newtonian fluid, the ratio of τ and $\dot{\gamma}$ is constant, and we have defined this ratio as viscosity. With a non-Newtonian fluid, this ratio is not constant but is a function of shear rate, and we must replace Eq. (5.25) with the equation:

$$\tau = (\eta)_{\dot{\gamma}}\dot{\gamma} \qquad (5.30)$$

In other words, the shear stress equals a viscosity at a specified shear rate times the shear rate itself. This viscosity function, $(\eta)_{\dot{\gamma}}$, is the ratio of τ and $\dot{\gamma}$ at a specific

Fig. 5.25 Time-dependent fluids

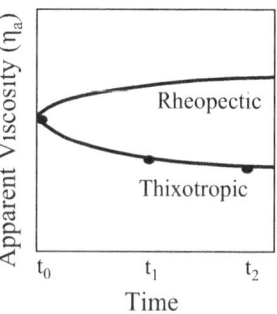

Fig. 5.26 Time-dependent
shear thinning

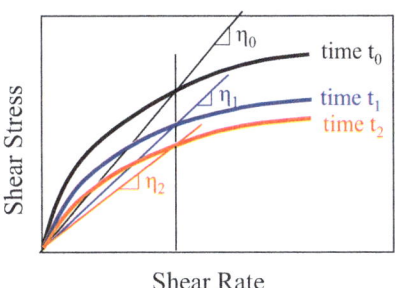

value of $\dot{\gamma}$ and is called the **apparent viscosity** (η_a) of the fluid at that value of $\dot{\gamma}$, and Eq. (5.30) may be rewritten as follows:

$$Apparent\ viscosity, \eta_a = \frac{\tau}{\dot{\gamma}} \qquad (5.31)$$

To be meaningful, therefore, the value of apparent viscosity must always accompany the shear rate at which it was determined. In Fig. 5.27, the curved line represents the behavior of a non-Newtonian fluid (pseudoplastic in this case), the slopes of the straight lines (η_{a1}) and (η_{a2}) represent the apparent viscosities of the fluid for two shear rates, $\dot{\gamma}_1$ and $\dot{\gamma}_2$. Note that in the case of a pseudoplastic fluid, the apparent viscosities decrease with increased shear, i.e., the fluid becomes thinner.

Check Your Understanding

9. Draw flow curve for a dilatant fluid, show the apparent viscosities at two shear rates and verify that the apparent viscosity is higher at the higher shear rate. ◀

Fig. 5.27 Apparent
viscosities

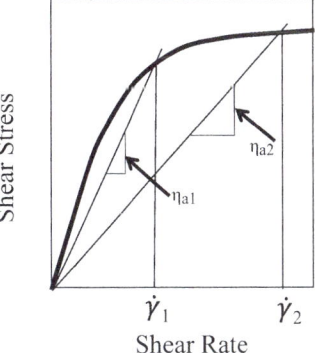

(D) The Power Law Model

The relationship represented by Eq. (5.31) merely states that apparent viscosity changes with shear rate and shear stress, but it does not tell us the nature of the change. For many fluids, the relationship between shear stress and shear rate can be approximated with the following relationship, known as the Ostwald de Waele or simply the power law model:

$$\tau = K \dot{\gamma}^n \tag{5.32}$$

where K = the coefficient of consistency (SI units Pa-s^n), n = the flow behavior index (dimensionless), a measure of deviation from Newtonian flow.

The meanings of the constants K and n are illustrated in Fig. 5.28. The three curves in that figure are all plotted from Eq. (5.32) using the same value of K, which represents the initial slope of each curve. The shape of the curve is determined by the value of n.

- If $n = 1$, Eq. (5.32) reduces to Eq. (5.26b), the curve is a straight line, and the fluid obeys Newton's law. In that case, $K = \eta$ or viscosity.
- If $n > 1$, the curve bends upward and the fluid is dilatant.
- If $n < 1$, the curve bends downward, and the fluid is pseudoplastic.

Using the power law, we can obtain an expression for apparent viscosity. To do this, factor $\dot{\gamma}$ out of Eq. (5.32) to obtain:

$$\tau = \left(K\dot{\gamma}^{n-1}\right)\dot{\gamma} \tag{5.33}$$

Comparing this with Eq. (5.31), we can see that, for a power law fluid,

$$Apparent\ viscosity, \eta_a = K\dot{\gamma}^{n-1} \tag{5.34}$$

Example

11. The τ versus $\dot{\gamma}$ curve for a fluid fits the equation: $\tau = 1.12\ \dot{\gamma}^{0.85}$ What shear stress is required to maintain a shear rate of 0.2 per second? What is the

Fig. 5.28 Power law plots and classification of fluids

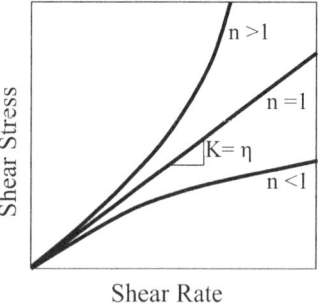

Shear Stress

n >1

n =1

K= η

n <1

Shear Rate

apparent viscosity of the fluid at this shear rate? What is its apparent viscosity at 0.3 s⁻¹?

Solution: *To maintain a shear rate of 0.2 s⁻¹, the shear stress must be*

$$\tau = 1.12 \, (0.2)^{0.85} = 0.285 \text{ Pa}$$

This means that opposing shear forces of 0.285 N must be exerted for every square meter of fluid. The apparent viscosities of the fluid at 0.2 and 0.3 s⁻¹ are

$$\eta_a = 1.12 \, (0.2)^{0.85-1} = 1.43 \text{ Pa}$$
$$\eta_a = 1.12 \, (0.3)^{0.85-1} = 1.20 \text{ Pa}$$

We see that the apparent viscosity decreases with increasing shear rate. ◄

Example

12. *Using the shear stress and shear rate data of a non-Newtonian fluid shown below, determine its consistency coefficient (K) and flow behavior index (n). What type of fluid is it?*

τ (Pa)	1.0	4.0	9.5	34.5
$\dot{\gamma}$ (s⁻¹)	0.5	3.5	10.0	53.0

Solution: *The data may be fitted to the power law model by either transforming the data into logs and then plotting on a linear graph or plotting the data directly on a log-log graph and then determining the values of n and K from the slope and intercept, as shown in* Fig. 5.29. *The value of n is less than one (0.81), and thus, it is a pseudoplastic or shear-thinning fluid.* ◄

(E) **Equations for Plastics**

A Bingham plastic requires a slightly different equation If you examine Fig. 5.24, you will note that a Bingham plastic plots as a straight line with a positive intercept. It behaves like a Newtonian fluid with an intercept equal to the yield stress. It therefore has the equation:

$$\tau = \tau_0 + \eta \, \dot{\gamma} \tag{5.35}$$

where τ_0 = the y intercept = the yield stress, i.e., the stress required before any flow takes place.

Fig. 5.29 Solution of
Example 12

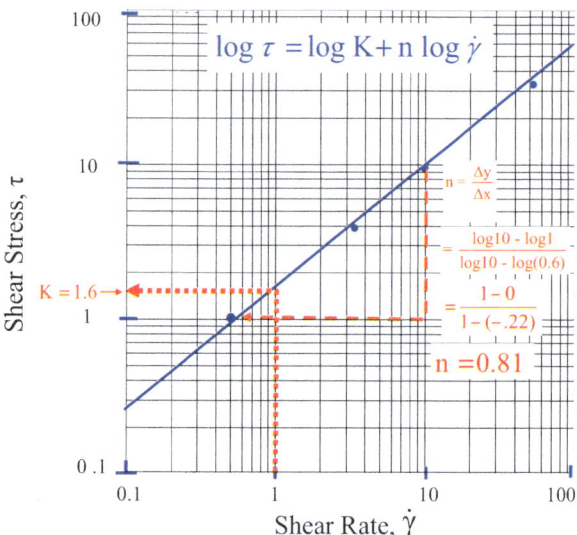

$\eta =$ the slope of the line after this threshold has been exceeded and represents viscosity. This is shown in Fig. 5.30.

This flow pattern of a shear-thinning, non-Bingham plastic can be approximated with the Herschel-Bulkley model, commonly written as:

$$\tau = \tau_0 + K\,\dot{\gamma}^n \qquad (5.36)$$

where K and n are the same parameters as defined earlier.

5.2.6   Temperature Effects on Flow Properties

At a constant temperature and pressure, the flow behavior of a fluid remains constant. Gases and liquids respond differently to changes in both temperature and pressure.

Fig. 5.30 Plastic flow plots

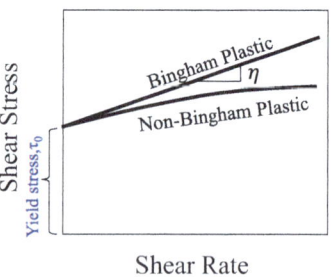

(A) **Gases**
- As the temperature of a gas increases, the kinetic energy and thus the viscosity of the gas increases linearly. This happens because flowing layers exchange momentum through collisions. A higher temperature leads to higher intermolecular collisions and thus greater viscosity.
- As the pressure increases, the intermolecular distance within a gas decreases. This leads to an increase in intermolecular forces and higher viscosity.

(B) **Liquids**
- Unlike gases, liquids decrease in viscosity as the temperature increases. This happens because as the temperature increases, the constituent molecules start vibrating faster. This weakens the cohesive interactions among them and leads to a loss of viscosity. Remember that hot fudge sauces flow better than cold sauces.
- The change due to temperature is quite noticeable and nonlinear, following a logarithmic relationship described by an **Arrhenius-type** equation.

$$\eta = \beta e^{\left(\frac{E_a}{RT}\right)} \tag{5.37}$$

where β is a constant, E_a = activation energy, R = the gas constant, and T = the absolute temperature.

The above equation is the familiar Arrhenius equation but without a negative sign in the exponent. The Arrhenius equation is used to describe the effect of temperature on the reaction rate constants. The exponent sign is reversed because chemical reaction rates increase with increasing temperatures, but the viscosity of liquids is well known to decrease when it is heated. The meaning of the activation energy (E_a) in the two systems is also different. In chemistry, it is the minimum energy needed for the reaction to occur. In flow systems, it is the energy barrier that must be surmounted for a molecule to "squeeze" by its neighbors when the temperature is raised. A lower E_a would thus imply a more efficient flow at higher temperatures.

Taking the logarithm of each side of Eq. (5.37), we obtain:

$$\ln(\eta) = \ln(\beta) + \left(\frac{E_a}{R}\right)\frac{1}{T} \text{ or } \log(\eta) = \log(\beta) + \left(\frac{E_a}{2.303R}\right)\frac{1}{T} \tag{5.38}$$

Equation (5.38) is the equation of a straight line. If you measure viscosities over a range of temperatures and plot the log of the viscosities versus the reciprocal of the temperature, the plot should yield a straight line with intercept equal to log (β) and slope equal to ($E_a/2.303R$). Once the values of β and E_a are established, a specific equation for the temperature dependence of the material under evaluation is generated for the prediction of viscosity at other temperatures within the tested range.

- The effect of pressure on liquid viscosity is negligible up to approximately 40 atm. since liquids are relatively incompressible.

5.2.7   Importance of Fluid Flow Behavior

Foods are multiphasic and complex materials, both compositionally and structurally. They exhibit a range of flow behaviors, such as low viscosity in milk or juice and viscoelasticity in dough (gluten). Many consumers perceive sensory properties such as creaminess, stickiness, and crispiness are largely related to flow properties. Additionally, the flow of liquid foods during processing and manufacturing or the ease with which they can be poured into or drained out of containers is largely determined by their flow characteristics. Some of the typical examples are summarized below:

(A) Product design implications:
  - High viscosity in liquid food is often associated with "richness" and "creaminess." Generally, humans can detect differences in viscosity as little as 1 mPa-s. This poses a challenge in developing lower fat foods.
  - Many frostings are designed to have a yield stress (i.e., Bingham behavior) greater than the stress induced by its weight so that the frosting will not run off the cake.
  - Flow (rheological) properties characterize the desirable pouring and spreading behavior of products and serve as useful tools for formulating new products for specific applications.

Check Your Understanding

10. A 1 mm thick layer of mayonnaise ($\eta = 20$ P) is applied to a slice of bread at a rate of 100 mm/s. Estimate the stress imposed on the bread. ◄

(B) Process design implications:
  - Bingham fluids will not completely drain from the walls of a tank, causing significant loss of such products.
  - During mixing, the viscosity of a pseudoplastic material will become lower near the impeller tip but may change very little toward the shaft; thus, no or very poor mixing may result.
  - Fluid-velocity profiles change with the type of flow behavior. This has very significant implications in process design, food quality control, and safety assurance, as you will see discussed later.

- Highly viscous fluid flow is laminar, while it is turbulent for low viscosity fluids. Consequently, mixing is reduced during pipeline flows and compositions or properties may not blend out.
- Pressure drops (friction losses, discussed later) in lines and fittings are high for highly viscous products, and thus, larger pumps are often needed. Additional higher power dissipation may cause the temperature to rise, which may change the product characteristics.

5.2.8 Fluid Viscosity and Rheology

Viscosity as a liquid property is not a good metric to quantify the flow behavior of suspensions, emulsions, gels, and soft solids such as cheeses. The study of how fluids and soft solids deform and flow is called rheology. It is more than just the study of a liquid's viscosity and is extensively used to understand and quantify the textural and mechanical properties of foods and in the design of processing operations. Figure 5.31 summarizes the rheological classification of materials. In addition to what has been discussed thus far, note the following:

- **Liquid.** A liquid is a material that undergoes an irreversible strain when subjected to a shear stress. The increasing viscous strain is what we know as flowing. When the stress is removed, the fluid remains in its new shape because the energy has been transformed.
- **Solid**. A solid shows elastic behavior and follows Hooke's law, as described in Chap. 2. When subjected to a shear stress, it deforms, but when the stress is removed, it partially or fully returns to its original shape and size.
- **Viscoelastic material**. Most materials of practical importance are viscoelastic in that they show both liquid-like (viscous) and solid-like (elastic) behaviors. When stressed, it will deform, and if immediately released, it may return to nearly its original shape. Under continued stress, however, some flow will take place, and

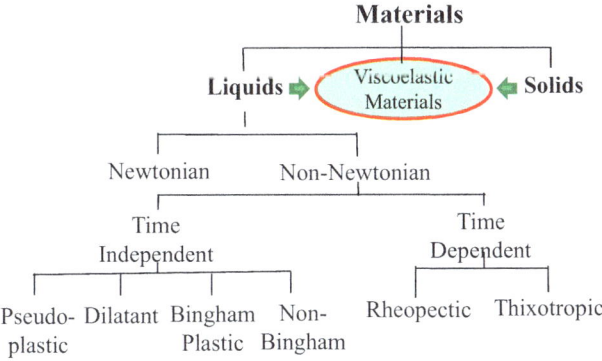

Fig. 5.31 Rheological classification of materials

when released, it will only partially return to its original shape. A vast majority of food materials show viscoelastic behavior. If you squeeze a block of cheese, for example, it partly rebounds when released. The longer you hold the cheese, the less it will rebound. The viscoelasticity of food materials not only affects the processing technology but also determines the perception of biting and chewing. Viscoelasticity is characterized by measuring the elastic (spring-like) storage modulus (G'), viscous (flow-like) loss modulus (G''), and their ratio (tan δ).

Many times, it is difficult to decide whether a particular food is a solid or a liquid since they exhibit some properties of both and are termed viscoelastic. In such situations, these are evaluated in terms of a nondimensional number called the **Deborah number (D_e)**. It is the ratio of the relaxation time (λ) of the sample, i.e., how long it takes a material to return to its original state after undergoing stress, and the observation time (t). The characteristic relaxation time (λ) of a material can be experimentally determined. It is nearly zero for liquids and infinite for ideal solids. A high Deborah number (λ/t) defines a solid-like behavior, and a low number indicates a more liquid-like material. You can find these and other related materials discussed in standard textbooks on rheology.

5.2.9 Fluid Flow Patterns

In 1883, Osborn Reynolds studied the movement of fluids through a pipe. To do this, he set up a pipe so that he could inject a dye into the flowing fluid, as shown in Fig. 5.32. By sliding the injector tube up and down, he was able to inject the dye into different parts of the moving stream. Using this equipment, he studied the relationship between fluid velocity and fluid flow pattern.

Through a series of experiments, he observed the following:

- At low fluid velocities, the dye followed a straight path parallel to the axis of the pipe, as shown in Fig. 5.32.
- At low fluid velocities, dye movement was rapid in the center (Fig. 5.32) and progressively slower toward the side of the pipe, as shown in Fig. 5.33. This suggested a flow pattern in which layers of fluid slide over each other in the manner already described in this chapter. We call this **laminar flow**. Such a flow pattern produces little mixing. As the average velocity of the fluid increased, a point was reached when the dye began to move chaotically with much lateral

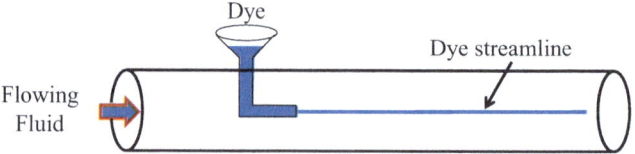

Fig. 5.32 Laminar flow (straight path parallel to pipe axis)

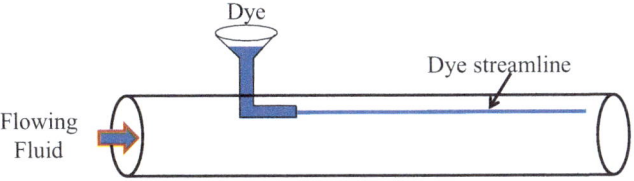

Fig. 5.33 Laminar flow slower toward the side of pipe

movement and breaking up of the dye stream, as shown in Fig. 5.34. This flow
pattern results in mixing of the fluid and is called **turbulent flow.**

(A) **Reynolds Number**

By experimenting with a variety of fluids at different velocities, Reynolds
determined that the transition from laminar to turbulent flow was affected by
the following conditions:

- **Laminar flow** takes place at **low velocities,** while **turbulent flow** takes place
  at **higher velocities.**
- The transition takes place at <u>lower velocities</u> with fluids of **higher density.**
- The transition takes place at <u>lower velocities</u> in pipes of **greater diameter.**
  Larger diameter pipes provide more room for lateral movement and are more
  conducive to turbulent flow.
- The transition takes place at higher velocities with fluids **of greater viscosity.** High-viscosity fluids such as thick oils stay in laminar flow longer than
  low viscosity fluids such as water.

Reynolds found that he could combine these 4 factors into a single number
that predicted well whether a fluid would exhibit laminar or turbulent flow. This
number, called the **Reynolds number (Re),** is computed as follows:

$$Re = \frac{\rho d <v>}{\eta} \tag{5.39}$$

where ρ — density of the fluid, d = inside diameter of the pipe, $<v>$ = average
velocity of the fluid, and η = viscosity of the fluid.

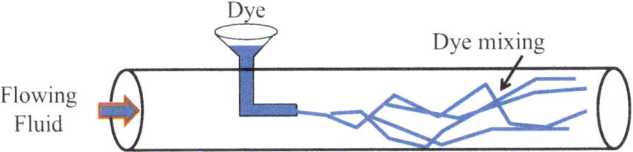

Fig. 5.34 Turbulent flow

- A system with a Reynolds number less than 2100 will exhibit laminar flow.
- A system with a Reynolds number greater than 4000 will exhibit turbulent flow.
- A system with a Reynolds number between 2100 and 4000 tends to be unstable as it goes through the **transition** from laminar to turbulent flow. The transition also depends on the nature of the system. With very special precautions, laminar flow can be maintained up to Re = 50,000, but this is not of practical use.

An analysis of the dimensions of the four properties that appear in the Reynolds number shows that they cancel out and that the number is dimensionless. For example, using the SI set of units, it can be observed as follows:

$$Re\,[=]\ \frac{\left(\frac{kg}{m^3}\right)(m)\left(\frac{m}{s}\right)}{\frac{kg}{m \times s}} = 1 \tag{5.40}$$

Example

13. *A pipe with an inside diameter of 3.2 cm carries water at 20 °C at a rate of 0.05 m³/min. Is the flow laminar or turbulent?*
    ***Solution:** The average velocity of the water is:*

$$<v> = \left(0.05\,\frac{m^3}{min}\right)\left(\frac{1\ min}{60\ s}\right)\left(\frac{1}{\pi\frac{0.032^2}{4}\,m^2}\right) = 1.04\ m/s$$

At 20 °C, water has a density of 1000 kg/m³ and a viscosity of 1.005 cP or 1.005 × 10⁻³ Pa-s. The Reynolds number is

$$Re = \frac{\rho d <v>}{\eta} = \frac{\left(1000\,\frac{kg}{m^3}\right)\ (0.032\ m)\ \left(1.04\,\frac{m}{s}\right)}{\left(1.005 \times 10^{-3}\,\frac{kg}{ms}\right)} = 3.31 \times 10^4$$

Since 3.31 × 10⁴ is greater than 4000, we conclude that the water in this pipe is turbulent. ◄

Check Your Understanding

11. In Example 13, the water is heated to 60 °C, causing its viscosity to drop to 0.469 cP. Compute the Reynolds number and verify that the water should now show turbulent flow. ◄

(B) **Physical Meaning of the Reynolds Number**

When water travels through a channel, the velocity varies across the width of the channel. Usually, the velocity is greatest in the center, so the molecules have the greatest momentum there. However, some of this momentum is passed to the slower-moving molecules, dragging them along with the faster ones. You have seen two mechanisms by which momentum is passed between molecules:

- **Viscous Transport**. When flow is laminar, momentum is passed from layer to layer in the fluid by the frictional forces between the layers. We call this viscous transport of momentum. We can also say that viscous forces are transporting momentum.
- **Molecular Transport**. When flow is turbulent, molecules move laterally and in so doing carry momentum to different layers. We call this a molecular transport of momentum. Since the molecules carry kinetic energy, we can also say that kinetic or inertial forces are transporting momentum.

Viscous transport predominates when the Reynolds number is low, and molecular transport predominates when it is high. This suggests that the Reynolds number might represent a ratio of

$$\text{Re} = \frac{\text{Molecular Momentum Transport}}{\text{Viscous Momentum Transport}} = \frac{\text{Kinetic or Inertial Forces}}{\text{Viscous Forces}} \quad (5.41)$$

The above interpretation is very helpful in many practical applications. For example, high Reynolds numbers indicate that the inertial forces are more significant than the viscous (friction) forces, and thus the viscosity effects may be neglected. In such situations, the flow is assumed to approximate an inviscid flow, making some analyses of practical utility simpler.

We will now demonstrate that the Reynolds number can indeed be interpreted as shown in Eq. (5.41). First, we define the inertial or kinetic forces of a fluid as its mass (m) times velocity (v) squared divided by some, yet undefined, characteristic dimension, L.

$$F_L \approx \frac{m <v>^2}{L} \quad (5.42)$$

Then, we define viscous forces, the forces associated with shear stress, as a function of η.

$$F_v \approx \frac{\eta <v>}{L} A \quad (5.43)$$

You can easily verify that both equations have units of force and can be used to show that the Reynolds number is the ratio of these forces. Taking the ratio of Eq. (5.42) to Eq. (5.43), we have:

$$\frac{F_L}{F_V} = \frac{\frac{m<v>^2}{L}}{\frac{\eta<v>A}{L}} = \left(\frac{m}{A}\right)\left(\frac{<v>}{\eta}\right) \quad (5.44)$$

Because the numerator and the denominator both have units of force, this ratio is dimensionless. However, mass equals density times volume, and volume equals area times length, so we can write:

$$\left(\frac{m}{A}\right) = \frac{\rho V}{A} = \frac{\rho AL}{A} = \rho L \tag{5.45}$$

Substituting (5.45) into (5.44), we have:

$$\frac{F_L}{F_V} = \left(\frac{m}{A}\right)\left(\frac{<v>}{\eta}\right) = \frac{\rho L <v>}{\eta} \tag{5.46}$$

To date, we have been rather vague about the meaning of L except to specify that it has the unit of length. If we let L be the diameter of the pipe, the ratio becomes the equation for the Reynolds number.

$$\frac{F_L}{F_V} = \frac{\rho d <v>}{\eta} = \text{Re} \tag{5.47}$$

The above analysis has established a reasonable physical meaning for the Reynolds number that is consistent with its relationship to flow patterns and is worth remembering.

(C) Reynolds Number: Various Geometries

The equation for the Reynolds number given above applies to fluid flowing through a pipe with a circular cross-section. For pipes with noncircular cross-sections, the Reynolds number is computed with the equation:

$$\text{Re} = \frac{\rho d_e <v>}{\eta} \tag{5.48}$$

where d_e = the equivalent diameter which is 4 times the hydraulic radius (r_h) defined as:

$$r_h = \text{the hydraulic radius} = \frac{\text{Cross SectionArea}}{\text{Wetted Perimeter}} \tag{5.49}$$

Check Your Understanding

12. Show that for a filled circular pipe, the equivalent diameter reduces to the diameter of the pipe. ◀

Example

14. *Compute the equivalent diameters for the geometries illustrated in* Fig. 5.35.

    Solution: The wetted perimeters are shown here as a heavy line. Note that only surfaces where fluid contacts solid walls are considered part of the wetted perimeter. Any interface between fluid and air is ignored. The equivalent diameter becomes:

A. *Completely filled rectangle of sides L_1 and L_2*

$$Cross\text{-}section\ area = L_1\,L_2,\ \ldots, Perimeter = 2(L_1 + L_2), d_e = 4r_h$$

$$= 4\left(\frac{L_1 L_2}{2(L_1 + L_2)}\right) = \frac{2L_1 L_2}{L_1 + L_2}$$

B. *Half-filled circular trough*

$$Cross\ section\ area = \frac{\pi \frac{d^2}{4}}{2} = \pi \frac{d^2}{8}, Perimeter = \frac{\pi d}{2}, d_e = 4r_h = 4\left(\frac{\pi \frac{d^2}{8}}{\pi \frac{d}{2}}\right) = d \quad \blacktriangleleft$$

(C) Reynolds Number for Non-Newtonian Fluids

The Reynolds number defined above contains viscosity as a factor and is, therefore, only applicable to Newtonian fluids. Recall that the viscous behavior of power law fluids can be described by Eq. (5.32). Using K (the consistency coefficient) and n (the flow behavior index), a **generalized Reynolds number** has been defined that is applicable to any power law fluid as follows:

$$G\,Re = \frac{\rho d^n <v>^{2-n}}{K 8^{n-1}\left(\frac{3n+1}{4n}\right)^n} \tag{5.50}$$

Fig. 5.35 Different flow geometries

A. Rectangular conduit wetted on all 4 surfaces

B. Semicircular trough wetted on bottom

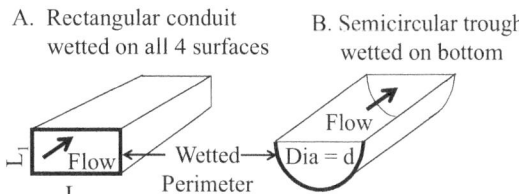

It is important to recognize that for a Newtonian fluid, where $n = 1$ and $K = \eta$, the generalized Reynolds number reduces to the ordinary Reynolds number, as shown below:

$$G\,Re = \frac{\rho d^1 <v>^{2-1}}{K 8^{1-1}\left(\frac{3(1)+1}{4(1)}\right)^1} = \frac{\rho d^1 <v>^1}{K 8^0 \left(\frac{4}{4}\right)^1} = \frac{\rho d <v>}{K} = \frac{\rho d <v>}{\eta} = Re$$

Example

15. *The viscous properties of a fluid are found to fit the power law with*

$$n = 0.85, K = 0.0024 \frac{Ns^{0.85}}{m^2}, and \rho = 950 \ kg/m^3$$

What is the generalized Reynolds number when this fluid is flowing at a rate of 1.4 g/s through a pipe with an inside diameter of 3.0 cm?
Solution: The average velocity of the fluid is:

$$\left(0.0014\frac{kg}{s}\right)\left(\frac{1 \ m^3}{950 \ kg}\right)\left(\frac{1}{\pi(0.015m)^2}\right) = 2.09 \times 10^{-3}\frac{m}{s}$$

The generalized Reynolds number is

$$G\,Re = \frac{\left(950\frac{kg}{m3}\right)(0.03m)^{0.85}\left(2.09 \times 10^{-3}\frac{m}{s}\right)^{2-0.85}}{\left(0.0024\frac{Ns^{0.85}}{m^2}\right)8^{0.85-1}\left(\frac{3(0.85)+1}{4(0.85)}\right)^{0.85}} = \frac{0.0399\frac{kg}{ms^{0.85-2}}}{0.00182\frac{kg}{ms^{0.85-2}}} = 21.9$$

Flow will be laminar. Note that the units have fractional exponents but still cancel to make the generalized Reynolds number dimensionless. ◄

5.2.10 Mass, Energy, and Momentum Balances for Fluid Flows

Flow Classification: Fluid flow is classified as ***internal*** when it flows inside a bounded surface such as a pipe, duct, or open channel and ***external*** when it flows unbounded over a surface, Fig. 5.36a. In this chapter, we will focus on internal flow driven by a pressure difference. Flow of liquid foods and water in pipes during processing, blood flow in arteries and veins, and oil and gas distribution via a network of pipelines are examples of internal flow systems. Since pipes with a circular cross-section can withstand larger pressure differences between the inside and the outside and are easier to clean, they are preferred for heating, cooling, and transportation of liquids.

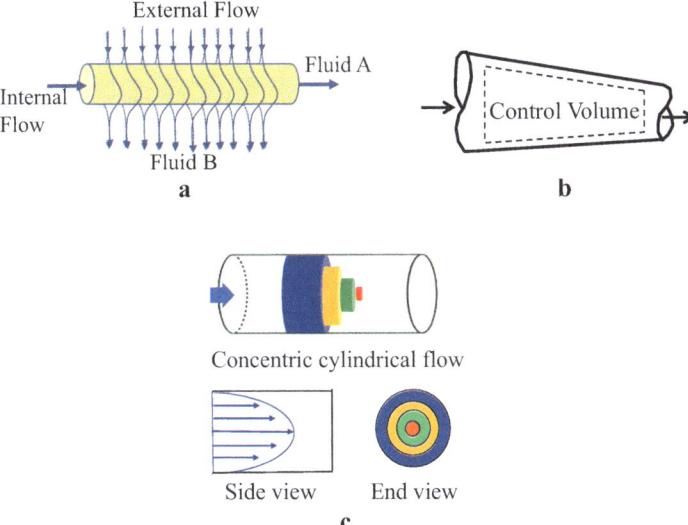

Fig. 5.36 (**a**) Internal vs. external flow. (**b**) Control volume concept. (**c**) Pipe flow velocity profile views

Flow Analysis: We are interested in mass, energy, and momentum balances and their effects on fluid flows. This is facilitated by understanding and performing the following analyses:

- **Overall (or Macroscopic) Balances**: When we are interested in an overall characteristic of flow and not in the detailed structure of the flow field, e.g., average velocity, we do an overall balance.
- **Differential (or Microscopic) Balances**: When our interest lies in knowing the detailed structure of the flow field, e.g., velocity distributions (profiles), we perform a differential balance.
- **Control Volume Approach:** The balances are better characterized if the basic laws of conservation of mass, energy, and momentum are performed on a control volume. A control volume is a fixed region of the system in space through which the fluid flows, as shown in Fig. 5.36b. We use the control volume approach instead of the system-surroundings analysis discussed previously for two reasons:
  - For a system-surroundings formulation, it will be necessary to identify and follow a fixed mass of fluid all the time, which is not practical. In the control volume, we focus on a fixed region,
  - Frequently, we also need to learn about the effects of the fluid motion within the space where the flow is occurring, such as a pipe or surface, which is better addressed by a control volume approach.
- **Velocity Distribution and Average Velocity:** The velocity of flow in a pipe generally varies across the cross-section, zero at the wall because of the no-slip

condition and maximum toward the center, as shown in Fig. 5.36c for flow in a pipe along with its side and end views.

In incompressible fluid flow, it is more desirable to work with an average velocity since it remains constant when the cross-section of the flow channel is constant.

The average velocity of an incompressible fluid (constant density) at some stream wide cross-section is then computed by applying the conservation of mass principle:

$$<v> = \frac{1}{A} \iint v dA \qquad (5.51)$$

where dA is an area element, v is the velocity in that element, and A is the total area. Using the above equation, the average velocity can thus be obtained from the velocity profile via integration. A simple system in the below example illustrates the principle.

Example

16. *A fluid flowing in a 5.0 cm × 5.0 cm ducts develops a velocity profile shown in* Fig. 5.37 *in m/s. Determine the average velocity.*
    Solution: *The average velocity of the fluid is obtained by adding velocity in each of the small elements and then dividing by the total area as shown below:*

$$\frac{1}{A} \iint v dA = \frac{1}{A}[v_{11}(dA) + v_{12}(dA) + v_{13}(dA) + \cdots + v_{54}(dA) + v_{55}(dA)]$$

$$= \frac{1}{25}[2(1) + 2(1) + 2(1) + 2(1) + \cdots + 4(1) + 4(1) + \cdots + 6(1)] \text{m/s}$$

$$= \frac{1}{25}[2(16) + 4(8) + 6(1)] \text{m/s} = \frac{1}{25}[70] \text{m/s} = 2.8 \text{ m/s} \qquad \blacktriangleleft$$

(A) Mass Balance and the Continuity Equation

Consider a control volume element with the flow details of a fluid shown in Fig. 5.38. If we assume that velocity (v) is normal to the surface and carries fluid

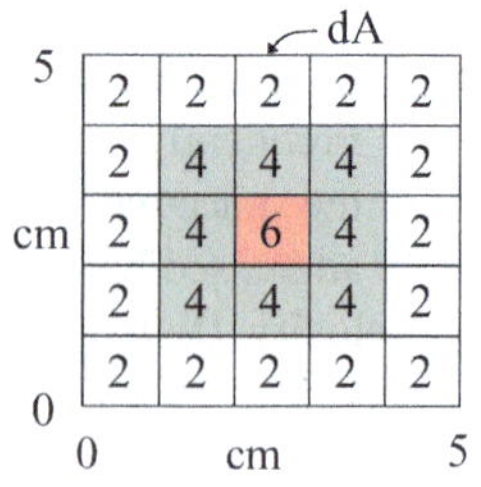

Fig. 5.37 Local velocities in a duct

Fig. 5.38 Macroscopic mass balance

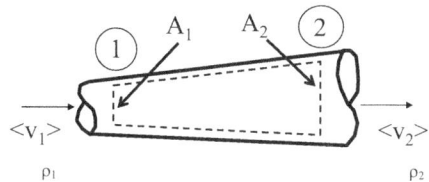

across, then the law of conservation of mass written in terms of mass balance for the control volume with no generation term becomes:

Rate of mass input $=$ Rate of mass accumulation $+$ Rate of output

$$\rho_1 <v_1>A_1 = \frac{d\dot{m}}{dt} + \rho_2 <v_2>A_2 \tag{5.52}$$

At **steady state,** there is no accumulation of mass, and the above equation reduces to:

$$\rho_1 <v_1>A_1 = \rho_2 <v_2>A_2 \tag{5.53}$$

The above Eq. (5.53) is known as the **continuity equation,** written in terms of mass flow rate $(\dot{m})$ since $\dot{m} = \rho <v>A$. Additionally, since the fluid velocity times the cross-sectional area of flow equals the volumetric flow rate $(\dot{V})$, Eq. (5.53) may be written as:

$$\rho_1 \dot{V}_1 = \rho_2 \dot{V}_2 \tag{5.54}$$

For incompressible fluids, ρ is constant; then, $\dot{V}_1 = \dot{V}_2$ or $<v_1>A_1 = <v_2>A_2$. It is therefore easy to note:

$$\frac{<v_1>}{<v_2>} = \frac{A_2}{A_1} \tag{5.55}$$

i.e., in steady-state flow of an incompressible fluid, if the area increases, the velocity must correspondingly decrease and vice versa.

Example

17. *A fluid with a density of 1005 kg/m³ is flowing steadily at a velocity of 2.0 m/ s in a pipe of 10.0 cm internal diameter. If the pipe gradually contracts to an internal diameter of 5.0 cm, find the new flow velocity.*
   Solution: From the continuity equation for steady-state flow, mass flow rate $(\dot{m})$ remains constant throughout the pipe, then for the two sections:
   $\dot{m}_1 = \dot{m}_2$ or $\rho_1 v_1 A_1 = \rho_2 v_2 A_2$ *and if the density is constant:*

$$v_2 = v_1 \left(\frac{A_1}{A_2}\right) = \left(\frac{2.0m}{s}\right)\left[\frac{(0.1)^2}{(0.05)^2}\right] = 8.0 \; m/s \; \text{i.e., } \textit{decreasing the diameter by}$$

half increased the velocity four times! ◀

(B) **Total Energy Balance**

Applying the principle of the conservation of energy to a control volume, we can also obtain a total energy balance for fluid flow, such as obtaining the continuity equation by invoking the principle of conservation of mass. Consider a control volume, Fig. 5.39, through which a fluid is flowing at **steady state** between points 1 and 2 with areas A_1 and A_2 and fluid properties as shown. The fluid receives energy $\overline{q}$ and performs work $\overline{w}$.

Thus, if $\overline{E}_T$ is the total energy per unit mass of the fluid, the energy balance according to the first law of thermodynamics becomes:

$$\Delta \overline{E}_T = \overline{q} - \overline{w}_s \tag{5.56}$$

The total energy (E_T) consists of internal energy (E_i), pressure–volume energy ($P \overline{V}$), kinetic energy (E_k), and potential energy (E_p), i.e.,

$$\overline{E}_T = P\overline{V} + \overline{E}_i + \overline{E}_k + \overline{E}_p \tag{5.57}$$

We have learned earlier that internal energy (E) plus pressure–volume energy is enthalpy (H). The above equation can then be written in terms of per unit mass of the flowing fluid as:

$$\overline{E}_T = \overline{H} + \frac{1}{2}\langle v\rangle^2 + gh \tag{5.58}$$

where $\frac{1}{2}\langle v\rangle^2$ and gh are the kinetic energy and potential energy terms, respectively, expressed on a per unit mass basis. Thus, Eq. (5.56) can be written as:

Fig. 5.39 Overall energy balance

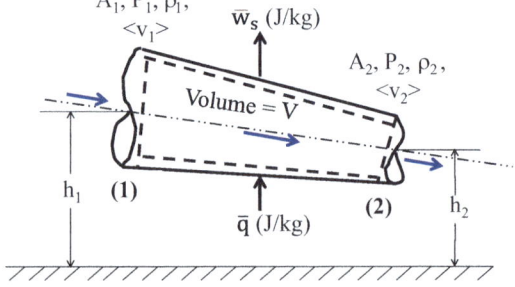

$$\left(\overline{H}_2 + \frac{1}{2}\langle v_2\rangle^2 + gh_2\right) - \left(\overline{H}_1 + \frac{1}{2}\langle v_1\rangle^2 + gh_1\right) = \overline{q} - \overline{w}_s \qquad (5.59)$$

The above equation can also be rearranged in terms of differences in the three types of energies on the left side of Eq. (5.59) as follows:

$$\left(\overline{H}_2 - \overline{H}_1\right) + \frac{1}{2}\left(\langle v_2\rangle^2 - \langle v_1\rangle^2\right) + g(h_2 - h_1) = \overline{q} - \overline{w}_s \qquad (5.60)$$

$$\text{Or,}\ \Delta\overline{H} + \frac{1}{2}\Delta < v >^2 + g\Delta h = \overline{q} - \overline{w}_s \qquad (5.61)$$

This is the first law of thermodynamics for a flow system and provides a total energy balance. In many situations, the kinetic and potential energy changes are insignificant, and there is no shaft work (w_s), i.e., no pump, no mixing, etc., and the energy balances may be simplified as:

$$\Delta\overline{H} = \overline{q} \qquad (5.62)$$

Kinetic Energy Correction Factor (α): The kinetic energy term in the above equations assumes that all mass is moving at the same average velocity. When velocity varies across the cross-section of a stream, the kinetic energy term can be corrected for the bulk motion through the introduction of a correction factor, α, to correct the kinetic energy term as follows:

$$\overline{E}_k = \frac{1}{2\alpha}\langle v\rangle^2 \qquad (5.63)$$

The value of the correction factor can be determined from the average velocity, $\alpha = \frac{<v>^3}{<v^3>}$, but is more commonly generalized based on the type of flow. The value of α is generally assumed to be 0.5 for laminar flows and 0.90–0.99 (~1) for turbulent flows. The overall energy balance Eq. (5.61) is thus modified as:

$$\Delta\overline{H} + \frac{1}{2\alpha}\Delta < v >^2 + g\Delta h = \overline{q} - \overline{w}_s \qquad (5.64)$$

(C) Mechanical Energy Balance (MEB) and the Bernoulli Equation

The total energy balance can be modified to include only the mechanical energy terms. Mechanical energy is energy that is either work or can be converted into work. It includes work (w_s), kinetic energy (E_k), potential energy (E_p), the pressure–volume term of enthalpy, and some of all energy lost due to friction (ΣF) but excludes heat (q) and internal energy (E_i), since these cannot be fully converted into work due to the second law of thermodynamics. To account for them in developing the MEB, we begin with the first law of thermodynamics applied to the flow of a fluid in the control volume shown Fig. 5.39. In the process of flowing from the inlet to the outlet, the fluid performs work, PdV, and loses energy due to friction, which is summed up as ΣF, and we can write the energy balance equation per unit mass basis as:

$$\Delta \overline{E}_i = \overline{q} - \overline{w}_s = \overline{q} - \int_{\overline{V}_1}^{\overline{V}_2} P \, d\overline{V} + \sum F \qquad (5.65)$$

We also know that

$$\Delta \overline{H} = \Delta \overline{E}_i + \Delta \left(P \overline{V}_0 \right) = \Delta E_i + \int_{\overline{V}_1}^{\overline{V}_2} P \, d\overline{V} + \int_{P_1}^{P_2} \overline{V} \, dP \qquad (5.66)$$

Substituting the value of $\Delta \overline{E}_i$ from Eq. (5.65) into (5.66), we obtain:

$$\Delta \overline{H} = \overline{q} - \int_{\overline{V}_1}^{\overline{V}_2} P \, d\overline{V} + \sum F + \int_{\overline{V}_1}^{\overline{V}_2} P \, d\overline{V} + \int_{P_1}^{P_2} \overline{V} \, dP \qquad (5.67)$$

$$\text{Or,} \ \Delta \overline{H} = \overline{q} + \sum F + \int_{P_1}^{P_2} \overline{V} \, dP \qquad (5.68a)$$

where $\overline{V} = $ specific volume $= \frac{1}{\rho}$ and for an incompressible fluid $\int_{P_1}^{P_2} \overline{V} \, dP = \frac{\Delta P}{\rho}$, thus:

$$\Delta \overline{H} = \overline{q} + \sum F + \frac{\Delta P}{\rho} \qquad (5.68b)$$

Now, if we substitute the above expression for $\Delta \overline{H}$ in the total energy balance, Eq. (5.64), we obtain:

$$\overline{q} + \sum F + \frac{\Delta P}{\rho} + g\Delta h + \frac{1}{2\alpha} \Delta <v>^2 = \overline{q} - \overline{w}_s \qquad (5.69)$$

$$\frac{\Delta P}{\rho} + g\Delta h + \frac{1}{2\alpha} \Delta <v>^2 + \sum F + \overline{w}_s = 0 \qquad (5.70)$$

Or, writing in terms of output = input:

$$\frac{P_2}{\rho} + gh_2 + \frac{1}{2\alpha} <v_2>^2 + \sum F + \overline{w}_s = \frac{P_1}{\rho} + gh_1 + \frac{1}{2\alpha} <v_1>^2 \qquad (5.71)$$

The above MEB equation is for a simple steady-state flow of an incompressible fluid at a constant temperature undergoing no phase change. All the terms in the above equations are easily measurable except the friction term, which is better obtained from the developed correlations discussed in Chap. 6. For some simple and small systems, it may be neglected.

Example

18. *Pasteurized juice ($\rho = 1100$ kg/m³) needs to be pumped up 10 m from a large storage tank to a filling line at a steady flow rate of 90 m³/h through a stainless-steel pipe with an internal diameter of 5.0 cm (Fig. 5.40). The energy loss due to friction is estimated to be 101 J/kg. Compute the pump horsepower required if the pump is assumed to be 90% efficient.*

Solution: *Volumetric rate of flow, $\dot{V} = 90$ m³/h $= 0.025$ m³/s*
Velocity in the pipe, $v = (0.025$ m³/s$)[4/(3.14\ (0.05)^2)] = 12.74$ m/s
Applying MEB between locations 1 and 2:

$$\frac{P_2}{\rho} + gh_2 + \frac{1}{2\alpha} <v_2>^2 + \sum F = \frac{P_1}{\rho} + gh_1 + \frac{1}{2\alpha} <v_1>^2 + \overline{w}_s$$

However, $P_2 = P_1 =$ atmospheric pressure; $h_2 = 10$ m, $h_1 = 0$ (datum)
$v_1 = 0$ (large tank), $v_2 = 12.74$ m/s; $\sum F_f = 101$ J/kg; $\alpha = 0.5$, assumed laminar

$$gh_2 + \frac{1}{2\alpha} <v_2>^2 + \sum F = \overline{w}_s$$

$$\overline{w}_s = \left(9.81 m/s^2\right)(10.0m) + \frac{1}{2(0.5)}(12.74\ m/s)^2 + \left(101.0\frac{J}{kg}\right) = 361.41\ J/kg$$

Power $= [(361.41$ J/kg$)\ (0.025$ m³/s$)(1100$ kg/m³$)]/0.90 = 11.0$ kJ/s $= 11.0$ kW
Since 1 Horsepower (h.p.) $= 0.746$ kW, $\overline{w}_s = 14.8$ h.p. ◄

The Bernoulli equation: In the very special case of fluid flow along a streamline, where no energy is exchanged via shaft work ($w_s = 0$) and there is no loss of energy due to friction ($\sum F_f = 0$), i.e., the fluid is assumed to be inviscid, the mechanical energy balance simplifies further to what is known as the Bernoulli equation and Eq. (5.71) is written as:

Fig. 5.40 MEB on a pumping system

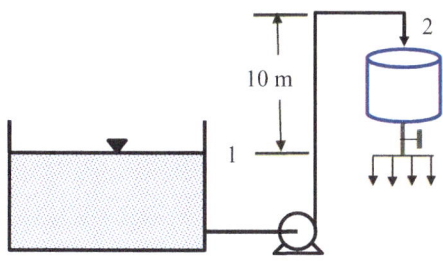

10 m

1

2

$$\frac{P_2}{\rho} + gh_2 + \frac{1}{2} <v_2>^2 = \frac{P_1}{\rho} + gh_1 + \frac{1}{2} <v_1>^2 \tag{5.72}$$

Bernoulli's equation is also known as the "Energy Equation" since it accounts for the changes in types of energy that may occur along a fluid path while the total energy remains constant. It embodies a principle that is named after Swiss mathematician and physicist Daniel Bernoulli, who first published this in his book Hydrodynamica in 1738. Qualitatively, it says that the sum of the pressure, potential, and kinetic energy in a fluid flow along a streamline must remain constant. The above Eq. (5.72) can also be summarized using the steady-state Euler equation, which says:

$$\frac{dP}{\rho} + g\,dh + v\,dv = 0 \tag{5.73}$$

If you have been following the derivation of the above equation, you will note that each term in the above equation has units of energy/mass. Often, it is more useful to write these equations in other units, such as pressure:

$$P_2 + \rho gh_2 + \frac{1}{2}\rho <v_2>^2 = P_1 + \rho gh_1 + \frac{1}{2}\rho <v_1>^2 \tag{5.74}$$

Or, in the units of height or head where each term has a specific meaning:

$$\underbrace{\frac{P_2}{\rho g}}_{\substack{\text{Pressure} \\ \text{(or Static)} \\ \text{Head}}} + \underbrace{h_2}_{\substack{\text{Elevation (or} \\ \text{Potential) Head}}} + \underbrace{\frac{1}{2g} <v_2>^2}_{\substack{\text{Velocity (or Dynamic)} \\ \text{Head}}} = \frac{P_1}{\rho g} + h_1 + \frac{1}{2g} <v_1>^2 \tag{5.75}$$

The sum of the three components is always constant, i.e., energy is conserved, ignoring the frictional losses. It is simple and practically very handy to approximate the balance between the pressure of a flowing fluid of low viscosity and its velocity and elevation. Many devices of utility ranging from aspirators, Bunsen burner, paint sprayers, venturi masks, etc., to thermocompressors, venturi-based mixers and airplane wing design are all based on the Bernoulli principle.

The velocity head expression identified in Eq. (5.75) is frequently utilized in many applications of direct interest to us, including sizing holes in a sparger and orifice opening, calculating leakage through small holes, and relating energy losses due to flow through valves and fittings.

15. Show that the equivalent head corresponding to a flow velocity of 5 m/s is 1.27 m. ◄

Example

19. *Estimate the time needed to drain a large tank (1.0 m id and 3 m high, open to the atmosphere) filled with water using a tap (5 cm id) located at the bottom of the tank, Fig. 5.41.*

   Solution: Assuming negligible friction loss, incompressible fluid and neglecting the velocity in the large tank since its diameter is too large compared to the tap diameter, the Bernoulli equation for the system between points 1 and 2 in Fig. 5.41 reduces to:

$$\frac{1}{2g} <v_2>^2 = h_1$$

Since $P_1 = P_2 = $ atmospheric pressure
$h_1 = $ datum and v_1 assumed negligible
By writing a mass balance on the rate of flow, we obtain

$$-A_1 \frac{dh}{dt} = <v_2> A_2$$

Substituting $<v_2>$ in terms of h_1 from Eq. (i), we obtain

$$-A_1 \frac{dh}{dt} = A_2 \sqrt{2gh_1} \ or -\frac{dh}{\sqrt{h}} = \frac{A_2}{A_1} \sqrt{2g} \, dt$$

Integrating the above into the limits of 3 and 0, we obtain

$$\int_3^0 -\frac{dh}{\sqrt{h}} = \frac{A_2}{A_1} \sqrt{2g} \int_0^t dt \ or \ 2\sqrt{h} = \frac{A_2}{A_1} \sqrt{2g} \, t$$

Fig. 5.41 Draining a tank

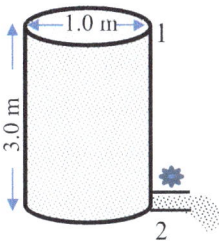

$$\text{or } t = \frac{A_1}{A_2}\frac{2\sqrt{h}}{\sqrt{2g}} = \frac{D_1^2}{D_2^2}\frac{2\sqrt{h}}{\sqrt{2g}} = \frac{(1m)^2}{(0.025)^2}\frac{2\sqrt{3m}}{\sqrt{2\left(9.81\frac{m}{s^2}\right)}} = 312.8\ s \qquad \blacktriangleleft$$

(D) Fully Developed Flow and Entrance Length

The velocity of a fluid entering a pipe from a reservoir will be nearly uniform (in **plug flow**), but as the fluid moves forward, viscous effects cause it to stick to the pipe wall (the **no-slip** condition), and a velocity gradient develops. The fluid in contact with the pipe wall has zero velocity, and a retarding shear stress on the fluid establishes a region of slower-moving fluid near the wall, called a **boundary layer**. The region of slower-moving fluid grows wider as the flow moves further downstream and the boundary layer expands to fill the entire pipe. To maintain the same average velocity (constant mass flow rate) at all cross-sections, the flow in the center (core flow) must accelerate as it moves downstream. The flow is fully developed when the shape of the velocity profile becomes the same at all cross-sections and flow characteristics do not change along the pipe length (Fig. 5.42). The shape of the velocity profile is discussed in details in the next section.

The distinction between laminar and turbulent flow applies only to the fully developed flow conditions. The entrance length (Le) is the distance from the entry point to where the flow is fully developed. The entrance length depends on whether the flow is laminar or turbulent and is related to the Reynolds number and the pipe diameter (d) as follows:

$$\text{Laminar flow} : (L_e/d) = 0.06\ \text{Re} \tag{5.76}$$

The longest practical entrance length corresponding to laminar flow (Re $= 2100$) is thus
$(L_e/d)\text{max} = 125.$

$$\text{Turbulent flow} : (L_e/d) = 4.4\ (\text{Re})^{1/6} \tag{5.77}$$

In most applications, Re ranges from 10^4 to 10^5, and on average, $(L_e/d) = 25$

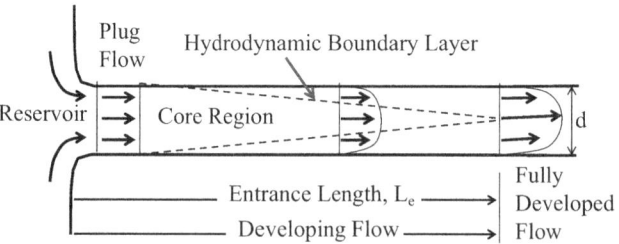

Fig. 5.42 Fully developed flow inside a pipe

(E) **Momentum Balance in Pipe Flow**

Like mass and energy balances, a macroscopic momentum balance can also be formulated on a control volume, as shown in Fig. 5.39, where a fluid of mass m flowing at velocity v will carry a momentum equal to mv. However, we are more interested in understanding what happens inside the control volume at a microscopic level. To accomplish this, we perform a differential momentum balance on a tiny size of the control volume, combine it with the concepts of viscosity and then provide answers to the four very important questions of engineering utility in designing many important food processing operations. They are (i) how is the shear stress (momentum flux) distributed within the flow field, (ii) what does the fluid flow profile look like and what is the maximum velocity, (iii) how can the pressure (energy) loss be estimated due to frictional resistance in the flow path, and finally (iv) how is the average velocity related to the maximum velocity? To accomplish this, consider a steady-state, one-dimensional, well-developed laminar flow of a Newtonian fluid inside a pipe, as shown in Fig. 5.43.

- **Shear stress or momentum flux distribution** during flow is obtained by writing the equilibrium force balance on the cylindrical control volume of fluid concentric with the pipe having length L and radius r, as shown in Fig. 5.43. The motion of the cylinder results from a difference in pressure (ΔP) at points 1 and 2, which in terms of net force acting in the x direction is given by:

$$F_x = \pi r^2 (\Delta P) \tag{5.78}$$

However, the motion of the fluid cylinder is resisted by a friction force acting on the surface and given by:

$$F_r = \tau_{rx}(2\pi r L) \tag{5.79}$$

At equilibrium, $F_x = F_r$, so equating the above two equations, we obtain:

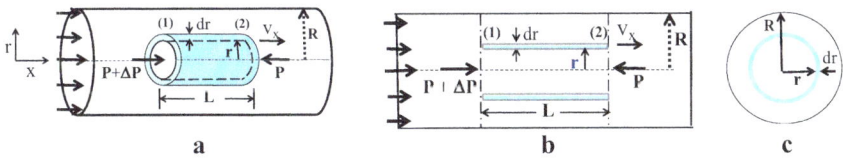

Fig. 5.43 Fluid flow and force balance on a control volume. (a) Inside a pipe (b) Front view (c) End view

$$\tau_{rx} = \left(\frac{\Delta P}{2L}\right) r \qquad (5.80)$$

Since at the wall of the tube, $r = R$, Eq. (5.80) is modified to obtain the shear stress at the wall as:

$$\tau_w = \left(\frac{\Delta P}{2L}\right) R \qquad (5.81)$$

Eliminating $(\Delta P / 2\,L)$ from Eqs. (5.80) and (5.81), we obtain:

$$\tau_{rx} = \tau_w \left(\frac{r}{R}\right) \qquad (5.82)$$

Equation (5.82) indicates a linear distribution of shear stress or momentum flux in the pipe. It is zero at the center ($r = 0$) and maximum at the wall ($r = R$), and the resulting distribution or profile is shown in Fig. 5.44.

- **The velocity profile** during flow of a fluid is obtained by recognizing that momentum flux during flow is related to viscosity as discussed earlier and expressed as:

$$\tau_{rx} = -\eta \frac{dv_x}{dr} \qquad (5.83)$$

The minus sign is used since the fluid velocity decreases as radius r increases. Rewriting the above equation in terms of velocity and substituting τ_{rx} in terms of ΔP, r and L from Eq. (5.80), we obtain:

Fig. 5.44 Shear stress and velocity profile for laminar flow in pipes

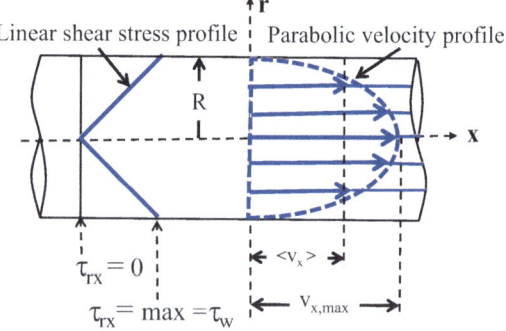

$$dv_x = -\left(\frac{\tau_{rx}}{\eta}\right)dr = -\left(\frac{\Delta P}{2L}\right)\left(\frac{r}{\eta}\right)dr \tag{5.84}$$

Now integrating the above equation and using the boundary condition at the wall, when $r = R$ and v_x is zero, to determine the constant of integration yields the equation for the velocity distribution as follows:

$$\int_{v_x}^{0} dv_x = -\left(\frac{\Delta P}{2\eta L}\right)\int_{r}^{R} rdr \text{ or } v_x = \frac{\Delta PR^2}{4\eta L}\left[1 - \left(\frac{r}{R}\right)^2\right] \tag{5.85}$$

By examining the above equation, the following important conclusions can be drawn:

- All the terms are constant except v_x and r, and the velocity in the x direction (v_x) varies as a square of the radial position (r), which represents an equation of a parabola. Thus, the velocity profile is parabolic, as shown in Fig. 5.44.
- At $r = R$, $v_x = 0$, i.e., velocity is zero at the wall.
- At $r = 0$, $v_x = v_{x,\max}$, and Eq. (5.85) gives:

$$v_{x,\max} = \frac{\Delta PR^2}{4\eta L} \tag{5.86}$$

• The pressure (energy) loss due to frictional resistance in the flow path is estimated beginning with the expression that defines average velocity and applying it to the small control volume shell of thickness dr and area $dA = 2\Pi rdr$, as shown in Fig. 5.43. It is then combined with the flow profile equation and integrated to obtain the final expression that relates pressure loss (ΔP) to average velocity ($<v>$), as shown below:

$$<v_x> = \frac{1}{\pi R^2}\int_{0}^{R} v_x(2\pi rdr) \tag{5.87a}$$

Substituting Eq. (5.85) in (5.87a):

$$<v_x> = \frac{1}{\pi R^2}\int_{0}^{R} \frac{\Delta P(R^2 - r^2)}{4\eta L}(2\pi rdr) = \frac{\Delta P}{2\eta LR^2}\int_{0}^{R}(R^2 - r^2)rdr \tag{5.87b}$$

Integrating the above equation between the indicated limits yields:

$$<v_x> = \frac{\Delta P}{2\eta L R^2} \left(\frac{1}{4}R^4\right) \tag{5.88}$$

and simplifying it gives the final equation that we are looking for:

$$<v_x> = \frac{\Delta P R^2}{8\eta L} \tag{5.89}$$

The above expression is known as the **Hagen-Poiseuille equation**. It relates the average velocity (Volumetric flux) to the pressure drop, fluid viscosity and pipe dimensions. It is used in several different applications, including the following:

– Experimental determination of viscosity, given pressure drop, flow rate, and tube (pipe) dimensions. This is generally done using a capillary tube to minimize experimental error in pressure measurement and maximize accuracy since, according to Eq. (5.89), the pressure drop is inversely related to the square of the tube radius.
– Estimating flow rates using pressure data, as shown below:

$$\dot{V} = \langle v \rangle A = \langle v \rangle \pi R^2 = \frac{\Delta P \pi R^4}{8\eta L} \tag{5.90}$$

• Finally, how the **average ($<v_x>$) and maximum ($v_{x,\text{max}}$) velocities** of a Newtonian fluid in a fully developed laminar flow are related can be easily deduced if we divide Eq. (5.86) by Eq. (5.89):

$$\frac{v_{x,\text{max}}}{<v_x>} = \frac{\left(\Delta P R^2\right)/(4\eta L)}{\left(\Delta P R^2\right)/(8\eta L)} = 2 \tag{5.91}$$

The maximum velocity is twice the average velocity, and this is a very useful finding for designing continuous flow operations in food manufacturing involving laminar flow of food showing Newtonian behavior. It becomes more complicated with turbulent flows. In such cases, it is estimated using empirical correlations such as a generally used expression, known as the Blasius 1/7th power law, shown below:

$$\frac{v_{x,\text{max}}}{<v_x>} = \left(1 - \frac{r}{R}\right)^{-1/7} \tag{5.92}$$

For turbulent flow with Re $> 10^5$, a reasonable approximation generally used is:

$$\frac{v_{x,\text{max}}}{<v_x>} = 1.2 \tag{5.93}$$

For non-Newtonian, power law fluids, the situation becomes even more complex. A graphical solution, shown in Fig. 5.45, is very helpful in relating average and maximum velocities, both in laminar and turbulent flows, as function of the generalized Reynolds number. For the laminar flow, however, it can also be estimated using the below equation:

$$\frac{v_{x,max}}{<v_x>} = \frac{3n+1}{n+1} \qquad (5.94)$$

where n is the flow behavior index of the fluid.

Practical Implications of Velocity Profiles: The velocity profile of fluids in pipe flow is quite complex. A simplified schematic of the laminar flow velocity profiles as a function of radius in a pipe is shown in Fig. 5.46. With the aid of Eq. (5.94), let us look at what happens to the ratio of maximum velocity to average velocity for Newtonian and non-Newtonian fluids in laminar flows:

• Dilatant fluids, $n > 1$

$\frac{v_{x,max}}{<v_x>} > 2$, the velocity profile is more like an elongated parabola, Fig. 5.46a.

• Newtonian Fluids, $n = 1$

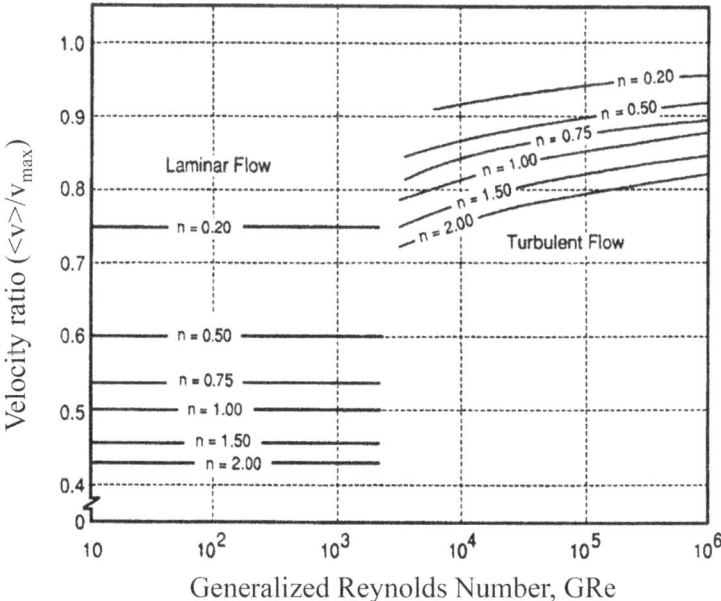

Generalized Reynolds Number, GRe

Fig. 5.45 Velocity ratio—Generalized Reynolds number correlations for power law fluids. (From V. A. Jones and J. A. Palmeri [1]. With permission)

Fig. 5.46 Axial velocity profiles of Newtonian and non-Newtonian fluids, and plug flow in a pipe

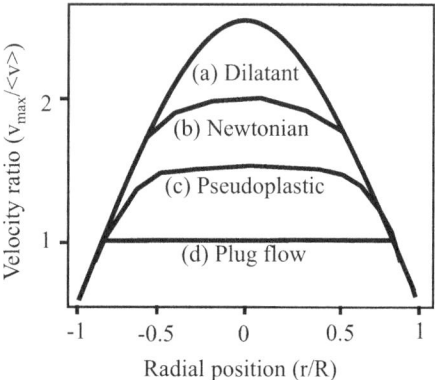

$\frac{v_{x,max}}{<v_x>} = \frac{3(1)+1}{1+1} = 2$, and the result is similar to what is shown in Eq. (5.91), i.e., in the parabolic velocity profile, the centerline velocity is twice that of the average velocity, as shown in Fig. 5.46b.

- Pseudoplastic, $n < 1$
  $\frac{v_{x,max}}{<v_x>} < 2$, the velocity profile mimics a flatter parabola, approximating the turbulent flow profile, Fig. 5.46c

Plug flow, Fig. 5.46d, occurs when all streams have the same velocity. It occurs due to slippage at the flow boundary.

Heat processing of liquid foods such as milk, fruit juices, liquid eggs, wines, beer, etc., on large commercial operations is performed in indirect contact heat exchangers with the food continuously flowing on one side and the heating medium such as hot water on the other side. In such situations, the greater the velocity difference between fluid streams at the center and at the wall, the greater the temperature difference in the food streams and the higher the quality difference. A flatter profile is highly desirable for more uniform heating and cooling operations.

Flash or high-temperature short-time (HTST) pasteurization is a continuous process that involves heat treatment of foods sufficient to kill pathogenic bacteria that are hazardous to human health and to reduce or eliminate spoilage microorganisms for shelf-life extension of the product. The process requires heating every particle of the product in a properly designed and operated pasteurization system to a specified temperature and holding at that temperature for the correspondingly specified time. The temperature and time requirements vary with the type of food. Some typical temperature–time requirements for HTST pasteurization of a few selected liquid products are given in Table 5.3.

For example, pasteurization of milk requires that all particles of the milk be heated to a temperature of at least 72 °C and then held continuously at that temperature for a minimum of 15 seconds. This is accomplished by heating the product to the desired temperature in a heat exchanger and then passing it through a holding tube at a rate such that the fastest moving particle will traverse the holding

Table 5.3 Recommended temperature and time for HTST pasteurization of selected products

Product	Temperature (°C)	Time (s)
Beer	72	20
Ice cream mix	80	20
Liquid eggs	80	25
Milk	72	15
Soft drinks	95	10

tube in the required holding time. Another heat exchanger is then used to cool the product down before it goes for packaging, as illustrated in Fig. 5.47.

The role of the velocity profile becomes critically important in the design of such continuous processing systems. We learned that all fluid streams in a pipe flow do not move at the same velocity and that a velocity distribution exists depending on the product characteristics and the type of flow. For example, in the case of a Newtonian fluid in laminar flow, the velocity of the stream at the center is maximum, and it is twice that of the average velocity. Therefore, for such a fluid, the theoretical length of the holding tube (L) needed to ensure the required hold time (t) must be based on the fastest traveling stream. It is predicated on the fact that if the fastest traveling stream gets the desired time-temperature treatment, then the rest of the streams will get more and that would add to the margin of safety of the process. The length of the holding tube required for a given duty is thus calculated as follows:

$$Holding\ tube\ length\ (L) = (v_{max})(t) \tag{5.95}$$

where $v_{max} = 2 <v>$, and the average velocity, $<v>$, may be obtained from mass ($\dot{m}$) or volumetric flow rate ($\dot{V}$) provided the pipe inside diameter (i.d.) and product density (ρ) values are available. The calculated length is then experimentally confirmed by measuring the time for an added trace substance (salt) to pass through the holding tube.

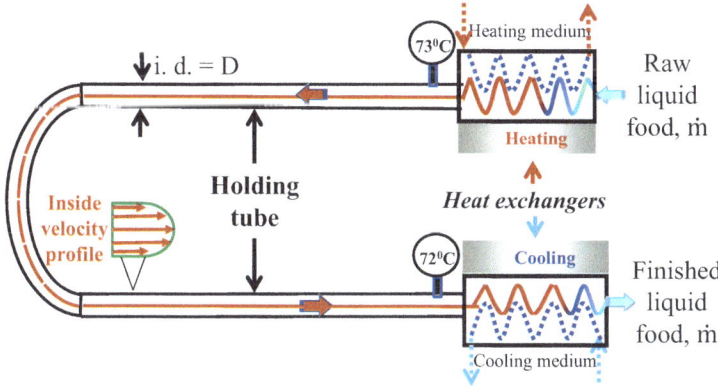

Fig. 5.47 Basics of a high-temperature short-time (HTST) pasteurization system

Example

20. *Compute the holding tube length needed for pasteurization of a Newtonian behaving beverage ($\rho = 1150$ kg/m³, $\eta = 3.0$ cP) at 80 °C for 20 s. The flow rate was 1000 kg/h, and the inside diameter of the holding tube was 6 cm.*
Solution: *we know that $\dot{m} = \rho <v> A$, then*

$$<v> = \frac{4\dot{m}}{\rho \pi D^2} = \frac{4\left(1000\frac{kg}{h} \cdot \frac{h}{3600s}\right)}{1150\frac{kg}{m^3}(3.14)(0.06m)^2} = 0.085\frac{m}{s} \text{ and}$$

$$Re = \frac{\rho D <v>}{\eta} = \frac{\left(1150\frac{kg}{m^3}\right)(0.06m)\left(\frac{0.085m}{s}\right)}{3x\ 10^{-3}\frac{kg}{m.s}} = 1955, \text{laminar flow,}$$

$$v_{max} = 2<v> = 2\left(0.085\frac{m}{s}\right) = 0.170\ m/s$$

The holding tube length, $L = (v_{max})(t) = (0.170)(20\ s) = 3.40\ m$ ◀

(F) Friction Loss in Fluid Flow

In the mechanical energy balance, Eq. (5.65), the term $\sum F$ was included to account for the loss of all energy due to friction, but nothing was said about the nature of this term. It is made up of various components of pipeline systems used in the processing and conveying of fluids, as shown in Fig. 5.48. The total energy loss depends on several factors, such as the fluid's viscosity and rate of flow as well as the diameter and roughness of the pipeline and whether the flow is laminar or turbulent. To explore how energy is dissipated *by friction in each of the four components of a pipeline set up identified in* Fig. 5.48 *and how to estimate it, we will begin with friction loss in pipes.*

$$\frac{P_1}{\rho} + gh_1 + \frac{1}{2\alpha}<v_1>^2 + \bar{W}_s = \frac{P_2}{\rho} + gh_2 + \frac{1}{2\alpha}<v_2>^2 + \left(\sum F\right)$$

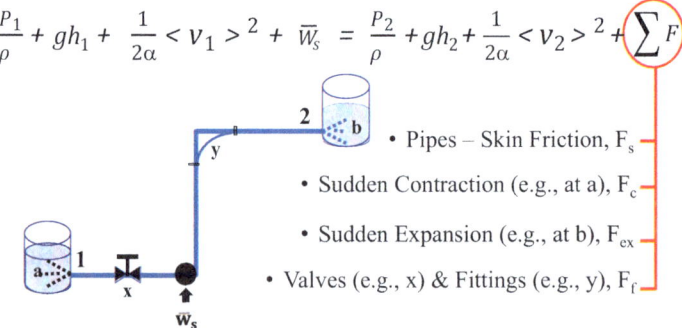

- Pipes – Skin Friction, F_s
- Sudden Contraction (e.g., at a), F_c
- Sudden Expansion (e.g., at b), F_{ex}
- Valves (e.g., x) & Fittings (e.g., y), F_f

Fig. 5.48 Mechanical energy balance and friction loss in pipeline systems

- For the mechanical energy loss in pipes due to friction, known as **skin friction** (F_s), consider a fluid of constant density flowing steadily in a horizontal pipe of uniform internal diameter, as shown in Fig. 5.49. Applying the MEB between locations 1 and 2, we obtain:

$$\frac{P_1}{\rho} + gh_1 + \frac{1}{2\alpha} <v_1>^2 + \overline{w}_s = \frac{P_2}{\rho} + gh_2 + \frac{1}{2\alpha} <v_2>^2 + F_s \qquad (5.96)$$

However, $<v_1> = <v_2>$ from the continuity equation; $h_1 = h_2$, horizontal pipe and $W_s = 0$, no mechanical energy added. Eq. (5.96) reduces to:

$$F_s = \frac{P_1 - P_2}{\rho} = \frac{\Delta P}{\rho} \qquad (5.97)$$

The above expression indicates that when a fluid flows in a pipe, the observed pressure drop along the pipe length is due to friction. It represents the mechanical energy that is irreversibly converted to thermal energy. This is only a part of the ΣF term in the MEB equation and is frequently a major component of the total loss of energy due to friction. For **laminar flow,** as discussed earlier, the Hagen-Poiseuille Eq. (5.89) relates the pressure drop to the average velocity and thus can also be rearranged to estimate the energy loss due to skin friction as follows:

$$F_s = \frac{\Delta P}{\rho} = \frac{8\eta L <v>}{\rho R^2} \qquad (5.98)$$

For turbulent flow, however, no such formulation as shown above is available to estimate Fs. A new generic parameter called the **Fanning friction factor (f)** *as defined* below is used for both turbulent and laminar flows.

$$f = \frac{\text{Drag force/wetted area}}{(\text{Density})(\text{Velocity head})} = \frac{\tau_w}{\frac{\rho<v>^2}{2}} = \frac{[(\Delta P)\pi R^2]/2\pi RL}{\frac{\rho<v>^2}{2}}$$

$$= \frac{R\Delta P}{L\rho <v>^2} \qquad (5.99)$$

Thus, for $D = 2R$

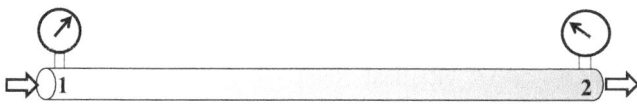

Fig. 5.49 Flow in a straight pipe and skin friction

$$F_s = \frac{\Delta P}{\rho} = 4f\left(\frac{L}{D}\right)\frac{<v>^2}{2} \tag{5.100}$$

The above equation, known as the **Fanning equation,** applies to any flow regime, turbulent or laminar. All we need is now some ways to estimate f, the Fanning friction factor. For **laminar flow** of Newtonian fluids, combining Eq. (5.99) and the Hagen-Poiseuille Eq. (5.89) yields:

$$f = \frac{16\eta}{\rho D <v>} = \frac{16}{Re} \tag{5.101}$$

The above equation indicates that the Fanning friction factor, f, depends on the Reynolds number only and can be easily calculated and used in Eq. (5.100) to estimate the friction loss in pipe flows.

For the **turbulent flow** of Newtonian fluids, however, in addition to Re, the friction factor f depends on the roughness of the pipe surface. Based on experimental data, a graphical friction factor chart as a function of the Re and relative surface roughness (ϵ/D), where ϵ is the average irregularity or depth of roughness, was prepared. Known as the Moody diagram, it is shown in Fig. 5.50. A similar chart is also available for non-Newtonian fluids flowing in a smooth pipe as a function of the GRe, Fig. 5.51.

A similar chart is also available for non-Newtonian fluids flowing in a smooth pipe as a function of the GRe, Fig. 5.51.

- The loss of energy due to **sudden contraction (F_c)** occurs when the flow cross-section changes abruptly, and the formation of eddies causes additional frictional loss. It is given as:

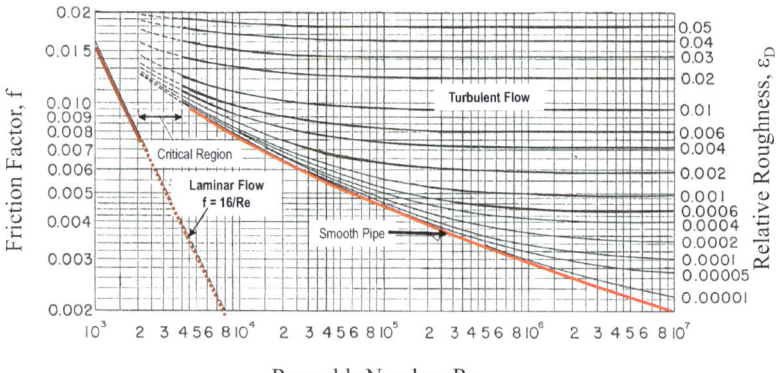

Fig. 5.50 Friction factors for incompressible Newtonian fluids in pipe flow (From L. F. Moody [2]. With permission)

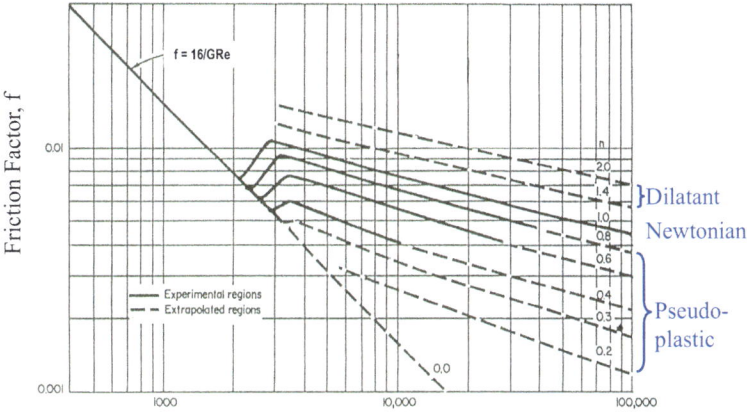

Generalized Reynolds Number, GRe

Fig. 5.51 Friction factors for non-Newtonian fluids in smooth pipe flow. (From A.B. Metzner and D.W. Dodge [3]. With permission)

$$F_c = 0.55\left(1 - \frac{A_2}{A_1}\right)\frac{<v_2>^2}{2\alpha} = K_c\frac{<v_2>^2}{2\alpha} \tag{5.102}$$

where $K_c = 0.55\left(1 - \frac{A_2}{A_1}\right)$ is the contraction coefficient. When $A_1 >>> A_2$, $K_c = 0.55$

- Similarly, for **sudden expansion (F_{ex})**, such as when a fluid suddenly enters a large diameter pipe or a tank, the jetting effect causes eddies to form, and loss of energy occurs. The friction loss in such cases can be calculated as:

$$F_{ex} = \left(1 - \frac{A_1}{A_2}\right)^2\frac{<v_1>^2}{2\alpha} = K_{ex}\frac{<v_1>^2}{2\alpha} \tag{5.103}$$

where $K_{ex} = \left(1 - \frac{A_1}{A_2}\right)^2$ is the expansion coefficient. When $A_2 >>> A_1$, $K_{ex} = 1$ and $F_{ex} = \frac{<v_1>^2}{2\alpha}$ (or velocity head)

In both Eqs. (5.102) and (5.103), α is 0.5 for laminar flow and 1.0 for turbulent flow, and the velocity used is the smaller area velocity (see Fig. 5.52).

- Pipe fittings such as valves, elbows, couplings, tees, etc. (see Fig. 5.55) resist flow and cause additional friction losses. These become significant when several of these are employed in short piping systems. The energy loss due to valves and fittings is estimated as follows:

$$F_f = K_f\frac{<v>^2}{2} \tag{5.104}$$

Fig. 5.52 Energy loss on
sudden contraction and
expansion

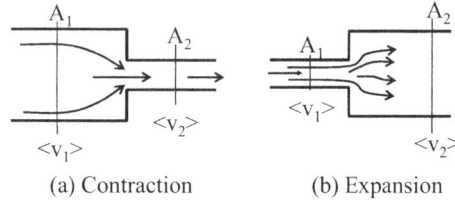

(a) Contraction   (b) Expansion

Table 5.4 Friction loss for turbulent flow through selected fittings and valves

Type of fitting or valve	Frictional loss (number of velocity heads) K_f
Elbow, 45°	0.35
Elbow, 90°	0.75
Bend, 180°	1.5
Tee	1
Return bend 180°	1.5
Coupling and Union	0.04
Gate valve	
Wide open	0.17
Half open	4.5
Globe valve	
Wide open	6.0
Half open	9.5

where K_f is the fitting loss coefficient and the experimental values for many types
of fittings and valves are given in Table 5.4, and $<v>$ is the average velocity
leading to the fitting or valve.

In summary, the sum of all frictional losses, $\sum F$, due to the straight pipe, sudden
contraction and expansion, and fittings and valves can be combined such that:

$$\sum F = 4f \left(\frac{L}{D}\right) \frac{<v>^2}{2} + K_c \frac{<v_2>^2}{2\alpha} + K_{\exp} \frac{<v_1>^2}{2\alpha} + K_f \frac{<v>^2}{2} \quad (5.105)$$

In the special case when the pipe diameter is constant and all the velocities are the
same, the above equation may be simplified to:

$$\sum F = \left(4f \left(\frac{L}{D}\right) + \frac{K_c}{\alpha} + \frac{K_{exp}}{\alpha} + K_f\right) \frac{<v>^2}{2} \tag{5.106}$$

Example

21. *Apple juice ($\rho = 1200$ kg/m³,$\eta = 1.2$ mPa·s) is pumped through a 50-mm internal diameter smooth pipe between two tanks at a steady rate of 5 kg/ s using the piping network shown in* Fig. 5.53. *The tanks are open to atmospheric pressure, and the level in each tank remains constant. Estimate the pumping power needed.*

Solution: *Writing MEB on the system and applying the given conditions, we obtain:*

Then, $\overline{w}_s = g\,(h_2 - h_1) + \Sigma F$

And $\Sigma F = F_c + F_s + F_{ex} + F_f$

Then, $\Sigma F = \left(K_c + 4f\frac{L}{D} + K_{ex} + K_f\right)\frac{<v>^2}{2}$

$$<v> = \frac{\dot{m}}{\rho A} = \frac{4\dot{m}}{\rho \pi D^2}$$

$$= [(4)(5\ kg/s)]/\left[\left(1200\frac{kg}{m^3}\right)(3.14)\left(50x\ 10^{-3}m\right)^2\right] = 2.12\ m/s$$

$$Re = \frac{\rho D <v>}{\eta} = \left(1200\ \frac{kg}{m^3}\right)\left(50x\ 10^{-3}m\right)\left(2.12\frac{m}{s}\right)\Big/\left(1.2\ x\ 10^{-3}\frac{m.s}{kg}\right)$$

$$= 1.5\ x\ 10^5 \Rightarrow Turbulent\ flow$$

For the above Re, the Moody diagram, Fig. 5.50, *for the smooth pipe shows the value of the Fanning friction factor (f) as 4.0 x 10⁻³. The other terms for the total friction are:*

$K_c = 0.55(1 - A_2/A_1) = 0.55$, large tank; $K_f = (0.75)\,(2) + 0.17 = 1.67$, two 90° *elbows and one fully open gate valve;* $K_{ex} = (1 - A_1/A_2)^2 = 1$

Then,

$\Sigma F = [(0.55) + (4)(4.0 \times 10^{-3})(5\ m + 2.5\ m + 0.5\ m)/(50 \times 10^{-3}m) + 1 +$ $1.67]\,(2.12\ m/s)^2/2.58\ m^2/s^2$

Fig. 5.53 Process schematic

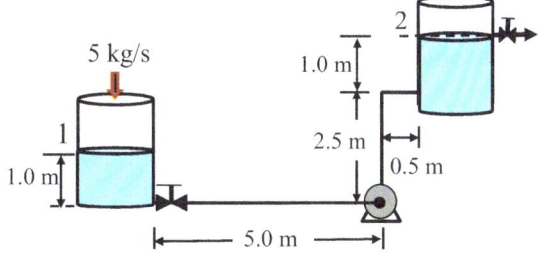

5 kg/s

1.0 m

2

1.0 m

2.5 m

0.5 m

1.0 m

5.0 m

$$\overline{w}_s = g\,(h_2 - h_1) + \Sigma F = \left(9.81\,\frac{m}{s^2}\right)(2.5\ m) + \left(2.58\,\frac{m^2}{s^2}\right) = 27.2\,\frac{m^2}{s^2}$$

$$Power,\, P = \overline{w}_s.\dot{m} = \left(27.2\,\frac{m^2}{s^2}\right)\left(5\,\frac{kg}{s}\right) = 136.0\ W \qquad \blacktriangleleft$$

(G) Pipes, Tubes, and Fittings

Pipes and tubes along with fittings and valves are extensively used for the transportation and processing of fluids. They are fabricated to meet the high hygienic standards of regulatory agencies, and various standards and specifications have evolved over the years. Stainless steel is the preferred material for food contact, and iron and copper are also used for the transport of water, steam, and other ancillary materials.

Tubes are thin-walled, cylindrical hollow conduits whose sizes are specified based on the outside diameter and wall thickness (gauge). In most applications, 16-gauge tubing is generally used in low-pressure systems. Pipes are thicker walled than tubes, and their size has a reference to a "nominal" (not actual) diameter. The term **sanitary pipe** (also called **O D tubing**) used in the food industry refers to stainless-steel tubing, where the size designation is based on the outside diameter (Fig. 5.54).

Iron pipes are designed for working under various pressures and are thus classified according to wall thickness in terms of **schedule number**. The schedule number, a dimensionless number, is estimated by a standard formula developed by the American Standard Association as shown below:

Schedule number

$$= \left(\frac{Internal\ working\ pressure}{Allowable\ stress\ the\ material\ can\ withstand\ under\ operation\ condition}\right)$$

$$\times\,1000$$

$$(5.107)$$

Table 5.5 shows the dimensions for **"50 mm" nominal diameter** iron pipes of different schedule numbers. Note that all have an outer diameter of 60.33 mm, but the inner diameter depends on the designed strength. A higher schedule number means higher pipe wall thickness.

Fig. 5.54 Size nomenclature for pipes and tubes

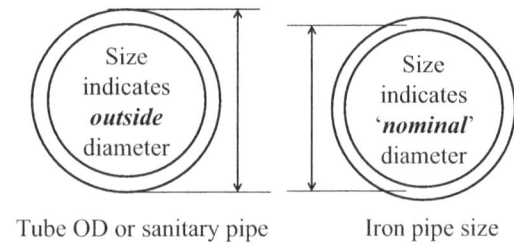

Tube OD or sanitary pipe   Iron pipe size

Table 5.5 Nominal 50 mm diameter iron pipe dimensions

Schedule number	Inner diameter (mm)	Wall thickness (mm)
5 S	57.03	1.65
10 S	54.79	2.77
80 S	49.25	5.54
160 S	42.85	8.74

Table 5.6 Steel pipe and sanitary pipe dimensions

Nominal size (in.)	Steel pipe (Sch. 40)		Sanitary pipe	
	ID in./(m)	OD in./(m)	ID in./(m)	OD in./(m)
0.5	0.622 (0.01579)	0.840 (0.02134)	–	–
0.75	0.824 (0.02093)	1.050 (0.02667)	–	–
1	1.049 (0.02664)	1.315 (0.03340)	0.902 (0.02291)	1.00 (0.0254)
1.5	1.610 (0.04089)	1.900 (0.04826)	1.402 (0.03561)	1.50 (0.0381)
2.0	2.067 (0.05250)	2.375 (0.06033)	1.870 (0.04749)	2.00 (0.0508)
2.5	2.469 (0.06271)	2.875 (0.07302)	2.370 (0.06019)	2.5 (0.0635)
3.0	3.068 (0.07793)	3.500 (0.08890)	2.870 (0.07289)	3.0 (0.0762)
4.0	4.026 (0.10226)	4.500 (0.11430)	3.834 (0.09739)	4.0 (0.1016)

ID = inside diameter, OD = outside diameter

Listed in Table 5.6 for comparison are schedule 40 steel pipe and sanitary pipe dimensions in the range of diameters most used in food processing operations.

Several different types of fittings, joint styles, and sealing techniques are used with sanitary pipelines to attach different equipment, change flow directions, reduce pipe size or split flows. Examples include Tri-clamp with Butt Weld, Bevel Seat, and H-Line, among others. A triclamp is the simplest and versatile type of joint design. It uses a gasket and a clamp for sealing the two flanged ends of the fittings. Bevel seat fittings are threaded with hexagonal union nuts and provide more rigid and secure connections. A bevel seat gasket is recommended to eliminate leakage when a metal seat-only connection is used. See Fig. 5.55 for some examples. Food-grade materials such as Teflon (PTFE), Viton, Buna, and silicone are used for gaskets in sanitary pipe joints.

Problems

5.1 Answer the following questions:
   (a) Following tooth extraction, dentists generally recommend not drinking through a straw. Why? Explain.
   (b) What is the main use of high-pressure processing in food science and engineering and why is it advantageous over thermal processing? List 5 foods that can benefit from high-pressure processing and 5 that cannot. Why is high-pressure processing also a thermal processing technique?

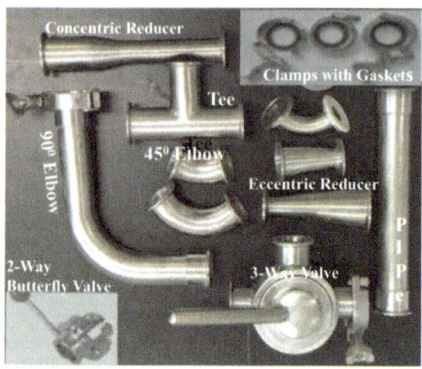
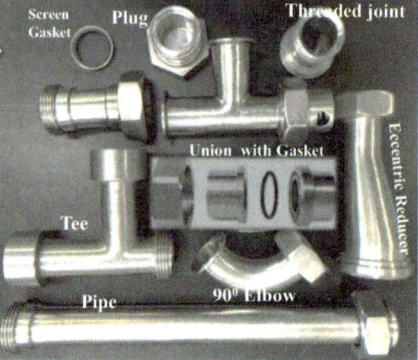

(a) Tri-clamp Buttweld   (b) Bevel Seat

Fig. 5.55 Sanitary fittings and sealing techniques using clamps and unions

5.2 A canning plant uses saturated steam in a continuous hydrostatic retort to process canned fruits, as shown. The retort needs to be maintained at 135 °C. If saturated steam is used as the heating medium, how high will the water head need to be in meters to process the cans? State your assumptions.

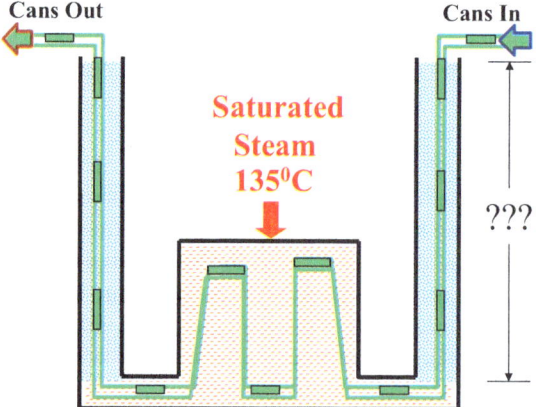

5.3 A U-tube is first half filled with water, and then an oil of unknown density is added to one leg until the system attains the configuration illustrated in the figure. What is the density of the oil?

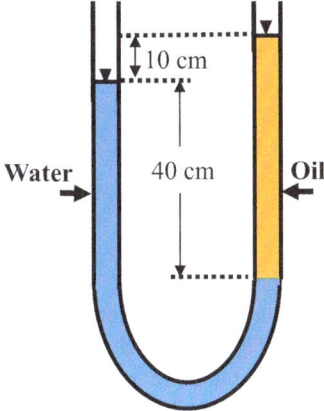

5.4 A U-tube manometer filled with mercury (density 13,600 kg/m^3) and water (density 997 kg/m^3), as shown in the figure, is attached to a water line at point 1. If point 5 is open to the atmosphere, what is the pressure at points 1 and 2?

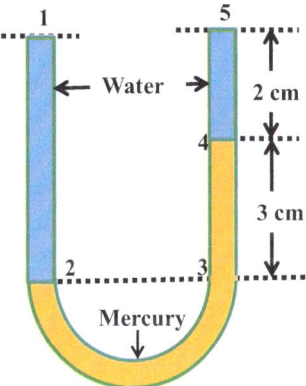

5.5 Brewed tea extract is pumped through a mixing tank and flows in a pipe with an inside diameter of 1.5 inches. If the volumetric flow rate is 7 ft^3/s, what is the velocity and the mass flow rate if the density of the extract is 60 lb$_m$/ft^3?

5.6 Alcohol (S.G. = 0.8) flowing at 8 m/s and water at 12 m/s enter pipes 1 and 2, respectively, and after mixing, thoroughly leave via pipe 3, as shown. Assuming the fluids to be incompressible, compute the exit velocity and density of the mixture exiting pipe 3.

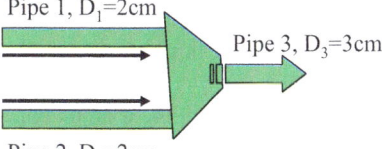

Pipe 1, D$_1$=2cm

Pipe 3, D$_3$=3cm

Pipe 2, D$_2$=2cm

5.7 A water line with an internal radius of 6.5 × 10^{-3} m is connected to a shower head that has 10 holes. The velocity of the water in the line is 1.2 m/s. (a) What is the volumetric flow rate in the line, and (b) at what velocity does the water come out from each hole in the shower head (effective hole radius $= 0.5$ mm) in the head?

1.2 m/s

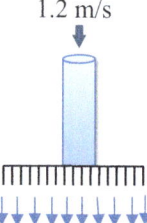

5.8 A manufacturing operation uses two counterrotating rollers operating at different speeds, 10 RPM and 15 RPM, to shell nuts of approximately 1.5 mm in size commercially. The rollers have a diameter of 1 m and are 1.5 cm apart from each other. Calculate the shear rate imposed on the nuts.

5.9 The shear stress vs shear rate data of a material are tabulated below.
  (a) Plot the shear stress vs shear rate and determine the flow behavior exhibited by the material.
  (b) Determine the yield stress and viscosity of the material.

Shear rate (1/s)	Shear stress (Pa)
0.25	9.3
0.63	9.8
0.98	10.2
1.23	10.6
2.5	12.2

5.10 Use the data shown in the table below for a liquid food to graph its shear stress vs. shear rate and determine the product viscosity. Is the fluid Newtonian or non-Newtonian? Why?

Shear stress (Pa)	Shear rate (s^{-1})
0	0
2	6.4
5	16
6	19.2

5.11 The shear stress and shear rate data of a power law liquid food are tabulated below.
  (a) Plot the above data in two different ways: (i) using linear graph paper and (ii) using log-log graph paper.
  (b) From each of the above graphs determine: (i) the values of the consistency coefficient (K) and flow behavior index (n), (ii) the type of flow behavior exhibited by the food, and (iii) the apparent viscosity at a shear

rate of 10 s^{-1} by using the shear stress–shear rate data and the power law model.

Shear stress (Pa)	Shear rate (s^{-1})
0.24	0.1
1.5	1.0
4.09	3.5
9.5	10
34.3	50

5.12 The effect of temperature on the viscosity of a liquid food follows Arrhenius-type behavior, and its experimental data are tabulated below.
   (a) Plot the above data in two different ways: (i) using linear graph paper and (ii) using semilog graph paper
   (b) From each of the above graphs, estimate and tabulate the activation energy and viscosity of the food at 37 °C.

Temp (°C)	Viscosity (cP)
10	44
30	15
50	5.5
80	1.25
95	0.7

5.13 The flow properties of olive oil ($\rho_{\text{olive oil}} = 930$ kg/m^3) are listed as: $\tau_{\text{olive oil}} = 0.025$ Pa, $\dot{\gamma}_{\text{olive oil}} = 0.21$ s^{-1}, $n = 0.85$. Determine (a) the consistency coefficient and (b) the generalized Reynolds number when the oil is flowing at a rate of 1.4 kg/s through a pipe of 3.0 cm internal diameter.

5.14 To produce chocolate-coated strawberries, a strawberry is tested by dipping it into melted chocolate having a density of 1050 kg/m^3 and a yield stress of 14 Pa. Assuming that the strawberry is a cone with a smooth surface, determine the thickness and the mass of the chocolate coating. The coated surface area of the strawberry may be assumed to be 35 cm^2.

5.15 Tomato sauce ($\rho = 1200$ kg/m^3) with a yield stress of 70 Pa is stored in a cylindrical tank of diameter 1 m and height 3 m. Assuming the tank is completely full, estimate: (a) the amount of tomato sauce that will be left on the sidewall of the tank after it is drained and (b) the percent of the total stored volume of the mayonnaise that will be lost?

5.16 Develop expressions for the Reynolds number for the following cases: (a) flow rate is often given as mass flow rate, $\dot{m}$ (kg/s), express the Reynolds number in terms of $\dot{m}$ and (b) flow rate is also given as volumetric flow rate, V (m^3/s), express the Reynolds number in terms of V.

5.17 An unknown fluid ($\rho = 1030$ kg/m^3) flows inside a pipe (I. D.$=10$ cm) at an average velocity of 2 m/s. The viscosity of the fluid is 0.0045 Pa-s. Calculate (a) the Reynolds number for the fluid and determine what kind of flow the fluid exhibits in the pipe and (b) the minimum velocity required to make the flow turbulent.

5.18 In an oil extraction facility, soybean oil flows at 2 L/s through a pipe of 0.127 m I. D. for a horizontal distance of 10 m. Find the pressure differential required across the pipe to maintain the above flow.

5.19 Pasteurization of apple cider ($\eta = 1.5$ cP, S.G. $= 1.01$) requires heating at 71 $^\circ$ C for 20 s. For a continuous flow pasteurization system for this product, compute the length of a 15 cm I. D. holding tube that would be required for a flow rate of 500 gal/h.

5.20 For a concentrated fruit juice ($k = 13$ Pa.s$^{0.75}$, $n = 0.75$, $r = 1560$ kg/m^3) to be pumped through a holding tube (4" I.D.) of a concentric tube pasteurizer at a rate of 1000 kg/min, determine (a) the maximum velocity of the flow and (b) if the fluid is to have a minimum residence time of 15 s in the pasteurizer, determine whether the flow rate is adequate for a holding tube that is 20 m long.

5.21 Water enters a horizontal channel with a rectangular cross-section. The width and height of this channel are 1 m and 0.5 m, respectively. The mass flow rate of water is 1 kg/s.
   (a) What is the equivalent hydraulic diameter of this channel?
   (b) What is the average velocity of water inside the channel?
   (c) Calculate the Reynolds number and indicate if the flow is turbulent or laminar?
   (d) Estimate the entrance length for the flow to be fully developed.

5.22 Orange juice ($\rho_{orange\ juice} = 1085$ kg/m^3) flows at a rate of 350 L/min in a 5 cm-diameter pipe at a pressure of 72 kPa. If the diameter of the pipe changes to 3 cm, calculate the change in pressure. Assume the flow is turbulent and the friction loss is negligible.

5.23 A 55% aqueous sugar solution flows through a tube at an average velocity of 0.6 m/s. The I.D. of the tube is 0.0254 m. The solution has a density of 1260 kg/m^3 and a viscosity of 0.032 Pa·s. Calculate the pressure gradient along the tube.

5.24 Olive oil at 30 °C ($\rho = 920$ kg/m^3, $h = 4.06 \times 10^{-2}$ Pa·s) is pumped through a 5 cm I. D. commercial steel pipe at a flow rate of 0.3 m^3/s. Determine the friction loss in a 15 m long section of the pipe.

5.25 A liquid with a density of 765 kg/m^3 and viscosity of 1.73 cP flows through a horizontal straight pipe at an average velocity of 3.9 m/s. The commercial steel pipe is 2" nominal size, schedule 40, $\in = 4.6 \times 10^{-5}$ m. For a pipe length of 53 m, calculate (a) the friction loss F_s (J/kg) and (b) the additional friction loss if a 90° elbow, two wide open globe valves and a 45° elbow are added to the pipeline (J/kg).

Bibliography

1. Jones VA, Palmeri JA (1976) Prediction of holding times for continuous thermal processing of power law fluids. J Food Sci 41(5):1233
2. Moody LF (1944) Friction factors for pipe flow. Trans ASME 66:671
3. Metzner AB, Dodge DW (1959) Turbulent flow of non-Newtonian systems. AIChE J 5:189
4. Barnes HA, Hutton JF, Walters K (1989) Introduction to rheology. Elsevier, Amsterdam
5. Herschel WH, Bulkley R (1926) Konsistenzmessungen von gummibenzollusungen. Kolloid-Zeitschr 39:291
6. Kittridge CP, Rowley DS (1957) Resistance coefficient for laminar and turbulent flow through one-half-inch valves and fittings. Trans ASME 79:1759
7. Steffe JF (1996) Rheological methods in food process engineering.2nd edn. Freeman Press, East Lansing, MI

Fluid Mechanics: Applications

6

In this chapter, we will focus on some selected applications of fluid mechanics. Food processing and manufacturing operations involve handling various types of fluids, including liquids, gases, steam, and fluidized materials. They follow many of the same laws and exhibit similar resistance to flow and are thus grouped together for analysis and process design purposes. Based on their unifying theory, engineering operations are often categorized into unit operations such as heating, cooling, drying, sheeting, and homogenization. The fluid flow properties and design of the handling and transport system in each unit operation and how fluids are moved from one operation to another control the resulting quality, safety, and consistency of the final products. Thus, a good understanding of the design considerations is needed. Although specific requirements depend on the product characteristics and the processes involved, elimination of dead spots where microorganisms could grow, ease of clean-in-place (CIP) and overall operational efficiency apply to all food handling systems. In Chap. 5, we discussed the basics of the flow of fluids. In this chapter, we will learn how those principles are utilized in selecting and setting up equipment and pipeline networks that are used to process and transport fluids such as liquid foods and related materials.

6.1 Pumps and Pumping Systems

Fluid handling and transport systems consisting of pipelines, valves, fittings, and tanks are used to move fluids in industrial operations, including food processing and manufacturing. If flow cannot be achieved by gravity alone, pumps are needed. Pumps are one of the most ubiquitous mechanical devices around us. For example, our heart is a pump. Our automobiles have a fuel pump, a coolant pump, a steering pump, a windshield pump, and transmission oil pump. A network of pipes and pumps transport water to our homes that we take for granted. It is estimated that approximately 20% of all the world's electrical energy is used in industrial pumping operations. In general, liquids are transported by pumps, while fans, blowers, and compressors are used to move gases. Liquids containing particulate materials such as soups and slurries and liquid–gas mixtures often require special handling systems.

The basic components needed to design a pumping system include the following:

- Types of pumps: Centrifugal (also called dynamic) or positive displacement pumps
- Types of drive elements: electric motors, pneumatic system, or engines
- Types of pipes, valves, fittings, and instrumentation to carry and control the fluid

The final choice depends on the fluid flow rate for the task to be performed and the viscosity, density, temperature, and energy needed (expressed in the units of pressure head) to meet the process requirements. While the mass and energy balances determine the flow rate and temperature, to account for possible variations in the process, a safety margin of 5–20% is generally added to the material-balance flow rate.

In food processing operations where cleanliness and safety are mandated, sanitary (easily cleanable) pipes, fittings, and pumps are used. Their sanitation standards are specified by local and federal regulatory agencies. Various private organizations also provide voluntary standards and guidance for good manufacturing practices to ensure compliance with the required standards.

6.1.1 Pumps: Types and Description

Pumps are basically devices designed to convert electrical/mechanical energy into pressure energy. They are classified based on the principle by which they add energy to the fluid and fall into two major categories, which are further divided into subcategories (Fig. 6.1).

A. **Positive Displacement Pumps:** Also known as PD pumps, they add energy periodically by doing work on fluid-containing volumes to transfer them from the

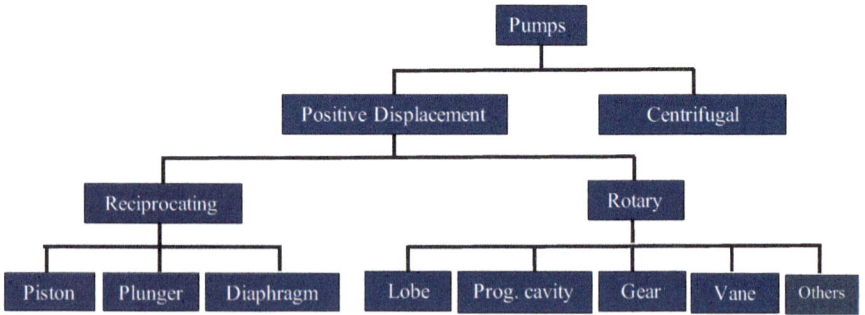

Fig. 6.1 Classification of pump types

suction to the discharge port at a constant rate, driven by pistons, diaphragms, screws, gears, lobes, vanes, etc.

- **Reciprocating positive displacement pumps** are also metering pumps since they deliver a precise volume of fluid per unit time, independent of the discharge pressure. They work by the repeated back-and-forth movement (strokes) of either a piston, plunger, or diaphragm. The flow rate depends on the stroke length and frequency. These pumps are generally used in low flow, high-pressure applications. They can handle many different fluids up to viscosities of 500 cP and can develop head pressures ranging from a few bars to over 1000 bars. Blockage or throttling of the discharge section is thus not tolerated, and caution is needed in operating them. The construction of reciprocating pumps consists of two basic parts: the pump head and the drive element (Fig. 6.2). These types of pumps are further divided into the following subtypes:

Piston and Plunger pumps: As indicated by the name, the reciprocating motion of a piston or a plunger draws the fluid into a cylinder through the suction check valve during the intake stroke and then forces it out through the outlet check valve

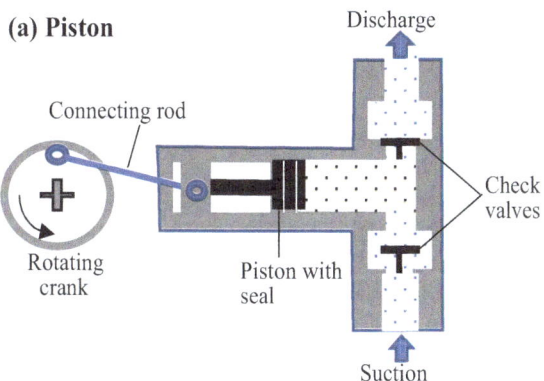

Fig. 6.2 Reciprocating pumps

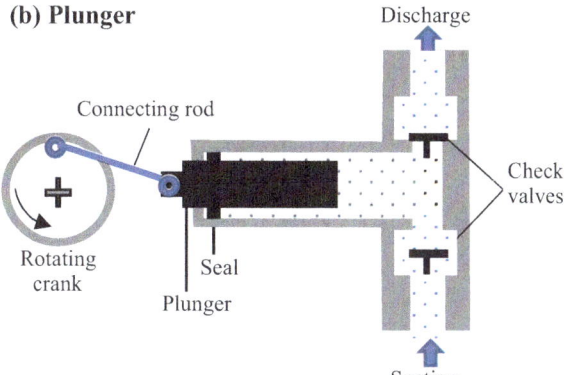

on the discharge stroke. The piston/plunger moves back and forth by a crankshaft (or cam) attached to a drive unit (generally electric). Although both the piston and plunger pumps have self-priming capability and work in the same way (Fig. 6.2), the major difference between them is the placement of high-pressure seals (also known as O-rings). The seal in a piston pump is attached to the piston and thus moves with the piston and wears out faster. Although both types are used in high-pressure applications, piston pumps are good for lower pressures, at a maximum of 100 bar. In plunger pumps, the plunger moves through a stationary seal into a sealed space filled with the fluid and thus can tolerate much higher pressures. Due to their dynamically operating seals, some back leakage does occur. These pumps are not very suitable for liquids with suspended solids, as they can damage the valves. They are also not recommended for use with high-viscosity liquids above 400 cP. However, there are many uses of such pumps in the food industry. For example, milk homogenizers use 3 to 7 piston pumps driven by an eccentric shaft through connecting rods and a sanitary head upon which a homogenizing valve is mounted.

Diaphragm pumps are fitted with a flexible membrane (diaphragm) to move fluids (Fig. 6.3). By repeatedly expanding and compressing the diaphragm in a reciprocating action, the volume of the pumping chamber draws the fluid into the pump and then pushes some fluid out. Their pump heads are hermetically sealed, which makes them highly suitable for pumping slurries and high viscosity, shear-sensitive fluids. Diaphragm pumps are also self-priming and work well in intermittent operations.

All reciprocating pumps deliver a pulsating discharge flow due to their cyclic action. During compression, the fluid accelerates and slows down during the suction phase. Although pulsation cannot be eliminated, it can be minimized by using two (duplex) or more (triplex, quadplex) pistons, plungers, or diaphragms working in tandem between suction and discharge modes (Fig. 6.4).

Fig. 6.3 A diaphragm pump

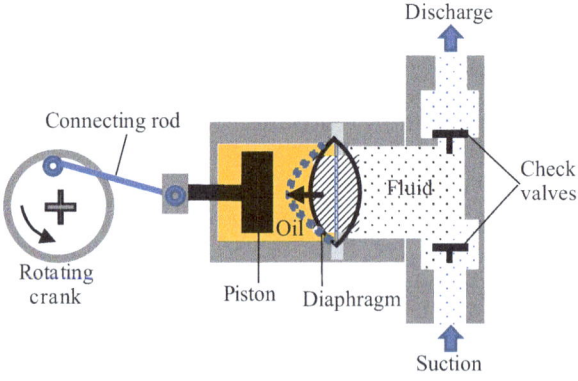

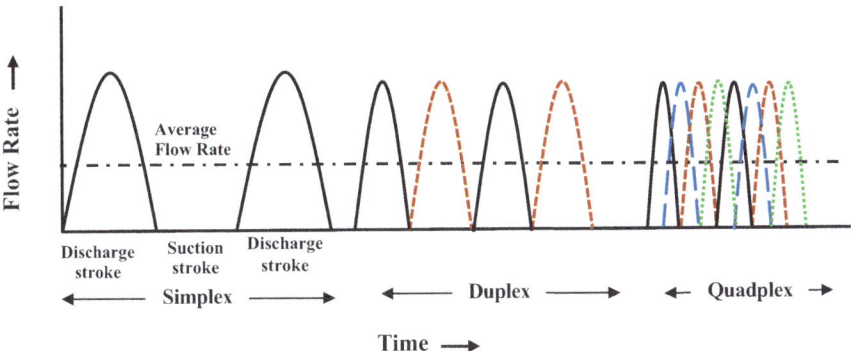

Fig. 6.4 Flow characteristics of reciprocating pumps

• **Rotary positive displacement pumps** use rotating elements such as lobe, cavity, gear, screw, vane, or peristaltic to trap a fixed volume of the liquid and then push it out of the pump chamber into the discharge port, generating lower-pulsating flow than their reciprocating counterparts. They are also more economical than piston, plunger, or diaphragm pumps, can handle fluids of very high viscosity, and have self-priming capabilities. As these are positive displacement pumps, for safety reasons, they should not be operated with a throttled or closed discharge. Some of the most commonly used rotary pumps in the food industry are described below. In practice, they all need to be coupled with a drive unit such as motors and gearboxes (shown only in some of the diagrams below).

Lobe pumps have two rotors, usually with 2–3 lobes each closed in the pumping chamber. The rotor surfaces create sealing with very tight tolerance, and a vacuum is created at the inlet when the rotors are rotated by a drive (electric motor) and the liquid is drawn into the cavity. The liquid then moves along the periphery of the chamber and is pushed out through the outlet (Fig. 6.5). These pumps are not ideal for water-like viscosities because they lose their pressure stiffness, and some slippage occurs between the rotor and the Fig. 6.5 casing, especially at pressures higher than 2–3 MPa. However, at low pressures, they have almost 100% volumetric efficiency (no slip) when the viscosity exceeds approximately 300 cP. Their sanitary design and the gentle treatment of the product of high viscosity allows this type of pump to be widely used for pumping cream with a high fat content, cultured milk products, curd/whey mixtures, juices with pulp, fruits and vegetable purees, etc. They can also handle liquids that contain small to medium size particulates and entrained vapor.

    Progressive cavity pumps, also known as eccentric-screw, helical rotor or Mono (Moyno, named after the French inventor, MoiNeau) pumps, consist of a helical metal rotor and an elastomer stator with a double helix internal cavity, sealed in a pumping chamber. The rotor helix is eccentric to the axis of rotation. This rotor-stator combination creates a shifting space to capture fluid into, progressivly push it forward, and then expel it through the discharge port. Its

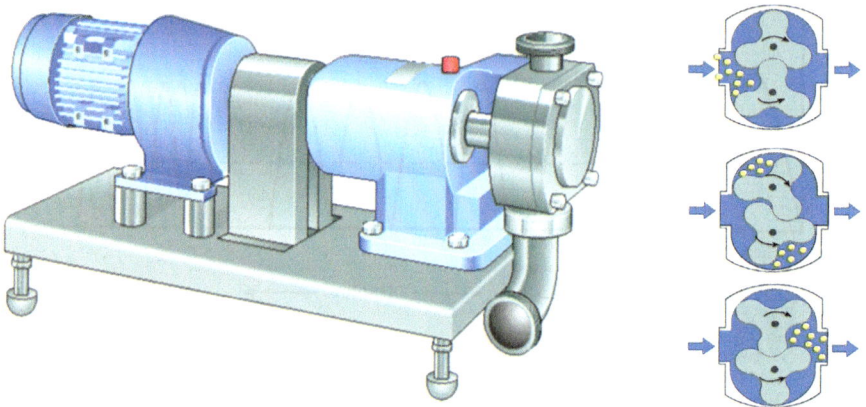

Fig. 6.5 A lobe pump assembled with drive and its cutaway section of head. From *Dairy Processing Handbook* © *Tetra Pak*

continuous seal minimizes pulsation, and the stator is immediately damaged if the pump is run dry. They are very versatile and used for both metering and transport. They work well in many pumping applications with shear-sensitive fluids, pastes, and abrasive slurries at discharge pressures up to 5 MPa at 7000 liters per minute. In the food industry, various products, such as apple sauce, milk concentrate, soft cheese, yogurt, peanut butter, jam & jelly, corn syrup, tomato soup and puree, salad dressing, etc., are routinely handled using pc pumps (Fig. 6.6).

Gear Pumps provide pumping action via two intermeshing gears encased in housing with tight clearance between the tips of the gears and the housing surface. One of the gears drives the other gear, and when the teeth are out of mesh, liquid flows between the teeth and the housing. It is then pushed out of the discharge port when the teeth come into mesh again. They are self-priming and provide pulse-free flow. These pumps are used in applications requiring accurate metering (dosing) of liquids at high pressures. In the food industry, such pumps find

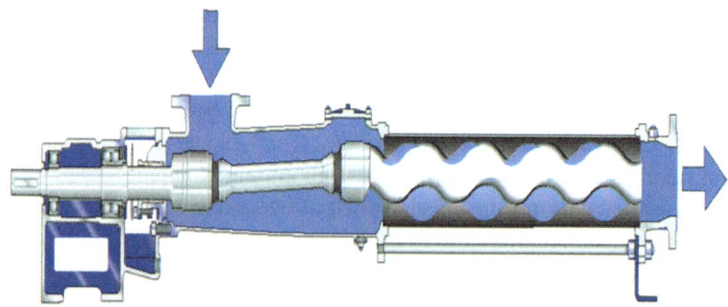

Fig. 6.6 An eccentric-screw or progressive cavity pump. From Dairy Processing Handbook ©Tetra Pak

applications in the handling of chocolate, cacao, fats and oils, and pet foods, among many others (Fig. 6.7).

Vane pumps consist of an eccentric rotor to which either spring-loaded or supple, rubber type flexible vanes are attached that rotate inside the pump cylinder (Fig. 6.8a & b). The vanes confirm to shape of the internal wall and help seal and maintain contact with the pump casing. Pumping is accomplished by the expanding and contracting volumes between the rotor and vanes where fluid enters and leaves the housing. These pumps are not steam sterilizable and thus not considered sanitary. Their applications are limited to pumping clean liquids such as wine, oils, syrup, honey, molasses and are used in coffee machines, drink dispensers, etc. These pumps are lightweight, compact, and low maintenance but not suitable for very high-pressure applications.

Other pumps included in the rotary positive displacement category are **screw and peristaltic pumps**. Screw pumps move liquid by capturing it in the spaces between the meshing screw threads, similar to gear pumps, except that the fluid is pushed axially as the screws mesh. An extruder is a screw pump that is extensively used in manufacturing many kinds of foods, such as breakfast cereals,

Fig. 6.7 A gear pump

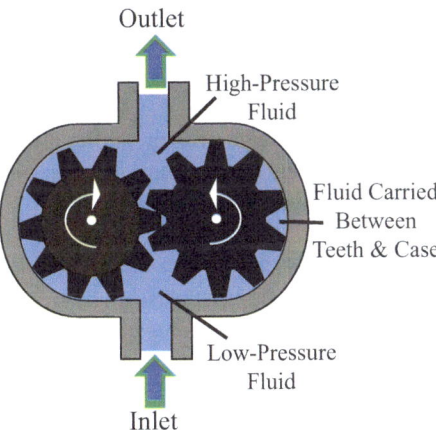

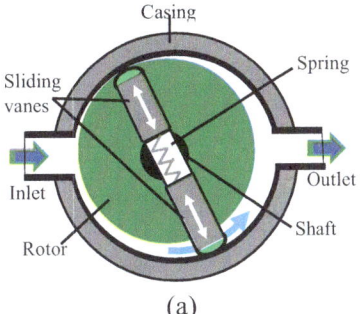
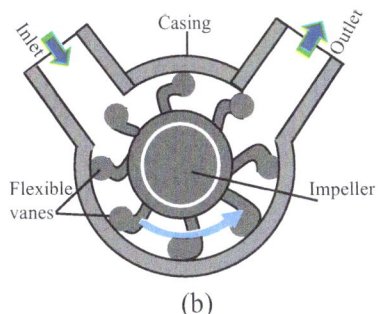

(a) (b)

Fig. 6.8 Vane pumps: (**a**) Sliding vane pump (**b**) Flexible vane pump

snacks, pasta, and pet foods. **Peristaltic pumps,** based on the principle of peristalsis (e.g., muscle contractions to move foods in our alimentary canal), use a flexible tube that contains the liquid and is compressed between a series of moving rollers inside the pump casing. As the roller moves over the tube, a vacuum is created that draws the liquid into the tube, and the next roller forms a "pillow" and pushes the liquid in front toward the outlet (Fig. 6.9). These pumps are ideal for pumping sterile and reactive fluids since they are in contact with the inside surface of the tube only. They are extensively used in research and development laboratories. In the food and pharmaceutical industries, these pumps find applications in dispensing, filtration, dialysis of liquids, and bioprocessing, albeit on a small scale (up to ~50 L/min and <1 MPa). However, they do produce high-amplitude, low-frequency flow pulsations and may have a short tube life. The flow rate of a peristaltic pump may be estimated as:

Volume per min. = (Pillow volume) × (Number of rollers) × (Pump RPM)

The flow rate of pumps is controlled by regulating the speed of the drive elements. Throttling the outlet of a positive displacement pump will increase the pressure dramatically and dangerously, and thus, it is important to avoid any possibility of restricting the flow out of such pumps.

Check Your Understanding

1. If the discharge pressure at the outlet of a positive displacement is increased, the power requirement for the pump will (a) increase, (b) decrease, or (c) remain unchanged. ◄

B. **Centrifugal Pumps:** Also known as dynamic pumps, they continuously add energy to the fluid being pumped via its centrifugal acceleration and subsequently transform the kinetic energy into pressure energy. This is different

Fig. 6.9 A peristaltic pump

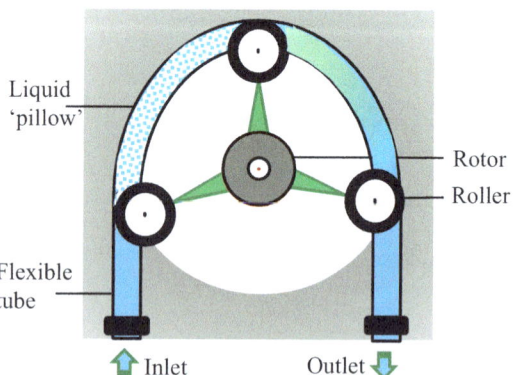

from the way positive displacement pumps function. As you may have noted, positive displacement pumps generate flow that results in pressure. Centrifugal pumps, on the other hand, create pressure that translates into flow. The two main parts of centrifugal pumps are the impeller (bladed rotor, vanes) and the volute (casing), which surrounds the impeller. In these devices, the fluid is pulled into the eye (center) of the spinning impeller axially, acquires kinetic energy due to rotation of the curved impeller blades, and moves in a tangential and radical direction toward the periphery, where it attains the same velocity as the impeller tips. As you may recall from elementary physics, when the diameter or the rotational velocity (rpm) of the impeller increases, the tip speed increases as well, and thus, more kinetic energy is added to the fluid. Upon leaving the impeller, the liquid goes to the casing and then travels to a discharge port either via a volute (diffuser) of increasing diameter or via a nozzle (shaped like a divergent cone with an angle of 7–10 degrees) or both, depending on the design, where the high velocity is gradually reduced and kinetic energy is converted to equivalent pressure energy, Fig. 6.10, approximating the Bernoulli principle. The pressure generated as a result, expressed in terms of the height of the liquid or

Fig. 6.10 A centrifugal pump coupled with motor and its cutaway section of head

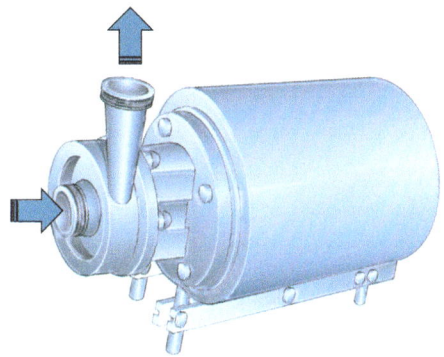

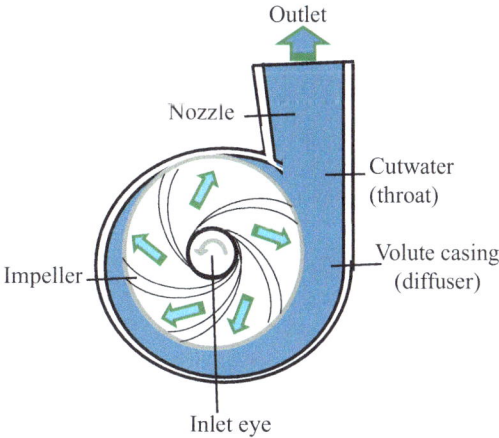

head (h), may be estimated from the liquid velocity (v) at the impeller periphery as follows:

$$h = \frac{v^2}{2g} \tag{6.1}$$

Because of their relatively simple engineering, centrifugal pumps are less costly and economical to operate and maintain. However, these pumps are generally not self-priming and require venting and priming for smooth operations. They are well suited for handling low-viscosity (<300 cP) liquids at high flow rates and moderate pressures. For more viscous liquids, pumps with larger inlet sizes may be used. Centrifugal pumps are the most commonly used pumps and are the workhorse of the processing industry.

How the three (flow rate, developed head, and power) most important characteristics of a centrifugal pump change with impeller speed have been mathematically developed into what is known as **Affinity Laws** (or Pump Laws). They are applicable to geometrically similar pumps and fans. These laws can be used to determine, for example, how a given centrifugal pump's performance will change if its impeller speed is changed while the diameter is fixed or the diameter is changed at a fixed speed. These laws are expressed as follows:

$$\dot{V}_2 = \dot{V}_1 \left(\frac{N_2}{N_1}\right)_D, \quad \dot{V}_2 = \dot{V}_1 \left(\frac{D_2}{D_1}\right)_N \tag{6.2}$$

$$h_2 = h_1 \left(\frac{N_2}{N_1}\right)_D^2, \quad h_2 = h_1 \left(\frac{D_2}{D_1}\right)_N^2 \tag{6.3}$$

$$\Phi_2 = \Phi_1 \left(\frac{N_2}{N_1}\right)_D^3, \quad \Phi_2 = \Phi_1 \left(\frac{D_2}{D_1}\right)_N^3 \tag{6.4}$$

where $\dot{V}$ is the volumetric flow rate, N is the impeller speed (rpm), "h" is the head, Φ is the power, and D is the impeller diameter.

Check Your Understanding

2. What impeller diameter will be needed to develop a head of 20 m by a centrifugal pump running at 1800 rpm? ◄

C. **Positive Displacement vs Centrifugal Pump Selection Criteria:** Although proper selection of a sanitary pump for any given application requires careful considerations and evaluation of the operational requirements, process safety and cost, a simple decision process based on product and process characteristics shown in Fig. 6.11 below provides a good starting point.

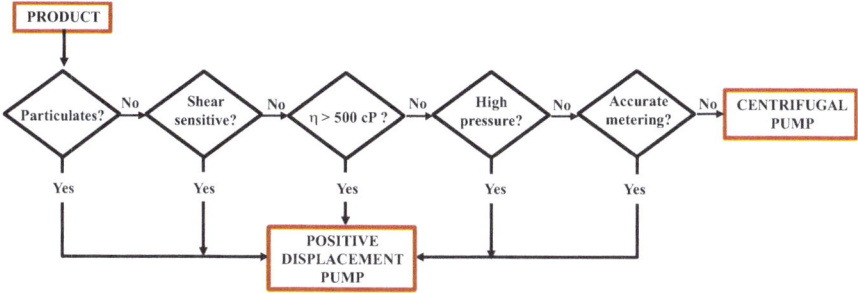

Fig. 6.11 Sanitary pump selection criteria (After T.J. Bowser [1])

6.1.2 Pumps: Sizing and Operation

Before selecting a pump for a particular application, pumping system design must be completed. To do so requires a good understanding of the following pump-related operational details.

A. **Pump sizing** for a given task involves matching the flow rate and total dynamic head required for the process with the flow and pressure rating of a pump provided by the manufacturer. As discussed above, the head (h) is the most common method used to express the energy that a pump imparts to the fluid being pumped. A simple example of a pumping system is given in Fig. 6.12, where h_s is the suction head, h_d is the discharge head, and the difference between the two is the total static head (Δh). For the pump to transport the liquid from position 1 to position 2, it will have to develop the head more than what will be needed just to overcome the static head. The pump will need to overcome differences in pressures and velocities between positions 1 and 2 as well as the loss of energy due to friction. To account for all such different forms of energy that will be needed, we resort to our now familiar MEB. Assuming turbulent flow and writing an MEB in terms of energy (or equivalent head) the pump will require to transport fluid to its destination, and calling it the **total dynamic head (TDH)**, we obtain:

$$\frac{P_2}{\rho} + gh_d + \frac{1}{2} <v_2>^2 + \sum F = \frac{P_1}{\rho} + gh_s + \frac{1}{2} <v_1>^2 + \overline{w}_s \quad (6.5a)$$

$$\text{Or,} \quad \frac{\overline{w}_s}{g} = \frac{P_2 - P_1}{\rho g} + \frac{<v_2>^2 - <v_1>^2}{2g} + (h_d - h_s) + \frac{\sum F}{g} \quad (6.5b)$$

$$\text{Or,} \quad TDH = \frac{P_2 - P_1}{\rho g} + \frac{<v_2>^2 - <v_1>^2}{2g} + (h_d - h_s) + \frac{\sum F}{g} \quad (6.5c)$$

where subscripts 1 and 2 refer to process start (source) and end (destination); P is the pressure; $<v>$ is the average velocity; h is the elevation, and ΣF is all the pressure losses due to friction in the system. Generally, a more elaborate network

Fig. 6.12 A simple pumping system

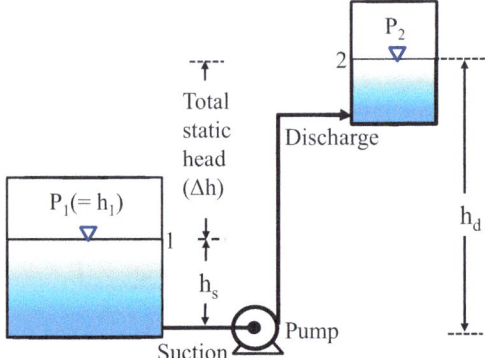

of pipes, fittings, valves, and instrumentation is needed to perform a specific job and contribute to friction losses. The above equation may be abbreviated as follows:

$$TDH = \frac{\Delta P}{\rho g} + \frac{\Delta <v>^2}{2g} + \Delta h + \frac{\sum F}{g} \tag{6.6}$$

Using Eq. (6.6), the total dynamic head (TDH) of the pumping system is computed for different flow rates and plotted. Remember, each flow rate will result in different friction loss. The resultant flow rate vs. total dynamic head (TDH) plot is known as the **system curve**, Fig. 6.13. It provides information on the total dynamic head needed for different flow rates and helps select the right pump. Additionally, Fig. 6.13 shows the **pump curves** of two centrifugal pumps, each operating at a different rpm (low and high). Such curves, also called pump characteristic curves, are provided by pump manufacturers. The intersection of the **pump curve** and the **system curve** represents the **operating point** for the pump. For Flow1, for example, the system curve intersects with the High RPM

Fig. 6.13 System and pump curves for sizing a pump

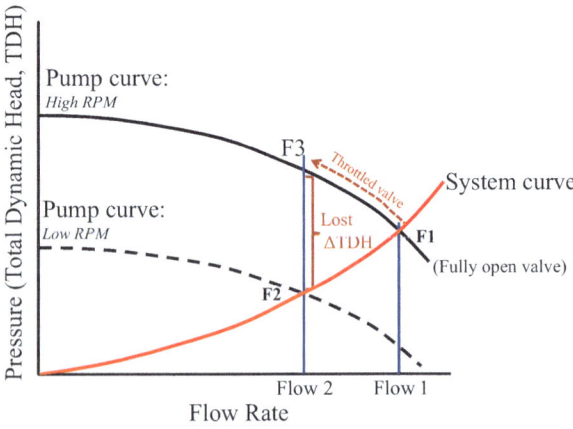

pump at F1, and for Flow2, it intersects with the Low RPM pump at F2. For constant flows, these selections will be optimum. In applications where flow rate requirements vary over time, a strategy to control the flow is needed. For a fixed speed centrifugal pump, it can be done either by recycling some of the flow back to the suction side or by throttling the flow by a valve in the discharge line. As noted in Fig. 6.13, throttling Flow1 to Flow2 using the high RPM will make the operating point F1 move along the pump curve to the new operating point F3. Compare this with using the lower RPM curve to attain the same flow rate F2 and thus save energy (lost ΔTDH). A variable speed pump may thus be preferred in such situations, although other considerations such as efficiency, reliability, and maintenance cost will also affect the final choice.

B. **The power** required to pump a fluid, called **fluid power** (P_F), represents the rate of energy delivered by the pump to the fluid and can be calculated from an MEB on the system of interest or it can be estimated from its head-flow rate curve at the operating point of the pump, Fig. 6.13, by the equation below:

$$P_F = \dot{m}\, g\, (TDH) \tag{6.7}$$

where $\dot{m}$ is the mass flow rate and TDH is the corresponding total dynamic head. The actual power draw from the source, called **the break power (P_B),** is the actual power taken by the pump drive and depends on the drive **efficiency (E_p)**, as shown in Fig. 6.14. The relationship between P_B and P_F is:

$$P_B = \frac{P_F}{E_p} \tag{6.8}$$

The brake power or power input to a pump is greater than the fluid power or actual output to the fluid being pumped due to the losses incurred in the pump. The ratio of these two values indicates the pump efficiency.

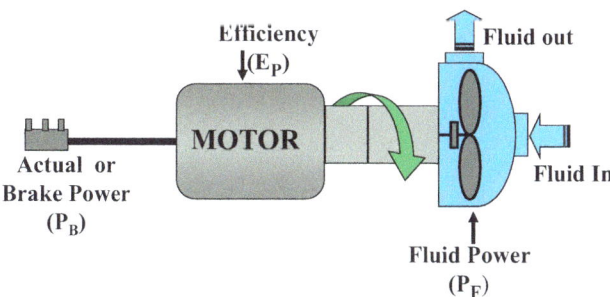

Fig. 6.14 Pump power nomenclature

Example

1. *Canola oil (specific gravity 0.9) is pumped at 10.0 L/s through a uniform pipe system with a 10.0 cm internal diameter to a height of 10.0 m above the level in the tank (Fig. 6.15). The pump's rated efficiency is 69%. The total friction loss in the pipe system is 25.0 J/kg. If the flow is assumed to be turbulent, determine the actual power used by the pump.*

   Solution:

   Using Eq. (6.7), $P_F = \dot{m}\, g\, TDH$

$$P_F = \left(10.0\frac{L}{s}\right)\left(0.9x\,1\frac{kg}{L}\right)\left(9.81\frac{m}{s^2}\right)TDH$$

   For TDH, using Eq. (6.6):

$$TDH = \frac{\Delta P}{\rho g} + \frac{\Delta <v>^2}{2g} + \Delta h + \frac{\Sigma F}{g}$$

   Now, $\Delta P = 0$, since both ends open to the atmosphere, $<v_1> = 0$, constant level in the tank

$$<v_2> = \left[(10.0\ L/s)\left(1\ m^3/1000\ L\right)\right]/\left[(3.14)\left(0.1\ m\right)^2/4\right] = 1.27\ m/s$$

$$TDH = 0 + \frac{\left(1.27\frac{m}{s}\right)^2}{(2)\left(9.81\frac{m}{s^2}\right)} + 10.0\ m + \frac{25\frac{J}{kg}}{9.81\frac{m}{s^2}} = 0.08 + 10.0 + 2.55 = 12.63\ m$$

$$P_F = \left(10\frac{L}{s}\right)\left(0.9\,\times\,1\frac{kg}{L}\right)\left(9.81\frac{m}{s^2}\right)(12.63\ m) = 1115.1\ J/s$$

   Then the actual or break power $P_B = 1115.1/0.69 = 1616.1\ W$ ◀

Fig. 6.15 Power required for pumping

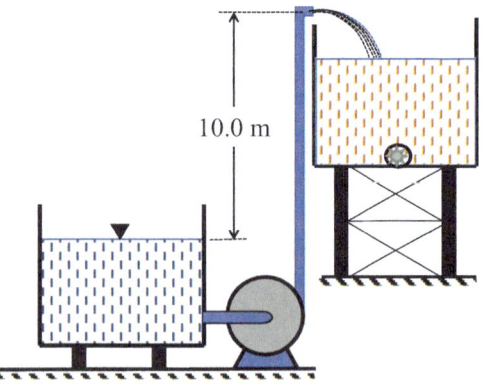

10.0 m

C. **Cavitation and Net Positive Suction Head (NPSH)** are two interrelated issues of significant importance in the design of any pumping system. Cavitation is the conversion of some of the liquid being pumped into vapor bubbles and their subsequent collapse within a pump. Vapor bubbles begin to form inside the eye of the pump impeller when the absolute pressure at the inlet to a pump falls below the vapor pressure of the liquid. These bubbles then move to the impeller periphery, where they start to collapse abruptly and cause implosions (inward bursting) due to higher pressure. The collapse of the bubbles creates liquid microjets with hammering noise, such as pumping gravel and signaling cavitation. In addition to creating noise and vibrations, they pit and damage the impeller to the extent that they may lead to equipment damage, process shutdown, and personnel injury. Cavitation can be eliminated simply by increasing the pressure of the liquid at the suction point of the pump, generally expressed as the **net positive suction head (NPSH)**. The pump design and characteristics determine the **net positive suction head required (NPSHR)** for smooth operation of the pump without cavitation, and the manufacturers include this information as well in their pump performance curves (Fig. 6.17). It is obtained by actual testing performed by the pump manufacturer. It is interesting to note that short-lived cavitation becomes inevitable in certain situations, such as emptying a tank with a pump where the NPSH goes down well below the required value for smooth operation of the pump.

Referring to the simple pumping system shown in Fig. 6.12, the net positive suction head available (NPSHA) at the pump inlet depends on the following:

- The pressure at the surface of the liquid, P_1 (or h_1 as head).
- The suction head, h_s
- The velocity head, h_v
- The friction loss in the suction line, h_f
- The vapor pressure of the liquid at the pumping temperature, h_{vap}. It may be estimated or obtained from the literature.

The above factors are combined into the following expression for NPSHA, with all values expressed in units of fluid head.

$$NPSHA = h_1 + h_s + h_v - h_f - h_{vap} \qquad (6.9)$$

Thus, to avoid cavitation, **NPSHA > NPSHR**. As a margin of safety, the NPSHA is recommended to be 1 to 2.5 times higher than the NPSHR, depending on the pump type, its design, and application. Should NPSHA be insufficient to eliminate cavitation, a few operational options may be considered to increase it. In Example 1 above, we saw that the contributions of the velocity head and the liquid vapor pressure to NPSHA are insignificant in comparison with other factors; thus, they may not be worthwhile to pursue and are ignored. Although vapor pressure can be lowered by cooling, it will still not be very effective. An examination of Eq. (6.11) suggests that the following remedial action may be helpful:

- Increase the source pressure. However, this may not be very practical and doable.
- Increase the static head by increasing the level of fluid in the source vessel, lowering the pump elevation, or increasing the source vessel elevation.
- Reduce suction side frictional loss by decreasing the length and increasing the diameter of the inlet pipe.

Example

2. *As shown in Fig. 6.16, water at 20 °C is pumped from the tank at a velocity of 2 m/s with a pump with that requires an NPSHR of 10 m. The friction loss due to the elbow, valve, and pipe on the suction side is estimated to be equivalent to 1 m of head pressure. Does the pump have sufficient NPSHA to prevent cavitation?*

    Solution: *Using Eq. (6.9)*

    $h_l = 10.3$ m *(atm. pressure, open to atmosphere)*, $h_s = 3.0$ m

    $h_v = \frac{v^2}{2g} = (4 \ m^2/s^2)/(2)(9.81 \ m/s^2) = 0.20$ m, $h_f = 1.0$ m *(given)*

    $h_{vap} = 0.031$ atm *(from steam table)* $= 0.32$ m

    Therefore, NPHSA $= 10.3$ m $+ 3.0$ m $+ 0.2$ m $- 1.0$ m $- 0.3$ m $= 12.2$ m is adequate to prevent cavitation. ◀

D. **The Pump Curves** (also known as pump performance curves or pump characteristic curves) are pressure (head) vs capacity (flow rate) graphs created by the pump manufacturers based on actual evaluation (Fig. 6.17). The pump curves depend on the geometry and velocity of the pump and are fixed and cannot be changed. They are needed in the selection and operation of pumps. For centrifugal pumps, as the flow rate increases, additional loss of energy within the pump casing and impeller occurs due to friction and eddy currents, which cause the

Fig. 6.16 Pump NPSHA
estimation

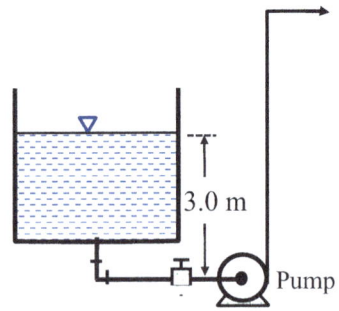

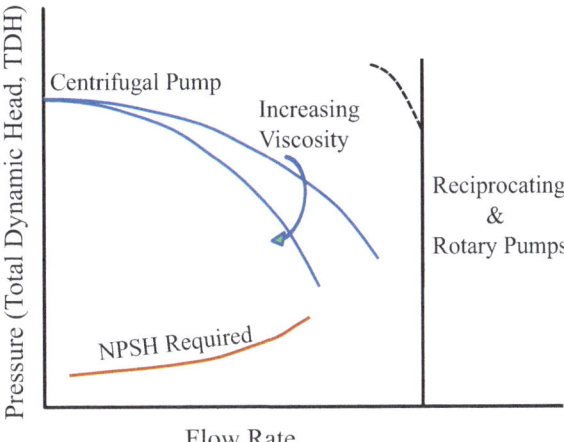

Fig. 6.17 Pump performance curves

produced head to decrease. For this reason, centrifugal pumps are not very efficient for pumping high-viscosity liquids.

Reciprocating and rotary PD pumps, on the other hand, exhibit very steep, almost vertical characteristics. The dashed line reflects the effect of slippage from discharge that reduces the performance of PD pumps at very high pressures. Typically, pump curves also include other information such as pump efficiency, power requirements, and NPSHR as a function of flow rate. Pump curves are indeed essential to select and source the best pump for any application.

E. **Other Considerations:** Once the flow and power rating of a pump have been matched with the flow rate and power required for a given system, the final choice on the type of pump to be used is dictated by the product characteristics and the processes involved. One of the important issues to consider is product integrity, especially when handling viscous and shear-sensitive products. Churning of high-fat cream, breakdown of emulsions, disintegration of suspended solids, and many other factors that lead to quality loss may result due to improperly chosen pumps. Other important considerations in the final selection of a pump include sanitary design and ability to clean-in-place (CIP) and sterilize-in-place (SIP).

In addition to transporting fluids, pumps are also used to add a precisely controlled quantity of one liquid to another. This is called metering or dosing and requires a flow tolerance in the range of 0.25 to 3%. This is not achievable with centrifugal pumps since their flow rates vary nonlinearly with pressure. Reciprocating pumps for tighter flow rate tolerance and rotary positive pumps for other applications are generally used.

6.2   Fluid Flow Measurements

Flow rate measurements are necessary to ensure that processes are running as designed and that each component of the system is operating smoothly and cost-effectively. In the food processing industry, monitoring and controlling the flow rate is especially critical because of the stringent quality control and safety requirements. An inaccurate flow rate may pose serious problems to product quality and result in production loss.

Various types of flow meters are available to measure the volumetric or mass flow rate of a fluid. Many of these devices provide flow rate information by determining the fluid velocity from pressure differential measurements. A common feature of these devices is that they do not require any external power to operate, and a few very common units are discussed below. However, before that, we need to understand the types of pressures that a flowing fluid exhibits and how they can be utilized to obtain flow rate data. In Chap. 5, we discussed the principle behind hydrostatic pressure measurements. The static or hydrostatic pressure is the pressure exerted by a fluid at rest. In a flowing fluid, the static pressure is measured relative to the moving streamline. A pressure transmitter traveling with the fluid will indicate free-stream static pressure at its location. It can also be measured through a flat opening in the pipe parallel to the flow. In Fig. 6.18, h_1 is the static pressure at location 1. The total pressure (also known as **stagnation** or Pitot pressure) is the pressure exerted when fluid stops moving or stagnates in the pipe. At this point, all kinetic energy is converted into pressure and added to static pressure to generate total pressure. It is measured at right angles to the flow direction in a low turbulence location, represented by location 2 in Fig. 6.18. The difference between the total pressure and static pressure is the dynamic pressure d. It measures the kinetic energy of a fluid, which appears as dynamic or velocity head in the above diagram and can be easily converted into velocity, based on the Bernoulli principle ($h = v^2/2\,g$).

Fig. 6.18 Types of pressures in fluid flow

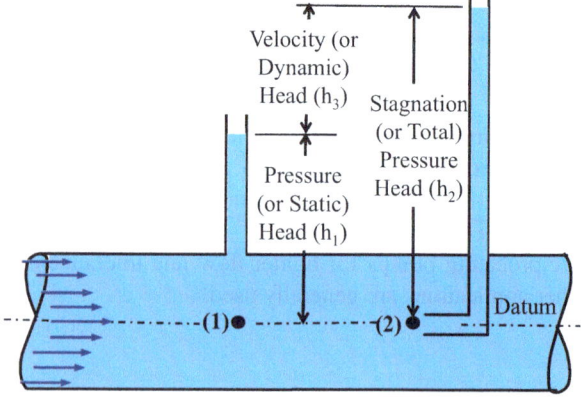

6.2.1   Fluid Flow Measurement Devices

Many simple flow measuring devices are based on creating a pressure differential by inserting an obstruction in the path of fluid flow. The measured pressure differential is then converted into velocity by invoking the Bernoulli principle and applying empirical correction to account for losses due to friction. Some of these devices include the following:

A. The **Pitot Tube,** invented by Henri Pitot in 1732, is a simple and inexpensive sensor widely used to measure local velocity during fluid flow. It measures the dynamic pressure (velocity head) as the difference between the total (stagnation) and static pressures in terms of the liquid head using a differential pressure meter such as a manometer (Fig. 6.19). The static tube opens parallel, and the impact tube opens perpendicular to the direction of flow. The fluid enters at point 2, the stagnation point, and becomes stationary ($v_2 = 0$) and indicates the total pressure. The static tube, on the other hand, measures the static pressure upstream of point 2. This differential pressure between points 1 and 2 is then converted into local or point velocity by application of the Bernoulli principle (pressure energy + kinetic energy + potential energy = constant along a streamline). Applying it at points 1 and 2:

$$\frac{P_2}{\rho} + gh_2 + \frac{1}{2}v_2{}^2 = \frac{P_1}{\rho} + gh_1 + \frac{1}{2}v_1{}^2 \tag{6.10}$$

Since $h_1 = h_2$ (assumed datum) and $v_2 = 0$, the above expression reduces to:

$$\frac{P_2}{\rho} = \frac{P_1}{\rho} + \frac{1}{2}v_1{}^2 \tag{6.11}$$

$$\text{Or,} \quad v_1 = \sqrt{\frac{2\,(P_2 - P_1)}{\rho}} = \sqrt{\frac{2\Delta P}{\rho}} \tag{6.12}$$

To measure the pressure difference ΔP, a manometer, as shown in Fig. 6.19, filled with a fluid of density ρ_m, is used, then invoking the hydrostatic principles discussed in Chap. 5, Eq. (6.12) can be written as:

Fig. 6.19 A Pitot tube

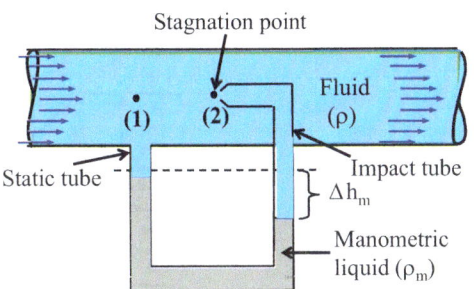

$$v_1 = \sqrt{\frac{2\,(\rho_m\,g\Delta h_m)}{\rho}} \tag{6.13}$$

To correct for the assumption of inviscid fluid inherent in the Bernoulli equation, a dimensionless coefficient, called the discharge coefficient, C, ranging in value from 0.98 to 1.0 is included in the final expression. Additionally, if the density of the flowing fluid (ρ) is not negligible compared to the manometric fluid density (ρ_m), Eq. (6.13) is modified as follows:

$$v_1 = C\sqrt{\frac{2\,(\rho_m - \rho)\,g\Delta h_m}{\rho}} \tag{6.14}$$

The point velocity may be measured at several locations in a pipe flow to obtain an average velocity, which can then be multiplied by the cross-sectional area of the pipe to obtain the volumetric flow rate.

B. The **Venturi Meter** was developed by G.B. Venturi in 1797 in the form of a short tube with a converging-diverging, throat-like passage that increases flow velocity and reduces the pressure correspondingly. Gradual contraction and expansion minimize frictional losses and provide a better estimate of the average cross-sectional velocity, determined from the pressure drop between two points (Fig. 6.20).

Similar to the Pitot tube analysis, writing the Bernoulli equation for points 1 and 2 for the average velocity and using the same datum, Eq. (6.10) becomes:

$$\frac{P_2}{\rho} + \frac{1}{2} <v_2>^2 = \frac{P_1}{\rho} + \frac{1}{2} <v_1>^2 \tag{6.15}$$

For the steady-state flow of fluid with constant density (ρ) through areas A_1 and A_2, the continuity equation gives:

$$<v_1>A_1 = <v_2>A_2, \text{or} <v_2> = v_1(d_1/d_2)^2. \tag{6.16}$$

where d_1 and d_2 are the respective diameters. Substituting for $<v_2>$ in Eq. (6.15) and solving for $<v_1>$, we obtain:

Fig. 6.20 A Venturi meter

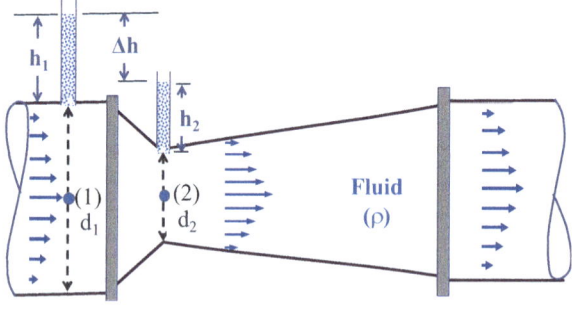

$$<v_1> = \sqrt{\frac{2\,(P_2 - P_1)}{\rho\left[\left(\frac{d_1}{d_2}\right)^4 - 1\right]}} \tag{6.17}$$

The pressure difference can be replaced with the observed differential head (Δh) to obtain:

$$<v_1> = C_v \sqrt{\frac{2\,g\Delta h}{\left[\left(\frac{d_1}{d_2}\right)^4 - 1\right]}} \tag{6.18}$$

A discharge coefficient (C_v) has been introduced as a correction factor to account for the loss of energy due to friction. The value of C_v is provided by the manufacturer, or the unit may be experimentally calibrated.

If a manometer is used to measure the pressure differential, as discussed earlier in the Pitot tube section, Eq. (6.17) will need to be modified as:

$$<v_1> = C_v \sqrt{\frac{2\,\rho_m g\,\Delta h_m}{\rho\left[\left(\frac{d_1}{d_2}\right)^4 - 1\right]}} \tag{6.19}$$

When the densities of the flowing fluid (ρ) are not negligible compared to the manometric fluid density (ρ_m), Eq. (6.19) is modified as follows:

$$<v_1> = C_v \sqrt{\frac{2\,(\rho_m - \rho)\,g\,\Delta h_m}{\rho\left[\left(\frac{d_1}{d_2}\right)^4 - 1\right]}} \tag{6.20}$$

C. **A nozzle flow meter** is another device for measuring the flow of fluids. It is more compact than a Venturi meter, but the friction loss is higher. It has a high coefficient of discharge, C_n. It is used in applications such as fluid-containing suspended solids or high pressure and temperature steam flows. The analysis and estimation of the average velocity at location 1 is very similar to what was discussed in the Venturi meter section. The final equation, in terms of the observed Δh, Fig. 6.21, is thus:

Fig. 6.21 A nozzle flow meter

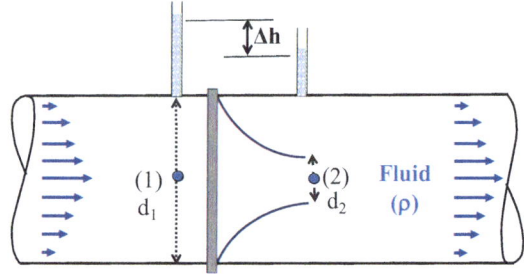

$$<v_1> = C_n \sqrt{\frac{2\, g\Delta h}{\left[\left(\frac{d_1}{d_2}\right)^4 - 1\right]}} \qquad (6.21)$$

Alternatively, if a manometer is employed to measure the pressure difference, it is modified as follows:

$$<v_1> = C_n \sqrt{\frac{2\, (\rho_m - \rho)\, g\, \Delta h_m}{\rho \left[\left(\frac{d_1}{d_2}\right)^4 - 1\right]}} \qquad (6.22)$$

D. **Orifice Meter**, Fig. 6.22, is the simplest flow measuring meter. It consists of a flat plate with a circular hole drilled in the middle. It is very compact and simple to construct and install. One pressure is measured upstream from the orifice plate and another just downstream, and the differential pressure is used, as discussed earlier, to estimate the average flow velocity, which may then be conveted into mass or volumetric flow rates. For an orifice meter, Eqs. (6.21) and (6.12) are also applicable with their own discharge coefficient, C_o, inserted into them, as shown below:

Fig. 6.22 An orifice meter

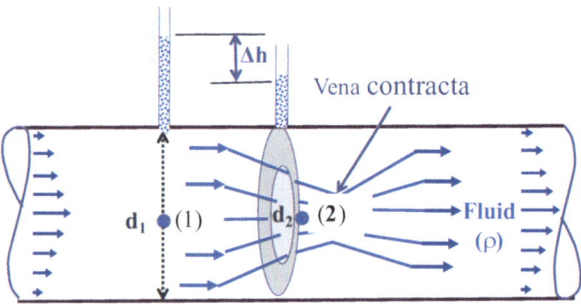

$$< v_1 > = C_o \sqrt{\frac{2\,g\Delta h}{\left[\left(\frac{d_1}{d_2}\right)^4 - 1\right]}} \qquad (6.23)$$

Similar to the orifice meter, if a manometer is employed to measure the pressure difference, the above equation becomes:

$$< v_1 > = C_o \sqrt{\frac{2\,(\rho_m - \rho)\,g\,\Delta h_m}{\rho\left[\left(\frac{d_1}{d_2}\right)^4 - 1\right]}} \qquad (6.24)$$

An interesting fact about the orifice meters is that the fluid passing through an orifice edge is not able to turnaround as eddies; it converges to a minimum diameter stream, called **vena contracta**, downstream from the orifice. The location of vena contracta depends on the ratio of the orifice diameter to the pipe diameter. The velocity of the fluid is highest, while the pressure is lowest at this point.

E. **Rotameter,** also known as a variable area meter, is a volumetric flow rate measuring device that was first introduced in Germany by Karl Kueppers in 1908. It consists of a tapered glass, plastic or metal tube enclosing a small float and has a scale that has been calibrated for reading flow rate directly. The float has a density higher than the fluid, so it does not float on top of the fluid at rest and is made of metal, plastic, or glass in different shapes and sizes. The fluid flows through a constriction of variable annular area between the float and the tapered tube wall, Fig. 6.23. This makes the float move up and down the tube in response to forces acting on it: the velocity head of the moving fluid and the buoyancy acting upward, and the downward gravitational force exerted by the weight of the float. The float stabilizes when the upward and downward forces are in equilibrium. The float movement varies linearly with the flow rate of the fluid in the tapered tube. When the flow rate changes, the float moves up or down in proportion to the flow rate. The changing annular area changes the pressure differential across the float, and it moves to a new position to be in equilibrium again. Since gravity plays a critical role in the correct operation of rotameters, they must be vertically oriented and mounted. The industrial utility of rotameters is numerous since they provide reliable service over a wide range of flows.

6.3 External Flow and Flow Past Immersed Objects

Thus far, we have been concerned with fluid mechanics applications related to momentum transfer and friction losses for fluid flow inside pipes and conduits, called internal flow, but fluid flow outside immersed objects, called external flow, is equally important. Some common examples of such flows include flow past

Fig. 6.23 A variable-area
flow meter (Rotameter)

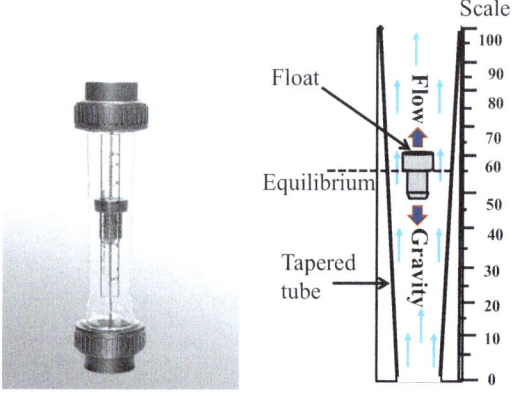

particles during settling and floatation (e.g., sedimentation, creaming), flow of steam
or hot water past heat exchanger tubes, flow of hot air past drying beds, and flow of
liquids past solid materials during filtration, among many others.

A fluid moving over a surface exerts a drag force on the surface because of
friction caused by viscosity and comes to a complete stop at the interface because of
the no-slip condition. A velocity profile with a boundary layer develops, as shown in
Fig. 6.24. We learned in Chap. 5 that in flow over a pipe surface, the Fanning friction
factor, f, defined by Eq. (5.100), is related to the drag force as follows:

$$f = \frac{F_D/A}{\frac{\rho <v>^2}{2}} \qquad (6.25)$$

where F_D is the drag force. By knowing f, the drag force exerted by the fluid on the
solid can be estimated. It is called **wall drag or skin drag**. This results from the
transfer of momentum perpendicular to the direction of flow and equals the shear
stress at the wall. In flows past immersed objects, in addition to skin or wall drag,
another drag called **form drag** exists because the fluid changes directions to pass
around the object, and the boundary layer separates from the object and forms large
eddies in the back (Fig. 6.25). The drag thus becomes form dependent. In such cases,

Fig. 6.24 Velocity profile
during fluid flow over an
object

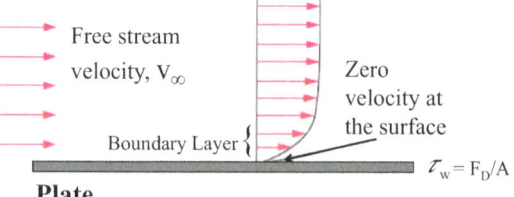

Fig. 6.25 Flow past a sphere

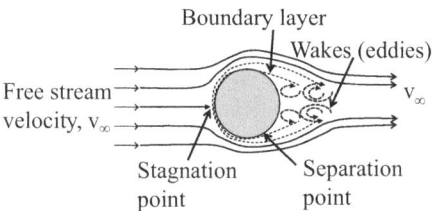

the total drag equals **skin drag + form drag,** and the object offers more resistance to flow.

Similar to the Fanning friction factor, a drag coefficient, C_D, is defined for immersed objects as:

$$C_D = \frac{F_D/A_p}{\rho \frac{v_\infty^2}{2}} \tag{6.26}$$

where F_D = total drag force, A_p = projected area, e.g., for a sphere of diameter D_p, $A_p = \frac{\pi D_p^2}{4}$, v_∞ = free-stream velocity and C_D = drag coefficient (dimensionless), which varies as a complex function of Re and the object shape. Correlations of C_D and Re for several geometries are available in the literature. From the knowledge of C_D, the total drag force may be computed as follows:

$$F_D = C_D \rho \frac{v_\infty^2}{2} \cdot A_p \tag{6.27}$$

In the laminar flow regime, when Re < 1, the experimental drag force for a sphere, G.G. Stokes (1851) showed, is proportional to the flow velocity, fluid viscosity, and sphere radius. It is given as:

$$F_D = 3 \pi \eta D_p v_\infty \tag{6.28}$$

Then, the drag coefficient is simplified to

$$C_D = \frac{F_D/A_p}{\rho \frac{v_\infty^2}{2}} = \frac{3\pi \mu D_p v_\infty}{\rho \frac{v_\infty^2}{2} A_p} = \frac{6\pi r v_\infty}{\rho v_\infty^2 \frac{\pi D_p^2}{4}} = \frac{24\eta}{\rho v_\infty D_p} = \frac{24}{Re} \tag{6.29}$$

The Reynolds number for a solid object immersed in a flowing fluid is defined as:

$$Re = \frac{\rho D_p v_\infty}{\eta} \tag{6.30}$$

6.3.1   Terminal Velocity of a Particle in a Fluid

The rising (creaming) or falling (settling) velocity of a particle suspended in a fluid, such as fat globules rising to the top in a milk container or solids settling out of suspension in a brewery tank, depends on several factors: the size of the particle, its density, and the viscosity of the liquid it is traveling through. Consider all the forces acting on such a suspended spherical particle in a fluid, as shown in Fig. 6.26. The particle will move upward if it is lighter than the fluid and downward if it is heavier. Starting from rest, it will accelerate until the net force acting on it is zero and then will begin to move with a constant velocity, called the **terminal velocity,** as shown in Fig. 6.27. The force balance on the particle is such that the downward gravitational force (FG) must be equal to upward buoyant (F_B) and drag (F_D) forces, i.e.,

$$F_G = F_B + F_D \tag{6.31}$$

$$\text{Or, } mg = mg\frac{\rho}{\rho_P} + F_D \quad \text{or} \quad F_D = mg\left(1 - \frac{\rho}{\rho_P}\right) \tag{6.32}$$

When Re < 1, Stokes' law applies, and Eq. (6.32) may be written as:

$$6\pi\mu r_p v = \left(\frac{4}{3}\pi r_p{}^3\right)\rho_p g\left(\frac{\rho_p - \rho}{\rho_p}\right) \tag{6.33}$$

which simplifies to:

Fig. 6.26 Force balance on a particle suspended in a fluid

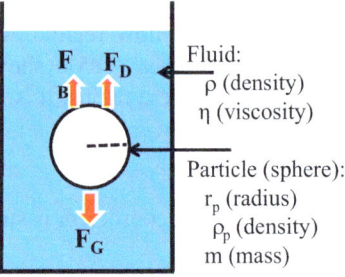

Fluid:
ρ (density)
η (viscosity)

Particle (sphere):
r_p (radius)
ρ_p (density)
m (mass)

Fig. 6.27 Terminal velocity of a settling particle

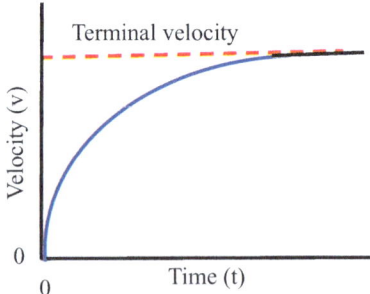

$$v = \frac{2}{9} r_p^2 \frac{(\rho_p - \rho)g}{\eta} \qquad (6.34)$$

The above equation expresses what is known as **Stokes' law,** which is indeed based on the equation for the drag experienced by a sphere first proposed by Stokes as described earlier and applies strictly to the creeping flow regime where the Reynolds number is less than unity. As shown in Eq. (6.34), it relates the terminal velocity of a sphere directly to its radius (squared) and inversely to the viscosity of the fluid it is moving through. Thus, the greatest impact on settling/rising velocity is that of the particle's radius. Doubling the radius, for instance, would cause a fourfold increase in the rising/falling velocity of the particle. This relationship is used for determining the rising or settling velocity of particles suspended in fluids, such as separating suspended impurities from incoming raw material or air streams, ice crystals from mother liquor, cream from whole milk, etc. It is also used in the measurement of liquid viscosity using falling-ball viscometers.

Check Your Understanding

3. For settling solid particles from a liquid with a terminal velocity (also known as the "v" factor) of 20 cm/hour, everything else being identical, which of the below two choices would you recommend and why? (Fig. 6.28)
   (a) a 1-meter-deep tank, or
   (b) a 1-meter-deep tank but divide it into 5 strata, each 20 cm deep. ◀

Fig. 6.28 Settling tank setups

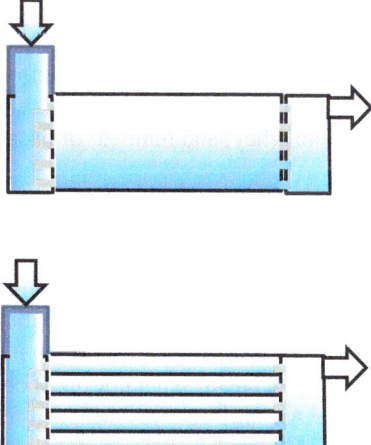

Gravity-based separation is slow and not very efficient for many industrial practices since gravitational force is a constant of nature. A modified version of Stokes' equation is used in centrifugal separations that replaces the "g" factor with centrifugal forces that can be easily manipulated, as described below.

A. **Centrifugal separation** uses a centrifuge, which is a device that spins at high speeds and causes denser materials in fluids to migrate outward in the rotating field faster than they would under the force of normal gravity. In circular motion, the acceleration due to the angular velocity is

$$a_c = \omega^2 R \tag{6.35}$$

where a_c is the acceleration from the centrifugal force in m/s^2, ω is the speed of revolution or angular velocity in radians per second, and R is the radial distance from the center of rotation. If the rotational velocity is expressed as N rpm, the above expression is written as:

$$a_c = \left(\frac{2\pi N}{60}\right)^2 R \tag{6.36}$$

For centrifugal separation of particles in the Stokes' law range, the terminal settling velocity in a centrifuge of radius R and spinning at N rpm is obtained by substituting Eq. (6.36) to replace the gravitational constant in Eq. (6.34).

$$v = \frac{2}{9} r_p^2 \frac{(\rho_p - \rho)}{\eta} \left(\frac{2\pi N}{60}\right)^2 R \rightarrow v = r_p^2 N^2 R \frac{(\rho_p - \rho)}{410\eta} \tag{6.37}$$

Centrifuges are extensively used for a variety of purposes to separate liquids and solids from solutions, suspensions, or slurries in numerous applications. Some examples of its utility in the food industry include sugar and oil refining, dewatering of salads, removal of cream from milk, clarification of fruit juices, recovery of coffee solubles, cell (yeast) harvesting, separation of yogurt curd from whey, and removal of bacteria from milk (bactofugation).

Check Your Understanding

4. If an average milk fat globule has a diameter of 6 μm and the density of fat is 950 kg/m^3 and if the skim milk density is 1050 kg/m^3 and viscosity is 2 cP, calculate the rate of separation of fat globules:
(a) under the influence of gravity
(b) in a cream separator of 20 cm diameter bowl running at 8600 rpm
5. A centrifuge was designed to separate yeast cells (assumed spherical with a 10.0 μm diameter and density of 1250 kg/m^3) from fermentation broth with water-like properties. The centrifuge spins perpendicular to the axis of rotation at 500 rpm. The distance between the surface of the liquid and the axis of rotation is 5 cm, and the distance from the bottom of the cylinder to

that axis is 15 cm. Estimate the time it would take to have a complete separation. ◄

B. **Relative centrifugal force (RCF)** (or Centrifugal Number, N_c) is used to compare the force of acceleration generated by different centrifuges operating at different rpm. It is expressed as multiples of the Earth's gravitational field (g), e.g., 100 g = 100 RCF. It is computed as follows:

For a particle of mass m, the gravitational force is $F_G = mg$, and if it spins at N rpm at a radial distance R from the center, then:

$$\text{Centrifugal force, } F_c = m\omega^2 R = m\left(\frac{2\pi N}{60}\right)^2 R \tag{6.38}$$

The magnitude of the force developed in a centrifuge compared to the gravitational force, expressed as equivalent to the g force or centrifugal number (N_c), becomes:

$$N_c = \frac{\omega^2 R}{g} = \left(\frac{2\pi N}{60}\right)^2 \frac{R}{g} = 0.00112RN^2 \tag{6.39}$$

Commercial centrifuges operate in the range of $N_c \cong 200$ (basket centrifuges) to $N_c \cong 360{,}000$ (ultracentrifuges)

C. **Centrifuges** are classified according to the centrifugal number and range from laboratory-scale models to industrial units. The latter cover a wide range of throughputs or solid concentrations in suspensions that can be handled. The industrial centrifuges most commonly used in food processing include the following types:

- **Filter centrifuges** are designed for solid–liquid separations and have a perforated screen bowl media (cloth, wire screen, or plate) located at the periphery for the fluid to exit while retaining the solids. The filtered liquid is collected in the bowl and drains out, Fig. 6.29. The separated solids are periodically removed by different designed mechanisms. Applications include washing and cleaning salad greens, separation of sugar crystals from syrup, removal of precipitate from supernatant, and production of pulp-free orange juice, among many others. Subtypes include desludgers, which are used for high-solid suspensions, and polishers, which are used for lower-solid suspensions.

- **Disk bowl centrifuges**, also known as conical plate centrifuges or disk stack separators, consist of a stack of conical shaped disks that rotate on a vertical axis. The discs have holes that serve as the feed channels for the incoming feed liquid to be separated. These disks, up to approximately 200 in number with an outer diameter of 20 to 30 cm and angle of inclination between 45°–60°, are spaced approximately 0.5 to 1 mm apart. They divide the feed stream into thin strata to speed up separation. Due to the centrifugal force, the heavy phase

Fig. 6.29 Sectional view of a filter centrifuge

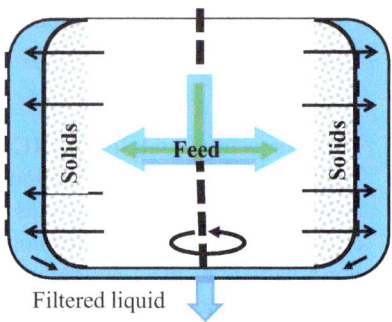

Filtered liquid

along with any present solids travel outward. The solids collect at the wall while the heavy liquid is pushed upward to the discharge outlet. The "self-cleaning or self-desludging" bowl opens intermittently and discharges the separated solids. The light phase liquid travels inwardly to the appropriate discharge outlet. More than 80% of the separation is known to occur right at the holes where the feed is distributed between the disks. For this reason, the location of the holes in the disc becomes important. When the holes are located toward the periphery of the disc, the light phase having to travel the longest distance under the influence of the centrifugal force gets separated with higher purity. The reverse is true when the holes are located closer to the axis of rotation. Disk bowl centrifuges are very versatile and can be used in batch or continuous separations. Figure 6.30 shows how a disk bowl operates during the separation of cream and some suspended solids from milk. The bowl rotation speed is generally in the range of 6000 to 10,000 revolutions per minute, which generates an N_c of 10,000 to 15,000.

The rate of cream separation in the Stokes' law regime is governed by Eq. (6.37), which can also be written as:

$$v = c\, r_p^2 N^2 R \qquad (6.40)$$

where $c = (\rho_p - \rho)/410\eta$. It represents the properties of milk and fat, contributing to the rate of separation in opposite directions. The value of c is maximum at approximately 35–40 °C due to an increase in the density difference between the serum and fat of milk and a decrease in viscosity. It is a preferred temperature range frequently used to attain maximum separation efficiency.

A schematic of the principle for direct in-line standardization of cream and milk using a centrifugal separator along with a homogenizer is shown in Fig. 6.31.

Cyclone separators are another very common, nonmechanical, and least expensive device for the separation of small solid particles from fluids. It consists of a vertical cylinder attached to a conical bottom (Fig. 6.32). The air-borne particle suspension enters a tangential inlet in the cylindrical top and

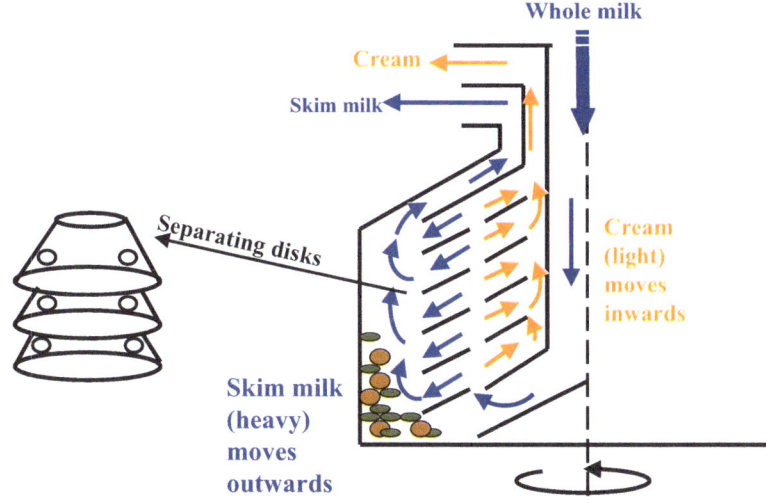

Fig. 6.30 Sectional view of a disk bowl centrifuge

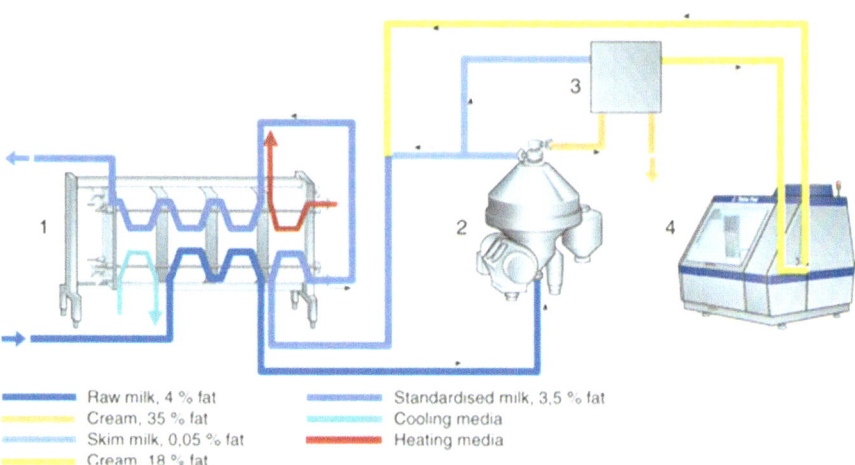

Fig. 6.31 A pasteurization unit with cream separator and homogenizer. (1) Heat exchanger, (2) Centrifugal separator, (3) Automatic fat standardization device, (4) Homogenizer. From *Dairy Processing Handbook* ©*Tetra Pak*

develops centrifugal force due to vortexing motion, which throws heavier particles radially toward the wall. However, it is not very effective if the particle size is <10 μ. Cyclone separators are an integral part of any spray drying unit, as shown in Fig. 6.33, both for collecting dry powder and for conveying and bagging it. In addition to spray drying, these separators are

Fig. 6.32 A cyclone
separator with top view

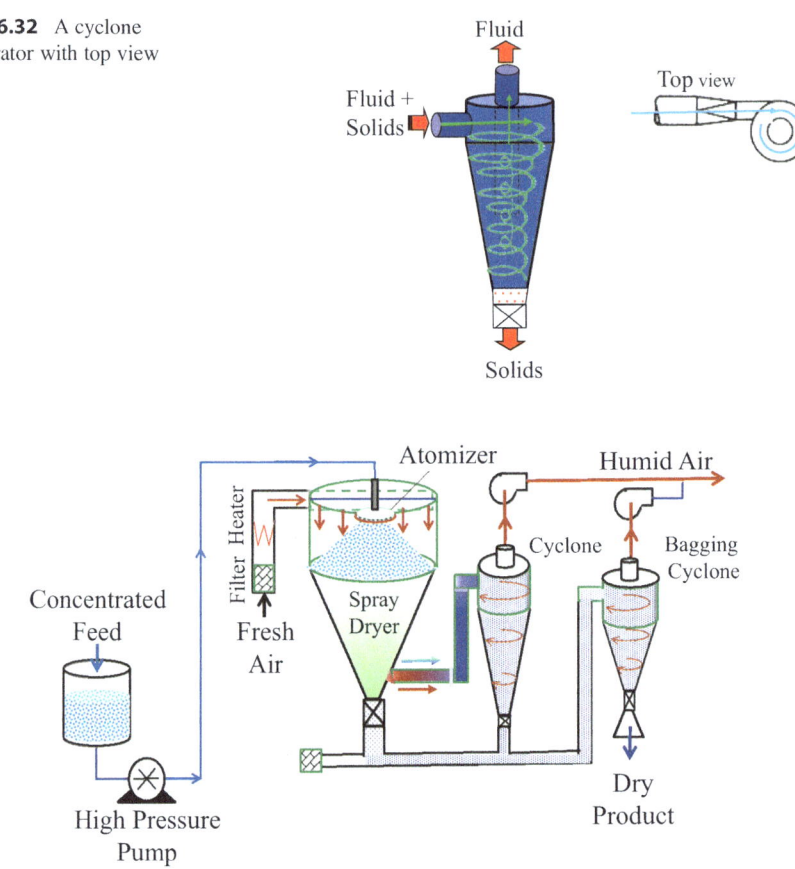

Fig. 6.33 Cyclone separators as part of a spray dryer and pneumatic conveying system

used in many applications in the food industry, such as pneumatic conveying, milling, and grinding operations. The Centrifugal Number developed in a cyclone ranges from 5 for large, low-velocity units to 2500 in small, high-velocity systems. Cyclones used for the separation of solids from liquids, such as in oil refining, are called **hydrocyclones.** These kinds of units also find applications in French fries and potato chips plants for starch recovery from cutting water and from wastewater. In the wheat and corn starch industries, they are used for both the concentration and washing of crude starch suspensions.

6.3.2   Fluid Flow in Packed and Fluidized Beds

There are other good examples of fluid flow passing through immersed objects, such as packed and fluidized bed systems (Fig. 6.34), that are widely used in the food

Fig. 6.34 Packed Bed (**a**) and Fluidized Bed (**b**)

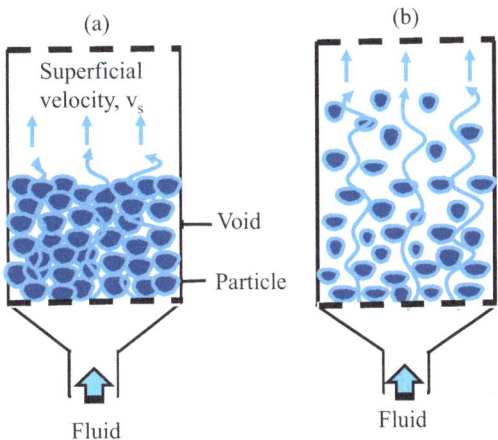

industry in various processes, such as filtration, distillation, drying, surface coating, individual quick freezing (IQF), biocatalysis by immobilized enzymes or microbial cells, and valorization of waste streams. In these systems, a bed or column is packed with particles of different shapes and sizes, and a fluid is introduced through a perforated distribution plate located at the bottom of the container holding the material. Drying moist foods such as grains, coffee beans, extrudates, etc. using thin-layer beds is also a technologically relevant process widely practiced in the food industry. It is somewhat more complex and involves simultaneous heat and mass transfer between the particles and the drying fluid passing through the porous bed.

In food processing, the food elements themselves are considered packing materials, and different types of packing materials are employed in processes such as distillation and gas absorption and are described in more advanced textbooks. A large surface area of contact between the packed particles and the fluid helps speed up heat and mass transfer processes and catalytic reactions. In practice, the important design expressions needed to estimate the size of the equipment for a given duty include the drop in pressure for a fluid flowing through a packing. In this discussion, the focus will be on packing materials that are uniform and spherical in shape.

The flow of fluid past submerged particles in **a packed bed** occurs via a tortuous path through channels in and around the bed material in a complex pattern. In such situations, a complete analytical solution is not easy to obtain, and experimental data generated for a variety of different systems are correlated and used. Some of the important characteristics of a packed bed are as follows:

(A) **Porosity**: A measure of the volume not occupied by the bed material and through which flow occurs is porosity (ε), defined as:

$$\varepsilon = \frac{volume\ of\ voids\ in\ bed}{Total\ volume\ of\ bed\ (voids + bed\ material)} \tag{6.41}$$

For uniform spherical bed material, ε is approximately 0.4. The average interstitial velocity (v_i) in the void space channels is related to the superficial velocity (v_s) as follows:

$$v_i = \frac{v_s}{\varepsilon} = \frac{\dot{V}}{A\varepsilon} \tag{6.42}$$

where superficial velocity (v_s) is obtained from the volumetric flow rate divided by the cross-sectional area of the empty container ($\dot{V}/A$).

(B) **Pressure drop:** As mentioned above, the most important issue in designing a packed bed unit is indeed the pressure drop required for the fluid to flow through the bed channel at a specified flow rate. For laminar flow, this can be estimated from the Hagen-Poiseuille equation for fluid with viscosity η with modifications for velocity (v) and replacing diameter with the hydraulic radius, as shown below:

$$\frac{\Delta P}{\Delta L} = \frac{32\,\eta v_i}{D^2} = \frac{32\,\eta\left(\frac{v_s}{\varepsilon}\right)}{(4r_H)^2} \tag{6.43}$$

where ΔL is the bed height and the hydraulic radius (r_H) for flow in bed channels is defined as:

$$
\begin{aligned}
r_H &= \frac{Total\ volume\ available\ for\ flow}{Total\ wetted\ surface\ of\ solids} \\
&= \frac{Void\ volume\ void/Bed\ volume}{Wetted\ surface/Bed\ volume} = \frac{\varepsilon}{a}
\end{aligned} \tag{6.44}
$$

In the above definition, a is related to the specific surface area a_s follows:

$$a = a_s(1 - \varepsilon) \tag{6.45}$$

where a_s is the particle surface area/particle volume, which for spherical particles of diameter D_p is $6/D_p$. Thus, the hydraulic radius for such particles becomes:

$$r_H = \frac{\varepsilon}{6(1 - \varepsilon)}\,D_P \tag{6.46}$$

Combining Eqs. (6.43) and (6.46), the pressure drop expression becomes:

$$\frac{\Delta P}{\Delta L} = \frac{32\,\eta\left(\frac{v_s}{\varepsilon}\right)}{(4r_H)^2} = \frac{72\,\eta v_s(1 - \varepsilon)^2}{\varepsilon^3\,D_p^2} \tag{6.47}$$

In real situations, ΔL is larger because of the tortuosity of the flow channels, and based on experimental data, the constant has been modified to 150, and the new equation is known as the **Blake-Kozeny** equation:

$$\frac{\Delta P}{\Delta L} = \frac{150 \, \eta v_s (1 - \varepsilon)^2}{\varepsilon^3 \, D_p^2} \tag{6.48}$$

The above equation works well for laminar flow with Reynolds numbers less than 10 and void fractions less than 0.5.

The Reynolds number for flow in a packed bed (Re_p) is similarly modified as:

$$Re_p = \frac{\rho(4r_H)(v_s/\varepsilon)}{\eta} = \frac{2 \, \rho D_P \, v_s}{3(1-\varepsilon)\eta} \tag{6.49}$$

Frequently, the constant 2/3 is omitted in calculating the Reynolds number. The values of Re_p to characterize turbulent vs. laminar flow in packed beds are hard to define because of the complexity of the flow involved, and the suggested guidelines are $Re_p > 100$, turbulent flow; $Re_p < 10$, laminar flow.

An empirical correlation for laminar flow in packed bed, introduced by French engineer, P.G. Darcy, now known as **Darcy's law**, can also be used to estimate the pressure drop:

$$\frac{\Delta P}{\Delta L} = \frac{v_s \eta}{\beta} \tag{6.50}$$

where β is the specific permeability coefficient, which depends only on the properties of the bed. It is expressed in units of cm^2 or Darcy units (1 Darcy = 10^{-8} cm^2), and some values of β for various packing materials are available in the literature.

Using the Fanning friction factor (f) to estimate the pressure drop for turbulent flow and performing analyses such as laminar flow analysis, it can be shown that for the packed bed:

$$\frac{\Delta P}{\Delta L} = \frac{3f \, \rho \, v_s^2 \, (1 - \varepsilon)}{\varepsilon^3 \, D_P} \tag{6.51}$$

For highly turbulent flow, $Re_p > 1000$, the friction factor assumes a constant value, 3f is replaced with 1.75, and the final equation, known as the **Burk-Plummer** equation, is used:

$$\frac{\Delta P}{\Delta L} = \frac{1.75 \, \rho \, v_s^2 \, (1 - \varepsilon)}{\varepsilon^3 \, D_P} \tag{6.52}$$

The above equations are useful in predicting the pressure that is needed to force a fluid flow through a packed bed at a given velocity.

6. Show that the pressure drop in a packed bed linearly increases with height. ◀

A **fluidized bed** works on the principle of fluidization, a process where fluid flows at such a high velocity that the particles attain a dynamic fluid-like state and the mixture resembles a boiling liquid (Fig. 6.33b). At low fluid velocity, the particles in a packed bed remain stationary because the pressure drop is not large enough to overcome the gravitational force on the particles. As the fluid velocity is increased, the friction forces (pressure drop times the cross-sectional area) and buoyancy (negligible) acting upward counterbalance the net gravitational force, and the particles begin to move and fluidize. The fluid superficial velocity at this point is known as the **minimum fluidization velocity** (v_{mf}), and it is one of the most important parameters in designing a fluidized bed system.

The **Ergun equation** is generally used for the calculation of pressure drop through a fluidized bed containing spherical particles. The Ergun equation is given as:

$$\frac{\Delta P}{\Delta L} = \frac{v_s (1 - \varepsilon)}{\varepsilon^3} \left(\frac{150 \, \eta \, (1 - \varepsilon)}{D_p^2} \right) + 1.75 \, \frac{\rho \, v_s}{D_p} \tag{6.53}$$

The above equation indicates that the pressure drop depends on the packing material size, fluid viscosity, and fluid density. The first term in the above equation accounts for the pressure drop due to friction. The second term, on the other hand, represents the pressure drop due to a sudden change in the flowing fluid direction and is negligible when $Re_p < 10$. The pressure drop needed for v_{mf} is obtained from the above expression by setting $v_s = v_{mf}$. For laminar flow, $Re_p < 10$, and the minimum fluidization velocity (v_{mf}) is also approximated by using the Kozeny-Carmen equation shown below:

$$v_{mf} = (D_P)^2 \, g \, \frac{(\rho_p - \rho) \, \varepsilon^3}{150 \, \eta \, (1 - \varepsilon)} \tag{6.54}$$

If the velocity becomes too high and the pressure drop approaches zero, particle breakage, entrainment and carryover from the fluidized bed leads to product loss. Some of the advantages of fluidized bed processes include high heat and mass transport between the particles and hot or cold air and ideal for processing temperature solids. However, sticky or too wet materials are not easy to handle and vigorous movement of particle in hot air drying may lead to generation of static electricity which can be dangerous in explosive atmosphere.

6.4 Agitation and Mixing of Fluids

Agitation is a process of providing bulk motion to a body of fluid to accomplish physical changes, chemical changes and/or increase rates of transport processes and includes operations such as dispersion, mixing, heating, cooling, and many others. It is nearly impossible to transform varied formulations into final products of defined quality unless the components are well mixed. Mixing reduces nonuniformity in composition, properties, or temperature in a process. For example, all containers of soups, drinks, salad dressings, etc. are expected to be uniform in composition with the same amount of each ingredient.

It is generally accomplished in a container in which the liquid is circulated and subjected to a desired level of shear with an agitator of different geometry and design, such as a paddle, impeller, propeller, or turbine attached to a power source such as a motor (Fig. 6.35). Many tanks are fitted internally with long, flat plates attached to the side called baffles. Baffles reduce radial swirling of the fluid generated by the mixing device and promote axial (top to bottom) mixing, especially when the viscosity is low.

During mixing, mechanical energy is required to rotate the agitator, which in turn transmits this energy to the fluid. Rotation of the agitator causes both bulk motion and fine-scale eddies in the fluid. The velocity gradients between the agitator and the sides and bottom of the container cause mixing and dispersion of the contents.

6.4.1 Commonly Used Agitator Designs

The design of mixing systems is dictated by the types of fluids involved. The mechanisms controlling the mixing of miscible fluids are very different from the dispersion of one immiscible material into another. The types of fluids (Newtonian vs. non-Newtonian) and quantity to be mixed control the energy

Fig. 6.35 Mixing tank with baffles

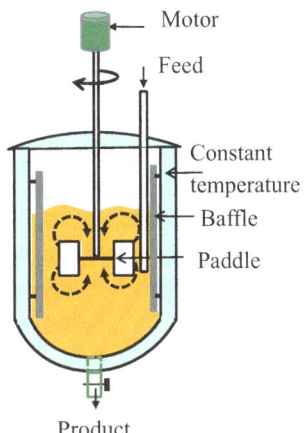

requirement to achieve the desired result. Different types of agitators have been designed to meet various needs and may be categorized into the following types (Fig. 6.36). The flow pattern created in the tank depends on the agitator design, geometric proportion of the tank and agitator, and the properties of the fluids to be mixed. Correct selection of an agitation device requires an understanding of the mechanism of their operation. A propeller produces fluid motion parallel to the axis of rotation linearly, i.e., it propels fluids by an extrinsic action. On the other hand, in an impeller the fluid moves to the center and is then impelled, or "thrown" out in some axial but mostly in radial direction movements.

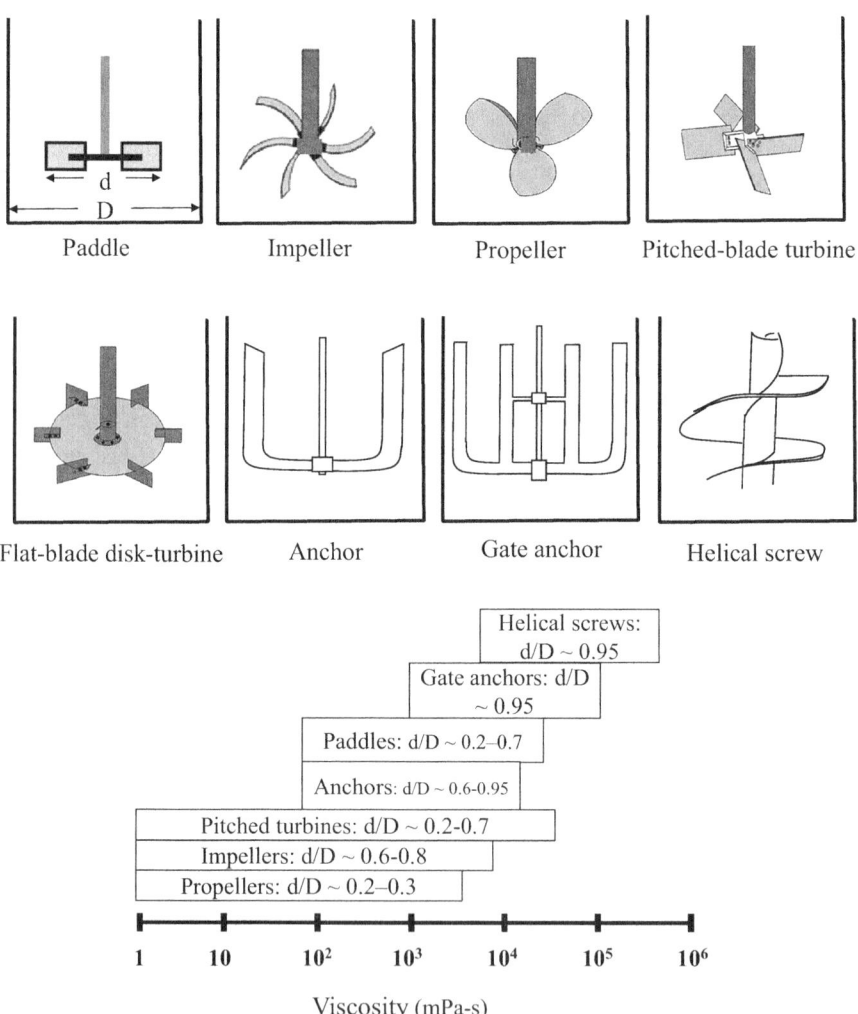

Fig. 6.36 Common agitators: design and viscosity ranges

A. **Radial flow agitators** discharge a fluid in a horizontal direction to the vessel wall since their blades are parallel to the axis, e.g., flat-blade paddle, impeller agitators.
B. **Axial Flow Agitators** have blades tilted at an angle of less than 90° to the plane of rotation and create a flow parallel to the axis (shaft), top to bottom, e.g., propeller and turbine agitators.
C. **Mixed flow agitators** create both axial and radial flow patterns, e.g., pitched turbine agitators.

Several factors dictate the choice of an agitator for the best results. These include viscosity, miscibility, and characteristics of the ingredient to be mixed. Food products ranging in viscosity from a few cP (e.g., water-like products) to more than a million cP (e.g., purees, slurries, pastes) are common and require agitators ranging from simple propeller type to scaped surface mixers. Hard-to-mix ingredients require high-shear propeller-type mixers. To maintain the integrity of delicate and shear-sensitive ingredients such as cut fruits and vegetables, gentle lifting and folding type agitators are needed. Due to their gentle motion, anchor agitators are mostly used in reactors. Gate anchor agitators are employed to mix materials of high viscosity at low shear. Helical screw mixers are most useful for mixing highly viscous liquids. A good understanding of the product and process requirements may be used to design agitators dedicated to specific applications.

Agitators are often designed to undergo two different types of motions:

A. **Single-motion agitators** use one shaft and turn continuously in one direction. They are configured to work with different types of tanks and mixing vessels of different shapes (cone, flat-bottom, hemispherical, etc.)
B. **Double-motion agitators** have two (inner and outer) eccentric shafts that rotate in opposite directions and provide good mixing.

For viscosities less than 2–5 Pa-s, baffles are needed in the tank. If the vessel is unbaffled, swirling and vortexing may result. With baffles, a strong top-to-bottom current is produced. The flow pattern in the baffled and unbaffled tanks is illustrated in Fig. 6.37.

6.4.2 Agitator Tip Speed: Reynolds Number and Power Number

If the agitator is rotating at N (revolution per second), the angular velocity is $2\pi N$ rad/s. If the diameter of the tip is d_o, then the speed of the tip is $\pi d_o N$ m/s. Additionally, since the characteristic velocity is the tip speed, the Reynolds number for agitation is defined as:

$$\text{Re}' = \frac{\rho d_o^2 N}{\eta} \tag{6.55}$$

Fig. 6.37 Flow pattern in an unbaffled (**a**) and a baffled (**b**) tank

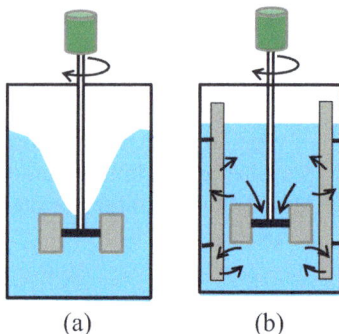

(a)   (b)

This Reynolds number is sometimes referred to as the agitator/impeller Reynolds number (Re′).

Re′ < 10   Laminar flow
Re′ = 10 − 10^4 Transition region
Re′ > 10^4 Turbulent flow

The energy input controls the quality of mixing. The power consumption is a function of the fluid density and viscosity, rotational speed, and impeller diameter. These variables are combined into a dimensionless number called the power number (N_p) and are defined as:

$$N_P = \frac{P}{\rho d_0^5 N^3} \tag{6.56}$$

where P is the mechanical power input to the fluid and ρ is the fluid density. The value of N_p for mixed flow impellers such as pitched blade turbines has been found to range from 0.8 to 2.0.

6.5   Measurement of Viscosity

Several different types of viscometers have been designed to measure the viscosity of fluids. In the case of simple, one-dimensional flow, viscosity is calculated from the measured ratio of shear stress and shear rate. During measurements, either the stress or strain rate is controlled, and the fluid viscosity determines the value of the other. Under normal experimental conditions, achieving the desired one-dimensional flow is not possible, and different errors come into play. To address these issues, various corrections have been suggested (see additional reading list).

The viscometers are classified (Fig. 6.38) based on how the flow takes place in the instrument. Additionally, vibrational (ultrasonic) viscometers are also available and used for high-viscosity fluids.

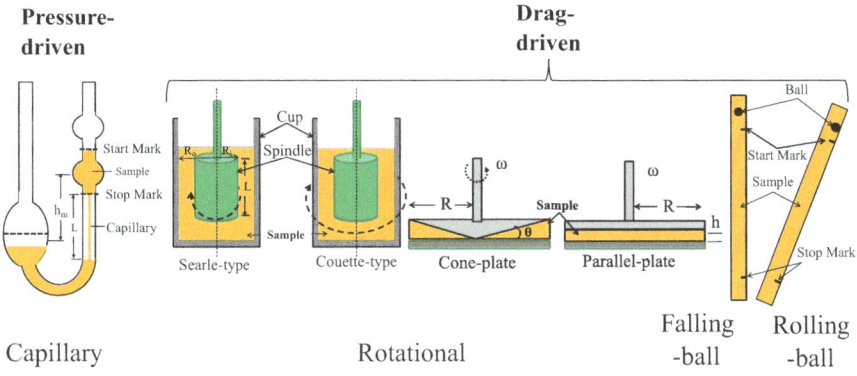

Fig. 6.38 Common types of viscometers

6.5.1 Capillary Viscometer

The classical Hagen-Poiseuille Eq. (5.89) discussed in Chap. 5 provides the basis for the measurement of the viscosity of Newtonian fluids in laminar flow using a capillary viscometer.

$$\eta = \frac{\Delta P \cdot \pi \cdot R^4}{8 L \dot{V}} \tag{6.57}$$

where $\dot{V}$ is the volumetric flow rate, ΔP is the pressure drop across the capillary, R is the capillary radius, η is the fluid viscosity, and L is the capillary length. For a capillary viscometer (Cannon Fenske type), shown in Fig. 6.38, the mean static fluid height difference between two liquid reservoirs, h_m, the pressure drop ΔP is $\rho g h_m$, and if volume V of the fluid flows through the capillary in time t, then $\dot{V}$ is V/t. The Eq. (6.57) can be written as:

$$\eta = \left(\frac{\pi R^4 \rho g h_m}{8 L V} \right) t \tag{6.58}$$

Alternatively, the kinematic viscosity ν ($-\eta/\rho$) may also be computed as follows:

$$\frac{\eta}{\rho} = \nu = \left(\frac{\pi R^4 g h_m}{8 L V} \right) t \tag{6.59}$$

When the fluid volume V drains every time between the start and stop marks, the mean height h_m is fixed, and the terms in parentheses become constant. Equation (6.59) may be simply written as:

$$v = \left(\frac{\pi \cdot R^4 g \, h_{\mathrm{m}}}{8 \, L \, V} \right) t \ \text{or} \ v = b \cdot t \tag{6.60}$$

where $b = \left(\frac{\pi R^4 g h_{\mathrm{m}}}{8LV} \right)$ is a viscometer constant, which can be simply determined by determining the efflux time for a fluid of known kinematic viscosity, as shown below:

$$b = \frac{v_{\mathrm{known}}}{t_{\mathrm{known}}} \tag{6.61}$$

Alternatively, the viscometer constant b can also be determined from the values of the parameters that define it. Using the value of b, the kinematic viscosity of a test fluid (unknown) can be determined by just measuring its efflux time and plugging it in Eq. (6.60). The capillary entrance and exit effects (called the end effects) become important if the efflux time is less than 200 s. In such cases, a modified version of Eq. (6.60) is utilized (see [2]).

6.5.2 Coaxial-Cylinder Rotational Viscometer

The concentric-cylinder or coaxial-cylinder viscometer is one type of rotational viscometer commonly used in rheological studies (Fig. 6.38). It consists of a cup containing the sample and a concentric cylinder (bob) that is placed in the sample to be tested. A motor drive provides a significantly stronger force for rotation than the Earth's gravitational force used in capillary viscometers discussed above, enabling characterization of more viscous liquids. Depending on whether the bob or the cup is rotated, there are two types of rotational viscometers:

- The **Searle** type in which the bob rotates inside an immovable cup filled with sample. Most rotational viscometers are based on the Searle principle of rotating the inner cylinder developed by the British physicist George Searle in 1912.
- The **Couette** type, where the sample-filled cup is rotated around the fixed bob. This idea was proposed and developed by the French physicist Maurice Marie Couette in 1888.

In principle, the torque T, required to rotate either the cup or the cylinder at a given number of revolutions per minute (N), is measured to obtain the flow curve for the material filled in the gap between the concentric cylinders. The shear stress (τ) exerted in the angular direction θ by a fluid surface at a radius R on the fluid in the region $R_{\mathrm{i}} < R < R_{\mathrm{o}}$, where R_{i} is the radius of the inner cylinder and R_{o} is the radius of the outer cylinder. Now, the torque M exerted at the radial position R is given by

$$M = F \cdot R \tag{6.62}$$

where F is the force exerted. From the definition of shear stress, τ, and substituting for the force from Eq. (6.62),

$$\tau = \frac{F}{A} = \left(\frac{M}{R}\right)\left(\frac{1}{2\pi RL}\right) \tag{6.63}$$

where A is the surface area at radius R, and L is the length. Therefore,

$$\tau = \frac{M}{2\pi LR^2} \tag{6.64}$$

The shear rate ($\dot{\gamma}$) as a function of angular velocity, ω ($= 2\,\pi\,N$), for a rotational system is:

$$\dot{\gamma} = R\frac{d\omega}{dR} \tag{6.65}$$

Remembering the definition of viscosity as the ratio of shear stress and shear rate, Eqs. (6.63) and (6.64) are equated to obtain the following relationship:

$$\frac{M}{2\pi\eta LR^2} = R\frac{d\omega}{dR} \tag{6.66}$$

Now, separating the variables and integrating the above equation with the boundary conditions $\omega = 0$ at R_0 and $\omega = \omega_i = 2\pi N$ at R_i, we obtain the final expression for the determination of viscosity as follows:

$$\int_0^{\omega_i} d\omega = \frac{M}{2\pi\eta L}\int_{R_0}^{R_i} R^{-3}dR \rightarrow \eta = \frac{M}{8\,\pi^2\,N\,L}\left(\frac{1}{R_i^2} - \frac{1}{R_o^2}\right) \tag{6.67}$$

In situations where the outer cylinder is a large container, it is reasonable to assume that $R_o >>> R_i$. On ignoring R_o, Eq. (6.67) is simplified as follows:

$$\eta = \frac{M}{8\,\pi^2\,N\,L\,R_i^2} \tag{6.68}$$

6.5.3 Cone-Plate and Parallel-Plate Rotational Viscometers

Cone-plate and parallel-plate viscometers are the most commonly used instruments for the measurement of non-Newtonian fluid properties. The sample is contained in a defined gap between the fixed plate and the rotating bob and undergoes uniform shear. The top is shaped as a cone or as a plate (Fig. 6.38). For cones, angles of $<4°$ are needed, but $1°$ is preferred, and the cone radius is generally between 10 mm and 100 mm.

As before, the shear stress and shear rate equation for the cone-and-plate geometry in terms of spherical coordinates can be written as:

$$\tau = \frac{3M}{2\pi R^3} \text{ and } \dot{\gamma} = \frac{\omega}{\theta} \tag{6.69}$$

where θ and R are the cone angle and radius, respectively. The angular velocity is ω ($= 2\pi N$). The viscosity is then given as:

$$\eta = \frac{3 M \theta}{4 \pi^2 N R^3} \tag{6.70}$$

Many viscometers are designed to permit the use of both cone-and-plate and parallel-plate geometries. For parallel plates, a gap between 0.5 mm and 3 mm is recommended, and the radius should be an order of magnitude larger than the gap. In this configuration, the shear rate depends on the gap between the plates (h) and the radial distance from the axis of rotation (r). The shear is thus not uniform and is expressed as a function of radial location:

$$\dot{\gamma}(r) = \frac{r \, \omega}{h} \tag{6.71}$$

For Newtonian fluids, the viscosity can be determined following integration over the plate area as

$$\eta = \frac{3 M h}{4 \pi^2 N R^4} \tag{6.72}$$

Since the gap between the plates is open to the side, care should be taken to avoid low-viscosity liquids from flowing away and thus distorting the results.

6.5.4 Falling-Ball and Rolling-Ball Viscometers

In these types of devices, a glass or steel ball of known mass and density is made to fall or roll through a tube filled with sample of the fluid being measured. The angle of inclination of the tube is adjusted to control the velocity of the ball to avoid turbulent flow and thus erroneous results. Rolling-ball viscometers have inclination angles between 10° and 80°, while the falling-ball viscometers' angle of inclination is between 80° and 90°. The viscosity is determined by monitoring the time of fall through a measured distance in the fluid at a constant temperature under the influence of gravity. Electrical sensors are used to determine the fall time for opaque samples, which permits determination of the terminal velocity (v) for use with the rearranged Stokes' equation discussed earlier, Eq. (6.34), to estimate the fluid viscosity as shown below:

$$\eta = \frac{2}{9} g r_p^2 \frac{(\rho_p - \rho)}{v} \tag{6.73}$$

Due to its simple design, this type of viscometer is frequently used. However, a finite container volume and the resulting interaction of the ball with the container wall and ends affect the terminal velocity and require modifications to Eq. (6.73) and some suggested empirical corrections are available from the literature.

Problems

6.1 A centrifugal pump is operating at 2000 rpm against a 28 m head. If the speed of the pump is halved, calculate the new velocity and the developed head.

6.2 You are designing a pipeline system using a centrifugal pump with an impeller diameter of 1500 mm to transport milk at a mass flow rate of 6 kg/s and the energy requirement to pump is 15 kJ/kg. If the pump efficiency is 90%, compute the brake power needed. However, the supplier can only provide a centrifugal pump with an impeller diameter of 2500 mm. If you decide to use the 2500 mm impeller diameter pump, calculate the power the new pump will require.

6.3 Compute the impeller diameter needed to develop a head of 20 m by a centrifugal pump running at 2500 rpm.

6.4 A storage tank is located 30 m above a body of water. If the pipe has an inside diameter of 5 cm, what power would have to be supplied to the fluid to pump the water into the tank at a rate of 0.03 m³/s? If the pump has an efficiency of 70%, what will be the power input required by the pump? Assume no energy loss due to friction in the connecting pipe.

6.5 Water flows through a pipe with 1" I. D. at a mass flow rate of 10 kg/s. The total length of the pipe is 20 m. There are two 45° elbows, one gate valve half open, and one swing check valve between the inlet and outlet. Determine the equivalent length of the straight pipe due to the friction loss caused by the fittings and valves. Assume a Fanning friction factor of 0.005.

6.6 Corn oil (S. G = 0.8) is being pumped at 10.22 m³/h through a pipe system of 30.48 mm I.D. to a height of 6.1 m above the initial level in the tank. The level in the tank drops at a rate of 61.0 mm/s. The total friction loss in the pipe system is 74.7 J/kg. If the flow is assumed to be laminar, what is the horsepower of the pump required for this process?

6.7 Vegetable oil (Newtonian fluid) is pumped from a feed tank at a rate of 57.5 m³/h to a storage tank located 30 m away and at a level 5 m higher than the feed tank. A pipe of 3" i.d. is used, and the piping system includes four standard 90° elbows and two gate valves (one fully opened and the other half opened). The viscosity of the oil is 60 mPa·s, and the density is 920 kg/m³. Determine the pumping power needed. Assume the pipe is smooth.

6.8 To pump milk ($\rho = 1150$ kg/m³; $\eta = 2 \times 10^{-3}$ kg/m·s) from a storage tank located at ground level to a process vessel situated 10.0 m above ground level, a 200 m long, 75 mm I.D. pipeline is used. The friction loss for the associated valves, fitting and entrance effects equals 10 times the velocity heads (based on the flow in the 75 mm I.D. pipe).

A centrifugal pump with the following characteristics is available for this job:

Flow rate (m^3/h)	10.0	15.1	19.5	25.2	30.0
Head (m)	26.1	25.1	23.9	21.0	16.1

If this equipment is to be used for the above process, develop a system's curve of dynamic head versus flow rate using the flow rates given in the table above and specify the optimal flow rate for using this and the power required if the pump is 85% efficient.

6.9 Orange juice (S.G. = 1.2) in a tank is being pumped at 20.0 L/s through a uniform pipe system (10.0-cm I.D.) to 5.0 m above the liquid surface. The total friction loss in the pipe system is 36 J/kg. If the flow is assumed to be turbulent, calculate (a) the velocity of the flow, (b) the total dynamic head of the pumping system, and (c) if the efficiency of the pump is 75%, calculate the fluid power and the break power requirements.

6.10 Chocolate milk at 24 °C ($\rho = 1025$ kg/m^3) in a large tank at atmospheric pressure is pumped through the system at a rate of 3.465 kg/s. The gauge pressure at the end of the discharge line is 217 kPa. The discharge height is 3.66 m, and the pump suction is 0.91 m above the level in the reservoir. The discharge line has an exit velocity of 2.1 m/s. The pressure drop by friction in the suction line is known to be 4.8 kPa and that in the discharge line is 33.1 kPa. If the vapor pressure of chocolate milk at 24 °C is 2.1 kPa and it is flowing turbulently, calculate (a) the differential head between the tank and the pump in kJ/kg and (b) if the pump manufacturer specifies an NPSHR of 3.048 m, what is the NPSHA, and will the pump be suitable for this service?

6.11 A Venturi meter is used to measure the velocity of water flowing through a pipe (I.D. = 75 cm). A manometer with mercury (S.G. = 13.5) is employed between the inlet and outlet of the nozzle, and the height difference in the manometer fluid level is 126 cm. If the coefficient of discharge of the Venturi meter is 0.9975 and the nozzle outlet diameter is 25 cm, calculate (a) the average velocity (in m/s) and (b) mass flow rate (in kg/s) of the water in the pipe.

6.12 Two identical particles ($D_p = 1$ mm) settle at terminal velocity in the Strokes flow regime, one in water ($\eta_w = 10^{-2}$ poise) and the other in glycerol ($\eta_{gly} = 800\, \eta_w$). Neglecting any buoyancy difference, how much slower or faster will the particle in glycerol fall?

6.13 Calculate the terminal velocity of snowflakes of 3 mm diameter falling through air at 0 °C. ($C_D = 2.5$, ρ_{air} (0 °C) = 1.3 kg/m^3, μ_{air} (0 °C) = 1.7×10^{-5} Pa·s, snow is on average 7% water by volume).

6.14 Corn starch has an average granular size of 10 μm and a particle density of 1500 kg/m^3. If corn starch is suspended in water, calculate (a) the velocity of gravitational sedimentation (in cm/h), (b) the Reynolds number of starch settling in the water, and c) if the starch suspension is centrifuged at 8000 rpm in a centrifuge with a 10 cm diameter of rotation, calculate the relative centrifugal force (RCF) and the velocity of separation.

6.15 Spherical particles ($\rho_p = 1500$ kg/m^3 and $d_p = 10$ μm) dispersed in water are harvested using a centrifuge consisting of several 20-cm long cylindrical tubes that are 85% full. The centrifuge operates at 1000 RPM, and the tubes spin perpendicular to the axis of rotation. During centrifugation, the distance between the top of the tube and the axis of rotation is 60 cm. How long would it take for all the particles to be completely separated?

6.16 A winery uses centrifugation to obtain clarified wine by separating impurities from wine of viscosity 2 cP and density 1.06 g/cm^3. The impure moieties can be assumed to be spheres with a diameter of 7.0 μm and density of 1.12 g/cm^3. The bowl of the separator has an outer diameter of 30 cm. If the centrifuge is operated at 600 rpm and the time it takes 90 min to separate the impurities, determine the distance between the surface of the liquid and the axis of rotation.

6.17 Milk ($\rho = 1025$ kg/m^3, $\eta = 3$ cP) flows at a rate of 10 ft^3/h through a 0.3 m diameter column packed with immobilized lactase-containing beads (diameter $= 1$ cm). If the bed porosity is 0.40, estimate (a) the superficial velocity, (b) the average interstitial velocity through the packed bed, and (c) the hydraulic radius of the beads.

6.18 An impeller stirrer (tip diameter $= 0.1$ m) is used for milk ($\rho = 1030$ kg/m^3, $\eta = 0.0012$ Pa s) agitation during heating to prevent burning. Calculate the speed at the impeller tip and the Reynolds number for agitation when the impeller is operating at 10 rpm.

6.19 To mix a liquid drink ($\rho = 1050$ kg/m^3, $\eta = 0.002$ Pa-s), an impeller stirrer is used to mix it during heating. The impeller diameter is 50 cm. You are told that to ensure adequate mixing and heat transfer within the fluid, the flow must be turbulent at 10% radial distance from the impeller center. Determine the minimum impeller speed in rpm and the Reynolds number at the tip.

6.20 A viscosity test is performed on a liquid food product with a density of 1200 kg/m^3 by dropping a steel ball with a density of 7000 kg/m^3 and a radius of 8 mm into a container filled with the food product. The ball attains a terminal velocity of 0.15 m/s. Compute the viscosity of the food. Verify whether Stroke's law is valid in this case.

6.21 The kinematic viscosity of canola oil ($\rho = 850$ kg/m^3) was measured at 25 °C using a capillary viscometer with a 40-mm long and 1 mm diameter capillary and reported to be 78 mm^2/s. The flow rate through the capillary was 5 mm^3/s. Determine the (a) dynamic viscosity of canola oil, (b) flow velocity of oil, (c) pressure drop across the capillary, and (d) head in mm required to produce the flow rate.

6.22 (A) You designed a new capillary viscometer and used a standard fluid of 1 cSt (centistokes) kinematic viscosity to calibrate it. It took 200 s for this fluid to drain from the start mark to the stop mark at 30 °C. Using this viscometer, you want to measure the viscosity of a new beverage ($\rho_m =$

1030 kg/m^3) that drains in 250 s at the same temperature. Determine the viscosity of the new beverage.

(B) To confirm your results, you use a falling-ball viscometer with a spherical bead ($\rho_s = 1.05$ g/cm^3) of 2 mm radius. The ball travels a distance of 10 cm between the start and stop marks in 750 ms. Compute the viscosity of the new beverage and indicate which of the two methods seems more reliable to you and why?

Bibliography

1. Bowser TJ (2006) Sanitary pump selection. OSU Food Technology Facts, FAPC-108
2. Van Wazer JR, Lyons JW, Kim KY, Colwell RE (1963) Viscosity and flow measurement. Interscience, New York
3. Oklahoma State University Cooperative Extension Service, Stillwater, OK
4. Macosko CW (1994) Rheology: principles, measurements, and applications. VCH, New York
5. Walters K (1975) Rheometry. Wiley, New York

Heat Transfer: Steady-State Conduction

Heat transfer involves the flow of thermal energy between objects due to temperature differences. It is an integral part of food processing and preservation techniques. Heating to make foods easier to digest and safe to eat has been practiced since time immemorial. It also helps extend their shelf life and makes them taste better. To achieve this, food processes require that heat be transferred into and out of the food. Some examples include:

- **Blanching**: At the start of processing, vegetables are subjected to moderate heat for a short period of time to stop enzymatic action that would otherwise reduce the quality of the food.
- **Cooking**: Many foods are heated to change their physical and chemical properties and make them more palatable and/or digestible.
- **Sterilizing**: In canning and certain other processes, food is subjected to high heat for sufficient time to destroy virtually all spores produced by pathogenic microorganisms to minimize the health hazard and produce foods that are stable for long periods of time.
- **Pasteurization**: Milk and certain other foods are subjected to enough heat to destroy vegetative cells of pathogenic organisms without radically changing their sensory characteristics. Not all organisms are destroyed by this process.
- **Evaporation and Drying**: The production of concentrated foods such as orange juice as well as dry foods requires the evaporation of water. This requires the addition of heat to supply the latent heat of water.
- **Freezing and Refrigeration**: These processes are used to preserve many foods and require first that heat be removed from food and second that new heat not be added. The first step requires refrigeration units. The second requires adequate insulation.

In each of these processes, it is necessary to control the temperature of the food and the rate of heating or cooling. In the chapters on thermodynamics, we focused on

© The Author(s), under exclusive license to Springer Nature Switzerland AG 2024
S. S. H. Rizvi, *Food Engineering Principles and Practices*,
https://doi.org/10.1007/978-3-031-34123-6_7

the quantity of energy that entered and left a system and on the direction the energy would move. Thermodynamics makes no mention of the speed of these energy changes. In this chapter and those that follow, we will examine the ways in which heat is transferred from one material or location to another and the factors that affect the rate of transfer.

7.1   Heat-Transfer Methods

Imagine that you have filled a pot with water and placed it on a support above the burner of an electric stove, as shown in Fig. 7.1. When the stove is turned on, **radiant** heat from the burner heats the bottom of the pot. This heat is **conducted** from molecule to molecule through the wall of the pot until it reaches the water. As the water heats, the warmer water tends to rise while the colder water sinks, resulting in a **natural convection** that distributes the heat throughout the pot. Should you choose to stir the water you will be transferring heat by **forced convection**. The methods of heat transfer described in this example are defined as follows:

* **Conduction** is the transfer of heat from molecule to molecule within a material and requires no movement of the material. Conduction is the only method of transfer within solids and plays some role in heat transfer within liquids and gases.
* **Convection** is the transfer of heat through a material by the mass movement of portions of the material. It is the principal form of heat transport in liquids and gases. **Natural convection** occurs when there is uneven heating of fluids. Because most materials expand on heating, hotter regions are lighter than colder regions, so the hotter regions rise while the colder regions sink. When pumps, fans, or other means are employed to force movement, we call it **forced convection**.
* **Radiation** transfers heat from one object to another without heating the intervening material. You have felt the heat on your face when you stand before a fire or a hot stove. If you place a shield in front of your face, the heat immediately ceases, showing that the air around you has not been heated. This form of heat transport uses electromagnetic radiation, which we also observe as light, radio waves, X-rays, etc.

This chapter will explore heat transfer by conduction. The next chapters will treat convection and radiation.

Fig. 7.1 Modes of heat transfer

7.2 Conduction Basics

In this section, we examine the underlying mechanisms of conduction, introduce Fourier's first law of conduction, and compare the rates of conduction in various food and nonfood materials.

7.2.1 Mechanisms of Conduction

In the earlier study of thermodynamics, you learned that when heat enters a system, it increases the internal energy of the system's material. If no nuclear or chemical changes take place, this results in an increase in internal energy in the form of increased kinetic energy of the individual molecules. This kinetic energy appears as changes in translation, rotation, or vibration of the molecules or as changes in the energy levels of electrons. If this energy results in phase changes, it is called **latent heat or enthalpy of phase transformation**. If it results in a temperature increase that can be felt, it is called **sensible heat**. There are several mechanisms by which this molecular kinetic energy is conducted through a material.

- **Conduction in fluids**. If the molecules are free to move as in a liquid or gas, some will collide, and in such collisions, some energy will be passed from the faster molecule to the slower one, similar to energy passing from one billiard or croquet ball to another.
- **Conduction in solids**. In solids, the molecules are held in a lattice by electrostatic bonds that behave like the springs in a bedspring. Faster vibrating molecules will tug on to slower ones through these bonds and pass some of their energy.
- **Conduction in metals**. In metals, a large part of the heat energy is transferred by the movement of electrons from atom to atom.

Whether by collision, through lattice bonds, or by electron movement, there will always be a tendency for energy to move through a material from **hotter to colder** regions. This movement will continue until the average internal energy (and hence temperature) is the same in all parts of a system.

7.2.2 Symbols and Units

We are going to look at the factors that affect the rate at which thermal energy is conducted through various materials. It is common to use a dot over a letter to mean "rate per unit time." For example, when a material is flowing, the rate in mass per unit time, we have used the symbol $\dot{m}$. Similarly, we use the symbol q for thermal energy and $\dot{q}$ for the "rate of thermal energy transfer." Thus, the following expression will be used:

$$\dot{q} = \frac{dq}{dt} \tag{7.1}$$

7.2.3   Fourier's First Law of Conduction

We are particularly interested in how rapidly the thermal energy will move under different circumstances. In 1822, Fourier proposed two simple laws. The first describes the flow of heat under steady-state conditions and will be introduced here. The second describes the flow of heat under unsteady-state conditions and will be introduced in a later chapter.

To understand the first law, imagine a slab of metal, Styrofoam or any other material of thickness x and area A, such as the one in Fig. 7.2. To keep things simple, let the area of the slab be large compared to the thickness so that heat entering or leaving along the edges can be ignored. Imagine further that the entire area of each side of the slab is held at a uniform temperature.

- If the two sides are at the same temperature, no heat will flow through the slab. However, if the sides are at different temperatures, heat will flow from the hotter side to the colder side, as indicated by the "q" arrow in the figure.
- In Fig. 7.2, a line is drawn to represent the temperature through the slab. (For purposes of this profile, the vertical dimension represents the temperature rather than the height of the slab. See the coordinates in the lower right-hand corner of the figure.) We call this line a **temperature profile,** and it represents temperature as a function of distance ($T = f(x)$).
- The slope of the profile is a measure of the rate at which temperature changes with distance and is called the **temperature gradient**. It is represented by the derivative of the profile, dT/dx.
- After a period, if the surface temperatures remain constant and the material is uniform, the profile will develop a **constant slope,** as shown in the figure. In other words,

Fig. 7.2 Steady-state heat transfer

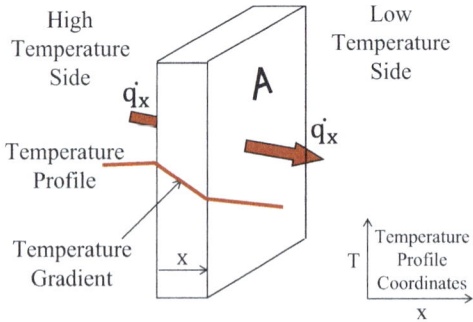

$$\frac{dT}{dx} = \text{constant} \tag{7.2}$$

- Heat transfer will then be in a **steady state**. Of course, since heat is constantly flowing through the slab, heat must be supplied to one side and removed from the other side if the steady state is to be maintained.
- Fourier observed that the rate of heat flow ($\dot{q}_x$) at any point in the slab was **proportional to the gradient**. In other words,

$$\dot{q}_x \propto \frac{dT}{dx} \tag{7.3}$$

- Furthermore, the rate is **proportional to the area** of the material that is at right angles to the direction of heat flow. In other words, a large sheet of material will pass heat faster than a small sheet, so

$$\dot{q}_x \propto A \tag{7.4}$$

- The rate also depends on the **nature of the material** through which the heat flows. Most metals conduct heat rapidly, while nonmetals conduct more slowly. Materials with a large amount of trapped air conduct heat very slowly.

Fourier's first law of conduction summarized these observations in the following equation:

$$\dot{q}_x = -kA\frac{dT}{dx} \tag{7.5}$$

where $\dot{q}_x$ = the rate of heat flow in the x direction in energy units per unit time.

A = the area of the material at right angles to the heat flow
dx = an infinitesimal distance in the x direction
dT = the difference in temperature across the distance dx
k = thermal conductivity, which is a proportionality constant whose value depends on the nature of the material

The negative sign in the equation indicates that the direction of heat flow is in the direction of decreasing temperature.

- **Heat Flux**

  Fourier's first law can be made independent of area if it is expressed in terms of heat flux (q'), which is defined as the rate of heat flow per unit area.

  $$q' = \frac{\dot{q}}{A} = -\frac{kA\frac{dT}{dx}}{A} = -k\frac{dT}{dx} \tag{7.6}$$

  The heat flux depends only on the thermal conductivity and temperature gradient.

7.2.4 Thermal Conductivity

Thermal conductivity (k) is a characteristic of each material and measures the rate of heat transfer as a function of area and temperature gradient. SI units for k can be derived by solving Fourier's law for k and writing units for the factors in this equation, we get:

$$k = \frac{\dot{q}_x}{A\frac{dT}{dx}} \; [=] \; \frac{J/s}{m^2(K/m)} = \frac{W}{m-K} \tag{7.7}$$

Since the rate depends on the temperature difference and not the actual temperature, the Celsius and Kelvin scales can be used interchangeably.

Example

1. *If a plate has a conductivity (k) of 400 W/m-K and a thickness of 1 m, then a temperature difference of one degree Kelvin will cause 400 Watts (400 Joules/s) of heat to pass through each square meter of the plate.* ◄

A. **Typical Thermal Conductivity Values: Thermal energy transport by conduction happens by two mechanisms: free electrons and lattice waves. These effects are additive and control the thermal conductivity of a material**. To obtain a feel for the values of the thermal conductivity coefficient, look over Table 7.1. In this table, notice the following:

- **Metals** have relatively high thermal conductivities because of the free movement of electrons with negligible lattice vibration. Silver and copper lead the list at approximately 400 Watts per meter per degree. Iron and mercury are much lower at 67 and 8.3, respectively. High thermal conductivity metals are used to fabricate equipment used for heating and cooling food, while low conductivity materials are generally used to insulate against heat losses.
- **Nonmetals and alloys** have conductivities near or below 1 Watt per meter per degree because they have few free electrons and conduction occurs by lattice vibrations.

Table 7.1 Thermal conductivity of some common materials

Material	W/m-K
Silver	422
Copper	391
Aluminum	202
Iron	67
Stainless steel	15
Mercury	8.3
Glass, wood	0.78
Many plastics	0.45
Cement	0.29
Pine	0.11
Cork	0.04
Air	0.021
Styrofoam	0.014

- **Water** has a conductivity of 0.60 W/m °C at 25 °C in the liquid state but 2.25 W/m °C in the solid state (ice at 0 °C). These values have a great effect on the rates of freezing and thawing of foods.
- **Oils** have a lower conductivity than water (approximately 0.2 W/m °C).
- **Gases** have a very low thermal conductivity (approximately 0.02 W/m °C), and materials that contain trapped air, such as cork and fiberglass, are particularly important, making them good insulators.
- Fluids in general have lower thermal conductivity than solids because of fluids have larger intermolecular spaces and the of intermolecular motion are more random.

B. **Effect of Temperature**: For most materials, thermal conductivities vary somewhat with temperature.
- **Gases**. For monatomic gases, assuming rigid spheres of mass m and diameter d, it has been shown that:

$$k \propto \frac{1}{d^2} \sqrt{\frac{T}{m}} \tag{7.8}$$

Thus the thermal conductivity of gases is proportional to the square root of temperature (up to 10 atm) and increases with increasing temperature.
- **Liquids**. The thermal conductivities of most liquids decrease with increasing temperature. For polar liquids such as water, the conductivities increase to a maximum and then decrease as the temperature increases.
- **Solids**. The thermal conductivities of solids vary with no set pattern. Conductivities of some increase and some decrease with increasing temperatures. As temperature increases, lattice vibration of molecules increases and caused the thermal conductivity to increase as is seen in alloys. The opposite happens in metals because the increase in lattice vibrations

Table 7.2 Thermophysical properties of food constituents

Product	Conductivity (k) W/m K	Specific heat (C_p) kJ/kg K	Density (ρ) kg/m^3	Thermal diffusivity (α) mm^2/s
Ice (0 °C)	2.25	2.11	920	1.10
Water (25 °C)	0.60	4.18	997	0.14
Protein	0.21	1.55	1272	0.09
Fat (Liquid + Solid)	0.18	1.95	913	0.10
Carbohydrate	0.22	1.42	1413	0.10
Ash	0.14	0.84	1731	0.10
Air	0.03	1.00	1.16	22.0

impedes the motion of free electrons and lowers their energy conducting ability.

C. **Estimating Conductivity of Foods**: In designing food processes, it is important to specify the time and temperature needed to produce a safe and palatable product. The selection of these parameters is greatly influenced by the thermal properties of the food. In general, most foods are poor conductors of heat, and their thermal conductivities depend on composition, moisture, temperature and structure. Thermal properties of various constituents of foods have been reported [1, 2] and a summary is given in Table 7.2. An estimate of a non-porous food's thermal conductivity at normal room temperatures may be obtained by computing a weighted average of the conductivities of its compositional constituents using the additivity principle, as given below:

$$k_f = 0.60\, x_w + 0.21\, x_p + 0.18\, x_f + 0.22\, x_c + 0.14\, x_a \tag{7.9}$$

where x_w, x_p, x_f, x_c and x_a are mass fractions of water, protein, fat carbohydrate and ash, respectively.

Thermal conductivities vary significantly with temperature and to obtain a more precise value for materials, Choi and Okos [3] reported the temperature dependence of the thermal conductivity of major food components in the temperature range of -40 to 150 °C as follows:

$$k_{ice} = 2.22 - 6.25 \times 10^{-3}\,T + 1.02 \times 10^{-4}\,T^2$$

$$k_{carb} = 2.01 \times 10^{-1} + 1.39 \times 10^{-3}\,T - 4.33 \times 10^{-6}\,T^2$$

$$k_{protein} = 1.79 \times 10^{-1} + 1.20 \times 10^{-3}\,T - 2.72 \times 10^{-6}\,T^2$$

$$k_{fat} = 1.81 \times 10^{-1} - 2.76 \times 10^{-3}\,T - 1.77 \times 10^{-7}\,T^2$$

$$k_{ash} = 3.30 \times 10^{-1} + 1.25 \times 10^{-3}\,T - 6.70 \times 10^{-6}\,T^2$$

$$k_{fiber} = 1.83 \times 10^{-1} + 1.76 \times 10^{-3}\,T - 3.17 \times 10^{-6}\,T^2$$

where temperature (T) is in Celsius and conductivity values are in W/m-K.

Example

2. *A food product has the following composition by weight: 10% protein, 15% fat, 23% carbohydrate, 2% ash and 50% water. Estimate its thermal conductivity.*
   Solution*: Multiplying the fractional composition for each component by the thermal conductivities read from Table 7.2, we obtain*:

$$0.1(0.21) + 0.15(0.18) + 0.23(0.22) + 0.02(0.14) + 0.50(0.60) = 0.40 \ W/mK \quad \blacktriangleleft$$

Note that factors such as the direction of fibers in solid foods like meat, fish, chicken, etc. significantly affect their conductivity. For such cases, the effective thermal conductivity depends on the heat flow direction through the material and estimation using parallel (Eq. 7.10a) and series (or perpendicular) (Eq. 7.10b) predictive models, like electrical resistance models, are used. In the parallel model, the effective thermal conductivity (k_{II}) is the sum of the thermal conductivities of the food components multiplied by their volume fraction:

$$k_{II} = \sum_{i=1}^{n} V_i k_i \tag{7.10a}$$

where V_i is volume fraction of component i. In the perpendicular series model, the effective thermal conductivity (k_{Σ})is the reciprocal of the sum of the volume fractions of the food constituents divided by their thermal conductivities:

$$\frac{1}{k_{\Sigma}} = \sum_{i=1}^{n} \frac{V_i}{k_i} \tag{7.10b}$$

The volume fractions of component i may be calculated from the mass fraction (x_i) and density (ρ_i) as follows:

$$V_i = \frac{x_i/\rho_i}{\sum_{i=1}^{n}(x_i/\rho_i)} \tag{7.10c}$$

The amount of trapped air as in dry powders and porous or expanded materials adds complexity to thermal conductivity modeling of foods. It is strongly influenced by degree of porosity, pore size and shape, convection within pores, etc. For a food consisting of a continuous and a dispersed phase, the following Maxwell-Eucken equations with air as the continuous phase (Eq. 7.11) and with air as the dispersed phase (Eq. 7.12) are often used to estimate effective thermal conductivity (k_e) values.

Table 7.3 Thermal conductivity of some food materials

Material	W/m-K	Btu/h ft °F
Turkey, frozen	1.68	0.97
Fish, frozen	1.22	0.71
Potatoes, frozen	1.09	0.63
Applesauce	0.69	0.40
Pears	0.60	0.35
Cantaloupe	0.57	0.33
Potatoes, raw	0.55	0.32
Milk, skim	0.54	0.31
Turkey, fresh	0.50	0.29
Pork, fresh	0.46	0.27
Butter	0.20	0.12
Peanut or soy oil	0.17	0.10

$$\text{For air as continuous phase}: k_e = k_c \left(\frac{2k_c + k_d - 2\varepsilon(k_c - k_d)}{2k_c + k_d + \varepsilon(k_c - k_d)} \right) \tag{7.11}$$

$$\text{And for air as dispersed phase}: k_e = k_d \left(\frac{2k_d + k_c - 2(1 - \varepsilon)(k_d - k_c)}{2k_d + k_c + (1 - \varepsilon)(k_d - k_c)} \right) \tag{7.12}$$

where k_c refers to air conductivity, k_d is the liquid/solid (condensed) phase conductivity and (ε) is the porosity (i.e., volume fraction of air, see Eq. 1.18).

D. **Typical Thermal Conductivities of Foods**: Table 7.3 gives the thermal conductivities for some typical food products. Notice the following:
- The thermal conductivities of most foods fall between the thermal conductivity values of water and air, the two components in food with the highest and lowest thermal conductivity.
- Fats and oils have low thermal conductivity, their increasing contents thus will have a decreasing effect on the thermal conductivity of foods.
- Except when there is phase transition like from solid to liquid (e.g., ice to water) or vice versa, the thermal conductivity of foods is not very strongly affected by change in temperature.
- For frozen foods, the fractions of ice and water are first calculated using the method mentioned earlier (see Eqs. 3.42a & 3.42b) and then their respective thermal conductivities are estimated and combined with the thermal conductivity of the remaining food components to obtain the final value of the food product.
- **Frozen foods**. Ice has a higher thermal conductivity than liquid water, which affects the rates at which foods freeze and thaw. Figure 7.3a shows a piece of food undergoing freezing. Since heat must be removed from the outer surface, freezing begins there. However, this increases the thermal conductivity at the

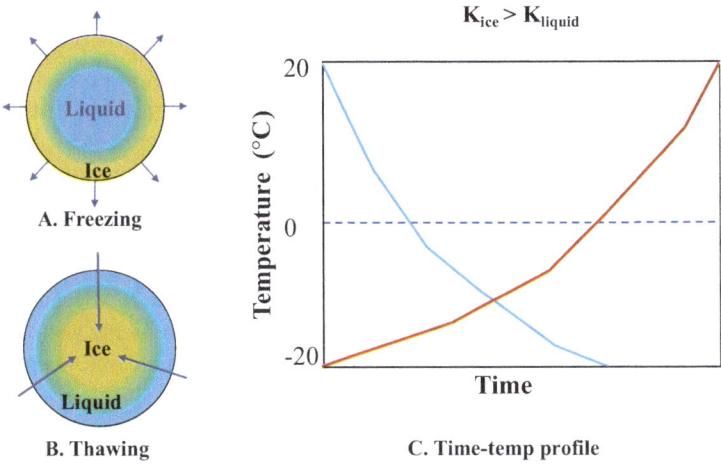

Fig. 7.3 Freezing versus thawing of foods

surface and hence the rate at which heat is removed. As more water is converted into ice, the freezing process speeds up.

During thawing, food also thaws from the outside as heat is added (Fig. 7.3b). The liquid layer that forms on the outside reduces thermal conductivity and acts as an insulator around the remaining ice. This slows down the heat removal with the result that thawing takes longer than freezing (Fig 7.3c). This becomes of very practical significance when large food containers are to be frozen or thawed.

While thermal conductivity determines how much heat flows in a material, there is another very important thermal property, called thermal diffusivity (α), which determines how rapidly heat will flow within the material. As discussed later in Chap. 9, thermal diffusivity controls the rate at which a material undergoes temperature change when heated or cooled. It is the ratio of a material's thermal conductivity (k) to its product of density (ρ) and specific heat at constant pressure (C_p). The thermal diffusivity of ice is approximately eight times higher than that of water and thus ice undergoes faster temperature change than water when exposed to identical temperature differentials.

Check Your Understanding

1. How is thermal diffusivity calculated? What is the thermal diffusivity of ice in m^2/s and how does it compare with the thermal diffusivity of water? ◄

7.3   Steady-State Heat Transfer Through a Slab

The next few sections will address conduction through a variety of geometric forms. The simplest situation occurs when heat is passing through a slab of uniform thickness that is made of uniform material. A flat griddle or a steak is a good example. If we know the thickness, area, and thermal conductivity of the slab, as well as the temperature on each side, we can use Fourier's law to determine the temperature profile through the slab and compute the rate at which heat flows.

Figure 7.4 shows a slab of area A with thickness $x_2 - x_1$. One side of the slab is at temperature T_1 and the other side is at a lower temperature T_2. Because of this temperature difference, heat flows through the slab, as shown by the vector $\dot{q}_x$. Our goals are to calculate the rate of heat flow and to determine the shape of the temperature profile through the slab.

7.3.1   Differential Slab

Rather than examining the flow through the entire slab, let us start with a simpler problem, namely, the heat flow through an infinitesimally thin slab called a **differential slab** that we imagine to be embedded in the real slab. The following are important properties of the differential slab:

- It can be placed at any **arbitrary location** along the thickness of the real slab and, in fact, represents an infinite number of similar slabs.
- It must be positioned **perpendicular** to the heat flow, $\dot{q}_x$.
- It has **differential thickness** dx, meaning that its thickness will be allowed to approach 0.
- For a simple slab, its **area is A**, the same as the real slab.

Fig. 7.4 Conduction through a slab

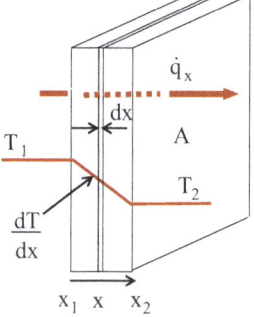

7.3.2   Energy Balance

The heat flowing through the real slab must pass through all arbitrary differential slabs that we select. Using the conservation of energy, the general equation for an energy balance through any differential slab is:

$$\begin{bmatrix} Rate\ of \\ energy\ in \end{bmatrix} + \begin{bmatrix} Rate\ of\ energy \\ generation \end{bmatrix} = \begin{bmatrix} Rate\ of\ energy \\ accumulation \end{bmatrix} + \begin{bmatrix} Rate\ of \\ energy\ out \end{bmatrix} \quad (7.13)$$

If no heat is generated in the slab, we can drop the generation term. Furthermore, if we assume a **steady state**, the temperature at any point will remain constant with time, so no heat can accumulate. Thus, if q is the thermal energy contained in the differential slab and t is time, we can write:

$$\frac{dq}{dt} = 0 \quad (7.14)$$

meaning that the quantity of thermal energy in the slab does not change with time. Under these conditions, the energy balance equation reduces to:

$$\begin{bmatrix} Rate\ of \\ energy\ in \end{bmatrix} = \begin{bmatrix} Rate\ of \\ energy\ out \end{bmatrix} \quad (7.15)$$

Thus, the heat entering the differential slab is always equal to the heat leaving. This means that we need to solve for just one value for heat flow through the differential slab. We call this $\dot{q}_x$.

By Fourier's first law, the magnitude of the heat flow through a differential slab is:

$$\dot{q}_x = -kA\frac{dT}{dx} \quad (7.16)$$

This is a differential equation that describes heat flow in terms of the rate of change in temperature (dT) with change in distance (dx) across the differential slab, which is simply the slope or gradient of the temperature profile.

Our first goal is to compute $\dot{q}_x$, the rate of heat flow through the slab in the x direction. To do this, we must rearrange Eq. (7.16), placing dx on the left with $\dot{q}_x$ and dT on the right with the other constants.

$$\dot{q}_x dx = -kAdT \quad (7.17)$$

We integrate this equation from one side of the slab to the other, taking x_1 and x_2 as limits for x and T_1 and T_2 as limits for T. Since $\dot{q}_x$, k, and A are constant and taken outside the integral, the above equation can be written and integrated as:

$$\dot{q}_x \int_{x_1}^{x_2} dx = -kA \int_{T_1}^{T_2} dT \tag{7.18}$$

Substituting the limits and solving for $\dot{q}_x$, we obtain:

$$\dot{q}_x = -kA\left(\frac{T_2 - T_1}{x_2 - x_1}\right) = kA\left(\frac{T_1 - T_2}{x_2 - x_1}\right) = kA\frac{\Delta T}{\Delta x} \tag{7.19}$$

If the area (A), thickness (Δx), and thermal conductivity (k) of the slab are known along with the temperature difference ($T_1 - T_2$) across the slab, Eq. (7.19) allows us to compute the heat flow ($\dot{q}_x$).

A. **The driving force and resistance** concept is very useful in understanding heat-transfer operations. In electricity, Ohm's law states that, in a circuit such as the one in Fig. 7.5, the current (I) passing through a resistor (R) equals the potential drop (voltage, V) across the resistor divided by the magnitude of the resistance (R), i.e., $I = E/R$. In other words, the rate of current flow is proportional to the driving force (voltage) and inversely proportional to the resistance.
Heat flow can be put in the same form if we take heat flow ($\dot{q}_x$) to be analogous to electric current (I) and temperature drop analogous to potential drop or voltage. Thermal resistance is defined as follows:

$$R = \frac{(x_1 - x_2)}{kA} = \frac{\Delta x}{kA}[=]\frac{^\circ C}{w} \text{ or } \frac{^\circ F - h}{Btu} \tag{7.20}$$

With this definition of thermal resistance, Eq. (7.19) becomes:

$$\dot{q}_x = kA\frac{\Delta T}{\Delta x} = \frac{\Delta T}{\left[\frac{\Delta x}{kA}\right]} = \frac{Driving\ Force\ (\Delta T)}{Resistance\ (R)} \tag{7.21}$$

This is the thermal equivalent of Ohm's law. Notice in the definition of thermal resistance (Eq. 7.21) that the resistance to flow:
- Increases with increasing thickness of the slab.
- Decreases with increasing slab area.
- Decreases with increasing thermal conductivity of the slab.

Fig. 7.5 An electric circuit analogy

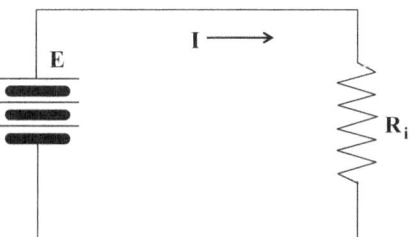

Table 7.4 Typical R-values of selected insulation materials

Material	R-value (ft^2·h·°F)/BTU	R-value (m^2·K/W)
Vacuum insulated panel (1 in. thick)	30.0	5.3
Brick (4 in. thick)	4.0	0.7
Fiberglass board (1 in. thick)	4.4	0.8
Cellulose fiber (1 in. thick)	3.7	0.6
Drywall (0.5 in. thick)	0.5	0.1

B. Thermal Insulation

Thermal insulation materials used for minimizing heat loss through pipes, walls, and other surfaces are characterized by their "R-value." It is defined as the ratio of the temperature difference across an insulator and the heat flux through it or R-value $(\Delta x/K) = (\Delta T)/(\dot{q}/A)$. The R-value is the reciprocal of the rate of heat flux through a material per degree temperature difference between its two sides.

The SI units for **R-values** are m^2·°C/W. In the USA R-values are given in units of ft^2·°F·h/Btu. The higher the R-value is, the less the heat flow and the better the insulation's effectiveness. Insulation for the home has R-values in the range of R–10 to R–30 in the American engineering system. Table 7.4 shows the R-values of some typical insulation materials.

Check Your Understanding

2. In still air ($k = 0.03$ W/m-K), it takes a food 3 hours to cool to 4 °C. How long will it take to cool the same temperature in water ($k = 0.6$ W/m-K) if heat transfer is assumed to occur by conduction only. ◄

7.3.3 Determination of the Temperature Profile

Our second goal is to determine the temperature profile, which is shown in Fig. 7.4 as a sloping line through the slab. The temperature profile expresses temperature (T) as a function of distance (x) into the slab. Thus, dT/dx in Eq. (7.3) must be the derivative of the temperature profile. Since integration is the inverse of differentiation, it follows that integrating Eq. (7.16) should yield the profile equation.

First separate the variables in Eq. (7.16), then integrate without limits (indefinite integral)

$$dT = -\frac{\dot{q}_x}{kA}dx \quad \text{or} \quad \int dT = -\int \frac{\dot{q}_x}{kA}dx \tag{7.22}$$

Because k and A are constants and, under steady-state conditions, $\dot{q}_x$ is also a constant, these factors can be brought outside the integral sign, and then integration is done, thus:

$$\int dT = -\frac{\dot{q}_x}{kA}\int dx \quad \text{or} \quad T = -\frac{\dot{q}_x}{kA}x + C \tag{7.23}$$

C in this solution is the usual arbitrary constant of integration that always occurs with an indefinite integral (one without limits). We evaluate such a constant by finding a set of conditions (called **boundary conditions**) in which both variables are known and substituting those values in the equation. Here, assume that x equals 0 on the left side of the slab, i.e., $x_1 = 0$. At this location, $T = T_1$. Substituting 0 for x and T_1 in Eq. (7.23) and solving for C, we obtain:

$$T_1 = -\frac{\dot{q}_x}{kA}(0) + C \quad \text{or} \quad T_1 = C \tag{7.24}$$

Substituting T_1 for C in Eq. (7.23) gives the equation for the temperature profile:

$$T = T_1 - \frac{\dot{q}_x}{kA}x \tag{7.25}$$

In the steady state, $\dot{q}_x$, k, and A are all constants, so $\dot{q}_x/kA$ must also be a constant. Thus, Eq. (7.25) is the equation for a straight line with intercept T_1 and slope $-\dot{q}_x/kA$. We can conclude that the temperature profile inside the slab is a straight line sloping downward, as shown in Fig. 7.4.

Example

3. A slab of iron is 40 cm wide × 70 cm high × 0.5 cm thick. One surface is at 21 °C, and the other is at 19 °C. What is the rate of heat flow through the slab in watts (Joules/second)? What is the temperature profile?
Solution: The area of the slab is 0.4 (0.7) = 0.28 m². From Table 7.1, we see that for iron, $k = 67$ W/m-°C. Substituting in Eq. (7.21) gives

$$\dot{q}_x = \left(67\frac{W}{m°C}\right)(0.28m^2)\left(\frac{21-19 \; °C}{0.005 \; m}\right) = 7504 \; W$$

Substituting into Eq. (7.25), the profile through this slab is given as

$$T = 21°C - \frac{(7504 \; W)}{\left(67\frac{W}{m°C}\right)(0.28 \; m^2)}(x)$$

$$T = 21°C - 400x$$

The temperatures at various distances into the slab are:

Distance from left side   *Temperature*
0.0 cm $21 - 400\,(0) = 21\,°C$
0.1 cm $21 - 400\,(0.001) = 20.6\,°C$
0.3 cm $21 - 400\,(0.003) = 19.8\,°C$
0.5 cm $21 - 400\,(0.005) = 19.0\,°C$ ◀

7.4   Steady-State Heat Conduction Through a Hollow Cylinder

Pipes are used extensively in industry to transport fluids from place to place. When the temperature of the fluid is greater or less than ambient, the problem of heat transfer with the surroundings becomes important. A pipe is basically a long hollow cylinder, so let us analyze that shape.

Figure 7.6 shows a hollow cylinder with inside radius r_i, outside radius r_o, and length L. The inside and outside surface areas are:

$$A_i = 2\pi r_i L \tag{7.26a}$$

$$A_o = 2\pi r_o L \tag{7.26b}$$

The inside wall of the cylinder is at temperature T_i and the outside wall is at temperature T_o. If $T_i > T_o$, heat will move outward through the wall. If $T_i < T_o$, heat will move inward through the wall. In either case, heat will tend to move parallel to the radii and at right angles to the surfaces.

7.4.1   The Classical Approach

The problem we face is that the area through which the heat passes continuously increases through the slab. A plausible approach would be to use an average area in the equation developed for the slab. If $A_m =$ mean area, Eq. (7.19) becomes:

Fig. 7.6 Heat conduction through a hollow cylinder

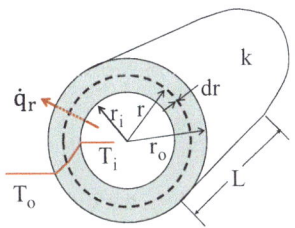

$$\dot{q}_r = kA_m \frac{T_i - T_o}{r_o - r_i} \tag{7.27}$$

However, how do we compute a suitable mean area? One possibility is:

$$A_m = \frac{A_i + A_o}{2} \tag{7.28}$$

As you will see shortly, this does not give the correct results, although, for thin-walled pipes, it is very close.

7.4.2   The Analytical Approach and the Differential Shell

A better approach is to apply calculus to the cylindrical geometry. Imagine a thin cylindrical shell at some arbitrary location within the wall of the real cylinder, as shown in Fig. 7.6. This shell will have some radius r and infinitesimal thickness dr. The area of this shell is equal to its circumference times its length or

$$A = 2\pi r L \tag{7.29}$$

The advantage of starting with the differential shell lies in the fact that it is so thin (approaching 0) that the inside and outside areas can be equal. Thus, the differential equation for this shell is the same as for a flat slab. This is the type of simplification that calculus allows.

A. **Energy balance is** written on the system when there is no heat generated within the wall of the cylinder and a steady state has been reached, both the heat generation term and the heat accumulation term will drop out of the energy balance as they did with the slab. Thus, we can again deal with a single heat flow through the shell that we call q_r. The subscript reminds us that the heat flow is along the radii.
B. **The differential equation** for the rate of heat flow through the shell, Fig. 7.6, by Fourier's first law is written as:

$$\dot{q}_r = -kA \frac{dT}{dr} \tag{7.30}$$

Unlike the case of the slab, however, the area of successive shells changes with r according to Eq. (7.29). Substituting this for A, Fourier's Eq. (7.5) becomes

$$\dot{q}_r = -k(2\pi r L) \frac{dT}{dr} \tag{7.31}$$

which is the differential equation for the slope (gradient) of the temperature profile through the wall of the cylinder. To determine the heat flow ($\dot{q}_r$), we first rearrange Eq. (7.31) and then integrate r from r_i to r_o and T from T_i to T_o

$$\dot{q}_r \left(\frac{dr}{r} \right) = -k(2\pi L)dT \quad \text{or} \quad \dot{q}_r \ln(r) \big|_{r_i}^{r_o} = -k(2\pi L)T \big|_{T_i}^{T_o} \qquad (7.32)$$

Substituting limits in this solution, we obtain:

$$\dot{q}_r[\ln(r_o) - \ln(r_i)] = -k(2\pi L)[T_o - T_i] \quad \text{or} \quad \dot{q}_r = 2k\pi L \left[\frac{T_i - T_o}{\ln(r_o) - \ln(r_i)} \right] \quad (7.33)$$

Finally, since $(\ln x - \ln y) = \ln (x/y)$, we can rewrite this equation as:

$$\dot{q}_r = 2k\pi L \left[\frac{T_i - T_o}{\ln \left(\frac{r_o}{r_i} \right)} \right] \qquad (7.34)$$

This is the analytical solution for the rate of heat flow through a cylinder wall.

C. **Resistance in a Hollow Cylinder**

If we rewrite Eq. (7.34) as:

$$\dot{q}_r = \frac{T_i - T_o}{\left[\frac{\ln \left(\frac{r_o}{r_i} \right)}{2k\pi L} \right]} = \frac{Driving\ force\ (\Delta T)}{Thermal\ resistance\ (R)} \qquad (7.35)$$

where the resistance through the wall of a hollow cylinder is defined as:

$$R = \frac{\ln \left(\frac{r_o}{r_i} \right)}{2k\pi L} \qquad (7.36)$$

It is interesting to note that the resistance of a hollow cylinder depends on the ratio of the outside radius to the inside radius and not on the actual magnitude of these radii. This means that a pipe with a 1 cm OD and ¾ cm ID will lose or gain heat at exactly the same rate as a pipe with a 4 cm Od and a 3 cm Id. Is this reasonable? Consider that as a pipe gets larger, its area gets larger, and this should reduce the resistance. However, for the same ratio, as the diameter increases, the wall thickness increases, which should increase the resistance. The effects of area and thickness cancel each other out, and only the ratio of radii is important.

7.4.3   The Log Mean Area

In the classical approach (Eq. 7.27), $\dot{q}_x$ was computed in terms of a mean area. If that equation is compared to Eq. (7.34), we can determine the proper way to compute this mean area. Setting Eq. (7.27) equal to Eq. (7.34) and rearranging:

$$k\left[\frac{A_M}{(r_o - r_i)}\right](T_i - T_o) = k\left[\frac{2\pi L}{\ln\left(\frac{r_o}{r_i}\right)}\right](T_i - T_o) \tag{7.37}$$

Dividing this equation by $k(T_i - T_o)$, we find that:

$$\frac{A_M}{r_o - r_i} = \frac{2\pi L}{\ln\left(\frac{r_o}{r_i}\right)} \tag{7.38}$$

If we solve Eq. (7.38) for A_M and then multiply the ratio in the denominator by $(2\pi L/2\pi L)$, we obtain:

$$A_M = \frac{2\pi L(r_o - r_i)}{\ln\left(\frac{r_o}{r_i}\right)} = \frac{2\pi L r_o - 2\pi L r_i}{\ln\left(\frac{2\pi L r_o}{2\pi L r_i}\right)} \tag{7.39}$$

However, $2\pi L r_o$ and $2\pi L r_i$ are the expressions for the outside and inside areas of the cylinder. Thus, Eq. (7.39) reduces to:

$$A_M = \frac{A_o - A_i}{\ln\left(\frac{A_o}{A_i}\right)} = A_{LM} = log\ mean\ area \tag{7.40}$$

This expression is called the **Log Mean Area** (A_{LM}) and is the correct mean to use in Eq. (7.27), which now becomes:

$$\dot{q}_r = kA_{LM}\frac{(T_i - T_o)}{(r_o - r_i)} \tag{7.41}$$

It turns out to be very easy to remember the order of subscripts in Eq. (7.41). In the numerator, "i" comes before "o", and in the denominator, "o" is before "i". If you reverse the subscripts in both the numerator and the denominator, the area is unchanged, so just remember to write the subscripts in opposite order in the numerator and the denominator.

7.4.4   The Temperature Profile

When we determined the profile equation for the slab, we integrated without limits (indefinite integral) and then substituted known boundary conditions to determine the constant of integration. We can do that here, but the same results can be achieved by integrating between limits if we use variables for upper limits. Specifically, we will use the constant r_i and the variable r for the radius limits and the constant T_i and the variable T for the temperature limits.

$$\int_{T_i}^{T} dT = -\frac{\dot{q}_r}{2k\pi L}\int_{r_i}^{r}\frac{dr}{r} \tag{7.42}$$

Integrating, substituting the limits, and solving for T give the profile equation:

$$T = T_i - \frac{\dot{q}_r}{2k\pi L}\ln\left(\frac{r}{r_i}\right) \tag{7.43a}$$

$$or, \quad T = T_i - \frac{2.303\,\dot{q}_r}{2k\pi L}\log\left(\frac{r}{r_i}\right) \tag{7.43b}$$

We see from this equation that the temperature is a linear function of the natural log of the radius. Figure 7.7 shows a plot of this equation compared to a linear plot. Note that the profile is slightly steeper near the inner wall of the cylinder, where the smaller area produces a greater resistance to heat flow. Additionally, note that, except where walls are very thick compared to the radius of the cylinder, the deviation from a straight line is small.

Example

4. *A 5-foot length iron pipe has an outside diameter (OD) of 1" and an inside diameter (ID) of 0.62 in. It carries hot water that raises the inside surface of the pipe to 130°F. If the outside surface of the pipe is at 25 °F, how much heat is lost through the walls of the pipe? Assume that the velocity of the water is large enough so that there is a negligible temperature drop along the pipe. What is the profile through the pipe wall?*

Fig. 7.7 Temperature profile through a cylinder wall

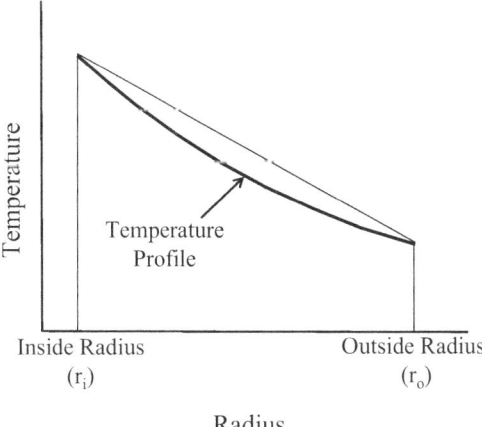

$$\textbf{\textit{Solution}} : \dot{q}_r = \left(39\,\frac{btu}{h\,ft\,°F}\right)2\pi(5ft)\left[\frac{(130\,°F - 25\,°F)}{\ln\left(\frac{1.00\,in}{0.62\,in}\right)}\right] = 2.69 \times 10^5\,\frac{btu}{h}$$

Solution: (Alternative)

$$A_i = \pi\frac{(0.62\,in)}{(12\,in/ft)}(5ft) = 0.812ft^2 \quad A_o = \pi\frac{(1.00\,in)}{(12\,in/ft)}(5ft) = 1.309ft^2$$

$$A_{LM} = \frac{(1.309ft^2 - 0.812ft^2)}{\ln\left(\frac{1.309ft^2}{0.812ft^2}\right)} = 1.04ft^2$$

$$\dot{q}_r = \left(39\,\frac{btu}{h\,ft\,°F}\right)(1.04ft^2)\left[\frac{(130\,°F - 25\,°F)}{\left(\frac{1}{24} - \frac{0.62}{24}\,ft\right)}\right] = 2.69 \times 10^5\,\frac{btu}{h}$$

$$\textbf{\textit{Profile}} : T = 130 - \left(\frac{2.69 \times 10^5\,\frac{Btu}{h}}{2\left(39\,\frac{Btu}{h\,ft\,°F}\right)\pi(5ft)}\right)\,\ln\left(\frac{r}{0.62}\right)$$

$$T = 130 - 219.5\,\ln\left(\frac{r}{0.62}\right)$$

Although the original data for this problem only warrant 2 significant figures in the result, we have retained 4 in the slope for plotting purposes only. This is necessary to make the profile pass through both T_1 and T_2. Except for that, we should report a slope of 2.2×10^2. The curve in Fig. 7.7 is a plot of this last equation. ◄

7.5   Slabs in Series

In the previous section, we developed the equation for heat flow through a slab made of a single material. More commonly, heat must pass through a series of materials. For example, the walls of a refrigerator can be thought of as slabs made up of a layer of steel, a layer of insulation, another layer of steel, and a layer of glaze. In this section, we will develop the equation for heat flow through a series such as this. In the next section, we will do the same for a multilayered pipe.

7.5.1 Conductor Geometry

Figure 7.8 shows a slab consisting of three layers A, B, and C, each W m wide and H m high. The thermal conductivities of the layers are k_A, k_B, and k_C, respectively. The thicknesses of the layers are $(x_2 - x_1)$, $(x_3 - x_2)$, and $(x_4 - x_3)$. The temperatures at the surfaces of the layers are T_1, T_2, T_3, and T_4, but only T_1 and T_4 are known. The problem is to determine the rate of heat flow ($\dot{q}_x$) through the slab and the temperature profile, including the unknown temperatures T_2 and T_3.

7.5.2 Energy Balance

Again, we assume no heat generation and a steady state so that at any point in the slab,

$$\text{Heat In} = \text{Heat Out}.$$

Since this is true at any point in the slab, we can say that $\dot{q}_x$ = constant throughout the slab. This allows us to apply Eq. (7.20) to each layer, and since $\dot{q}_x$ is the same in each layer, equating these equations, we obtain:

$$\dot{q}_x = k_A A \frac{T_1 - T_2}{x_2 - x_1} = k_B A \frac{T_2 - T_3}{x_3 - x_2} = k_C A \frac{T_3 - T_4}{x_4 - x_3} \tag{7.44}$$

Solving for the change in temperature in each layer:

$$T_1 - T_2 = \frac{\dot{q}_x}{A}\left(\frac{x_2 - x_1}{k_A}\right) \tag{7.45a}$$

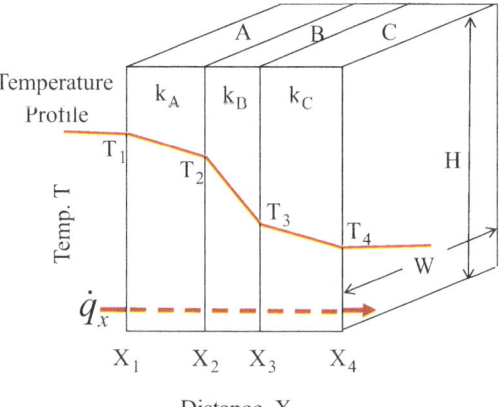

Fig. 7.8 Conduction through a multilayered slab

$$T_2 - T_3 = \frac{\dot{q}_x}{A}\left(\frac{x_3 - x_2}{k_B}\right) \tag{7.45b}$$

$$T_3 - T_4 = \frac{\dot{q}_x}{A}\left(\frac{x_4 - x_3}{k_C}\right) \tag{7.45c}$$

Once $\dot{q}_x$ is known, these equations will allow you to compute the temperature drop across each layer. For the moment, note that temperature drops are directly proportional to layer thickness ($x_{i+1} - x_i$) and inversely proportional to conductivity (k). In other words, there will be a greater drop across a good insulator than across a good conductor. (Which is why an insulated house is warmer in winter than an uninsulated house.)

Adding these three Eqs. (7.45a, 7.45b, and 7.45c) gives:

$$T_1 - T_4 = \frac{\dot{q}_x}{A}\left(\frac{x_2 - x_1}{k_A} + \frac{x_3 - x_2}{k_B} + \frac{x_4 - x_3}{k_C}\right) \tag{7.46}$$

Solving for heat flow ($\dot{q}_x$) gives:

$$\dot{q}_x = \frac{(T_1 - T_4)}{\left(\frac{x_2 - x_1}{Ak_A} + \frac{x_3 - x_2}{Ak_B} + \frac{x_4 - x_3}{Ak_C}\right)} \tag{7.47}$$

Note that each term in the denominator fits the definition for thermal resistance (Eq. 7.21). This allows us to rewrite the equation as

$$\dot{q}_x = \frac{T_1 - T_4}{(R_A + R_B + R_C)} = \frac{(\Delta T)_{\text{overall}}}{\Sigma R_i} = \frac{Overall\ Driving\ Force}{Sum\ of\ all\ Resistances} \tag{7.48}$$

Again, heat flow equals a driving force (temperature difference) divided by a resistance. Similar to the electric resistors shown in Fig. 7.9, the thermal resistances in series also contribute to the total resistance.

Fig. 7.9 A series electric circuit

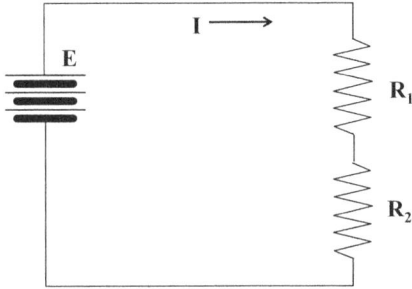

Example

5. *An old-fashioned refrigerator has a 2 by 4 ft door made of a 0.4" pine board, a 2" layer of cork, and a 0.7" pine board. The inside surface is at 40 °F, and the outside surface is at 70 °F. How many Btu's of heat pass through the door each hour and what are the temperatures on two cork surfaces?*

Solution: The conductivity values given in Table 7.1 *and their conversion into imperial units gives* $k_{Pine} = 0.11$ *W/mK* $= 0.065$ *Btu/h ft °F and* $k_{Cork} = 0.04$ *W/mK* $= 0.026$ *Btu/h ft °F. The thermal resistance offered by each layer is (Fig. 7.10):*

$$R_{Pine} = \frac{\frac{0.4}{12} ft}{(8\,ft^2)\left(0.065\,\frac{btu}{h\,ft\,°F}\right)} = \frac{0.0641\,°F}{\frac{btu}{h}}, R_{Cork} = \frac{\frac{2}{12} ft}{(8ft^2)\left(0.026\,\frac{btu}{h\,ft\,°F}\right)}$$

$$= \frac{0.8013\,°F}{btu/h}$$

$$R_{Pine} = \frac{\frac{0.7}{12} ft}{(8ft^2)\left(0.065\,\frac{btu}{h\,ft\,°F}\right)} = \frac{0.1122\,°F}{btu/h},$$

$$R_{Pine} + R_{Cork} + R_{pine} = 0.978\,°F/btu/h$$

The heat lost through the door is thus:

$$\dot{q}_x = \frac{(70 - 40\,°F)}{\left(0.978\,\frac{°F}{btu/h}\right)} = 30.7\,\frac{btu}{h}$$

The temperature drops across the layers are:

$$T_1 - T_2 = \left(30.67\,\frac{btu}{h}\right)\frac{\left(\frac{0.4}{12} ft\right)}{(8\,ft^2)\left(0.065\,\frac{btu}{hft\,°F}\right)} = 1.97\,°F$$

Fig. 7.10 Temperature profile through an insulated door

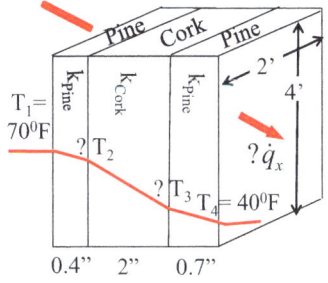

$$T_2 - T_3 = \left(30.67\frac{btu}{h}\right)\frac{\left(\frac{2}{12}ft\right)}{\left(8\,ft^2\right)\left(0.026\frac{btu}{hrft\,°F}\right)} = 24.58\,°F$$

$$T_3 - T_4 = \left(30.67\frac{btu}{h}\right)\frac{\left(\frac{0.7}{12}ft\right)}{\left(8\,ft^2\right)\left(0.065\frac{btu}{hrft\,°F}\right)} = 3.44\,°F$$

Therefore, the temperatures on the cork surfaces are:

$$T_2 = 70 - 1.97 = 68.03\,°F \text{ and } T_3 = 40 + 3.44 = 43.44\,°F \quad \blacktriangleleft$$

In the building trade, insulating materials such as fiberglass are rated with an R-value that indicates the resistance of a wall to the transmission of heat. R-values are given in terms of heat flux (q'), i.e., heat transmitted per square foot. The total R-value of a wall is equal to the sum of the R-values of the various layers. For example, a wall consisting of two layers of 3/4″ plywood ($R = 1.2$ per inch) separated by 3.5″ of cellulose insulation ($R = 3.3$ per inch) would have an R-value of: 0.75 (1.2) + 3.5 (3.3) + 0.75 (1.2) = 12.4. On a day when the temperature is 70 °F inside and 20 °F outside, the heat flux would be

$$q' = \frac{70 - 20(°F)}{12.4\left(\frac{°F}{\frac{btu}{h*ft^2}}\right)} = 4.0\frac{btu}{h*ft^2}$$

In other words, 4.0 Btu of thermal energy will be lost each hour through each square foot of the wall area. Calculations such as these are used to size furnaces and cold storage rooms, buildings, etc.

Fig. 7.11 Temperature profile in a laminate

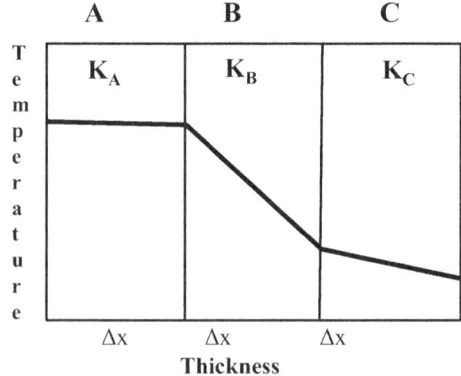

4. At steady state, the temperature profile in a laminated composite material appears as shown in Fig. 7.11. Which material has the highest thermal conductivity? Which material offers the controlling resistance to heat flow? ◄

7.6   Multilayered Cylinder

Another case of thermal resistance in series occurs with multilayered cylinders such as an insulated pipe.

7.6.1   Conductor Geometry

Figure 7.12 shows a hollow cylinder consisting of three layers with thermal conductivities k_A, k_B, and k_C and boundaries at radii r_1, r_2, r_3, and r_4. The temperatures at the boundaries are T_1, T_2, T_3, and T_4. Again, only T_1 and T_4 are known. The problem is to calculate the rate of heat flow through the cylinder wall and to determine the temperature profile through the wall of the cylinder, including the temperatures T_2 and T_3.

7.6.2   Energy Balance

No heat is generated, and we assume a steady state, as with the multilayered slab, so $\dot{q}_r$ must be the same through all layers, allowing us to write Eq. (7.41) for each layer and equate them to each other.

Fig. 7.12 Heat conduction through a multilayer cylinder

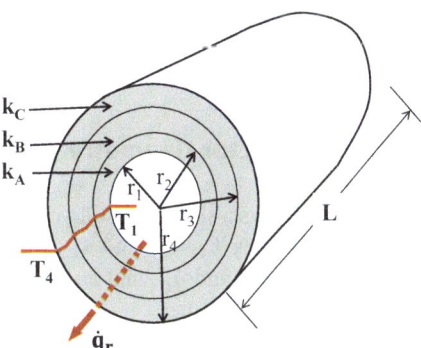

$$\dot{q}_r = k_A A_{\text{LM, } A} \frac{T_1 - T_2}{r_2 - r_1} = k_B A_{\text{LM, } B} \frac{T_2 - T_3}{r_3 - r_2} = k_C A_{\text{LM, } C} \frac{T_3 - T_4}{r_4 - r_3} \qquad (7.49)$$

Solving for each temperature drop, we obtain:

$$T_1 - T_2 = \dot{q}_r \frac{r_2 - r_1}{A_{\text{LM,} A} k_A} \qquad (7.50\text{a})$$

$$T_2 - T_3 = \dot{q}_r \frac{r_3 - r_2}{A_{\text{LM,} B} k_B} \qquad (7.50\text{b})$$

$$T_3 - T_4 = \dot{q}_r \frac{r_4 - r_3}{A_{\text{LM,} C} k_C} \qquad (7.50\text{c})$$

These equations describe the profile through each layer. Adding these equations gives:

$$T_1 - T_4 = \dot{q}_r \left[\frac{r_2 - r_1}{A_{\text{LM,} A} k_A} + \frac{r_3 - r_2}{A_{\text{LM,} B} k_B} + \frac{r_4 - r_3}{A_{\text{LM,} C} k_C} \right] \qquad (7.50\text{d})$$

$$\text{Solving for } \dot{q}_r, \text{ yield}: \ \dot{q}_r = \frac{T_1 - T_4}{\left[\frac{r_2 - r_1}{A_{\text{LM,} A} k_A} + \frac{r_3 - r_2}{A_{\text{LM,} B} k_B} + \frac{r_4 - r_3}{A_{\text{LM,} C} k_C} \right]} \qquad (7.50\text{e})$$

For the general case, it may be written as:

$$\dot{q}_r = \frac{T_1 - T_4}{(R_A + R_B + R_C)} = \frac{(\Delta T)_{overall}}{\Sigma R_i} = \frac{Overall\ Driving\ Force}{Sum\ of\ All\ Resistances} \qquad (7.51)$$

7.7　Conduction in Parallel

When a slab consists of two or more materials arranged side by side, the heat flowing through each one will be independent of the others. Under these conditions, the total heat flow equals the sum of the heat flow through each material. Let us see how you can compute the effective resistance of such a slab and the total heat flowing through it.

7.7.1　Conductor Geometry

Figure 7.13 shows a slab consisting of 3 different materials with areas A_A, A_B, and A_C and conductivities k_A, k_B, and k_C, respectively. The heat flowing through these sections is $\dot{q}_A$, $\dot{q}_B$, and $\dot{q}_C$, and as you can see, these heats are passing through the slab parallel to each other.

Fig. 7.13 Heat conduction in parallel

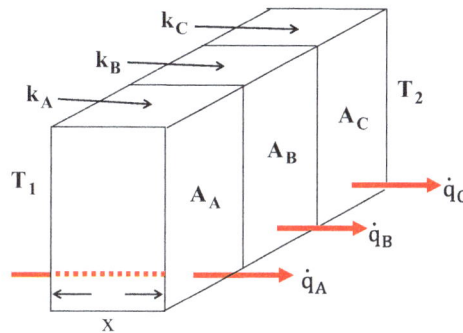

7.7.2 Energy Balance

The total heat flowing through the slab is equal to the sum of the individual heats:

$$\dot{q}_T = \dot{q}_A + \dot{q}_B + \dot{q}_C \tag{7.52}$$

$$\dot{q}_T = k_A A_A \frac{T_1 - T_2}{\Delta x} + k_B A_B \frac{T_1 - T_2}{\Delta x} + k_C A_C \frac{T_1 - T_2}{\Delta x} \tag{7.53}$$

$$\dot{q}_T = (T_1 - T_2) \left[\frac{1}{\frac{\Delta x}{k_A A_A}} + \frac{1}{\frac{\Delta x}{k_B A_B}} + \frac{1}{\frac{\Delta x}{k_C A_C}} \right] \tag{7.54}$$

$$\dot{q}_T = (T_1 - T_2) \left[\frac{1}{R_A} + \frac{1}{R_B} + \frac{1}{R_C} \right] \tag{7.55}$$

From this equation, we see that, similar to the electrical resistors in parallel in Fig. 7.14, the total resistance of a set of parallel resistances is the sum of the reciprocal of individual resistance.

Fig. 7.14 A parallel electric circuit

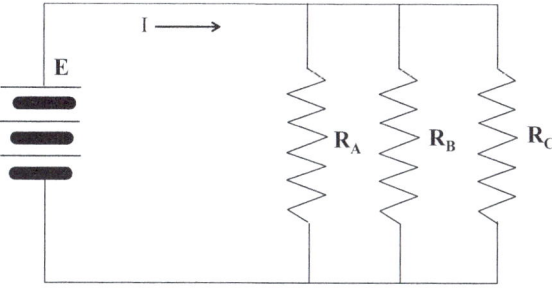

Problems

7.1 One side of a 1.5 cm thick slice of lean meat (7.8 wt.% fat, 1.5 wt.% ash, 19.0 wt.% protein, 71.7 wt.% water) contacts a hot pan that is maintained at 80°C while the other side remains at 25°C. Assuming steady-state heat transfer, draw the equivalent thermal resistance circuit and determine the rate of heat flux through the meat slice.

7.2 At steady state, the heat flow through a 0.6 cm thick aluminum sheet is 111408 W. One side of the sheet is maintained at 22 °C while the other is maintained at 18 °C. Determine (a) the area of the sheet, (b) the temperature profile, and (c) the temperature at 0.2 cm, 0.4 cm, and 0.6 cm from the 22 °C surface of the sheet.

7.3 To experimentally measure the thermal conductivity of a new food, it was formed into a 4 cm × 4 cm and 1 cm thick block. The edges of the block were insulated, and heat was supplied continuously to one side of the block at a rate of 0.4 W. At steady state, the temperature of both sides was measured with thermocouples and found to be 30 °C and 25 °C, respectively. Calculate the thermal resistance and thermal conductivity of the food. Given the composition of the food as 75 wt.% moisture, 19 wt.% protein, 2.5 wt.% fat, 2.2 wt.% carbohydrates, and 1.3 wt.% ash by weight, estimate the thermal conductivity of the product using the composition data and compare the reliability of the results.

7.4 (a) A formulated product has the following composition on dry weight basis: 70.0 wt.% carbohydrates, 15.0 wt.% fat, 1.0 wt.% protein, and 2.0 wt.% ash. Determine the thermal conductivity of the product if 100 g of wet product contains 60 g of water.

(b) The above food product is packed into an annular cylinder with the temperatures at the outer and inner walls maintained at constant values. The inner and outer walls are of negligible thickness and their radii are 5 cm and 11 cm, respectively. The temperature of the packed food product is monitored at different radial distances inside the cylinder with the help of thermocouples. At steady state, for a flux of 150 W/m^2, the temperature at various radial positions was recorded. Estimate the product's thermal conductivity.

Radial distance, r (cm)	Temperature, T (°C)
6.0	6.8
7.3	14.8
8.6	19.0
9.8	25.0
10.7	29.2

7.5 Single strength juice flows through a stainless-steel pipe of 50-mm inside diameter and 5 m long with a wall thickness is 2 mm. The outside surface temperature is 80 °C and the inside wall temperature is 85 °C. The thermal conductivity of stainless steel is 16 W/m-°C. Calculate the heat losses assuming steady-state flow.

7.6 Show that the rate of heat transfer through a spherical shell (thermal conductivity k) with inner radius r_1 and outer radius r_2 is given by the following expression:

$\dot{q}_r = \frac{4\pi k r_1 r_2 (T_1 - T_2)}{r_2 - r_1}$, where T_1 and T_2 are the inside and outside temperatures.

7.7 What length of cylindrical shell will conduct heat at the same rate as a spherical shell, both having the same internal radius and external radius of 1 m and 2 m, respectively. Given that the cylinder and sphere are made of the same material and the inside temperature (T_1) and outside temperature (T_2) are also the same for both the spherical and cylindrical shells.

7.8 A refrigerated room is designed with an outer 70 mm thick wall of cement (($k = 0.29$ W/m-K) and an inner wall of maple wood (15 mm thick, $k = 0.172$ W/mK), with the space in between (120 mm) filled with polyurethane foam ($k = 0.025$ W/mK). If the inner wall temperature is 4 °C and the outer wall is maintained at the ambient air temperature of 34 °C, estimate the rate of heat loss per unit area.

7.9 A rigid plastic cooler is built with high density polyethylene (HDPE) ($k = 0.51$ W/m-K) as a rigid shell and polyurethane foam ($k = 0.025$ W/m-K) serving as an insulator while it fills the inside of the rigid HDPE walls. You are trying to use this cooler to keep your beverages at a cool 0 °C while the ambient summer temperature is 38 °C. If the HDPE shell thickness is 30 mm and the polyurethane insulation is 140 mm thick, determine the heat flux through the walls of the rigid plastic cooler. Assume heat may only transfer through the walls of this cooler.

7.10 Three heat conducting slabs of materials A, B, and C are joined together in parallel. Each slab is 5 cm thick and has an area of 1.5 m^2. The thermal conductivities of A, B, and C are 0.29, 0.04, and 0.11 W/m-K, respectively. If the temperature at one surface is maintained at 40 °C, calculate the steady-state temperature at the other surface, given that heat flows through the slab at a rate of 500.7 W.

Bibliography

1. Hermans F (1979) The thermal diffusivity of foods, Thesis, University of Leven, The Netherlands

2. Miles CA, van Beek G, Veerkamp CH (1983) Calculation of thermophysical properties of foods. In: R Jowitt, F Escher, B Hallström, HF Th Meffert, WEL Spiess, G Vos, (Eds.), Physical Properties of Foods. Essex, England: Applied Science Publishers Ltd.
3. Choi Y, Okos MR (1986) Effects of temperature and composition on the thermal properties of foods. In: Le M. Maguer & P Jelen (Eds.), Food Engineering and Process Applications. Elsevier New York, NY.

Heat Transfer: Steady-State Convection

8

In Chap. 7, we introduced the idea of heat transport, concentrating on steady-state conduction through various geometric forms. In this chapter, we will look at heat transfer by convection and combinations of convection and conduction.

The major problem in dealing with convection is to estimate the rate at which heat is transferred from one location to another due to the movement of fluid. Fluid flow with accompanying energy changes makes exact analysis very challenging, and empirical methods are frequently used in engineering analysis of convection heat-transfer processes, as you will undoubtedly note later in this chapter.

The following are some examples that illustrate how extremely important convective heat transfer is in industrial processes:

- Air movement within a refrigerator transports heat to the cooling coils. On the outside, air movement removes heat from the exterior condenser coils to the room.
- Convection in an oven removes heat from the heating element and transfers it to the food being cooked.
- Within a soup pot, convection removes heat from the bottom of the pot and distributes it throughout the soup.
- Convection in the cooling system of your car removes heat from the engine block and transfers it to the radiator. Convection of outside air removes heat from the radiator.
- In a heat exchanger used to process food, the convection of steam is used to transport heat to one side of the separating walls and then it is convection of the food that removes heat from the other side of walls.

© The Author(s), under exclusive license to Springer Nature Switzerland AG 2024
S. S. H. Rizvi, *Food Engineering Principles and Practices*,
https://doi.org/10.1007/978-3-031-34123-6_8

8.1   The Mechanisms of Convection

In the absence of nuclear or chemical reactions and phase changes, an increase in the internal energy of a system results in greater molecular movement, which generally moves the molecules farther apart. This increases the volume and decreases the density of the material. If the material is a liquid or a gas and the heating is not uniform, the warmer, less dense regions will tend to rise, and the colder, denser regions will tend to sink. If heating takes place at the top and cooling at the bottom, the resulting temperature distribution will be stable, and no movement of material will take place. On the other hand, if heating occurs at the bottom or cooling occurs at the top, the upward movement of warm fluid and downward movement of cool fluid will result in **natural convection** (Fig. 8.1a). While flow is entirely driven by the temperature difference in natural convection, it can also be driven using pumps, fans, stirrers, or any other device that moves a fluid, which is then called **forced convection** (Fig. 8.1b).

8.2   Newton's Law of Cooling

Figure 8.2 shows an object surrounded by fluid that is flowing past it, either by natural or forced convection. The object is at temperature T_w, while the fluid is at temperature T_f. If $T_w > T_f$, heat will be transferred from the object to the fluid, with the result that the object will be cooled. Newton proposed that the rate of cooling was:

- Proportional to the area (A) of the object.
- Proportional to the difference in temperature (T) between the object and the fluid. These relationships are expressed in the equation below and is known as **Newton's law of convective heat transfer**:

Fig. 8.1 Natural convection (**a**) and forced convection (**b**)

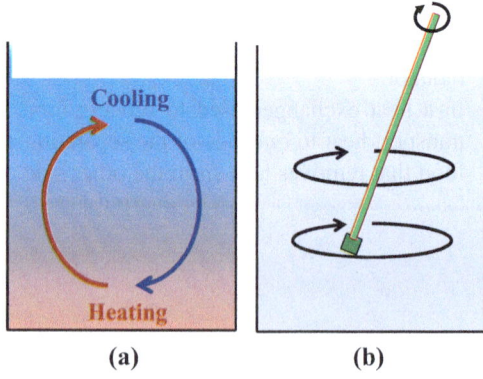

Cooling

Heating

(a)   (b)

Fig. 8.2 Convection heat transfer

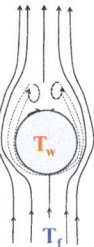

Table 8.1 Ranges for the convective heat-transfer coefficient

Mechanism	h_c (W/m²K)	h_c (Btu/hrft²°R)
Still air	0.5–4	0.1–0.7
Moving air	2–10	0.4–2
Moving water	50–3000	10–500
Condensing steam	1000–5000	200–900

$$\dot{q} = h_c A \left(T_w - T_f \right) = h_c A (\Delta T) \tag{8.1}$$

where $\dot{q}$ = rate of heat transfer, h_c = a proportionality constant, A = surface area of the object, T_w = object surface temperature, and T_f = fluid temperature.

The proportionality constant (h_c) in the above equation is called the **convective heat-transfer coefficient** and has units of energy per unit time per unit area per degree (W/m²K).

The value for h_c is usually determined experimentally and depends on several factors, including the nature of the fluid, the flow properties of the stream, the geometry of the system and the temperature difference. It is thus a system property Table 8.1 lists some approximate ranges for h_c for various heat-transfer mechanisms.

8.3   Heat Transfer at the Interface

When a fluid flows past a solid object, a stagnant film of fluid develops next to the solid. Heat exchange between the fluid and the solid takes place through this boundary film largely by conduction. However, since the thickness and behavior of the film cannot be measured exactly, it is not practical to apply Fourier's law. Therefore, Newton's law of cooling is used instead.

Although we speak of the boundary layer as a stagnant film, there is no distinct division between it and the free-flowing stream. Figure 8.3a shows a velocity profile of the fluid flowing over a hot surface. At the surface, the fluid velocity will be zero. Above the surface, the velocity will increase for a short distance and then become uniform. The fluid layer in which the velocity is changing is called the **hydrodynamic boundary layer**. The layer beyond uniform velocity is called the **free**

Fig. 8.3 Hydrodynamic and thermal boundary layers

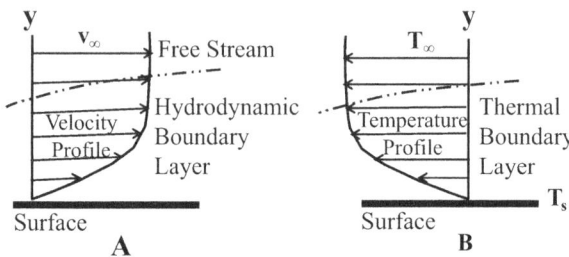

stream. There is no exact point of separation between these layers, it depends on several factors.

Figure 8.3b shows a temperature profile of the same fluid stream. At the surface, the fluid temperature is the same as the surface temperature. As you move away from the surface, the temperature gradually changes and then levels off. The layer of decreasing temperature is called the **thermal boundary layer** and corresponds to the hydrodynamic boundary layer. Thus, what we call a stagnant film is truly a zone of transition from totally stagnant to free-flowing fluid.

The ΔT in Newton's law of cooling refers to the difference in temperature between the free stream and the surface, i.e., the temperature difference across the boundary layer. The vagueness of the layer is the reason that the effect of thickness (Δx) is lumped into h rather than stated explicitly in the equation as it is with conduction.

8.4   k Versus h_c

The equations for Fourier's first law and Newton's law of cooling look similar.

Fourier's Law: $\dot{q} = kA \frac{(T_1 - T_2)}{\Delta x}$ and Newton's Law: $\dot{q} = h_c A (T_1 - T_2)$

- In each case, we see that the rate of heat transfer is proportional to the temperature difference. In conduction, it is the difference in temperature across a well-defined slab. In convection, it is the temperature difference across a poorly defined film of fluid adjacent to an object.
- In the conduction equation, the thickness of a slab (x) is a factor explicitly shown in the equation. In the convection equation, the thickness is not measurable, and its effect is included in the constant h_c.
- The value of k depends primarily on the type of material and to a lesser extent on its temperature. The value of h_c depends on various flow and heat-transfer characteristics of the fluid as well as the shape and size of the solid.

8.5   Determination of h_c

To use Newton's law of cooling to solve problems, it is necessary to know the value of the convective heat-transfer coefficient (h_c). Unfortunately, this is not always easy to determine.

8.5.1   Factors Affecting h_c

Convective heat transfer concerns itself with the transfer of heat from a freely flowing fluid to an adjacent solid. As we described above, this transfer takes place across a thin, stagnant film of the fluid that forms along the surface of the solid. The rate of this transfer was shown to depend on the following system characteristics:

- The size and shape of the surface
- The thermal conductivity, specific heat, viscosity, and density of the fluid in the film
- The velocity of the fluid stream
- The temperature difference across the film

With these many factors, predicting h_c for any set of system characteristics becomes a complex undertaking and requires considerable intuitive understanding of the system.

8.5.2   Approaches to Estimate h_c

There are several ways of determining the relationship between h_c and the other system variables:

- The value of h_c can be determined experimentally for each situation, and the findings can be tabulated. Since h_c depends on so many factors, running an experiment for every combination of factors is a formidable job.
- It would simplify things if we had an equation that predicted h_c from the characteristics of the system. Such an equation can be derived analytically from the physical laws of flow, heat transfer, etc., using the methods of calculus and algebra. Unfortunately, the number of factors involved makes the derivation quite complicated, and this approach is not generally used.
- A function predicting h_c from the other characteristics of the system can be developed by a technique called dimensional analysis. Although both seek a function that predicts h_c, this method and the analytical method go about it quite differently. While the analytical approach derives an equation based on physical theory, the dimensional approach ignores theory and concentrates on finding a function that is dimensionally homogeneous, i.e., every term has the same units. Experiments are then performed, and the equation is adjusted to fit the

data that are obtained. This is the approach that has been most successfully used to solve heat-transfer problems, and we will devote several sections of this chapter to it.

8.5.3 Dimensionless Numbers

The equation we are seeking turns out to look something like this:

$$h_c = 0.023 \frac{k_{\text{fluid}}}{d} Re^{0.8} Pr^{0.4} \qquad (8.2)$$

This makes it difficult to achieve homogeneous equation; what does one do with meters$^{0.8}$, for example. What happens to the units when the exponent is changed to a new fractional value?

One way to address this problem is to combine factors to form dimensionless numbers. We can then raise these to any power we wish without troubling the effect on dimensions. In fact, an equation that is composed entirely of dimensionless numbers is always homogeneous.

You are already familiar with one dimensionless number, the Reynolds number. We can show that this is also dimensionless by showing that all units cancel each other out.

Check Your Understanding

1. Verify that each of the following is a dimensionless number.

$$\text{Nusselt Number} = \frac{h_c d}{k}$$

$$\text{Prandtl Number} = \frac{C_p \eta}{k} \qquad \blacktriangleleft$$

8.6 Dimensional Analysis

Since Newton's time, calculus has been used to develop equations to describe physical phenomena. This approach usually involves setting up differential equations for, say, velocity and then solving them to obtain an equation for distance traveled. Experiments are then used to verify these equations and estimate their constants.

Another method for developing such equations is called **dimensional analysis**. This method is introduced in this section and compared with calculus by using both methods to derive the equation for the distance covered by a freely falling object at time t. In the next section, we use dimensional analysis to estimate the convective transfer coefficient.

8.6.1 Some Definitions

We start by defining some of the terms that are used in dimensional analysis.

- **Physical Equation** = algebraic expression that describes some physical phenomena. It will contain symbolic variables and constants for which numbers can be substituted.

Example

1. *The following equation describes the distance traveled in time t when an object is allowed to fall in a vacuum.*

$$x = \frac{1}{2}gt^2$$

where: x = distance (a variable),
g = acceleration due to gravity (a constant), and
t = time (a variable). ◄

To determine the distance traveled after 3 seconds of fall, substitute numbers as follows:

$$x = \frac{1}{2}\left(9.8\,\frac{m}{sec^2}\right)(3\ sec)^2 = 44.1\ m$$

- **Homogeneous Equation** = a physical equation in which every term has the same units.

Example

2. *The above equation for distance in free fall has two terms: "x" and "1/2 gt²".*
Both terms have the dimension of length, so the equation is homogeneous.
Similarly, in the equation v = gt, each term has dimensions of velocity, so the equation is homogeneous. However, if you add the two equations, a perfectly legal algebraic operation, the resulting equation, x + v = 1/2 gt² + gt, has two velocity and two distance terms and is, therefore, not homogeneous. ◄

Check Your Understanding

2. Write the Bernoulli equation with all units, then verify that it is homogeneous.

- **Dimension** is a qualitative physical characteristic on which quantitative measurements are made. A dimension does not depend on either the size of the measurements or the units used. ◄

Example

3. The measurements 10 inches, 25.4 cm, and 0.254 meters all involve different values and units, but all refer to the same quality of length. Length, then, is a dimension. ◄

In this discussion, dimensions will be represented by uppercase letters. For example, length will be represented by the letter L.

- **Fundamental Dimension.** In any dimensional system, there will be certain fundamental dimensions upon which all other dimensions are based. Some examples of fundamental dimensions are given in Table 8.2. The symbols shown represent the dimensions, not the quantities of the dimensions or the units used. Thus, we use "m" to represent the dimension "Mass," while "m" is used to represent the quantity of mass.
- **Derived Dimensions** are dimensions that are expressed as the product of powers of fundamental dimensions. For example, since L = the fundamental dimension "Length," L^2 represents the derived dimension "Area" and L^3 represents "Volume." Additionally, since t represents the fundamental dimension "Time," L/t or Lt^{-1} represents "Velocity" and L/t^2 or Lt^{-2} represents the derived dimension "Acceleration."

Table 8.3 lists some dimensions that are derived as follows:

$$\text{Force} = \text{mass x acceleration} = (m)(Lt^{-2}) = mLt^{-2}$$

$$\text{Energy} = \text{force x distance} = (mLt^{-2})(L) = mL^2t^{-2}$$

$$\text{Power} = \frac{\text{energy}}{\text{time}} = \frac{(mL^2t^{-2})}{t} = mL^2t^{-3}$$

$$\text{Pressure} = \frac{\text{force}}{\text{area}} = \frac{(mLt^{-2})}{L^2} = mL^{-1}t^{-2}$$

$$\text{Viscosity} = \text{pressure x time} = (mL^{-1}t^{-2})(t) = mL^{-1}t^{-1}$$

Table 8.2 Some fundamental dimensions

Dimension	Dimensional System			
	SI		Engineering	
	Symbol	Unit	Symbol	Unit
Length	L	meter	L	Ft
Mass	m	kg	m	lb_m
Time	t	sec	t	sec, hr
Temperature	T	K	T	°F

Table 8.3 Some derived dimensions

Dimension		SI		Engineering	
Name	Symbol	Dimension	Units	Dimension	Units
Area	A	L^2	m^2	L^2	ft^2
Volume	V	L^3	m^3	L^3	ft^3
Velocity	v	Lt^{-1}	m/sec	Lt^{-1}	ft/sec
Acceleration	a	Lt^{-2}	m/sec^2	Lt^{-2}	ft/sec^2
Force	f	mLt^{-2}	$N = kg\ m/sec^2$	F	lb_f
Energy	E	mL^2t^{-2}	$J = Nm$ $= kg\ m^2/sec^2$	FL	$ft\ lb_f$
Power	p	mL^2t^{-3}	$W = J/sec$ $= kg\ m^2/sec^3$	FLt^{-1}	$ft\ lb_f/sec$
Density	ρ	mL^{-3}	kg/m^3	mL^{-3}	lb_m/ft^3
Pressure	P	$mL^{-1}t^{-2}$	$Pa = N/$ $m^2 = kg/m\ sec^2$	FL^{-2}	lb_f/ft^2
Viscosity	η	$mL^{-1}t^{-1}$	$Pas = kg/m\ sec$	FTL^{-2}	$lb_f sec/ft^2$
Heat Capacity	C_p C_v	$L^2t^{-2}\,T^{-1}$	J/kgK	$HL^{-3}\,T^{-1}$	$Btu/lb_m\,°F$
Thermal Conductivity	k	$mLt^{-3}\,T^{-1}$	W/mK $= J/mK\ sec$ $= N/K\ sec$	$Ht^{-1}L^{-1}T^{-1}$	Btu/hr. $ft°F$
Convection Coefficient	h_c	$mt^{-3}\,T^{-1}$	W/m^2K $= J/m^2K\ sec$ $= N/mK\ sec$	$Ht^{-1}L^{-2}T^{-1}$	Btu/hr. $ft^2°F$
Thermal Diffusivity	α	L^2t^{-1}	m^2/sec	L^2t^{-1}	ft^2/sec

$$\text{Heat Capacity} = \frac{\text{energy}}{\text{mass x temperature change}} = \frac{\left(mL^2t^{-2}\right)}{(m)(\Delta T)} = L^2t^{-2}\Delta T^{-1}$$

$$\text{Thermal Conductivity} = \frac{\text{power}}{\text{area x }\frac{\text{temperature change}}{\text{distance}}} = \frac{\left(mL^2t^{-3}\right)}{\left(L^2\right)\left(\frac{\Delta T}{L}\right)} = mLt^{-3}\Delta T^{-1}$$

$$\text{Convective Coefficient} = \frac{\text{power}}{\text{area x temperature change}} - \frac{\left(mL^2t^{-3}\right)}{\left(L^2\right)(\Delta T)} = mt^{-3}\Delta T^{-1}$$

$$\text{Thermal Diffusivity} = \frac{\text{thermal conductivity}}{\text{density x heat capacity}} = \frac{\left(mLt^{-3}\Delta T^{-1}\right)}{\left(mL^{-3}\right)\left(L^2t^{-2}\Delta T^{-1}\right)} = L^2t^{-1}$$

- **Dimensional Equation** = an expression that gives the dimensions of a quantity. Unlike ordinary algebraic equations, you cannot substitute numbers into a dimensional equation. In these equations, square brackets are used to mean "dimension of." Some authors place the brackets around the physical expression and some around the equal signs.

Example

4. *The physical equation for density is:* $\rho = \frac{m}{V}$

   The dimensional equation for density can be written as $[\rho] = \frac{m}{V} = mV^{-1}$

   Either notation is read as "The dimension of ρ is m/V or m/V^{-1}." ◄

Check Your Understanding

1. Write the dimensional equation for the Bernoulli equation, then verify that it is homogeneous. ◄

8.6.2   Dimensionless Numbers

If a group of variables and constants are taken from a physical equation and combined by multiplying their powers and if, in the resulting number, all units cancel, the result is a dimensionless number.

Example

5. *The Reynolds number is a dimensionless number that is a product of powers of density, diameter, velocity, and viscosity.*

$$\pi = Re = \frac{\rho dv}{\eta} = \rho dv \eta^{-1}$$ ◄

This can be verified by canceling the units, thus

$$Re[=] \frac{\left(\frac{kg}{m^3}\right)(m)\left(\frac{m}{s}\right)}{\left(\frac{kg}{m.s}\right)}$$

Examination of this equation shows that kg cancels kg, m^3 cancels the three m's, and second cancels second, leaving the expression dimensionless. It can also be verified with a dimensional equation. The dimensions of the variables in the Reynolds number are:

$[\rho] = mL^{-3}$ (kg/m^3, for example)

$[d] = L$ (m, for example)

$[v] = Lt^{-1}$ (m/s, for example)

$[\eta] = mL^{-1}t^{-1}$ (Pas = kg/m s, for example)

Combining dimensions for this group of variables gives the dimensional equation for the Reynolds number.

$$[Re] = \left(mL^{-3}\right)(L)\left(Lt^{-1}\right)\left(mL^{-1}t^{-1}\right)^{-1}$$

Dimensional equations can be manipulated like ordinary algebraic equations. In particular, $(x^a)^b = x^{ab}$, so

$$[Re] = \left(mL^{-3}\right)(L)\left(Lt^{-1}\right)\left(m^{-1}L^1t^1\right)$$

Furthermore, $x^a x^b = x^{a+b}$, so

$$[Re] = m^{1-1}L^{-3+1+1+1}t^{-1+1}$$

From this, we see that all dimensions decrease to zero, so the dimensions disappear. This means that the Reynolds number is dimensionless and qualifies as a π.

$$[Re] = m^0 L^0 t^0 = 1$$

8.6.3 Buckingham's π Theorem

The pi theorem starts with the following assumptions:

- You have a homogeneous equation with n measurable quantities (variables) and dimensional constants (such as g, the acceleration of gravity). For generality, we call these values $(\alpha, \beta, \gamma, \delta, \ldots)$

Example

6. *The distance covered by a falling object is given by the equation* $y = \frac{1}{2}gt^2$ *The* $n = 3$ *dimensional constants and variables in this equation are* (y, g, t). ◀

- The equation can be expressed as a function in the form $f(\alpha, \beta, \gamma, \delta, \ldots) = 0$.

Example

7. *The falling object equation can be expressed as* $f(y, g, t) = \frac{1}{2}gt^2 - y = 0$ ◀

- The dimensions of $(\alpha, \beta, \gamma, \delta, \ldots)$ can be expressed in terms of r fundamental variables.

Example

8. *The dimensions of* (y, g, t) *can be expressed in terms of the* $r = 2$ *fundamental variables L (length) and T (time). Specifically, their dimensions are:*

[y] = L (meters, for example)
[g] = Lt^{-2} (meters per second per second, for example)
[t] = t (seconds, for example) ◄

The pi theorem states that, under these circumstances, it is possible to find $n-r$ numbers called $(\pi_1, \pi_2, \pi_3, \ldots \pi_{n-r})$ that have the following characteristics:

- The π's are independent products of the powers of the variables and constants (α, $\beta, \gamma, \delta, \ldots$) in the original function.

Example

9. *In the falling body equation, there are 3 quantities (x, g, and t) and 2 dimensions (L and T), so these can be 3–2 = 1 independent π. As we will show below, there is a possible π for this equation.*

$$\pi = \frac{y}{gt^2} = yg^{-1}t^{-2}$$

- *All π's are dimensionless.*

Example

10. *The dimensions of this π are: $[\pi] = L(Lt^{-2})^{-1}t^{-2} = L^{1-1}t^{2-2} = L^0t^0 = 1$*

- *They can be combined in a new function of the form $\phi(\pi_1, \pi_2, \pi_3 \ldots \pi_{n-r}) = 0$ ◄*

8.6.4 Derivation by Calculus

Before seeing how dimensional analysis can help us develop equations, let us see how it is done with calculus.

Example

11. *To derive the equation for a falling object, we can start with the assumption that all falling bodies accelerate at the same constant rate that we call g. Now acceleration is a rate of change in velocity per change in time, so we can start by writing the differential equation:*

$$acceleration = \frac{dv}{dt} = g$$

Integration of this equation gives an equation for velocity.

$$velocity = \int_0^v dv = \int_0^t g\,dt$$

$$velocity = v = gt$$

However, velocity is a rate of change in distance per change in time, so this equation can be rewritten as the differential equation:

$$velocity = v = \frac{dy}{dt} = gt$$

Integration of this equation gives the equation for distance that we have been seeking:

$$distance = \int_0^y dy = \int_0^t gt\,dt \quad or \quad y = \frac{1}{2}gt^2 \qquad \blacktriangleleft$$

8.6.5 Derivation by Dimensional Analysis

Now, let us use dimensional analyses to develop the same equation. Imagine that we do not know the equation for a falling object and that we are starting from scratch to derive it.

1. We begin by listing the quantities that we expect to be involved in describing the falling object. We know, of course, that they are distance (y), acceleration of gravity (g), and time (t). We must assume that these quantities are related by a function that can be put in the form: $f(y, g, t) = 0$.
2. Next, we select the fundamental dimensions that we will use for these quantities. These are length (L) and time (t).
3. Compute the number of π groups in the final solution. With 3 quantities and 2 fundamental dimensions, there must be only $3-2 = 1$ such π.
4. We set up this π as a product of powers of the selected quantities: $\pi = y^a g^b t^c$
5. We set up a dimensional equation for this π:

$$[\pi] = L^a \left(Lt^{-2}\right)^b t^c = L^{a+b} t^{-2b+c}$$

6. To make π dimensionless, it is necessary that the exponents in this last equation be 0. If they do this, the following equations will be true.
For the exponents of L: $a + b = 0$

For the exponents of t: $-2b + c = 0$

7. This gives 2 equations in 3 unknowns. They can only be solved if we select a value for one of the unknowns. Since we are looking for an equation for y, let us set the exponent of y (a) to 1. The two equations become:

$1 + b = 0$

$-2b + c = 0$

8. The solution to these equations is

$a = 1, \quad b = -1, \quad c = -2$

Substituting these values in the equation for π, we have: $\pi = yg^{-1}t^{-2} = \frac{y}{gt^2}$

9. The π theorem requires that there be a function of π's $\phi(\pi_1, \pi_2 \dots)$, which is equal to 0 for all values of the π's. In this example, there is only one π, so we are looking for a function:

$$\phi(\pi) = 0$$

Some possibilities are as follows:

$\phi(\pi) = 2\pi^2 - 3 \quad \phi(\pi) = \frac{\pi}{7} + 3 \quad \phi(\pi) = \pi - k$

Note that in each case, $\phi(\pi) = 0$ only if $\pi =$ a constant, so the equation is:

$$\pi = \text{constant} = \frac{y}{gt^2}$$

10. Solving this for y, we obtain $y = \text{constant}(gt^2)$

8.6.6   Experimental Confirmation

Before we can use this equation to predict distances, it must be confirmed by experiment. Furthermore, dimensional analysis provides no way to determine the value of any constants. In the above example, we know from the calculus derivation that the constant is 1/2. However, without that information, we must perform experiments to verify the equation and estimate its constants.

Example

11. *For the falling body example, we might perform the following experiment: take several balls of varying weights and drop them several distances, measuring the time of fall each time. Substitute these experimental times and distances into the π equation and compute the value of π. For example, $\pi = 5/(9.8 \times 1.032^2) = 0.480$. If all π values are the same, the data would support the derived equation, and the average of those values would be a good estimate of the unknown constant. Table 8.4 gives some possible data.* ◄

The following are some observations that can be made from these data:

Table 8.4 Falling body data

Experiment			Data		Computed
Drop	Weight	Distance	Time	Avg Time	π
1, 2	1 kg	5 m	1.04, 1.02	1.03 sec	0.480
3, 4		20 m	2.10, 2.06	2.08 sec	0.471
5, 6	5 kg	5 m	1.01, 1.01	1.02 sec	0.490
7, 8		20 m	2.05, 2.05	2.05 sec	0.485
9, 10	25 kg	5 m	1.03, 0.99	1.01 sec	0.500
11, 12		20 m	2.01, 2.03	2.02 sec	0.500
Average π					0.488

- The individual values of π are quite similar, which lends support to the idea that π is a constant.
- The average π of 0.488 is the best estimate of the "true" value of π. The equation then becomes: $y = 0.488gt^2$
- This value is quite close to the value of 1/2 obtained by the calculus approach.
- The variation that does exist in the results is of the same order of magnitude as the variation between duplicate measurements. This suggests that the variation is due to errors in measurement and not to a flaw in the equation or assumptions.
- However, a closer look at the results shows a consistent increase in π as the weight of the ball increases. This raises the suspicion that some factor, perhaps air resistance, has been overlooked. This suspicion cannot be tested with these data.

8.6.7   Summary

The above sections reviewed the use of calculus to analyze a phenomenon and develop an equation to describe it. It then showed how dimensional analysis, combined with experimentation, can be used to do the same thing. In cases such as this where we have an underlying theory, the calculus approach is simpler. In cases where we lack an underlying theory or where the theory requires measurements that are difficult to obtain, the approach of dimensional analysis provides a good alternative. In the next section, we will apply it to convective heat transfer.

8.7   The Nusselt Equation

In this section, we will use dimensional analysis to derive the Nusselt equation. This equation predicts values of the convective heat-transfer coefficient (h_c) by correlating it with other system characteristics.

8.7.1   The Variables

The first step in the derivation is to list the variables that we think will help predict h_c. Earlier, several system characteristics were listed. We can express these characteristics in terms of the following variables:

h_c = convective heat-transfer coefficient
d = diameter of the fluid stream. For a fluid flowing in a cylindrical pipe, this variable can represent both the size and shape characteristics of the system.
$\bar{v}$ = average velocity of the stream
ρ = density of the fluid
η = viscosity of the fluid
C_p = specific heat of the fluid
k = thermal conductivity of the fluid

We start with the assumption that there exists some function of these variables of the form:

$$f\left(h_c, d, \bar{v}, \rho, \eta, C_p, k\right) = 0$$

8.7.2   The Fundamental Dimensions

These variables can be expressed in terms of 4 fundamental dimensions: length (L), mass (m), time (t), and temperature (T). Note that for viscosity, the symbol η has been used. Specifically:

$$[h_c] = \frac{m}{t^3 T} \quad [d] = L \quad [\bar{v}] = \frac{L}{t} \quad [\rho] = \frac{m}{L^3} \quad [\eta] = \frac{m}{Lt} \quad [C_p] = \frac{L^2}{t^2 T} \quad [k] = \frac{mL}{t^3 T}$$

8.7.3   How Many π's

We start this analysis with 7 variables that can be expressed with 4 basic dimensions. The number of independent dimensionless numbers (π's) that we can find is

$$7 - 4 = 3$$

8.7.4   The Basic Equation for π

According to Buckingham's π theorem, each of the three π's will be a product of powers of the variables, thus:

$$\pi = h_c^a d^b \overline{v}^d \eta^e \rho^f C_p^g k^m$$

The dimensions of each π are, therefore,

$$[\pi] = \left(\frac{m}{t^3 T}\right)^a (L)^b \left(\frac{L}{t}\right)^d \left(\frac{m}{Lt}\right)^e \left(\frac{m}{L^3}\right)^f \left(\frac{L^2}{t^2 T}\right)^g \left(\frac{mL}{t^3 T}\right)^m$$

8.7.5 Combining Exponents

If we combine exponents of this dimensional equation, we obtain:

$$[\pi] = m^{a+e+f+m} \, L^{b+d-e-3f+2g+m} \, t^{-3a-d-e-2g-3m} \, T^{-a-g-m}$$

8.7.6 Simultaneous Equations

Our goal is to make each π dimensionless, so the final exponent of each dimension must be 0. Thus, the following set of equations must be true.

$$
\begin{array}{rrrrrrrr}
a & & +e & +f & & & +m & = & 0 \\
& b & +d & -e & -3f & +2g & +m & = & 0 \\
-3a & & -d & -e & & -2g & -3m & = & 0 \\
-a & & & & & -g & -m & = & 0
\end{array}
$$

8.7.7 The First π

There are 4 equations and 7 unknowns. To solve them, it is necessary to assign values to three of the variables. Since we eventually want to obtain an equation for h as a function of the other variables, i.e., $h_c = f(d, \overline{v}, \rho, \eta, C_p, k)$. It seems best to have h to the first power in one of the π's. This suggests setting a, the exponent of h, to 1. In addition, let us arbitrarily assign 0 to the exponents d and e. The simultaneous equations become:

$$
\begin{array}{rrrrrrrr}
1 & & +0 & +f & & & +m & = & 0 \\
& b+0 & -0 & -3f & +2g & +m & = & 0 \\
-3(1) & & -0 & -0 & & -2g & -3m & = & 0 \\
-1 & & & & & -g & -m & = & 0
\end{array}
$$

This reduces to:

$$
\begin{array}{rccr}
f & +m & = & -1 \\
b \quad -3f \quad +2\,g & +m & = & 0 \\
-2\,g & -3\,m & = & 3 \\
-g & -m & = & 1
\end{array}
$$

The solution to these equations is $b = 1, f = 0, g = 0$, and $m = -1$. Substituting these into the general equation for π, we obtain π_1.

$$
\pi_1 = h_c^1 d^1 \overline{v}^0 \eta^0 \rho^0 C_p^0 k^{-1} = \frac{h_c d}{k}
$$

This is the Nusselt number that was mentioned earlier. You already verified that it is dimensionless.

8.7.8   The Other π's

Since we only want h to appear once in the final equation, we should set a equal to 0 for both of the remaining values of π.

Check Your Understanding

2. Set $a = 0$, $b = 1$, and $g = 0$, solve the simultaneous equations and use the solutions to determine the expression for π_2. Verify that π_2 is the Reynolds number.
3. Set $a = 0$, $g = 1$, and $f = 0$, solve the simultaneous equations and use the solutions to determine the expression for π_2. Verify that π_2 is the Prandtl number mentioned earlier. ◀

8.7.9   The ϕ Function

The π theorem states that there should be a function of the form: $\phi(\pi_1, \pi_2, \pi_3) = 0$ or in this case:

$$
\phi\left(\frac{h_c d}{k}, \frac{d v \rho}{\eta}, \frac{C_p \eta}{k}\right) = 0
$$

One possible solution to this equation is:

$$
f_1\left(\frac{d v \rho}{\eta}\right) f_2\left(\frac{C_p \eta}{k}\right) - \frac{h_c d}{k} = 0
$$

where f_1 and f_2 are functions of the Reynolds and Prandtl numbers whose forms have yet to be determined. Solving this equation for the Nusselt number yields:

$$\frac{h_c d}{k} = f_1 \left(\frac{dv\rho}{\eta}\right) f_2 \left(\frac{C_p \eta}{k}\right)$$

8.7.10 The Nusselt Equation

Numerous experiments show that these functions should take the following form:

$$\left(\frac{h_c d}{k_{\text{fluid}}}\right) = C \left(\frac{dv\rho}{\eta}\right)^a \left(\frac{C_p \eta}{k_{\text{fluid}}}\right)^b \quad or \quad Nu = CRe^a Pr^b \qquad (8.3)$$

where C, a, and b are constants to be determined experimentally. This is the Nusselt equation. The expressions in parentheses are known as the **Nusselt (Nu), Reynolds (Re), and Prandtl (Pr)** numbers.

As an example, experiments with fully developed flow in tubes have led to the following specific equation:

$$Nu = 0.023 Re^{0.8} Pr^{0.4} \qquad (8.4)$$

8.7.11 Determining h_c

The above equation can be solved in terms of h_c as follows:

$$h_c = 0.023 \frac{k_{\text{fluid}}}{d} \left(\frac{dv\rho}{\eta}\right)^{0.8} \left(\frac{C_p \eta}{k_{\text{fluid}}}\right)^{0.4} \qquad (8.5)$$

The value of h_c is estimated when the values of the other variables are known.

8.8 Dimensionless Numbers and Their Interpretation

Many dimensionless numbers have been found to be useful in describing momentum and heat transport phenomena. All of them can be expressed as ratios, and frequently, these ratios point to a meaningful interpretation of the numbers. We have previously discussed the significance and interpretation of the Reynolds number. Let us look at the others.

8.8.1 The Nusselt Number

The Nusselt number is defined as:

$$\mathrm{Nu} = \frac{h_c d}{k_{\mathrm{fluid}}} \tag{8.6}$$

where d = diameter, h_c = convective heat-transfer coefficient, k_{fluid} = fluid thermal conductivity

Consider a layer of fluid of thickness x in contact with a surface. The bulk of the fluid is at temperature T_b, and the surface is at temperature T_w, so a temperature difference of $(T_w - T_b)$ exists across the layer. If the entire layer is moving, heat transfer occurs entirely by convection, and the heat flux is determined by Newton's law of cooling, i.e.,

$$q'_{\mathrm{conv}} = h_c(T_w - T_b) \tag{8.7}$$

On the other hand, if the entire layer is stagnant, heat transfer across the layer would be entirely by conduction and determined by Fourier's law, i.e.,

$$q'_{\mathrm{cond}} = \frac{k_{\mathrm{fluid}}(T_w - T_b)}{x} \tag{8.8}$$

Taking the ratio of these fluxes, we obtain:

$$\frac{q'_{\mathrm{conv}}}{q'_{\mathrm{cond}}} = \frac{h_c(T_w - T_b)}{k_{\mathrm{fluid}}\frac{T_w - T_b}{x}} = \frac{h_c x}{k_{\mathrm{fluid}}} = \mathrm{Nu}$$

Thus, the Nusselt number can be interpreted as

$$\mathrm{Nu} = \frac{\textit{Heat transfer by convection}}{\textit{Heat transfer by conduction}} \tag{8.9}$$

A large Nusselt number indicates very efficient convection. A Nusselt number of approximately 1 indicates very sluggish fluid flow, and heat transfer is slightly more efficient than pure conduction.

8.8.2   The Prandtl Number

The Prandtl number is defined as:

$$Pr = \frac{C_p \eta}{k_{\mathrm{fluid}}} \tag{8.10}$$

where C_p = Specific heat, η = viscosity, and k = thermal conductivity.

Multiplying the numerator and denominator by $1/\rho$ and arranging slightly, we obtain:

$$Pr = \frac{C_p \eta}{k_{fluid}} = \frac{\left(\frac{C_p \eta}{\rho}\right)}{\left(\frac{k_{fluid}}{\rho}\right)} = \frac{\left(\frac{\eta}{\rho}\right)}{\left(\frac{k_{fluid}}{C_p \rho}\right)} = \frac{\nu}{\alpha} = \frac{momentum\ diffusivity}{thermal\ diffusivity} \tag{8.11}$$

A large Prandtl number indicates that momentum transfer is dominant. A small Prandtl number indicates that heat transfer is dominant.

8.8.3 Additional Correlations for h_c

Many other correlations with further refinements have been developed to improve estimation of h_c, examples include:

- **Turbulent Flow:** The Dittus and Boelter equation is used to estimate the convective heat-transfer coefficient for fluids in turbulent flow inside clean, round pipes.

$$Nu = C(Re)^n (Pr)^m \left(\frac{\eta_b}{\eta_w}\right)^x \tag{8.12}$$

where $C = 0.15$–4.0; $n = 0.65$–0.85; $m = 0.30$–0.45; $x = 0.05$–0.2

All properties at bulk temperature, Fig. 8.3, e.g., $(T_{b1} + T_{b2})/2$, except η_w, which is at film temperature, e.g., $(T_{o1} + T_{o2})/2$, Fig. 8.4.

For highly turbulent flow, when $Re > 10^4$, $Pr = 0.7$–700 and $L/d \geq 10$, Eq. (8.12) is simplified to:

$$Nu = 0.023(Re)^{0.8}(Pr)^{\frac{1}{3}}\left(\frac{\eta_b}{\eta_w}\right)^{0.14} \tag{8.13a}$$

- **Laminar Flow:** A modified form of the **Dittus-Boelter equation** was proposed by **Sieder** and **Tate** for laminar flow in a pipe as follows:

$$Nu = 1.86\left(Re * Pr * \frac{d}{L}\right)^{\frac{1}{3}}\left(\frac{\eta_b}{\eta_w}\right)^{0.14} \tag{8.13b}$$

- For heating or cooling of a spherical object immersed in a flowing fluid, the following correlation is often used when $1 < Re < 70{,}000$ and $0.6 < Pr < 400$:

Fig. 8.4 Temperatures in a flowing stream

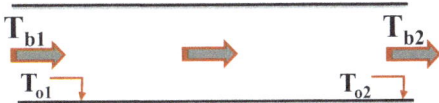

$$Nu = 2 + 0.6\,Re^{1/2}Pr^{1/3} \qquad\qquad (8.14)$$

Example

13. *Apples (average diameter 6 cm) at 20 °C are to be hydrocooled by immersion in a 0 °C water stream flowing at 0.1 m/s (Fig. 8.5). Estimate the convective heat transfer if the following correlation is found to apply to this system:*

$$Nu = 0.8(Re)^{0.3}(Pr)^{0.4}$$

Solution:
Given, $(h_c d/k_{fluid}) = 0.8\,[(\rho\,d < v>)/\eta]^{0.3}[(C_p\eta)/k_{fluid}]^{0.4}$
All the properties should be evaluated at the average temperature of $(20 + 0)/2 = 10\,°C.$
For water at $10°\ C; \rho = 1000\ kg/m^3; \eta = 1.3\ mPa - s, k_{water} = 0.6\ W/m - °\ C;$
$C_p = 4.2\ kJ/kg - °\ C\ Re = (\rho\,d < v>)/\eta = (1000\ kg/m^3)(0.06\ m)(0.1\ m/s)/(1.3 \times 10^{-3}\ kg/m - s) = 4615.4$

$$Pr = (C_p\eta)/k_{fluid} = (4.2 \times 10^3\ J/kg\text{-}°C)(1.3 \times 10^{-3}\ kg/m\text{-}s)/(0.6\ J/s\text{-}m\text{-}°C)$$
$$= 9.1$$

Using the above values, we obtain $\left(h_c\,d/k_{fluid}\right) = (0.8)(4615.4)^{0.3}(9.1)^{0.4} = 0.8 \times 12.6 \times 2.4 = 24.2$
Then, $h_c = (k_{fluid})(24.2)/d = (0.6\ W/m - °\ C)(24.2)/(0.06\ m) = 242.0\ W/m^2 - °\ C$ ◀

Next, we describe some of the most common types of processes that utilize combined conduction and convection heat-transfer principles in the processing and manufacturing of foods.

8.9   Combining Convection and Conduction

Convection and conduction frequently occur together in many food processing operations. For example, one type of heat exchanger used to pasteurize liquid foods consists of many parallel plates that separate thin parallel cavities. Hot water

Fig. 8.5 Hydrocooling by convection

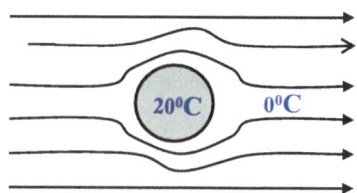

or steam moves through every other cavity by forced convection, bringing heat to each plate. This heat is conducted across the plate to cavities containing liquid food, where it is carried away by forced convection. Let us examine this process more closely.

8.9.1 Combined Heat Transfer Through a Flat Plate

Figure 8.6 shows the cross-section of a flat plate or slab that is conducting heat (q) from a hot side to a cold side. This plate has a thermal conductivity of k_A. A hot fluid flowing upward on one side of the plate brings heat. A cold fluid flows downward on the other side, carrying heat away. A thermal boundary layer forms on each side of the plate (dashed lines). These layers have convective heat-transfer coefficients of h_i and h_o, respectively.

A temperature profile is shown in the figure. The free-stream temperatures of the two fluids are T_1 and T_4, respectively. The temperatures at the surface of the plate are T_2 and T_3. At steady state, the profile through the plate will be linear, as shown. However, because of velocity shear through the boundary layers, the profiles may become parabolic, although the exact shape is hard to establish.

Heat flow through the plate is governed by fourier's law given as:

$$\dot{q}_p = k_A A \frac{T_2 - T_3}{\Delta X} \tag{8.15}$$

Using Newton's law of cooling, the heat flow through the boundary layers is:

$$\dot{q}_1 = h_i A (T_1 - T_2) \tag{8.16a}$$

$$\dot{q}_o = h_o A (T_3 - T_4) \tag{8.16b}$$

Fig. 8.6 Convection and conduction combined

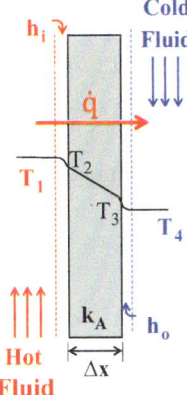

At steady state, the rate of heat flow ($\dot{q}$) will be the same through each layer, so we can equate these three heat flow equations as follows:

$$\dot{q} = h_i A (T_1 - T_2) = k_A A \frac{T_2 - T_3}{\Delta x} = h_o A (T_3 - T_4) \tag{8.17}$$

Now solving each of these three equations for the temperature difference,

$$T_1 - T_2 = \frac{\dot{q}}{h_i A} \tag{8.18a}$$

$$T_2 - T_3 = \frac{\dot{q}\Delta x}{k_A A} \tag{8.18b}$$

$$T_3 - T_4 = \frac{\dot{q}}{h_o A} \tag{8.18c}$$

and adding them up, we obtain

$$T_1 - T_4 = \dot{q} \left[\left(\frac{1}{h_i A} \right) + \left(\frac{\Delta x}{k_A A} \right) + \left(\frac{1}{h_o A} \right) \right] \tag{8.19}$$

and solving for $\dot{q}$ to obtain an overall equation for heat flow:

$$\dot{q} = \frac{T_1 - T_4}{\left(\frac{1}{h_i A} \right) + \left(\frac{\Delta x}{k_A A} \right) + \left(\frac{1}{h_o A} \right)} = \frac{(\Delta T)_{\text{overall}}}{\Sigma R} \tag{8.20}$$

In the above equation, we note that the driving force for the heat flow is the temperature difference between the free streams. The rate of heat flow ($\dot{q}$) equals this driving force divided by the sum of the three resistances, two computed using Newton's law of cooling and one computed using Fourier's law of conduction.

A layer with low resistance transfers a large amount of heat and vice versa. Thus, the ability of a layer to transfer heat is the reciprocal of its resistance. To express this idea mathematically, we need to define a new **overall heat-transfer coefficient** (U).

Multiplying and dividing Eq. (8.20) by area A and rewriting it, we obtain:

$$\dot{q} = \frac{A(\Delta T)_{\text{overall}}}{A \, \Sigma R} \tag{8.21}$$

and defining a new coefficient U as:

$$U = \frac{1}{A \, \Sigma R} = \frac{1}{\frac{1}{h_i} + \frac{\Delta x}{k_A} + \frac{1}{h_o}} \tag{8.22a}$$

or,

$$\frac{1}{u} = \frac{1}{h_i} + \frac{\Delta x}{k_A} + \frac{1}{h_o} \tag{8.22b}$$

Equation (8.21) in terms of U may be written as:

$$\dot{q} = UA\Delta T \tag{8.23}$$

Equation (8.23) is a very useful design equation. U is the **overall heat-transfer coefficient** with units of heat flux per degree difference. The units of U may be written as:

$$U[=]\frac{W}{m^2 K} \quad \text{or} \quad U[=]\frac{Btu}{hr - ft^2 - {}^0R}$$

This equation expresses the same relationship as Eq. (8.20) but does so in terms of the heat transmission capability of the system rather than its heat resistance.

8.9.2 Combined Heat Transfer Through a Hollow Cylinder

A similar equation can be developed for heat transfer through the walls of a hollow cylinder such as a pipe. Figure 8.7 shows a pipe through which a fluid flows at temperature T_4. Another fluid that might be steam, air, or water flows over the outside surface. Again, thermal boundary layers develop at each surface. The equation is similar to the one for the flat plate except for the areas involved.

$$\dot{q} = \frac{T_1 - T_4}{\frac{1}{h_i A_i} + \frac{r_o - r_i}{k A_{LM}} + \frac{1}{h_o A_o}} = \frac{T_1 - T_4}{\frac{1}{h_i A_i} + \frac{ln\left(\frac{r_o}{r_i}\right)}{2\pi k L} + \frac{1}{h_o A_o}} = \frac{(\Delta T)_{overall}}{\Sigma R} \tag{8.24}$$

where A_o = area outside of the cylinder = $2\pi r_o L$, A_{LM} = log-mean area of the cylinder, A_i = area inside of the cylinder = $2\pi r_i L$, and k = conductivity of the cylinder material.

In Eq. (8.23) above, we simplified the expression for the heat-transfer rate by introducing a parameter called **the overall heat-transfer coefficient, U.** In doing that for a cylinder, we are faced with the question of which area to use in determining the heat-transfer rate. As shown below, it may be defined using either the inside area (A_i) or the outside area (A_o). Multiplying and dividing Eq. (8.20) by A_i or A_o, we obtain:

Fig. 8.7 Convection and conduction in a hollow cylinder

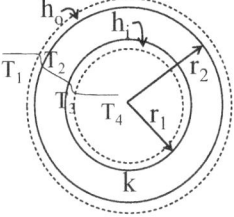

$$\dot{q} = \frac{(\Delta T)_{\text{overall}}}{\sum R} = \frac{A_i (\Delta T)_{\text{overall}}}{A_i \sum R} = \frac{A_o (\Delta T)_{\text{overall}}}{A_o \sum R} \qquad (8.25)$$

Now defining new U_i or U_o in terms of A_i or A_o, we can write:

$$U_i = \frac{1}{A_i \Sigma R} = \frac{1}{\left(\frac{1}{h_i} + \frac{A_i (r_o - r_i)}{k_m A_{LM}} + \frac{A_i}{A_o h_o} \right)} \quad \text{and} \quad U_o = \frac{1}{A_o \Sigma R}$$

$$= \frac{1}{\left(\frac{A_o}{A_i h_i} + \frac{A_o (r_o - r_i)}{k_m A_{LM}} + \frac{1}{h_o} \right)} \qquad (8.26a)$$

Alternatively, the above expressions may be written as follows:

$$\frac{1}{U_i} = \frac{1}{h_i} + \frac{A_i (r_o - r_i)}{k_m A_{LM}} + \frac{A_i}{A_o h_o} \quad \text{and} \quad \frac{1}{U_o} = \frac{A_o}{A_i h_i} + \frac{A_o (r_o - r_i)}{k_m A_{LM}} + \frac{1}{h_o} \qquad (8.26b)$$

These coefficients, U_i and U_o, have the following relationship to each other:

$$U_o = U_i \left(\frac{r_i}{r_o} \right) \qquad (8.27)$$

Using these definitions, we can write the heat-transfer equations as:

$$\dot{q} = U_i A_i (\Delta T)_{\text{overall}} = U_o A_o (\Delta T)_{\text{overall}} \qquad (8.28)$$

The following equations present two expressions for U, one based on flux through the inside area of the pipe and one based on flux through the outside area. They both should yield the same answer.

Example

14. A room at 20 °C has steel pipes (52.50 mm i.d., 60.33 mm o.d.) carrying steam at 120 °C. The convective heat-transfer coefficient for the steam on the inside is 6000 W/m²-C, while the coefficient for convection on the outside of the pipe is 50 W/m²-C. Calculate the overall heat-transfer coefficient based on the outside area (U_0) if the thermal conductivity of steel is 45 W/m-K. Compute the heat loss per meter length of the pipe.

Solution: The combined convection and conduction equation based on the outside area is:

$$\frac{1}{U_0} = \frac{A_0}{A_i h_i} + \frac{A_0 (r_0 - r_i)}{k_m A_{LM}} + \frac{1}{h_o} \quad \rightarrow \quad \frac{1}{U_0} = \frac{d_0}{d_i h_i} + \frac{d_0 (r_0 - r_i)}{k_m d_{LM}} + \frac{1}{h_o}$$

The log-mean diameter is: $d_{LM} = \dfrac{d_o - d_i}{\ln\left(\frac{d_o}{d_i}\right)} = \dfrac{60.33 - 52.50}{\ln\left(\frac{60.33}{52.50}\right)} = \dfrac{7.83}{0.14} = 55.9 \ mm$

$$\frac{1}{U_0} = \frac{0.06033}{(0.0525)(6000)} + \frac{(0.06033)(0.06033 - 0.0525)/2}{(45)(0.0559)} + \frac{1}{50} \rightarrow U_0$$

$$= 49.3 \ W/m^2 - {}^\circ C$$

For a $1-m$ long pipe: $A_o = \pi d_0 L = (3.14)\,(0.06033)\,(1) = 0.189 \ m^2$
    Then, $\dot{q} = U_0 A_0 \Delta T = (49.3)(0.189)(120 - 20) = 931.8 \ W$ ◄

8.9.3 Fouling of Heat-Transfer Surfaces

In heat exchangers, for example, fluids are passing over both sides of plates or pipes. These fluids may corrode plates and pipes, or they may deposit films on the surface. This is called **surface fouling**. Fouling introduces additional layers through which heat must pass and thus increases the total thermal resistance of the system. Practically all industrial heat exchangers undergo fouling after extended use and show a decrease in performance. Let us examine how fouling affects heat flux.

Figure 8.8 shows that when fouling occurs on both sides of a flat plate, there are a total of 5 resistances to impede the flow of heat.

For the plate shown in Fig. 8.8, the resistances are written as:

$$\Sigma R_i = \frac{1}{h_i A} + \frac{1}{h_{i,f} A} + \frac{\Delta x}{kA} + \frac{1}{h_{o,f} A} + \frac{1}{h_o A} \tag{8.29}$$

For fouling of a pipe, the resistances will be:

$$\Sigma R_i = \frac{1}{h_i A_i} + \frac{1}{h_{i,f} A_i} + \frac{\ln\left(\frac{r_o}{r_i}\right)}{2\pi L k} + \dots \tag{8.30}$$

where $h_{i,f}$ = the inside fouling film coefficient and $h_{o,f}$ = the outside fouling film coefficient.

Fig. 8.8 Fouling of a flat plate surface

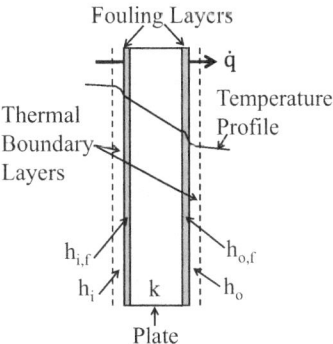

Table 8.5 Typical fouling factors

Type of fluid	R_f (m²K/W)
Boiler Feed Water	0.002
Steam	0.00009
Industrial Air	0.0004
Refrigerating Liquid	0.0002
Vegetable Oil	0.0006

Fouling cannot be determined by theory. It must be determined experimentally. Earlier, we defined U such that:

$$\Sigma R_i = \frac{1}{U_o A_o} = \frac{1}{U_i A_i} \tag{8.31}$$

If experiments are performed to determine U when the system is clean and again when the system is fouled, we can define a fouling factor (R_f) such that:

$$\frac{1}{U_{\text{fouled}}} = R_f + \frac{1}{U_{\text{clean}}} \tag{8.32}$$

Some typical values for R_f are given in Table 8.5.

Example

15. *The overall heat-transfer coefficient of a sanitary double-pipe heat exchanger based on the inside area (U_i) is 1000 W/m²-K when clean. After use to pasteurize liquid food for eight hours, U_i drops to 100 W/m²-K due to the deposition of foulants on the inside surface. What percent of the total resistance to heat transfer is due to fouling?*

Solution:

$$\frac{1}{U_c} = \frac{1}{1000} = 0.001 \text{ and } \frac{1}{U_f} = \frac{1}{100} = 0.01$$

Using Eq. (8.32), we can calculate the foulant resistance:

$$R_f = \frac{1}{U_f} - \frac{1}{U_c} = 0.01 - 0.001 = 0.009$$

Therefore, $R_f = (0.009/0.01)\ (100) = 90\%$ ◄

8.10   Heat Exchangers

Heat exchangers are devices used to transfer thermal energy from one medium to another, each at a different temperature. There are two main types of heat exchangers: direct and indirect. In direct heat exchangers, both media are in direct contact with each other; examples include direct steam injection systems or air-cooled water towers. In indirect heat exchangers, the two media are separated by a solid boundary (metal of high k). Indirect heat exchangers used by the food and other processing industries are of various types and designs. Some of the more common types used by the food industry include double-pipe, parallel plate, scraped surface, and shell-and-tube heat exchangers. The choice is dictated by fluid properties, design considerations, space, and budget. These exchangers are briefly described below.

8.10.1 Double-Pipe Heat Exchanger

Also known as a tube-in-tube heat exchanger, it is relatively simple in design, easy to operate and maintain, and low cost. It consists of a large diameter pipe with a concentric smaller diameter pipe inside the larger pipe, as shown in Fig. 8.9. One fluid flows in the inner pipe, and the other fluid flows in the annulus between the outer and inner pipes, with the inner pipe acting as a conductive barrier. Such types of heat exchangers could be configured as straight pipes or as coiled pipes and can be operated in a counter-current or co-current configuration, which results in different temperature profiles (Fig. 8.10).

Let us perform a quick analysis of the two types of heat flow profiles. The equation:

$$\dot{q} = UA\Delta T = U_i A_i \Delta T = U_o A_o \Delta T \qquad (8.33)$$

holds only at a particular point in the heat exchanger. If we make the assumptions that the flow (counter or co-current) is steady state with a constant U, heat exchange with the ambient is negligible and the specific heats of both fluids are constant and independent of temperature, then the properties of the fluids are as shown in the table below:

Fig. 8.9 A double-pipe heat exchanger

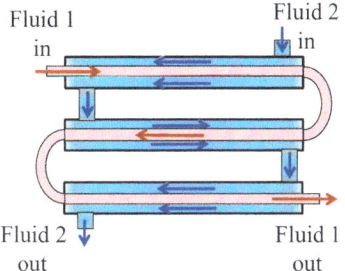

Fluid 1 in

Fluid 2 in

Fluid 2 out

Fluid 1 out

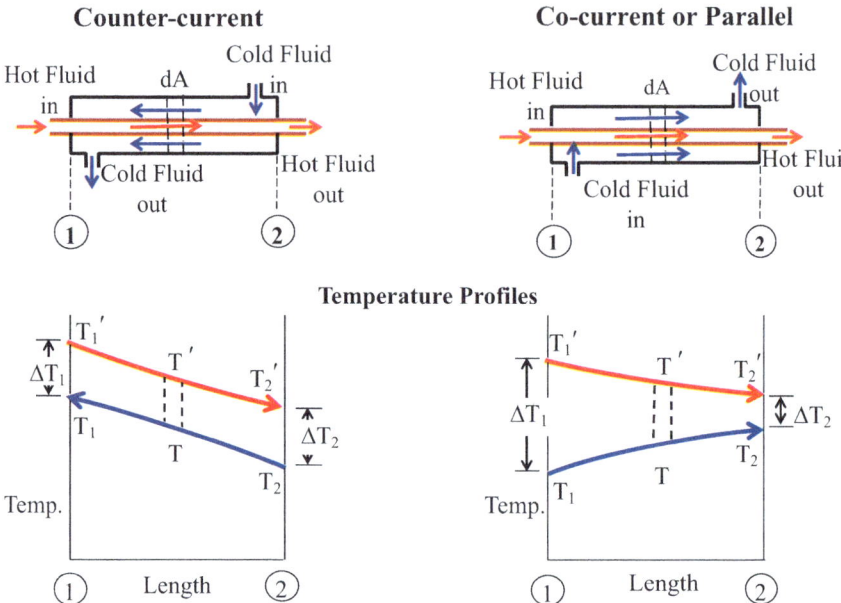

Fig. 8.10   Flow inside double-pipe heat exchangers and temperature profiles

Parameters	Hot Fluid	Cold Fluid
Specific Heat	Cp' J/kg-°C	Cp J/kg-°C
Flow rate	m' kg/s	m kg/s

Then, the amount of heat transferred between the two streams is

$$d\dot{q} = -m' \, Cp' \, dT' = m \, Cp \, dT \tag{8.34}$$

$$\text{Or,} \; dT' = \frac{d\dot{q}}{-m'Cp'} \quad \text{and} \quad dT = \frac{d\dot{q}}{mCp} \tag{8.35}$$

$$\text{Then,} \; dT'\text{-}dT = d(T'\text{-}T) = d\dot{q}\left[-\frac{1}{m'Cp'} - \frac{1}{mCp}\right] \tag{8.36}$$

At any point, e.g., dA:

$$d\dot{q} = UdA(T'\text{-}T) \quad \text{or} \quad \frac{d(T'\text{-}T)}{T'\text{-}T} = \text{-}UdA\left[\frac{1}{m'Cp'} + \frac{1}{mCp}\right] \tag{8.37}$$

Integrating the above equation between points (1) & (2):

$$\int_1^2 \frac{d(T'-T)}{(T'-T)} = -U\left(\frac{1}{m'Cp'} + \frac{1}{mCp}\right)\int_1^2 dA \quad \text{or} \quad \ln\frac{(T_2'-T_2)}{(T_1'-T_1)}$$

$$= -UA\left(\frac{1}{m'Cp'} + \frac{1}{mCp}\right) \tag{8.38}$$

However, $q = m'Cp'\left(T_1'-T_2'\right) = mCp(T_2-T_1)$,

Then, $m'Cp' = \frac{\dot{q}}{T_1'-T_2'}$ and $mCp = \frac{\dot{q}}{T_2-T_1}$ $\tag{8.39}$

Thus, $\ln\left(\frac{T_2'-T_2}{T_1'-T_1}\right) = -UA\left(\frac{T_1'-T_2'}{\dot{q}} + \frac{T_2-T_1}{\dot{q}}\right)$ or

$$\dot{q} = \frac{UA\left[-T_1' + T_2'-T_2 + T_1\right]}{\ln\left(\frac{T_2'-T_2}{T_1'-T_1}\right)} = UA\frac{\left[(T_2'-T_2)-(T_1'-T_1)\right]}{\ln\left(\frac{T_2'-T_2}{T_1'-T_1}\right)} \tag{8.40}$$

Also, $T_2'-T_2 = \Delta T_2$,   and $T_1'-T_1 = \Delta T_1$, then

$$\dot{q} = UA\frac{\Delta T_2 - \Delta T_1}{\ln\left(\frac{\Delta T_2}{\Delta T_1}\right)} \tag{8.41}$$

If $\frac{\Delta T_2 - \Delta T_1}{\ln\left(\frac{\Delta T_2}{\Delta T_1}\right)} = \Delta T_{LM}$, the log-mean temperature difference (LMTD), then the final equation can be written as:

$$\dot{q} = UA\Delta T_{LM} \tag{8.42}$$

This is a very useful equation for designing heat exchangers for a given duty.

Example

16. *Examine the temperature profiles of the two heat exchangers shown in Fig. 8.11 and compute their average and log-mean temperature differences. What are the salient features of the two configurations? Of the two, which configuration would you recommend for use in a processing operation and why? Under what circumstances would you recommend using the other configuration and why?*

   Solution: Counter-current flow (a):

$$\Delta T_{ave} = \frac{40 + 60}{2} = 50°C \quad and \quad \Delta T_{LM} = \frac{40 - 60}{\ln\left(\frac{40}{60}\right)} = \frac{-20}{-0.405} = 49.3°C$$

Co-current flow (b):

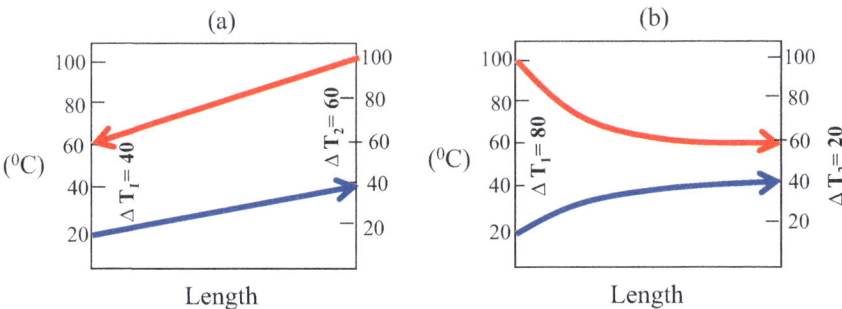

Fig. 8.11 Temperature profiles in co-current and counter-current flow heat exchangers

$$\Delta T_{ave} = \frac{80 + 20}{2} = 50°C \quad and \quad \Delta T_{LM} = \frac{(80 - 20)}{ln\left(\frac{80}{20}\right)} = \frac{60}{1.386} = 43.3°C$$

The above example shows the following:

- ΔT_{LM} depends on the flow pattern, while ΔT_{ave} does not.
- $(\Delta T_{LM})_{counter\text{-}current} > (\Delta T_{LM})_{co\text{-}current}.$
- *For the same $\dot{q}$ and U and using the equation $\dot{q} = UA\, \Delta T_{LM}$, it can be seen that*

$$A(\Delta T_{LM})_{countercurrent} = A(\Delta T_{LM})_{co\text{-}current}$$

However, $(\Delta T_{LM})_{counter\text{-}current} > (\Delta T_{LM})_{co\text{-}current}$, as shown above, thus,

$$(A)_{counter\text{-}current} < (A)_{co\text{-}current}$$

◄

It is thus clearly established that everything else being equal, the area required for the counter-current flow is less than that for the co-current flow, saving space and cost. Counter-current flow also provides a more uniform temperature difference between the two media over the total length of the heat exchanger and should thus be the preferred configuration. However, the co-current configuration is desirable when:

- The fluid to be heated is very viscous, and pumping is otherwise difficult. A higher ΔT at the entrance will reduce the viscosity quickly, increasing Re and thus h_c.
- The mean temperature of the pipe will be lower and thus minimize the differential thermal stress exerted on the pipes and joints.

4. Can ΔT_{LM} be the same in the co-current and counter-current flows for any special case?
5. Can ΔT_{LM} ever be the same as ΔT_{ave}?
6. Can ΔT_{LM} equal zero and, if so, how to handle it? ◄

8.10.2 Plate Heat Exchangers

A plate heat exchanger (PHE) is made of a pack of corrugated stainless-steel heat-transfer plates held together in a frame (Fig. 8.12). A channel is created between each pair of plates through which one fluid flows while another fluid flows through the adjacent channel.

Each plate is gasketed in a configuration that permits counter-current flow of the two media in a series of alternate channels (cavities), as shown in Fig. 8.13. Open

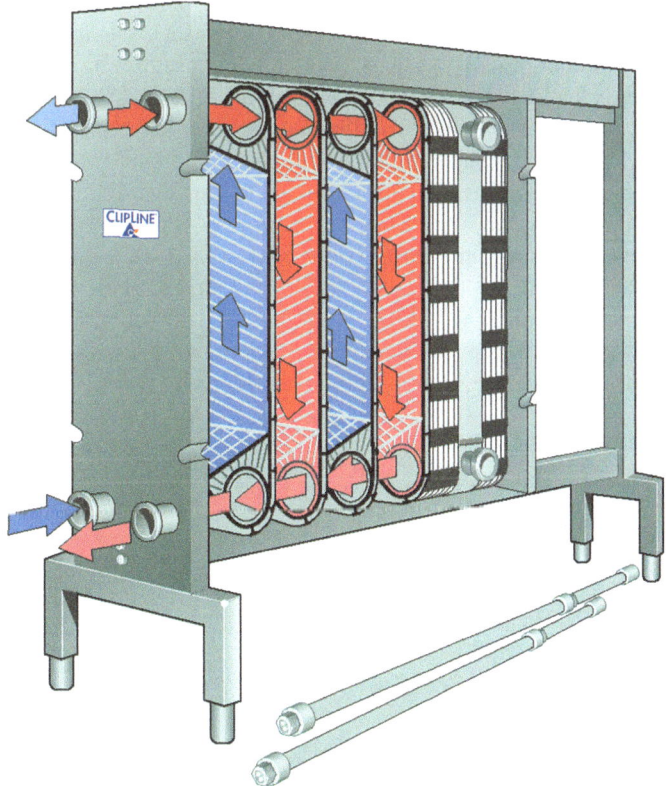

Fig. 8.12 Heat transfer in a plate heat exchanger. From Dairy Processing Handbook © Tetra Pak

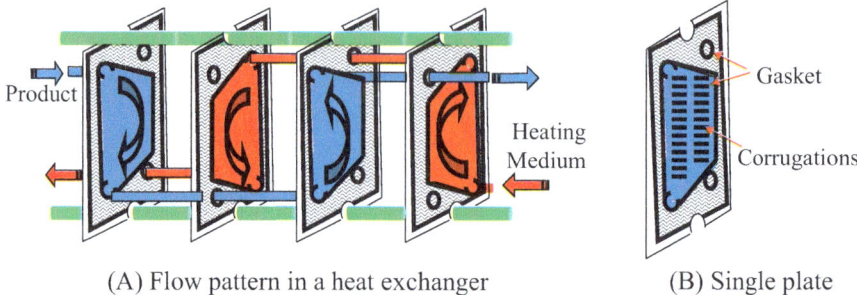

(A) Flow pattern in a heat exchanger   (B) Single plate

Fig. 8.13 Flow control using plate gaskets

and blind holes with different gasket configurations in each plate are utilized to generate varying liquid flow patterns. The corrugated plate design creates high turbulence, which leads to a high heat-transfer rate and less surface fouling. The overall heat-transfer coefficient for this kind of heat exchanger ranges between 5–8 kW/m²K. Frequently, several plate packs are put together as separate sections for heating, heat regeneration, and cooling in the same frame.

It is relatively easy to open, inspect, and clean plate heat exchangers for cleaning and hygienic control of products such as foods. Another useful feature of this type of heat exchanger is that it can be expanded to meet the heat-transfer requirements by adding plates. Using gaskets will limit the operating pressure and temperature, so other designs with welded plates exist that can operate in more extreme conditions but cannot be opened for inspection and cleaning, limiting their utility. In addition to corrugations, the narrow gaps between the plates help generate turbulence at low fluid velocities. To ensure uniformity of the narrow gap when the plate pack is compressed in the frame, supporting points are provided on the corrugations. The Reynolds number estimate for flow in the cavity between adjacent plates is performed using the concept of hydraulic diameter to estimate the equivalent diameter, as discussed before. Referring to Fig. 8.14,

$$D_e = 4\left(\frac{\text{Cross} - \text{sectional area}}{\text{Wetted Perimeter}}\right) = 4\frac{GW}{2(G+W)} = 4\frac{GW}{\sim 2W} = 2G \qquad (8.43)$$

where D_e is the equivalent diameter and G is the gap between the plates. W and L are the plate width and length, respectively.

Fig. 8.14 Equivalent diameter of a plate heat exchanger

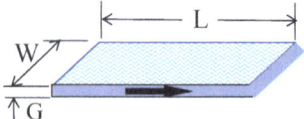

Plate heat exchangers are compact, low-cost, low-maintenance, 80–90% energy-efficient, and flexible capacity systems that are extensively used in the dairy and food industries.

8.10.3 Shell-and-Tube Heat Exchangers

As the name implies, shell-and-tube heat exchangers consist of a bundle of tubes enclosed within a cylindrical shell (Fig. 8.15). In such devices, one fluid flows through the tubes, and another fluid passes through the shell to exchange energy. The shell is also fitted with anti-swirl tank baffles to maintain the flow pattern and improve the system efficiency.

These heat exchangers are designed to have multiple passes and are named accordingly, such as 1–1, 1–2, or 1–6 heat exchangers. The first number refers to the number of shell passes, and the second number denotes the number of tube passes. Figure 8.15 illustrates two such heat exchangers. Overall, shell-and-tube heat exchangers are simpler in design, robust, and low in maintenance costs. They have a high thermal efficiency but require more space compared to a plate heat exchanger of equal duty.

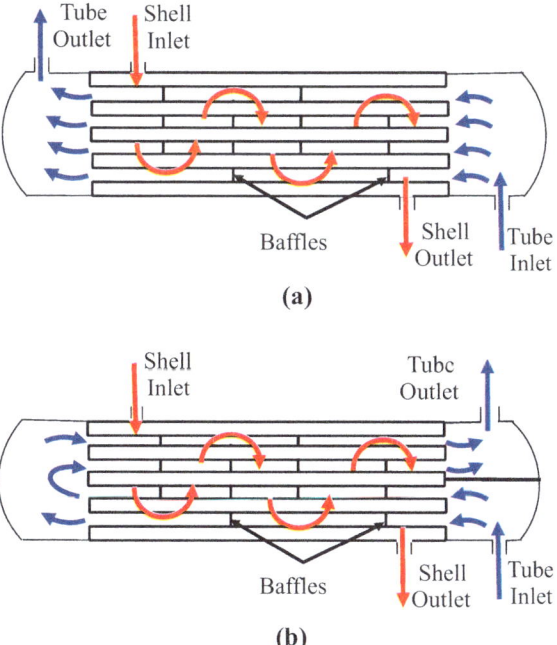

Fig. 8.15 Shell-and-tube heat exchangers: (**a**) 1 shell pass and 1 tube pass (1–1 exchanger), (**b**) 1 shell pass and 2 tube passes (1–2 exchanger)

8.10.4 Scraped Surface Heat Exchangers (also Called Votator Heat Exchangers)

These are used for heating/cooling viscous materials and liquid solid suspensions, e.g., margarine, O.J. concentrates and consist of a double-pipe heat exchanger with a jacketed cylinder containing steam/refrigerant. Internal rotating shaft is fitted with a wiper blade. The rotating scrapers or wiper blades continually scrape the surface, preventing localized overheating/freezing and giving rapid heat transfer (Figs. 8.16 and 8.17).

8.11   Designing Thermal Processes for Better Quality Food

Now that we have learned the basic principles of heat transfer and fluid flow, let us look at an example where these principles are used to design industrial scale processes to make finished products for the market. The application of heat energy is the most ubiquitous and effective method for processing a wide variety of foods. Although thermal processing indicates the transfer of heat energy from one object to another, in the food industry, thermal processing generally means any process in which the temperature of the product is elevated above ambient temperature. It includes processes such as blanching, pasteurization, and commercial sterilization.

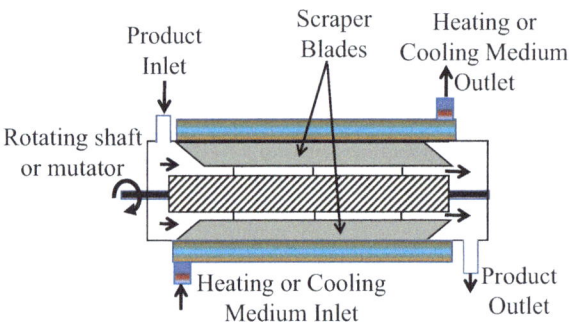

Fig. 8.16 Cross-sectional view of scraped surface heat exchanger

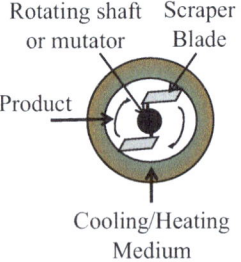

Fig. 8.17 End view of rotating scraper blades

The purpose of thermal processes is to extend the storage life of food while keeping it safe for human consumption. As heating is known to be injurious to the nutritional quality of foods, a derivative objective is to design thermal processes that also maximize nutrient retention. This can only be attained by a comprehensive understanding of knowledge from the basic disciplines of chemistry, physics, and microbiology and their integration by engineering to design intelligent processing operations, as discussed below.

8.11.1 Chemical and Microbiological Aspects of Food

It has been recognized for some time that the temperature dependence of the destruction of nutrients is markedly different from that of microorganisms. Listed in Table 8.6 are the ranges of first-order reaction rate constants for quality factors of food as well as for the types of microorganisms associated with foods. As may be observed from the shown values, nutrients and quality factors are considerably more resistant to thermal destruction than vegetative cells and some spores, ranging from one to six orders of magnitude. In the food industry, the rate constant (k) is represented by the D-value ($D = 2.303/k$), which for a first-order reaction commonly observed in foods becomes the time at a constant temperature for the concentration to decrease by one log cycle (or 90%).

Additionally, shown in the above table is activation energy, E_a, of selected items. The reaction rate constant with a larger E_a is known to increase rapidly with increasing temperature. The higher temperature dependence of reaction rates for vegetative cells and spores than for vitamins and quality factors does indeed offer an opportunity to intelligently pick a time–temperature combination that would achieve the desired microbial kill with minimum collateral damage to quality factors. Thus, high-temperature, short-time (HTST) processes should result in better quality products. Experimental data such as those plotted in Fig. 8.18 do indicate the efficacy of HTST processes. However, the underlying assumption of the process that every particle of the product receives the same lethal heat treatment makes it applicable to only convection heating liquid foods or liquid foods containing internally sterile particulates. This approach is not applicable to conduction heating foods or liquid foods containing internally nonsterile particulates. For such cases, more advanced optimization techniques are required.

Table 8.6 Approximate thermal resistance of selected food constituents and microorganisms

Constituent	Reaction Rate Constant, k, at 121 °C (min^{-1})	Activation Energy, E_a, (kcal/mole)
Vitamins	0.002–0.02	20–30
Quality Factors (Color, flavor, texture)	0.005–0.5	10–50
Vegetative cells	100–1000	100–150
Spores	0.5–20	50–80

Fig. 8.18 Bacterial thermal destruction curve and 1, 5, 20, and 50% destruction curves for thiamine (Data from Feliciotti and Esselen [1] and Lund [4])

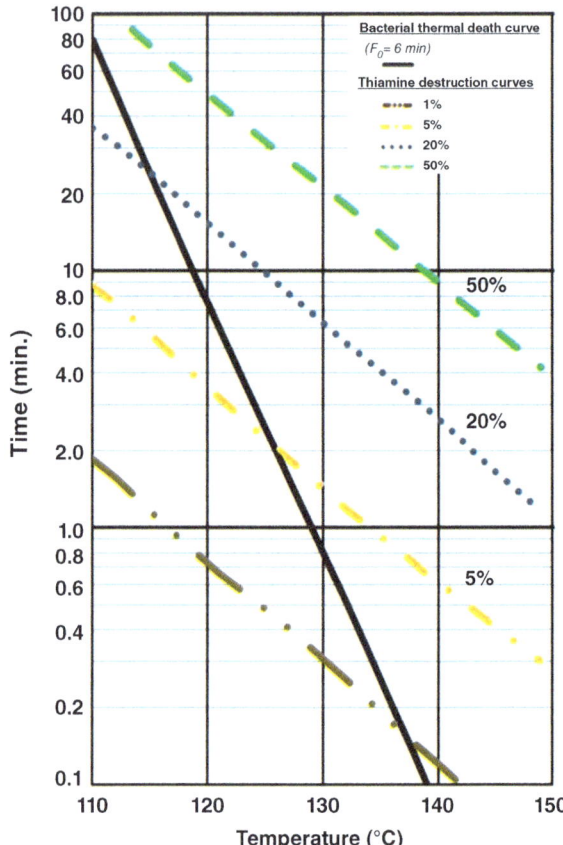

8.11.2 HTST Processing Systems

Based on the discussion presented in the previous section, it is reasonable to expect that HTST processing of liquid foods could be used to make better quality products. A wide range of food products, such as milk, ice cream mix, liquid eggs, fruit and vegetable juices, soups, wine, beer, and many other beverages, are indeed commercially processed using the HTST method. It is now an established, common process and is known as HTST pasteurization, named after Louis Pasteur, who in 1857 showed that souring of milk could be delayed by heating. Also known as flash pasteurization, a generically applicable definition of the process is given by the IDF as follows: "pasteurization is a process applied to a product with the objective of minimizing possible health hazards arising from pathogenic microorganisms associated with the product (milk) which is consistent with minimal chemical, physical and organoleptic changes in the product."

The first commercial HTST pasteurization system was approved in the USA in 1933 for milk, and it is now a widely accepted technology. The standard US protocol

Table 8.7 PMO requirements for pasteurization of milk	Temperature °C (°F)	Hold time
	63 (145)	30 min
	72 (161)	15 s
	89 (191)	1 s
	90 (194)	0.5 s
	94 (201)	0.1 s
	96 (204)	0.05 s
	100 (212)	0.01 s

for pasteurization of milk and related products is outlined in the Pasteurized Milk Ordinance (PMO) published by the Food and Drug Administration. A sample of the required temperature and hold time is shown in Table 8.7. The required time–temperature treatment is based on the destruction of *Coxiella burnetii,* the most heat-resistant, nonsporulating, and Q-fever-causing pathogen likely to be present in raw milk. For added safety, however, many HTST systems operate at higher temperatures with longer holding times (e.g., 75 °C for 25 s). International regulatory agencies such as Codex Alimentarius require a 5-log (99.999%) or greater reduction of *Mycobacterium tuberculosis, M. bovis,* and *C. burnetii* in whole milk (4% milk fat) following pasteurization.

The high-temperature short-time process for maximizing the quality of pasteurized liquid foods has been made possible only by the introduction of continuous processes using heat exchangers designed on sound engineering principles. The schematic of a typical pasteurization system using a plate heat exchanger is shown in Fig. 8.19, and its major components are briefly described below:

- Regeneration Section: Raw feed from a constant-level balance tank (to ensure a constant head to the feed pump) is pumped to the regeneration section of the heat exchanger, where it is preheated by the hot fluid stream of the product coming from the holding tube. A differential pressure > 14 kPa (2 psi) between the heated and raw product streams is maintained using a booster pump to eliminate the possibility of cross contamination or product dilution from raw to treated streams due to cracks or pinholes in the heat exchanger plates that may develop over time due to corrosion.
- Regeneration Efficiency: Recovery of energy in high-temperature thermal processing units is highly desirable to make the process more efficient. In the regen section, the heat from the hot stream leaving the holding tube (known as the starter stream) is used to reheat (regenerate) the incoming raw product stream instead of rejecting it to an external cooling medium. The effectiveness of regeneration is the ratio between actual heat transfer from the hot (starter) to cold stream and ideal or maximum heat transfer possible from the hot to cold stream. This can be estimated using the following steps:

In the regeneration section, heat received by raw feed = heat given up by the "starter" stream:

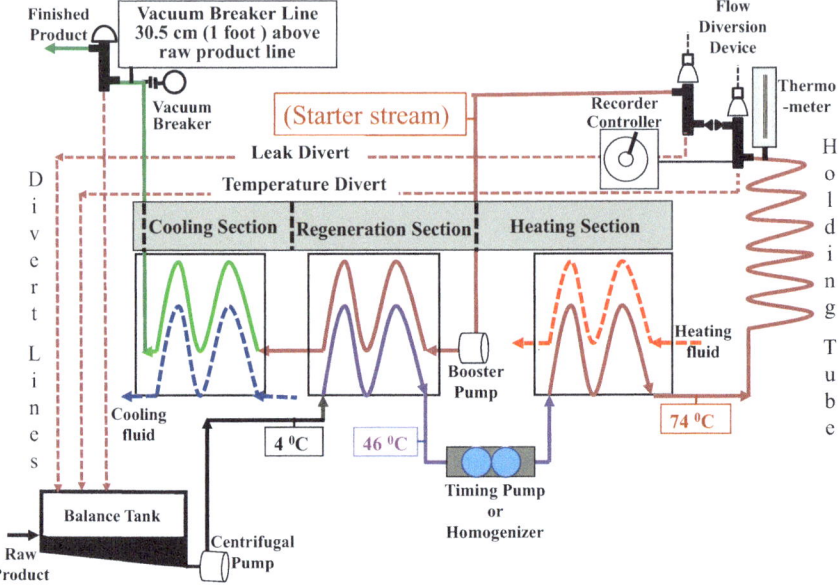

Fig. 8.19 Schematic of a HTST pasteurization system

$$\left(\dot{m}c_p\Delta T\right)_{\text{Feed}} = \left(\dot{m}c_p\Delta T\right)_{\text{Starter}} \qquad (8.44)$$

At steady state, the mass flow rate in each section of the system is constant, and if c_p does not change appreciably with temperature, then for 100% energy recovery under ideal conditions:

$$(\Delta T)_{\text{Feed}} = (\Delta T)_{\text{Starter.}} \qquad (8.45)$$

However, since this is not achievable in real systems due to the various types of energy losses that commonly occur, the actual regeneration efficiency ($\acute{\eta}$) is computed as a percentage of the maximum starter stream as shown below:

$$\acute{\eta} = \left\{ (\Delta T)_{\text{Feed in Regen.}} / (\Delta T)_{\text{Starter } Maximum} \right\} (100) \qquad (8.47)$$

Using the temperatures indicated for the process shown in Fig. 8.19, we obtain:

$$\acute{\eta} = \{(46-4)/(74-4)\}(100) = 60\%$$

• Heating Section: The preheated stream is then directed into the main heating section of the heat exchanger with the aid of a positive displacement (timing) pump, where it is heated to the required pasteurization temperature with some margin for safety. For products requiring homogenization, a homogenizer may also serve as a positive pump for controlled flow through the heating section and the holding tube.

Holding Tube: The hot stream flows through the holding tube where the residence time is designed to be at least equal to the stipulated hold time. As discussed in Chap. 5, even under turbulent flow in the holding tube, there will be a velocity distribution giving rise to a range of stream residence times. To ensure that all particles receive the minimum required heat treatment, the holding tube must be sized correctly, see 5.2.10 E and experimentally verified at regular intervals. The PMO also has the following two requirements for the HTST setup: (a) the process always flows against gravity such that in the case of power failure, the product drains backward, and (b) there must be a ¼ inch slope per foot of the holding tube to minimize the entrapment of air bubbles that may otherwise distort the hold time (Fig. 8.20).

Check Your Understanding

7. The hold time of a liquid food in a pasteurizer is 15 seconds at 72 °C. Compute the number of log reductions of *listeria monocytogenes* if its D_{72} value is 2.9 seconds.
8. Will a long holding tube with high velocity be better than a short holding tube with low velocity for HTST pasteurization? Why? ◄

- Flow Diversion Device: Located at the end of the holding tube, the FDD controls the direction of product flow. It diverts the product to forward flow only if the set temperature of the scheduled process is reached; otherwise, the flow is diverted back to the balance tank. To avoid flashing of the product in the divert line, FDD is located after the cooling section, especially for high-temperature and ultrahigh-temperature processing operations (Fig. 8.21). Should a pressure drop be detected between the raw and finished streams in the regen or cooling sections, the FDD automatically diverts the treated stream back to the balance tank.

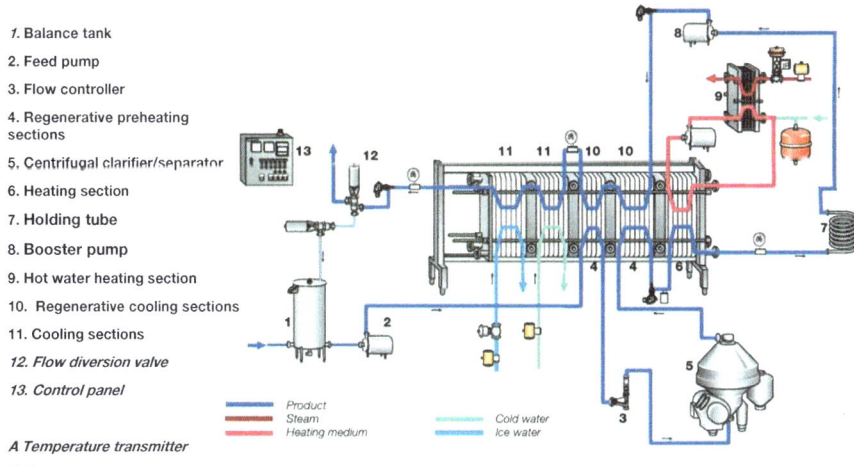

1. Balance tank
2. Feed pump
3. Flow controller
4. Regenerative preheating sections
5. Centrifugal clarifier/separator
6. Heating section
7. Holding tube
8. Booster pump
9. Hot water heating section
10. Regenerative cooling sections
11. Cooling sections
12. *Flow diversion valve*
13. *Control panel*

A Temperature transmitter

B Pressure gauge

— Product
— Steam
— Heating medium
— Cold water
— Ice water

Fig. 8.20 The complete HTST pasteurizer plant From Dairy Processing Handbook © Tetra Pak

Fig. 8.21 Holding tube with ¼ inch slope per foot and dual stem diversion device in a pasteurizer

- Cooling Section: After giving up some of its energy in the regen section, the product stream enters the cooling sections where it is further cooled to the final temperature by chilled water.
- Process Efficacy: Pasteurization processes are designed to kill at least 99.999% of pathogens (a 5-log reduction) and destroy yeasts, molds, and some spoilage microorganisms found in foods, and their efficacy is evaluated by testing the process using surrogate microorganisms. Milk pasteurization efficacy or contamination of pasteurized milk with raw milk is monitored indirectly by using alkaline phosphatase (ALP) enzyme as an indicator of safe quality. The ALP enzyme is present in all raw milks and is completely inactivated at the required time and temperature for pasteurization of milk. New phosphatase tests have been developed that can be performed in less than a minute and can detect raw milk contamination as low as 0.01%.
- Ultrahigh-Temperature (UHT) Processing: As a derivative of the HTST process, ultrahigh-temperature units for even shorter holding times with higher temperatures have been developed for making shelf-stable products and are in commercial use. It is important to remember that according to most local regulations, thermal processing systems must be approved, and their use must be authorized by the competent authority.

Problems

8.1 The surface temperature of a hot plate is 130 °C. When placed in a room where the ambient air temperature is 30 °C, the heat flux from the hot plate is estimated to be 800 W/m². Compute the convective heat-transfer coefficient.

8.2 A counter-current, double-pipe heat exchanger is designed to cool 300 kg/h of milk from 80 °C to 40 °C, using cooling water at 20 °C with a maximum allowable exit temperature of 35 °C. What surface area is required for the duty? Additional information: Overall heat transfer coefficients based on the inside area (Ui) = 140.84 W/m^2·K. Specific heats: milk = 4000 J/kg-K; water = 4180 J/kg-K. Heat exchanger: tube thickness = 2 mm, thermal conductivity = 20 W/m-K.

8.3 A double pane window (thermopane) consisting of two layers of glass, each 5.2 mm thick and separated by a stagnant 5 mm space of stagnant air is installed in a room maintained at 25 °C. If the outside temperature is -5 °C and the window is 0.5 m wide and 1.0 m high, estimate the heat loss through the window. The thermal conductivity of air is 0.021 W/m-K and that of glass is 0.78 W/m-K. The convective heat-transfer coefficients on the inside and outside surfaces of the window maybe assumed to be 10 W/m^2-K and 50 W/m^2-K, respectively.

8.4 Air at 95 °C is flowing at a rate of 0.5 m/s through a bed of peas for drying. The average diameter of the peas is 4 mm. Assuming the surface temperature to be 25 °C, estimate the convective heat-transfer coefficient. The air properties at the average air temperature of 60 °C are: $C_p = 1.02$ kJ/kg-°C; $\rho = 1.03$ kg/m^3; $k = 0.028$ W/m-°C; $\eta = 19.91*10^{-6}$ Pa-s.

8.5 Assuming the basic mechanism of heat transfer in a drum drier is conduction, a new drum drier is being designed for drying a liquid food from an initial total solids content of 12 wt.% to 4 wt.% final moisture content. An overall heat-transfer coefficient (U) of 1700 W/m^2-°C has been estimated for the product while the average temperature difference between the roller surface and the product is 85 °C. Estimate the roller surface area needed for a production rate of 20 kg product per hour. The enthalpy of vaporization of water is 2420 kJ/kg.

8.6 In a refrigeration system, a pipe of 1 cm inside diameter carries the refrigerant. The thermal conductivity of the pipe material is 45 W/m-K and the thickness of the pipe is 1 mm.

(a) What will be the heat gained by the refrigerant (in W/m) from the surroundings if the outside heat-transfer coefficient is 8.3 W/m^2-K, assuming that the inside surface temperature of the pipe is -10 °C and the ambient temperature is 30 °C?

(b) What is the heat gained if the pipe is wrapped with an insulation of thermal conductivity 0.1 W/m-K and thickness 1 mm? Is the heat transfer in this case higher or lower as compared to part (a)? Why?

8.7 In a concentric tube heat exchanger used for cooling a hot product after pasteurization, the outer tube carries cold water with inlet and outlet temperatures of T_{ci} and T_{co}, respectively. The inner tube carries the hot product with inlet and outlet temperatures of T_{hi} and T_{ho}, respectively. Draw a rough temperature vs. length profile for the heat exchanger fluids, for each of the following situations. Also in each case, determine if the profiles are feasible or not explain.

(a) Counter-current flow, $T_{ci} = 5$ ° C, $T_{co} = 28$ ° C, $T_{hi} = 50$ ° C, and $T_{ho} = 25$ ° C

(b) Co-current flow, $T_{ci} = 5\,^{\circ}$C, $T_{co} = 28\,^{\circ}$C, $T_{hi} = 50\,^{\circ}$C, and $T_{ho} = 25\,^{\circ}$C

(c) Co-current flow, $T_{ci} = 5\,^{\circ}$C, $T_{co} = 28\,^{\circ}$C, $T_{hi} = 60\,^{\circ}$C, and $T_{ho} = 17\,^{\circ}$C

(d) Counter-current flow, $T_{ci} = 5\,^{\circ}$C, $T_{co} = 28\,^{\circ}$C, $T_{hi} = 60\,^{\circ}$C, and $T_{ho} = 17\,^{\circ}$C

(e) Counter-current flow, $T_{ci} = 5\,^{\circ}$C, $T_{co} = 5\,^{\circ}$C, $T_{hi} = 60\,^{\circ}$C, and $T_{ho} = 25\,^{\circ}$C

8.8 A double-pipe heat exchanger consists of a 0.902 inch i.d. and $= 1.0$ inch o.d copper tube inside a 2-inch schedule 40 steel pipe. Water is flowing through the inner tube, while in the annular space, lube oil flows counter-current to the water. At a particular point in the exchanger, the bulk temperature of the oil is 350 °F, while the bulk water temperature at the same point is 95 °F. The heat-transfer coefficients at this point are 100 BTU/hr-ft²-F° for the oil and 400 for the water. (a) Calculate the heat-transfer rate at this point (BTU/hr). (b) What is the temperature (°F) at the outer surface of the copper tube? The thermal conductivity of copper is 215 BTU/hr-ft² (°F/ft).

8.9 A beverage is heated using a double-pipe heat exchanger from 20 to 60 °C. The product flows in the inner pipe which is 40 mm i.d. with a 5 mm thick wall. The heating medium is saturated steam condensing at 95 °C in the outer jacket. The product physical properties are as follows: Density: 1015 kg/m³ at 40 °C; Specific heat: 3.5 kJ/kg-K; Viscosity: 1.03 cP at 40 °C and 0.5 cP at 95 °C; Thermal conductivity: 1.2 W/m-K; Thermal conductivity of the pipe material: 50 W/m-K; The convective heat-transfer coefficient for the steam side is 9500 W/m²-K

   (a) Estimate the convective heat-transfer coefficient for a product flow rate of 10.0 m³/h

   (b) Compute the overall heat-transfer coefficients (U_i and U_o), based on both the inside and outside areas.

   (c) What length of the heat exchanger will be required to perform the heating duty in (a) above?

8.10 A plate heat exchanger was designed to pasteurize a product ($h_c = 85$ W/m²-K) with hot water ($h_c = 350$ W/m²-K). The design value for the heat exchanger area was 17 m², without considering fouling. Calculate the actual area required if a fouling resistance of 0.02 m²-K/W is anticipated.

8.11 Saturated steam at 120 °C flows inside a steel pipe ($k = 42.9$ W/mK) of 2.0 cm i.d. and 2.8 cm o.d. at an ambient air temperature of 20 °C. If the convective heat-transfer coefficients on the inner and outer pipe surfaces are taken as 5500 W/m²-K and 30 W/m²-K respectively, compute the heat loss (a) per meter of the pipe and (b) per meter of the pipe, if a 4 cm thick insulation ($k = 0.05$ W/m-K) is wrapped around outer surface.

8.12 A block of ice at 0 °C and 0.25 m on each side is sitting on a 7.6 thick layer of Styrofoam in a room at 22 °C. Assume the density of ice is 920 kg/m³.

   (a) Considering only convection, how much time will be needed to melt the ice in a room with no air movement ($h = 1$ W/m²-K)?

   (b) How much faster will it melt if you put a fan near the ice ($h = 5$ W/m²-K)?

   (c) What is the flow rate of heat through the Styrofoam ($k = 0.03$ W/m-K)

8.13 Calculate: (a) the <u>inner</u> overall heat-transfer coefficient for a 5 meter long, uninsulated pipe (5 cm ID, 7 cm OD) carrying oil. Given ($h_{oil} = 120$ W/m^2-K, $k_{pipe} = 237$ W/m-K, $h_{outer} = 25$ W/m^2-K), (b) the heat-transfer rate from the oil to the air if the temperature of the oil is 70 °C and the temperature of the air is 25 °C, and (c) the outer overall heat-transfer coefficient for this system. After several weeks of operation, it is found that the overall heat-transfer coefficient is reduced by 25% due to fouling, (d) what is the fouling resistance?

8.14 A pilot plant wants to use a new holding tube to pasteurize ice cream mix ($\rho = 63$ lb$_m$/ft^3, $K = 1.595*10^{-5}$ psi-s$^{0.7}$, $n = 0.7$). The holding tube consists of eight sections of 1 in OD stainless-steel tube, each 41 inches long, and joined together by seven 180° bends, each 15.5 inches long. Mix enters at the bottom of the holding tube which snakes up a total of 24 inches and the mix exits at the top.

(a) Determine the maximum legal flow rate of ice cream mix if the residence time in the holding tube must be 15 s. Assume laminar flow

(b) Calculate the pressure drop through the holding tube.

8.15 A food manufacturing company produces 1000 kg/h pasteurized apple cider daily in an 8-hour shift. The pasteurizer temperature setting is indicated on the diagram. Compute the regeneration efficiency of the process and estimate the energy savings per month if the cost of energy is $ 0.10 per kW-h.

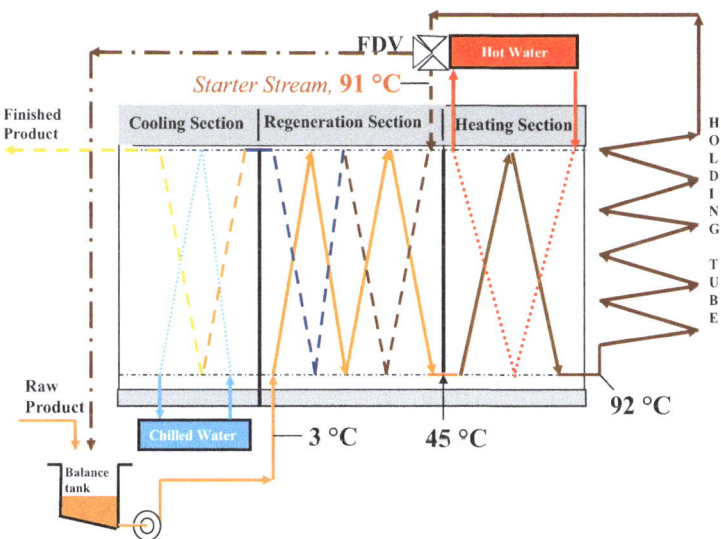

Regeneration in a HTST pasteurizer

Bibliography

1. Feliciotti E, Esselen WB (1957) Thermal destruction rates of thiamin in pureed meats and vegetables. Food Technol 11:77
2. International Dairy Federation. 1986. Monograph on Pasteurized Milk, International
3. Dairy Federation Bulletin 200.
4. Lund DB (1977) Maximizing nutrient retention. Food Technol 31(2):71
5. Omidi M, Farhadi M, Jafari M (2017) A comprehensive review on double pipe heat exchangers. Appl Therm Eng 110

Heat Transfer: Unsteady State

9

We have shown how heat travels through materials. However, the movement of heat does not necessarily imply a movement of temperature changes. In fact, in a steady state, heat flows with no temperature change with time at a particular point in a process. In this section, we will examine the unsteady state or transient heat transfer. Under unsteady-state conditions, the temperature within the material varies with time, and we will see what factors affect the migration of temperature changes through an object. All the heating transfer processes are initially under unsteady-state conditions.

Imagine a thick-walled pot filled with a liquid that has been sitting in a cool water bath for an extended period. An equilibrium state has been reached in which the water bath, the wall of the pot, and the liquid in the pot are all at the same temperature. Figure 9.1a shows the flat temperature profile that will exist through the wall of the pot. Clearly, no heat will flow through the wall.

Suppose the pot is suddenly plunged into a hot water bath. Heat will begin to penetrate the pot wall, and in a short timeframe, the temperature profile within the wall will resemble the one shown in Fig. 9.1b. The outer surface has been heated to the temperature of hot water, while most of the wall is still near the original temperature. This results in a steep gradient followed by a flatter gradient.

According to Fourier's first law, heat should move rapidly across the steep gradient and then slow down when the flatter gradient is reached. An energy balance at the point of gradient change will be:

$$\begin{bmatrix} Rate\ of \\ energy\ in \end{bmatrix} - \begin{bmatrix} Rate\ of \\ energy\ out \end{bmatrix} = \begin{bmatrix} Rate\ of\ energy \\ accumulation \end{bmatrix} \quad (9.1)$$

In other words, energy will accumulate at the point of gradient change with a resulting temperature rise. This will tend to straighten the temperature profile so that it resembles Fig. 9.1c. The rate of temperature rise with time (dT/dt) at any point in the wall will depend on three factors:

© The Author(s), under exclusive license to Springer Nature Switzerland AG 2024
S. S. H. Rizvi, *Food Engineering Principles and Practices*,
https://doi.org/10.1007/978-3-031-34123-6_9

- Thermal conductivity: The greater the conductivity (k), the faster heat will be brought to any given spot and the faster temperature will rise.

$$\frac{dT}{dt} \propto k \tag{9.2}$$

- Specific heat: The greater the specific heat (C_p) of the wall, the more heat is required to raise the temperature at any point and the slower the temperature will rise.

$$\frac{dT}{dt} \propto \frac{1}{C_p} \tag{9.3}$$

- Rate of gradient change: Since the rate of heat flow is proportional to the slope (gradient) of the temperature profile, a large change in slope at any point will result in a large difference between the heat into that point and the heat out of that point. Thus, the faster the slope changes, the faster the temperature will rise. We will show below that this change in slope is described by the second derivative of the temperature profile.

$$\frac{dT}{dt} \propto \text{Rate of change in} \frac{dT}{dx} = \frac{d^2T}{dx^2} \tag{9.4}$$

Eventually, a linear temperature profile will be established, as shown in Fig. 9.1c. Now, if the outside and inside temperatures remain the same, all the heat that enters the pot will flow through, and none will accumulate in the wall. Although the temperature will vary at different points in the wall, at any point in the wall, the temperature will remain constant with time, and we say that the wall will be in a steady state. Thus,

$$\frac{dT}{dt} = 0 \tag{9.5}$$

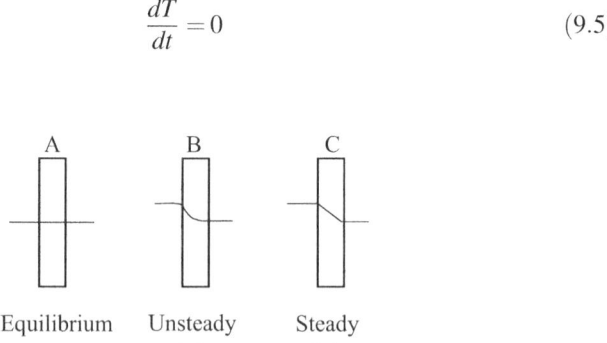

Fig. 9.1 Unsteady-state conduction

9.1   Unsteady-State Heat Transfer

A comparison of successive profiles in Fig. 9.1 shows that before attaining steady state, a temperature increase migrates through the wall of the container. This temperature migration should not be confused with the heat flow through the wall, although heat flow is necessary for temperature migration to take place. This migration of temperature is analogous to the diffusion of molecules in a solvent. For example, if a cube of sugar is dropped into a cup of coffee without stirring, sugar molecules will diffuse into the coffee. Initially, a steep concentration gradient will form near the sugar cube, such as the one in Fig. 9.1b. With passing time, this sugar concentration will migrate across the cup. Because of the similarity to molecular diffusion, we call the migration of temperature change thermal diffusion, and in the following sections, we will show that the thermal diffusion depends on thermal conductivity, specific heat, and the second derivative of the temperature profile.

It was previously noted that when a temperature profile is nonlinear, the temperature tends to rise most rapidly at the point where the temperature gradient changes most rapidly. In that case,

- The original function, $f(x)$, describes the temperature profile.
- The first derivative, $f'(x)$, describes the slope or gradient of the profile.
- The second derivative, $f''(x)$, describes the rate of change of the gradient of the profile. Thus, we expect that temperature will change most rapidly where the second derivative is greatest.

We will now derive the equation that verifies this expectation. Figure 9.2a shows a cross-section of a flat plate in which we have selected a small rectangular volume element at some arbitrary location. Figure 9.2b shows a three-dimensional enlargement of this tiny differential element whose edges have lengths Δx, Δy, and Δz. The faces of the element have areas equal to the products of the edges, i.e., $A_x = \Delta y \Delta z$, $A_y = \Delta x \Delta z$, and $A_z = \Delta x \Delta y$. Its volume will be the product of the three dimensions, i.e., $V = \Delta x \Delta y \Delta z$. The three-dimensional energy balance on the small element in Fig. 9.2b will help us analyze the heat-transfer mechanism. Heat flowing in the x direction $(\dot{q}_x)$ enters this element through one face and exits through

Fig. 9.2 Heat transfer in a differential element

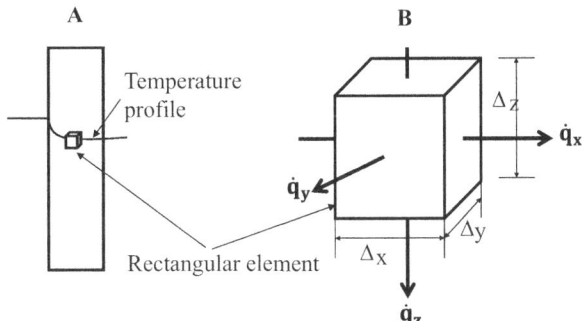

the opposite face, as shown in the figure. Assume that an unsteady state exists and that there is a nonlinear temperature profile through the element, as shown. This means that heat will enter the element with one gradient and exit with a less steep gradient. By Fourier's law, then, heat will enter faster than it leaves and the energy balance Eq. (9.1) may be written as:

$$-kA_x\left(\frac{dT}{dx}\right)_{in} + kA_x\left(\frac{dT}{dx}\right)_{out} = \dot{q}_x \tag{9.6a}$$

$$\text{Or, } \dot{q}_x = -kA_x\left[\left(\frac{dT}{dx}\right)_{in} - \left(\frac{dT}{dx}\right)_{out}\right] \tag{9.6b}$$

where $\dot{q}_x$ is the rate of energy accumulation and the expression in square brackets represents the difference in slope between the two sides of the tiny element. To make this derivation perfectly general, assume that the same thing is happening in the y and z directions, so

$$\dot{q}_y = -kA_y\left[\left(\frac{dT}{dy}\right)_{in} - \left(\frac{dT}{dy}\right)_{out}\right] \tag{9.7}$$

$$\dot{q}_z = -kA_z\left[\left(\frac{dT}{dz}\right)_{in} - \left(\frac{dT}{dz}\right)_{out}\right] \tag{9.8}$$

The total rate of heat accumulation in the element is the sum of these three heat flows, so we can write:

$$\text{Total accumulation } (\dot{q}) = \dot{q}_x + \dot{q}_y + \dot{q}_z \tag{9.9}$$

Substituting Eqs. (9.6a, 9.6b), (9.7), and (9.8) in (9.9) gives the complete three-dimensional energy balance:

$$\dot{q} = -k\left\{A_x\left[\left(\frac{\partial T}{\partial x}\right)_{in} - \left(\frac{\partial T}{\partial x}\right)_{out}\right] + A_y\left[\left(\frac{\partial T}{\partial y}\right)_{in} - \left(\frac{\partial T}{\partial y}\right)_{out}\right]\right. \tag{9.10}$$
$$\left. + A_z\left[\left(\frac{\partial T}{\partial z}\right)_{in} - \left(\frac{\partial T}{\partial z}\right)_{out}\right]\right\}$$

In this equation, the dependent variable (T) is a function of 4 independent variables $(x, y, z,$ and $t)$. We must therefore differentiate T separately for each independent variable, which results in partial derivatives.

9.1.1   Fourier's Second Law

The rate at which the temperature changes in the differential element depends on three factors:

- It is inversely proportional to the mass of the element, which equals its volume times its density ($\rho \Delta x \Delta y \Delta z$).
- It is inversely proportional to the heat capacity (specific heat) of the element (C_p) and.
- It is proportional to the rate of heat accumulation within the element ($\dot{q}$).

Thus, we can write:

$$\frac{\partial T}{\partial t} = \frac{\dot{q}}{C_p m} = \frac{\dot{q}}{C_p \rho \Delta x \Delta y \Delta z} \quad \text{or,} \quad (C_p \rho \Delta x \Delta y \Delta z)\frac{\partial T}{\partial t} = \dot{q} \tag{9.11}$$

Substituting Eq. 9.10 in 9.11 gives:

$$C_p \rho \Delta x \Delta y \Delta z \frac{\partial T}{\partial t} = -k\left\{ A_x\left[\left(\frac{\partial T}{\partial x}\right)_{in} - \left(\frac{\partial T}{\partial x}\right)_{out}\right] \right.$$

$$\left. + A_y\left[\left(\frac{\partial T}{\partial y}\right)_{in} - \left(\frac{\partial T}{\partial y}\right)_{out}\right] + A_z\left[\left(\frac{\partial T}{\partial z}\right)_{in} - \left(\frac{\partial T}{\partial z}\right)_{out}\right]\right\} \tag{9.12}$$

If we divide all terms by $\Delta x \Delta y \Delta z$, noting that $A_x = \Delta y \Delta z$, etc., we obtain:

$$\rho C_p \frac{\partial T}{\partial t}$$

$$= k\left\{\left[\frac{\left(\frac{\partial T}{\partial x}\right)_{out} - \left(\frac{\partial T}{\partial x}\right)_{in}}{\Delta x}\right] + \left[\frac{\left(\frac{\partial T}{\partial y}\right)_{out} - \left(\frac{\partial T}{\partial y}\right)_{in}}{\Delta y}\right] + \left[\frac{\left(\frac{\partial T}{\partial z}\right)_{out} - \left(\frac{\partial T}{\partial z}\right)_{in}}{\Delta z}\right]\right\} \tag{9.13}$$

Note that each of the three terms in square brackets is a change in the first derivative divided by Δx, Δy, and Δz. Taking the limit of these ratios as Δx, Δy, and Δz going to zero, the results will be the second derivatives of the temperature profile. Thus, the equation becomes:

$$\frac{\partial T}{\partial t} = \frac{k}{\rho C_p}\left[\frac{\partial^2 T}{\partial x^2} + \frac{\partial^2 T}{\partial y^2} + \frac{\partial^2 T}{\partial z^2}\right] \tag{9.14}$$

This equation is Fourier's second law, which verifies what we saw earlier that the rate of temperature change per unit volume is:

- Proportional to **conductivity** (k).
- Inversely proportional to **specific heat**. (Since the temperature change per unit mass is proportional to C_p, the change per unit volume is proportional to ρC_p.)
- Proportional to the rate of **gradient change**, i.e., the second derivative of the temperature profile.

9.1.2 Thermal Diffusivity

For any material, k, ρ, and C_p are all constants, and thus, the expression $k/\rho C_p$ is also a constant that is frequently represented by symbol α and is given the name "**thermal diffusivity.**" Substituting α, Eq. (9.14) becomes:

$$\frac{\partial T}{\partial t} = \alpha\left[\frac{\partial^2 T}{\partial x^2} + \frac{\partial^2 T}{\partial y^2} + \frac{\partial^2 T}{\partial z^2}\right] \tag{9.15}$$

Like thermal conductivity, it is characteristic of the material and reflects the fact that the rate at which a temperature change migrates through the material depends both on its conductivity and its heat capacity. Table 7.2 in Chap. 7 gives the thermal conductivity, specific heat (heat capacity), density, and thermal diffusivity for the principal constituents of food products. To summarize, it can be concluded that a temperature increase will penetrate a material at a rate that is proportional to its **thermal diffusivity** (α)

The units of α are

$$\alpha = \frac{k}{\rho C_p}[=]\frac{\left(\frac{Watts}{m\ K}\right)}{\left(\frac{kg}{m^3}\right)\left(\frac{Joules}{kg\ K}\right)} = \frac{m^2}{s} \quad Or \quad \alpha = \frac{k}{\rho C_p}[=]\frac{\left(\frac{Btu}{hr\ ft^0 R}\right)}{\left(\frac{lb_m}{ft^3}\right)\left(\frac{Btu}{lb_m\ {}^0R}\right)} = \frac{ft^2}{hr} \tag{9.16}$$

9.2   Governing Conduction Equations and Solutions

As presented in the preceding section, the governing equation for unsteady-state heat transfer involving no energy conversion (no heat generation) and having constant physical properties is the second-order partial differential equation, Eq. (9.15). It may be written in common coordinate systems such as Cartesian for products such as steaks and cookies, cylindrical for canned foods, and spherical for meatballs, as follows:

In Cartesian coordinates:

$$\frac{\partial T}{\partial t} = \alpha\left(\frac{\partial^2 T}{\partial x^2} + \frac{\partial^2 T}{\partial y^2} + \frac{\partial^2 T}{\partial z^2}\right) \tag{9.17}$$

In cylindrical coordinates:

$$\frac{\partial T}{\partial t} = \alpha \left[\frac{\partial^2 T}{\partial r^2} + \frac{1}{r} \left(\frac{\partial T}{\partial r} \right) + \frac{1}{r^2} \left(\frac{\partial^2 T}{\partial \theta^2} \right) + \frac{\partial^2 T}{\partial z^2} \right] \tag{9.18}$$

In spherical coordinates:

$$\frac{\partial T}{\partial t} = \alpha \left[\frac{1}{r} \frac{\partial^2 (rT)}{\partial r^2} + \frac{1}{r^2 Sin\theta} \frac{\partial}{\partial \theta} \left(Sin\theta \frac{\partial T}{\partial \theta} \right) + \frac{1}{r^2 Sin^2\theta} \frac{\partial^2 T}{\partial \phi^2} \right] \tag{9.19}$$

The above may be generically written as $\dfrac{\partial T}{\partial t} = \alpha \nabla^2 T$ $\qquad$ (9.20)

where ∇^2 (del 2) is the Laplace operator. It is used to generalize the second derivative to spaces of more than one dimensions. This abbreviation eliminates the need to describe each different geometry individually.

Solution to the above equation requires advanced analytical and numerical techniques and knowledge of the initial and boundary conditions. Analytical solutions for some geometries are available in the literature. In view of the complexity involved in these processes, some simpler solutions have been developed with the introduction of nondimensional parameters such as the Biot number (Bi), which simplify the mathematical formulation of the problem and will be discussed next.

9.2.1 The Biot Number (Bi)

A dimensionless number, named after the French physicist Jaen-Baptiste Biot (1774–1862), is used to evaluate the relative importance of internal and external resistances to heat transfer. It is a ratio of the heat-transfer resistance within (x_1/k_{solid}) and at the surface ($1/h_c$) of an object undergoing heat transfer, written as follows:

$$B_i = \frac{Internal\ resistance}{External\ or\ surface\ resistace} = \frac{\frac{x_1}{k_{solid}}}{\frac{1}{h_c}} = \frac{h_c x_1}{k_{solid}} \tag{9.21}$$

where h_c = convective heat-transfer coefficient, k_{solid} = thermal conductivity of the solid, and x_1 = characteristic dimension of the solid, which is calculated as shown below:

$$x_1 = \frac{Volume\ (V)}{Area\ (A)} \tag{9.22}$$

which for some of the conventional geometries becomes:

Slab (Flat plate) : (Thickness $= 2x_1$, length $= H$, and width $= W$), $\dfrac{V}{A}$

$$= \frac{WH2x_1}{2WH} = x_1 \tag{9.23}$$

$$\text{Cylinder : (Length} = L \text{ and radius} = r), \frac{V}{A} = \frac{\pi r^2 L}{2\pi r L} = \frac{r}{2} \tag{9.24}$$

$$\text{Sphere : (Radius} = r), \frac{V}{A} = \frac{\frac{4}{3}\pi r^3}{4\pi r^2} = \frac{r}{3} \tag{9.25}$$

As you may have noted, the Biot number (Bi) looks mathematically similar (hx_1/k) to the Nusselt number (Nu) we discussed in the previous chapter. Although both represent the ratio of heat transfer by convection to heat transfer by conduction, in Nu, the thermal conductivity is for the fluid (k_{fluid}), while in Bi, it is for the solid (k_{solid}) body.

Controlling resistance: Bi is very useful in evaluating the resistance that controls heat transfer. As shown in Fig. 9.3 and with the aid of Eq. (9.21), it may be stated that:

Bi < 0.1 indicates:

- k_{solid} is very high compared to h_c, Fig. 9.3a,
- Minimum temperature gradient exists within the object, internal temperature almost uniform.
- Heat flows in the object faster than it crosses the boundary layer.
- Boundary or surface resistance controls heat transfer, and internal resistance to heat transfer is negligible.

Bi > 40 suggests:

- h_c is very large compared to k_{solid}, Fig. 9.3b,
- A significant temperature gradient exists within the object, and a nonuniform internal temperature.
- Heat crosses the boundary faster than it flows within the object.

Fig. 9.3 Controlling resistance to heat transfer

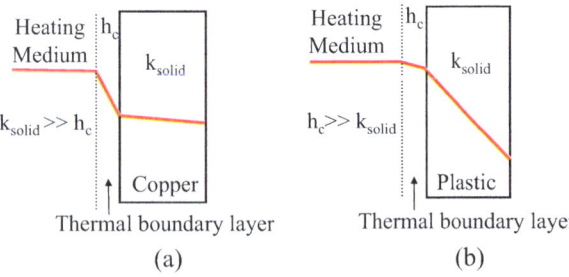

- Internal resistance controls heat transfer, and boundary resistance is negligible (surface temperature = heating (or cooling) medium temperature).

Bi = 0.1 to 40 implies:

- Both the internal and boundary resistances control heat transfer.

Check Your Understanding

1. Show that like the Reynolds number, the Bi number is also dimensionless. ◄

9.2.2   Negligible Internal Resistance & Lumped System Analysis (Bi <0.1)

A lumped system is one in which the dependence of temperature on position (spatial dependence) is negligible and thus disregarded. In such cases, the objects are observed to behave like a "lump," with almost uniform temperature at all times during the process. This may be the case for solids with very high thermal conductivity or fluids that are well mixed. In such situations, temperature is modeled as a function of **time only**. Examining Fig. 9.4, for any time t, we can write:

$$\text{Heat given up by the heating medium} : \dot{q} = h_c A (T_1 - T) \qquad (9.26)$$

$$\text{Heat gained by the object} : \dot{q} = m C_p \frac{dT}{dt} = \rho V C_p \frac{dT}{dt} \qquad (9.27)$$

Thus, equating the above two equations and separating the variables, we obtain:

$$\rho V C_p \frac{dT}{dt} = h_c A (T_1 - T) \text{ or } \frac{dT}{(T_1 - T)} = \frac{h_c A}{\rho V C_p} dt \qquad (9.28)$$

Integrating the above equation with the limits, we get:

Fig. 9.4 A lumped capacity system

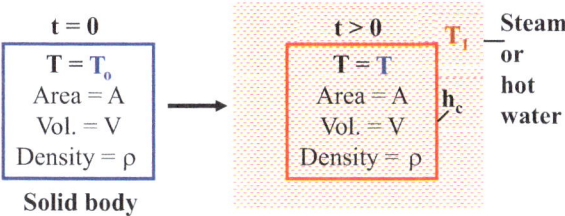

$$\int_{T=T_0}^{T=T} \frac{dT}{(T_1-T)} = \frac{h_cA}{\rho VC_p} \int_{t=0}^{t=t} dt \quad \text{Or} \quad -\ln\frac{T_1-T}{T_1-T_0}$$

$$= \frac{h_cA}{\rho VC_p} t \quad \text{or} \quad \frac{T_1-T}{T_1-T_0} = \exp\left[-\left(\frac{h_cA}{\rho VC_p}\right)t\right] \tag{9.29a}$$

For given system with constant properties, the above expression may be simplified to:

$$\frac{T_1-T}{T_1-T_0} = \exp(-kt) \tag{9.29b}$$

where $k = (h_c A)/(\rho VC_p)$. The temperature of the object during heating/cooling changes exponentially and approaches the medium temperature asymptotically, as illustrated in Fig. 9.5.

Eq. (9.29a) may also be written as:

$$\frac{T_1-T}{T_1-T_0} = \exp\left[-\left(\frac{h_c x_1}{k_{solid}}\right)\left(\frac{k_{solid}}{\rho C_p}\cdot\frac{t}{x_1^2}\right)\right] \tag{9.30}$$

Using the definitions of the Biot number and thermal diffusivity, Eq. (9.30) can now be expressed as:

$$\frac{T_1-T}{T_1-T_0} = \exp\left[-(B_i)\left(\frac{\alpha t}{x_1^2}\right)\right] \tag{9.31}$$

where $\frac{\alpha t}{x_1^2} = F_0$ (Fourier number), and Eq. (9.31) simplifies to:

$$\frac{T_1-T}{T_1-T_0} = \exp[-(B_i)(F_0)] \tag{9.32}$$

In the above expression, the Fourier number (F_0), named after Joseph Fourier, is an important dimensionless number used in heat transfer. To understand its physical significance, it may be written as:

Fig. 9.5 Temp profile for heating (**a**) and cooling (**b**)

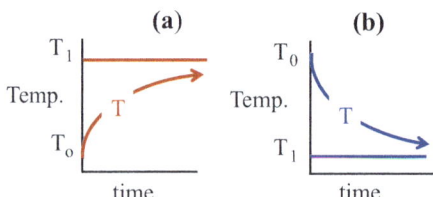

$$F_0 = \frac{\alpha\,t}{x_1^2} = \frac{kx_1^2\,(1/x_1)}{\left(\rho C_p x_1^3/t\right)}\,\frac{\Delta T}{\Delta T}$$

$$= \frac{Rate\ of\ heat\ conduction\ across\ x_1\ of\ an\ object}{Rate\ of\ heat\ storage\ in\ an\ object} = \frac{\dot{q}_{conducted}}{\dot{q}_{stored}} \quad (9.33)$$

As may be observed from the above expression, a large F_0 would indicate fast propagation of heat through an object.

In Eq. (9.32), the term $\frac{T_1-T}{T_1-T_0}$ is a dimensionless temperature ratio and indicates the unaccomplished temperature change. It is the ratio of the fraction of total accomplished temperature change and to that possible at any time. Its value ranges from zero to unity, as shown below:

$$\frac{T_1-T}{T_1-T_0} \Rightarrow \frac{T_1-T_0}{T_1-T_0} = 1 \text{ to } \frac{T_1-T_1}{T_1-T_0} = 0 \quad (9.34)$$

And thus indicates no change in the temperature (initial state) of an object to when it attains the heating/cooling medium temperature.

Example

1. *A bakery uses 2 m long, 1.5 m wide, and 0.4 cm thick aluminum sheets (density $= 2700$ kg/m^3, specific heat $= 0.9$ kJ/kg-K, and thermal conductivity $= 200$ W/m-K) for baking cookies in an oven set at 250 °C. Following baking, the sheets are placed in the kitchen where the temperature is 23 °C. Assuming a convective heat-transfer coefficient of 10 W/m^2-K, estimate the time required for the sheet to cool down to a handling temperature of 30 °C.*
   Solution:

$$Sheet\ area\ (A) = 2 \times (2.0 \times 1.5) = 6\ m^2\ and$$

$$Sheet\ volume\ (V) = (2.0 \times 1.5) \times \left(0.4 \times 10^{-2}\right)$$
$$= 1.2 \times 10^{-2} m^3$$

$$Bi = \frac{h_c V}{k_{solid} A} = \frac{(10) \times \left(1.2 \times 10^{-2}\right)}{(200) \times (6)} = 0.0001$$

$B_i < 0.1$, so the plate can be assumed to be a lumped system and then:

$$\frac{T_1 - T}{T_1 - T_0} = exp\left(-\frac{h_c A}{C_p \rho V}t\right) \quad or$$

$$\ln\left(\frac{23 - 30}{23 - 250}\right) = -\frac{(10) \times (6)}{(900) \times (2700) \times (1.2 \times 10^{-2})} \times t$$

$$t = 28.1 \ min.\quad \blacktriangleleft$$

Check Your Understanding

2. Define a lumped system. ◀

Example

2. To uniformly heat 250 kg cream soup ($C_p = 3.0$ kJ/kg-°C) in a jacketed and steam-heated kettle, agitators are being used. The heating surface area of the kettle is 1.7 m^2, and the overall heat-transfer coefficient is assumed to be 225 W/m^2-°C. If the initial temperature of the product is 21 °C and the internal surface temperature of the kettle is 95.7 °C, estimate the product temperature after 45 min of heating?

Solution:

Assuming a well-agitated system with uniform temperature that is a function of time only, it may be treated as a lumped system. Replacing the convective heat-transfer coefficient (h_c) with the overall heat-transfer coefficient (U) in Eq. (9.30) and substituting mass (m) for volume time density, we obtain:

$$\frac{T_1 - T}{T_1 - T_0} = exp\left[-\left(\frac{U\,A}{mC_p}\right)t\right]$$

Substituting the values of all the parameters except Temperature T, we obtain:

$$\frac{95.7 - T}{95.7 - 21} = exp\left(-\frac{225 \times 1.7}{250 \times 3000} \times 2700\right)$$

$$95.7 - T = 74.7 \times \ exp(-1.377) \ or \ T = 95.7 - 18.9 = 76.9°\,C \quad \blacktriangleleft$$

An overview of strategies for solving the governing equation for conduction heat transfer is illustrated in Fig. 9.6. Analytical solutions for various initial and boundary conditions are available in the literature and should be consulted as needed. It is

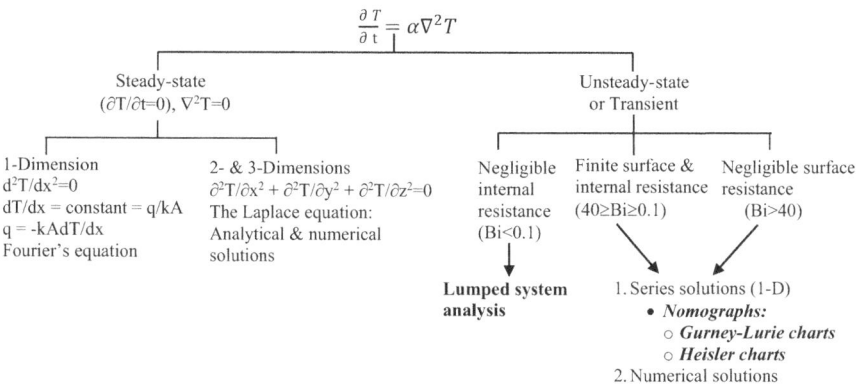

Fig. 9.6 Strategies for solving selected cases of heat transfer by conduction

worth recognizing that the analytical solutions for cases with finite surface and internal resistances typically involve infinite series.

This requires the evaluation of an infinite number of terms to determine the temperature at a given location and time. However, in many cases, the evaluation of just the first term is sufficient to estimate the time and temperature at a specified point. It is further simplified by assuming no variation in temperature in the y and z directions and considering conduction in the x direction only. This approach has been used for solving unsteady-state conduction in simple geometries such as plates or slabs and cylinders and spheres and is useful in mimicking various food shapes.

9.2.3 Finite Internal & Surface Resistances: $0.1 \leq Bi \leq 40$

The solution of the unsteady-state equation for an **infinite plate** (a flat plate in which the dimensions are infinite in two directions and finite in thickness) with uniform initial temperature and significant internal and surface resistances is as follows [4, 9]:

$$\frac{T_1 - T}{T_1 - T_0} = \sum_{n=1}^{\infty} \frac{2B_i}{\delta_n^2 + B_i^2 + B_i} \frac{\cos\left(\frac{x}{x_1} \delta_n\right)}{\cos(\delta_n)} \exp\left[\left(-\delta_n^2\right)(F_0)\right] \tag{9.35}$$

where the eigenvalues δ_n are the positive roots of the equation $\delta \tan \delta = B_i$.

9.2.4 Negligible Surface Resistance: $Bi > 40$

The solution of the unsteady-state equation for an **infinite plate** with uniform initial temperature and negligible surface resistance is given as:

$$\frac{T_1 - T}{T_1 - T_0} = \frac{4}{\pi} \sum_{n=0}^{\infty} \frac{(-1)^n}{2n+1} \cos\left(\frac{(2n+1)\pi x}{2x_1}\right) \exp\left[-\frac{(2n+1)^2 \pi^2}{4} F_0\right] \quad (9.36)$$

Expressions of the type shown above are also available for other simple geometries, such as infinite cylinders (a finite cylinder can be approximated as an infinite cylinder if its length is at least 10 times its diameter or if the two ends of the cylinder are insulated) and spheres. These series solution equations and their first-term approximations have been used to generate charts of the type shown in Figs.9.7, 9.8, 9.9, known as Gurney-Lurie charts. These charts for different geometries are plots of the Fourier number (F_0) on the x-axis against the dimensionless temperature, $(T_1-T)/(T_1-T_0)$, on the y-axis as a function of variables m and n. The variable m is the reciprocal of the Biot number with characteristic dimension x_1, as indicated in the figures, and n is the relative position within the object, equal to (x/x_1). For the surface of the object, $n = 1$ and $n = 0$ for the center.

These charts are very convenient for determining the temperatures at any position x in the object and at any time t. Alternatively, they may also be used for determining the time t for any location in an object to attain a given temperature. To use these charts, first, the m and n values for a given geometry are calculated, and then, using a known value of either the temperature ratio or Fourier's number, the value of the other one is obtained from the chart. The obtained value is then used to either estimate the temperature or the time. These graphs are very useful albeit approximate but serve as the base case for verification of other numerical and analytical solutions.

In many heat-transfer applications, it is more important to determine the center temperature of a body more accurately than what can be obtained from Gurney-Lurie charts. For such situations, another series of charts are available in the literature. Known as Heisler charts (not shown), these are also plots of dimensionless temperature as a function of Fourier number but only for the center temperature ($n = 0$).

Check Your Understanding

3. Newton's law for heating/cooling of solids by convection is related to which of the following cases: (a) $Bi < 0.1$, (b) $0.1 \leq Bi \leq 40$ or (c) $Bi > 40$ ◀

Example

3. *To reduce the temperature of freshly picked apples ($\rho = 900$ kg/m³, $C_p = 3500$ J/kg-°C, and k = 0.3 W/m-°C) from 25 °C to 5 °C, they are washed in a water stream at 4 °C and having a convective heat-transfer coefficient of 3500 W/m² °C. If the average diameter of the apples is 10 cm, how long would it take for the center of the apple to reach 5 °C?*

Solution: Assuming spherical geometry, the Biot number of the apple is:

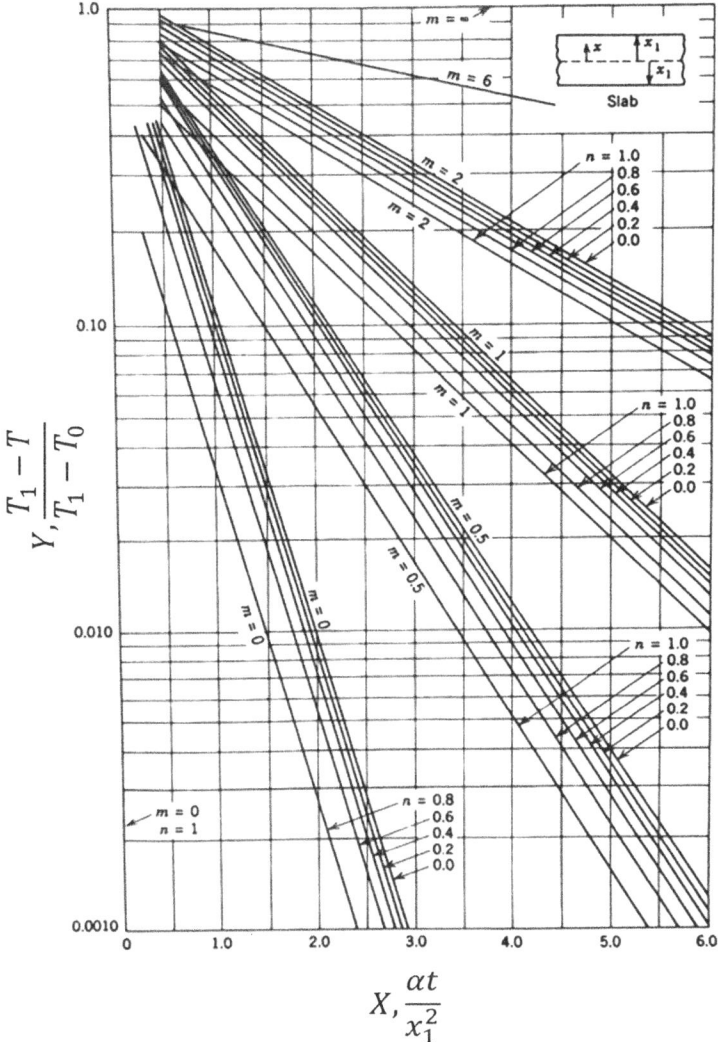

Fig. 9.7 Gurney-Lurie chart for unsteady state heat conduction in an **infinite flat plate:** $m = k/h\,x_1$ and $n = x/x_1$. From Welty et al. [8], with permission from Wiley & Sons, Inc.

$$Bi = \frac{h\left(\frac{r}{3}\right)}{k} = \frac{\left(3500\,\frac{W}{m^2\,{}^\circ C}\right)\left(\frac{0.05\,m}{3}\right)}{0.30\,\frac{W}{m\,{}^\circ C}} = 194.4 > 40$$

Since Bi > 40, internal resistance controls and series solutions (charts) may be used ◄

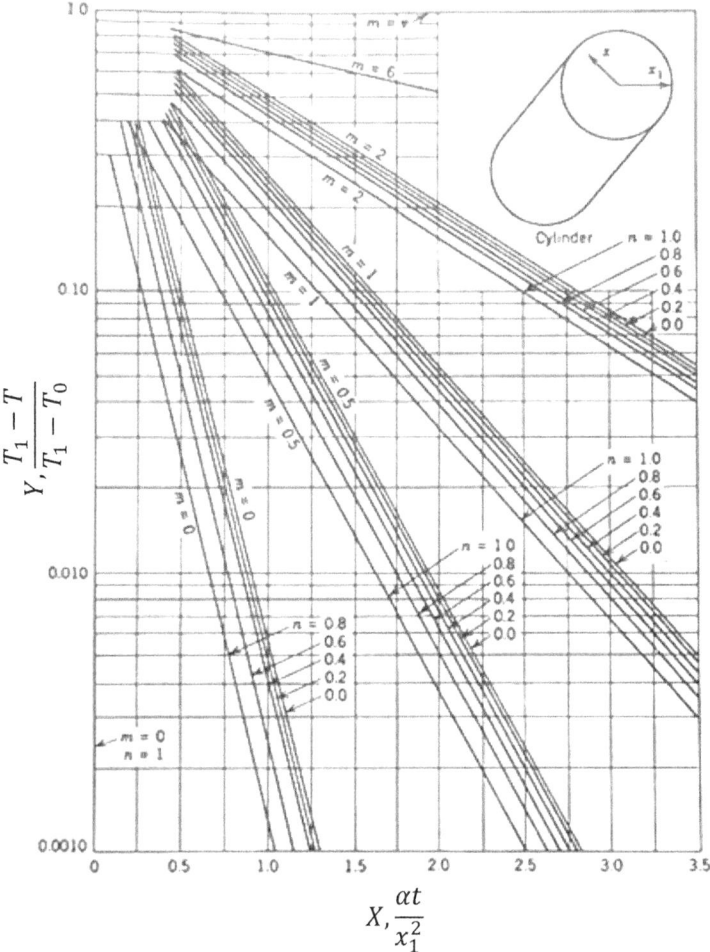

Fig. 9.8 Gurney-Lurie chart for unsteady-state heat conduction in an **infinite cylinder:** $m = k/h\,x_1$ and $n = x/x_1$. From Welty et al. [8], with permission from Wiley & Sons, Inc.

Time for the center temperature to reach 4 °C

$$m = \frac{k}{hx_1} = \frac{0.3\,\frac{\text{W}}{\text{m}\,°\text{C}}}{\left(3500\,\frac{\text{W}}{\text{m}^2\,°\text{C}}\right)(0.05\ \text{m})} = 0.002 \cong 0$$

$$n = \frac{x}{x_1} = \frac{0}{0.05\ \text{m}} \cong 0$$

$$Y = \frac{T_1 - T}{T_1 - T_0} = \frac{4 - 5}{4 - 25} = 0.048$$

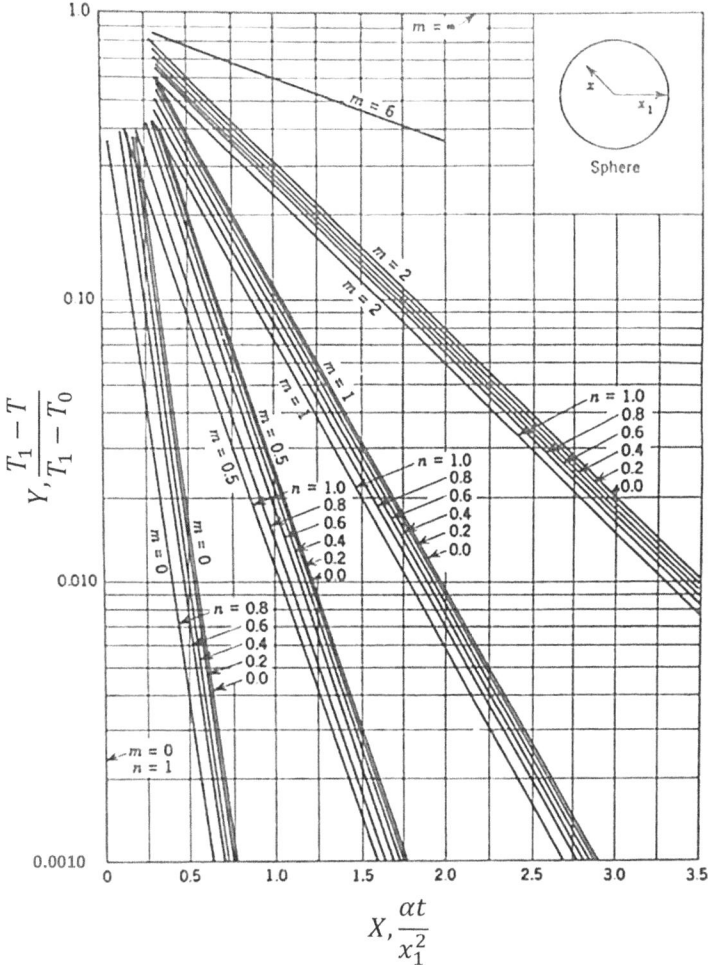

Fig. 9.9 Gurney-Lurie chart for unsteady-state heat conduction in a **sphere:** $m = k/h\, x_1$ and $n = x/$ x_1. From Welty et al. [8], with permission from Wiley & Sons, Inc.

From the Gurney-Lurie chart for sphere, for $Y = 0.048$ and $m = 0$, $X = 0.37$
Or, $0.37 = \frac{\alpha t}{x_1^2} = \frac{kt}{\rho C_p x_1^2}$

$$t = \frac{X \rho C_p x_1^2}{k} = \frac{(0.37)\left(900\,\frac{kg}{m^3}\right)\left(3500\,\frac{J}{kg\,^\circ C}\right)(0.05\text{ m})^2}{\left(0.30\,\frac{W}{m\,^\circ C}\right)} = 8633.3\text{ s} \cong 2.4\text{ hours}$$

4. *Cans (#1 Picnic: 2 11/16" × 4") measuring 68.1 mm in diameter and 101.6 mm high are filled with apple sauce (k = 0.750 W/m. K and α = 2 × 10^{-7} m^2/s) at 30 °C and vertically stacked 5-high in a retort. Steam at 120 °C is used for heating, and the*

heat-transfer coefficient is 4000 W/m²-K. Calculate the temperature at the center of the can in the middle of the stack after 40 min.

   Solution: *The cans in the middle of the stack are far enough from the two ends and may be assumed to be infinite cylinders*

$$Bi = \frac{h\left(\frac{r}{2}\right)}{k} = \frac{\left(4000 \frac{W}{m^2 \, ^\circ C}\right)(0.03405 \, m)/2}{0.75 \frac{W}{m \, ^\circ C}} = 90.8 > 40$$

Bi > 40 and therefore, internal resistance controls, use series solution (charts)

$$m = \frac{k}{hx_1} = \frac{0.75 \frac{W}{m \, ^\circ C}}{\left(4000 \frac{W}{m^2 \, ^\circ C}\right)(0.03405 \, m)} = 0.0055 \approx 0$$

$$n = \frac{x}{x_1} = \frac{0}{0.03405 \, m} = 0$$

$$X = \frac{\alpha t}{x_1^2} = \frac{\left(2 \times 10^{-7} \frac{m^2}{s}\right)(40 \times 60)}{(0.03405 \, m)^2} = 0.414 \text{ and from the chart, for } X = 0.414 \text{ and } m = 0,$$

$$Y = 0.09$$

$$Y = \frac{T_1 - T}{T_1 - T_0} = \frac{120 - T}{120 - 30} = 0.09 \rightarrow 120 - T = 90 \times 0.09 = 8.1 \rightarrow T = 111.9 \, ^\circ C$$

    4. Which of the following pair of dimensionless numbers are needed for analyzing the transient heat conduction problems in a flat plate, cylinder, and sphere:
(a) Re and Nu, (b) Bi and Fo, (c) Pr and Re, or (d) none of these. ◀

9.2.5 Finite Objects

Based on the analysis by Myers [6], the Gurney-Lurie and Heisler charts also enable estimation of temperatures for finite cylinders and slabs. In this approach, a finite cylinder of radius R and height X is considered a product of an infinite cylinder of radius R and an infinite slab of thickness X, as shown in Fig. 9.10. For given values of m, n, and F_o, the Y_X value for an infinite slab of thickness X is determined from the chart in Fig. 9.7. Using the same values of m, n, and F_o, the Y_R value for an infinite cylinder is then obtained from Fig. 9.8. Then,

$$Y_{XR} = (Y_X)(Y_R) \tag{9.37}$$

where $Y_{XR} = \left(\dfrac{T_1 - T}{T_1 - T_0}\right)_{\text{finite cylinder}}$, $Y_X = \left(\dfrac{T_1 - T}{T_1 - T_0}\right)_{\text{infinite slab}}$ and

$Y_R = \left(\dfrac{T_1 - T}{T_1 - T_0}\right)_{\text{infinite cylinder}}$

    The value of Y_{XR} is then used to compute the unknown temperature T. Similarly, a rectangular solid of length X, width Y, and thickness Z may be considered a product of three infinite slabs of thickness X, Y, and Z, so that

Fig. 9.10 Approximation of a finite cylinder

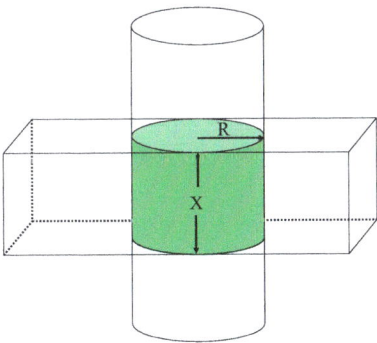

$$Y_{XYZ} = (Y_X)(Y_Y)(Y_Z) = \left(\frac{T_1 - T}{T_1 - T_0}\right)_{\text{finite slab}} \tag{9.38}$$

The unknown temperature T can then be computed using the value of Y_{XYZ}.

Example

5. A can with the code 301 × 400 is 3 1/16" in diameter and 4.00" high is filled with a food paste and is heated by steam from all sides. What is the temperature in the center of the can after 45 minutes of heating? The process parameters are given as follows: paste thermal conductivity $(k) = 0.480$ Btu/hr-ft-°F and thermal diffusivity $(\alpha) = 7.78 \times 10^{-3}$ ft²-hr; convective heat-transfer coefficient (h_c) of steam $= 800$ Btu/hr-ft²-°F; initial temperature of paste $= 85$ °F; steam temperature $= 240$ °F.

3.06"

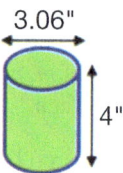

4"

Solution: Treat the finite can as a product of an infinite cylinder and an infinite slab.

First, treating the can as an infinite cylinder, diameter $3.06" = 0.255$ ft.

$$x_1 = R = \frac{0.255ft}{2} = 0.128ft$$

$$m = \frac{k}{hx_1} = \frac{\left(0.480\frac{Btu}{hr \cdot ft \cdot °F}\right)}{\left(800\frac{Btu}{hr \cdot ft^2 \cdot °F}\right)(0.128ft)} = 0.0047 \text{ (essentially 0)}$$

$$n = \frac{x}{x_1} = \frac{0}{0.128ft} = 0 \quad X = \frac{\alpha t}{x_1^2} = \frac{\left(7.78 \times 10^{-3}\frac{ft^2}{hr}\right)\left(\frac{45}{60}hr\right)}{(0.128ft)^2} = 0.356$$

From the Gurney-Laurie chart for an infinite cylinder, $Y_R = 0.2$

Second, treating the can as an infinite slab of thickness $4.00" = 0.333$ ft

$$x_1 = \frac{thickness}{2} = \frac{0.333ft}{2} = 0.167ft$$

$$m = \frac{k}{hx_1} = \frac{\left(0.480\frac{Btu}{hr.ft.°F}\right)}{\left(800\frac{Btu}{hr.ft^2.°F}\right)(0.167ft)} = 0.004 \;(essentially\; 0)$$

$$n = \frac{x}{x_1} = \frac{0\;ft}{0.167ft} = 0$$

$$X = \frac{\alpha t}{x_1^2} = \frac{\left(7.78 \times 10^{-3}\frac{ft^2}{hr}\right)\left(\frac{45}{60}hr\right)}{(0.167ft)^2} = 0.21$$

From the Gurney-Laurie chart for an infinite slab, $Y_X = 0.8$

The product of Y_R and Y_X can then be used to compute dimensionless temperature:

$$Y_{XR} = (Y_X)(Y_R) = \left(\frac{T_1 - T}{T_1 - T_0}\right) \; Or \; (0.8)(0.2) = \left(\frac{240 - T}{240 - 85}\right)$$

$$T = 240{-}0.16\,(155) = 215.2°F \qquad \blacktriangleleft$$

9.3 Thermal Processing: Canning/Appertization

Heating processing to make foods safe for consumption and/or extend their shelf life by minimizing pathogenic and spoilage microorganisms is an old and reliable method widely used in the food industry. Canning or Appertization, pioneered in the 1790s by a French confectioner, Nicolas Appert, is a more advanced thermal processing method employed to preserve the safety and wholesomeness of foods over long periods of storage at normal room temperatures. It is based on the application of heat, alone or in combination with pH and/or water activity, to attain commercial sterility by killing a defined number of log cycles of a microorganism of concern.

Designing these processes requires expertise in the disciplines of engineering and biological systems. The critical components that affect the process sterility of the canned products are the temperature and time of heating data for establishing the process schedule. To determine the process lethality, the product's temperature–time characteristics are determined by conducting heat penetration tests, which record the rate of change in temperature of the product inside the container at its slowest heating point (cold spot) while it is being heated. The cold spot within a container depends on the product properties, heating transfer mechanisms (conduction vs. convection), container size and type, and processing system (agitated vs. stationary) (Fig. 9.11). During thermal treatments in containers, the temperature inside the canned food

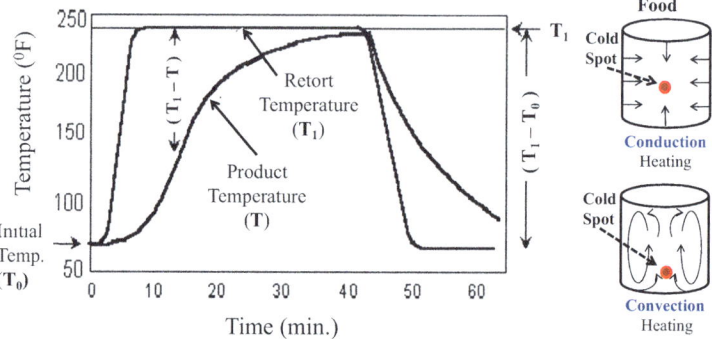

Fig. 9.11 Product "cold spot" and retort temperature profiles

slowly increases over the heating time and then slowly decreases as well over the subsequent cooling time. This presents the challenge of dealing with unsteady-state heat-transfer processes. The following two methods have been successfully developed to design and engineer thermal processing of foods that guarantee safety and quality.

9.3.1   The General Method

The general method is a useful graphical method for evaluating the sterility of a thermal process. It is based on a hypothetical thermal destruction that is parallel to the thermal resistance curve of the microorganism of interest and has an F value of 1 min. at 121.1 °C. It computes the total sterility of the process by integrating the lethal effects of a time/temperature history of a process at the slowest heating zone in the container. This permits comparison of the lethality of different processes, and this method is extensively used. However, it lacks predictive power for designing new process schedules.

9.3.2   The Ball Formula Method

First published by C. Olin Ball [1] this method permits the extrapolation of process time following time, and temperature data are obtained by a heat penetration test. It turns heat penetration data in the form of two parameters, f and j, and thus using parameter conversion procedures, new processes for the same product in containers of different sizes can be directly calculated. It successfully established the thermal process design procedure and is recognized as "a major milestone in the history of food technology," given the safety history of commercially canned products over the last several decades. It is based on the following approach to unsteady-state heat-transfer solutions.

For unsteady-state heat transfer, the governing partial differential equation in cylindrical coordinates is:

$$\frac{\partial T}{\partial t} = \alpha \left[\frac{\partial^2 T}{\partial r^2} + \frac{1}{r}\left(\frac{\partial T}{\partial r}\right) + \frac{1}{r^2}\left(\frac{\partial^2 T}{\partial \theta^2}\right) + \frac{\partial^2 T}{\partial z^2} \right] \tag{9.39}$$

The series solution for the center temperature of a sphere with both resistances to heat transfer is an infinite series solution given as:

$$\frac{T_1 - T}{T_1 - T_0} = \sum_{n=1}^{\bowtie} \frac{2(\sin\beta_i - \beta_i \cos\beta_i)}{\beta_i - \mathrm{Sin}\beta_i \cos B_i} \exp\left[-\frac{(\beta_i^2 \alpha t)}{x_1^2} \right] \frac{\mathrm{Sin}\left[\beta_i\left(\frac{x}{x_1}\right)\right]}{B_i\left(\frac{x}{x_1}\right)} \tag{9.40}$$

where x represents the radial location and β_i is the ith root of $Bi = 1 - \beta_i \cot \beta_i$

After a short heating time when $F_0 = \alpha t / x_1^2 \geq 2$, the series converges rapidly, and the first-term approximation provides a reasonable estimate. In such cases, the series becomes:

$$\frac{T_1 - T}{T_1 - T_0} = \exp\left[-\frac{(\beta_i^2 \alpha t)}{x_1^2} \right] \left[\left\{\frac{2(\sin\beta_i - \beta_i \cos\beta_i)}{\beta_i - \sin\beta_i \cos\beta_i}\right\} \left\{\frac{\sin\left[\beta_i\left(\frac{x}{x_1}\right)\right]}{B_i\left(\frac{x}{x_1}\right)}\right\} \right] \tag{9.41}$$

$$\mathrm{Or}, \log\frac{T_1 - T}{T_1 - T_0} = -\frac{(\beta_i^2 \alpha t)}{2.303 x_1^2} + \log\left[\left\{\frac{2(\sin\beta_i - \beta_i \cos\beta_i)}{\beta_i - \sin\beta_i \cos\beta_i}\right\} \left\{\frac{\sin\left[\beta_i\left(\frac{x}{x_1}\right)\right]}{B_i\left(\frac{x}{x_1}\right)}\right\} \right] \tag{9.42}$$

$$\mathrm{If}, \quad \frac{1}{f} = -\frac{(\beta_i^2 \alpha)}{2.303 x_1^2} \text{ and } j = \left[\left\{\frac{2(\sin\beta_i - \beta_i \cos\beta_i)}{\beta_i - \sin\beta_i \cos\beta_i}\right\} \left\{\frac{\sin\left[\beta_i\left(\frac{x}{x_1}\right)\right]}{B_i\left(\frac{x}{x_1}\right)}\right\} \right] \tag{9.43}$$

then, we can write:

$$\log\frac{T_1 - T}{T_1 - T_0} = -\frac{t}{f} + \log j \tag{9.44}$$

$$\mathrm{Or}, \log(T_1 - T) = -\frac{t}{f} + \log[j(T_1 - T_0)] \tag{9.45}$$

The above equation was used by Ball to develop a semi-analytical formula method to calculate the process time, or the lethality (F) imparted to products during thermal processing. According to Eq. (9.45), a log-linear (semilogarithmic) graph of (T_1–T) and time (t) will give a straight line. As shown in Fig. 9.12 for a conduction heating food in a container, after an initial lag period due to changes in the properties of the food with temperature, log (T_1–T) does decrease linearly with time. As

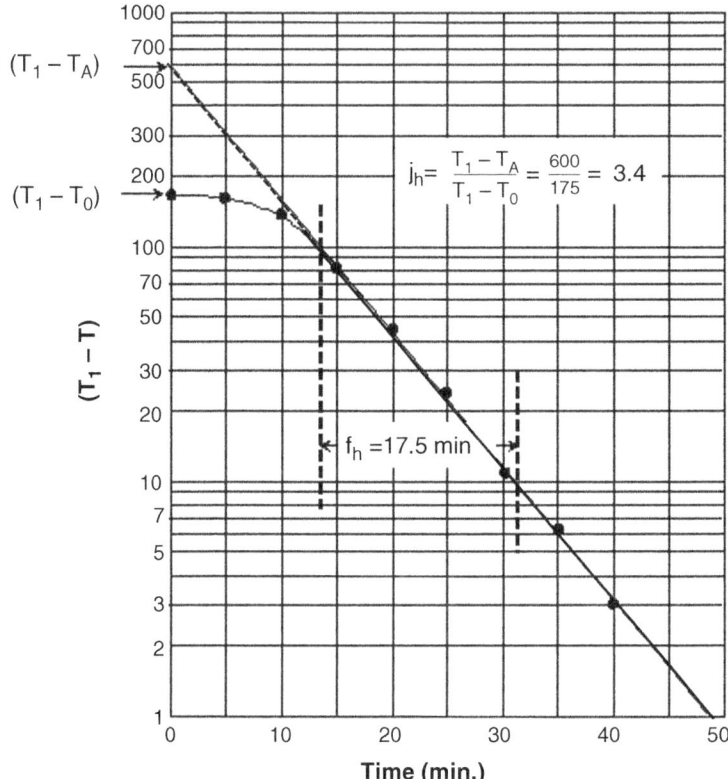

Fig. 9.12 Semilogarithmic temperature-time graph for heating canned food at temperature T_1

illustrated on the graph, f_h and j_h parameters, called heat penetration factors for heating, can then be evaluated from the straight-line segment. The parameter f_h is the time required for the straight line to pass through one log cycle or the time required for the temperature difference between the medium and product temperatures to decrease by 90%. And j_h is called the lag factor, which represents the lag time before the graph assumes a straight line and is calculated as a dimensionless ratio of $(T_1 - T_A)$ and $(T_1 - T_0)$. The value of $(T_1 - T_A)$ is obtained by extrapolation of the straight-line portion of the graph, see Fig. 9.12. Similar graph can also be constructed from the cooling time-temperature data and the corresponding f_c and j_c parameters for cooling are then similarly calculated.

The f and j parameters are obtained from the heat penetration (time-temperature) data for each product and its associated geometry. They are then used to calculate the lethality (F) in terms of equivalent time at a reference temperature (generally 121 °C/ 250 °F) of a thermal process such as canning. You will learn more about these methods in related food processing courses.

Problems

9.1 A meat ball with a radius of 1.0 in is at a uniform temperature of 240 °F. It is suddenly plunged into a medium whose temperature is held constant at 50 °F. Assuming a convective coefficient (h_c) of 5.0 W/m^{2}K, calculate the temperature of the ball after 2 hours. The average physical properties are $k = 0.48$ W/m-K, $\rho = 1240$ kg/m^3, and $C_p = 3.43$ kJ/kg-K.

9.2 A spherical steel ball of 2.5 in. diameter at an initial temperature of 60 °C is exposed to an ambient temperature of 4 °C with a convective heat-transfer coefficient of 30.9 W/m^2·K. (a) Calculate the time it takes for the temperature to reach 10 °C at the center of the ball, given that the thermal conductivity of steel is 7.7 W/m-K. Additionally, the specific heat and density of steel are 0.5 kJ/kg-K and 7800 kg/m^3, respectively. (b) Calculate the time required to obtain the cooling mentioned in part (a) if the material of the spherical ball is ground beef, which has a thermal conductivity of 0.49 W/m-K. The specific heat and density of ground beef are 3.43 kJ/kg-K and 1160 kg/m^3, respectively.

9.3 A food product with a characteristic dimension of 5 mm is uniformly heated to a temperature of 90 °C and then immersed in a cold fluid which is at 5 °C. The thermal diffusivity of the product is 1.1×10^{-6} m^2/s and the system has a Biot number of 0.05. Estimate the temperature of the product after 50 seconds.

9.4 Carrot puree ($\rho = 1040$ kg/m^3, $C_p = 4.0$ kJ/kg-°C)) at a uniform initial temperature of 25 °C is heated by steam in a stirred and jacketed kettle of 0.5 m in radius and 100 °C inside surface temperature. If the convective heat-transfer coefficient in the kettle is 4 kW/m^2-°C, find the temperature of the puree after 10 minutes of heating.

9.5 A bowl of soup at 95 °C is too hot to drink. When stirred and left for 2 minutes in a room at 25 °C, it has cooled to 80 °C in two minutes. How long will it take for the temperature to be just right for consumption at 50 °C?

9.6 A 2.5 cm thick salmon ($k = 0.40$ W/m-K and $\alpha = 5.0 \times 10^{-4}$ m^2/h) fillet at uniform initial temperature of 5 °C and with a large surface area is to be broiled in an oven set at 210 °C. Assuming a convective heat-transfer coefficient of 32.0 W/m^2-K, estimate the cooking time needed for the center temperature to reach 66.0 °C to make it safe for eating. The fillet may be treated as a large plate with heat transferring mostly from the top and bottom.

9.7 For thermal processing of beef sausage 50 mm in diameter and 0.3 m long in a retort, it takes one hour to reach the minimum recommended temperature of 70 °C at the center. The convective heat-transfer coefficient is known to be very high and most of the resistance to heat transfer has been found to reside internally. The thermal conductivity of the product is 0.5 W/m-K. How long will it take to reach the same internal temperature if the diameter is doubled?

9.8 A conduction-cooling food product with a density of 1000 kg/m^3, a specific heat of 4000 J/kg-°C, and a thermal conductivity of 0.4 W/m-°C was heated to 100 °C. The cooling of the product in a 4.0 cm high by 6 cm diameter steel can (1.0 mm wall thickness) is accomplished using cold water with a temperature of 0 °C and a convective heat-transfer coefficient of 20 W/m²-°C. What is the temperature at the geometric center of the can after 40 minutes?

9.9 A rectangular cheese block of 0.305 m by 0.457 m by 0.61 m is initially at 60 °C. It is suddenly immersed into a brine solution at 4 °C. Determine the temperature at the center of the block after 1 hour. The surface convective heat-transfer coefficient is 34 W/m²-K. The physical properties are $k = 0.3$ W/m-K and $\alpha = 0.0379$ m²/h.

9.10 The time and temperature data at the cold spot during retorting of a canned food are shown in the table below. The heating medium temperature was constant at 121 °C. Determine the heat penetration factors, f_h and j_h.

Time (min.)	Temperature (°C)
0	21
10	31
20	81
30	101
40	112
50	117

Bibliography

1. Ball, C.O. 1923. Thermal process time for canned food, Bull. 37. Vol. 7, Part 1. Nat'l. Res. Council, Washington, DC.
2. Gurney HP, Lurie J (1923) Ind. Eng. Chem. 15:1170
3. Ingersol et al (1948) Heat Conduction. McGraw-Hill Book, New York
4. Jaeger JC, Carslaw HS (1959) Conduction of heat in Solids. Oxford University Press, Oxford
5. McAdams WH (1948) Heat Transmission. McGraw-Hill Book, New York
6. Myers GE (1971) Analytical Methods in Heat Conduction Heat Transfer. McGraw-Hill, New York
7. Patashnik M (1953) A simplified procedure for thermal process evaluation. Food Technol 7:1
8. Welty JR, Wicks CE, Wilson RE (1969) Fundamentals of momentum, heat, and mass transfer. Wiley, New York
9. Yanniotis S (2008) Solving Problems in Food Engineering. Springer, New York

Heat Transfer: Radiation, Dielectric and Ohmic

As we learned in the previous chapters, in conventional thermal processes, heat is generated outside of the object to be heated and is then transferred to the material by conduction, convection, or a combination of the two mechanisms. Radiation is the third mode of energy transfer that occurs through electromagnetic waves without involving a medium. Light and heat are examples of types of radiations. In process engineering applications, all three modes of heat transfer are generally involved. Thermal processing technologies using specific radiation waves and Ohmic heating, where thermal energy is generated and absorbed directly by food materials, are finding increasingly new applications for targeted energy delivery to improve food product quality and achieve higher energy efficiencies.

10.1 Electromagnetic Radiation

Every physical body at temperatures above the absolute zero continuously radiates electromagnetic radiation due to vibrational and rotational movements of their atoms and molecules and carry energy. Electromagnetic radiation thus involves the flow of energy carrying photons (or quanta) traveling at the speed of light (2.998×10^8 m/s in vacuum) in a sinusoidal (wave-like) pattern. The collection of all electromagnetic (EM) radiation in the universe is known as the electromagnetic spectrum (EMS). As stored energy in motion with momentum, the electromagnetic spectrum consists of an electric field (E) and a magnetic field (V) at right angles to each other and moving in the travel direction (Fig. 10.1). Unlike mechanical waves such as sound waves, EM radiation does not necessarily require a medium to propagate and can travel through solids, liquids, gas, and vacuum. They suffer no attenuation when traveling in vacuum. However, the attenuation becomes progressively higher for air, liquids, and solids due to increasingly larger absorption and reflection by objects of higher density. Thus, the most efficient radiation heat transfer will occur in a

S. S. H. Rizvi, *Food Engineering Principles and Practices*,
https://doi.org/10.1007/978-3-031-34123-6_10

Fig. 10.1 Basic characteristics of an electromagnetic wave

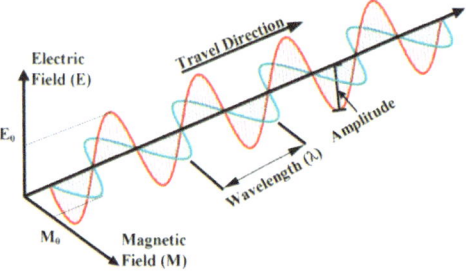

vacuum. EM wave fundamentals were first proposed by James Maxwell in 1864, and Heinrich Hertz in 1888 successfully produced the first EM waves.

The two types of waves, electric and magnetic, exist together and pervade the cosmos inseparably. They are characterized by their wavelength and frequency. The distance between two consecutive peaks (or crests) of a wave that are in phase is called the wavelength (λ), while the number of waves (cycles) that form in unit time is called the frequency (f). The latter is generally expressed as the number of wave cycles per second, or hertz (Hz), named in honor of German physicist Heinrich Hertz. The distance from the base to the top of the crest is called the amplitude (A). The wavelength and frequency are inversely proportional and related by the following equation:

$$\lambda = c/f \tag{10.1}$$

where c is the speed of light. The wave with the highest frequency will have the shortest wavelength. The EM spectrum spans wavelengths ranging from approximately 10^{-18} m to 100 km, which corresponds to frequencies decreasing from 3×10^{26} Hz to 3×10^{3} Hz.

As stated earlier, all physical objects, solid, liquid, or gas, with a temperature above absolute zero emit EM radiation. Electromagnetic radiation is also produced whenever an atom absorbs energy, which changes the velocity of one or more of its electrons due to acceleration and deceleration. The radiation intensity is a function of the temperature and surface characteristics of the object. It can take the form of different types of electromagnetic waves, such as thermal, UV, or others. The energy level (e) associated with each photon is given by the following equation:

$$e = hf = h\,(c/\lambda) \tag{10.2}$$

where h is the Plank's constant (6.626×10^{-34} J.s or J/Hz). Thus, shorter wavelength radiations such as X-rays and gamma-rays are more energetic and highly powerful. Based on the energy level of the photons, the electromagnetic spectrum is generally divided into seven basic regions. Arranged in order of increasing wavelength and decreasing energy and frequency, they are classified as gamma-rays, X-rays, ultraviolet (UV), visible light, infrared (IR), microwave, and radio waves, as shown in Fig. 10.2.

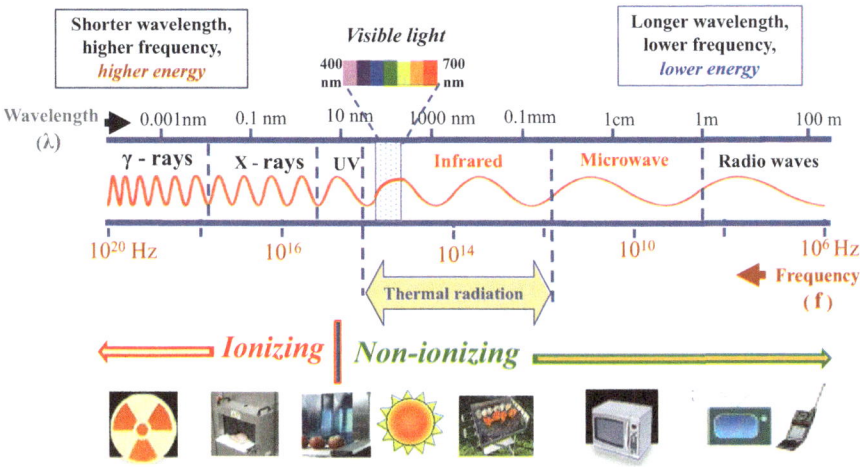

Fig. 10.2 Electromagnetic radiation spectrum

EM radiation is further divided into ionizing and nonionizing radiation at an ultraviolet wavelength of 100 nm. Ionizing radiation is a high-frequency and high-energy radiation, such as X-rays and gamma-rays, while nonionizing radiation is a low-frequency and low-energy radiation, such as infrared, microwave, and radio waves.

10.2   Ionizing Radiation and Irradiation of Food

Ionizing radiation possesses enough energy to eject tightly bound electrons from an atom or molecule and produce short-lived ions that quickly convert into free radicals. Such ions and radicals can react with biological materials to form hydrolytic products and cause damage to the cellular DNA that may lead to cell death. Ionizing radiation is advantageously utilized to kill microorganisms, inactivate enzymes, inhibit seed sprouting to extend the shelf life of foods and control tumor cells by radiation therapy.

Irradiation refers to material exposure to radiation. It is done using electromagnetic waves (γ-rays or X-rays) or electron beams (β-rays or cathode rays). The sources of ionizing radiation are either radioactive isotopes or electron accelerators. Cobalt-60 (^{60}Co) and Cesium-137 (^{137}Cs) are the main isotopes used for the generation of γ-rays. These isotopes are produced by bombarding stable natural elements with neutrons. Electron beams, on the other hand, are high-energy electrons produced by machines called electron accelerators that accelerate electrons to the desired energy level.

Radiation strength is characterized by the energy level and expressed in SI units of MeV (million electron volts or 1.6×10^{-13} J). Radiation absorption or dose, however, is expressed in the SI unit of Gray (Gy), which represents one Joule of energy absorbed per kg of the radiated material. Generally, doses in the range of 1–8 kGy for destroying nonspore-forming bacteria and 10–50 kGy for killing spore-forming bacteria are recommended. Irradiation as a technique to extend the shelf life of foods without collateral damage to their quality attributes started in the 1960s, and it is now practiced on a limited commercial scale. The high capital cost, stringent safety requirements, potential to misuse irradiation, and poor consumer acceptability of irradiated foods are some of the reasons why irradiated foods are not more commonly available.

Check Your Understanding

1. All objects radiate electromagnetic waves above what temperature and what is the most potent and strongest radiation that we deal with? ◀

10.3   Thermal Radiation

Radiation emitted from a material due to its heat is called **thermal radiation**. The thermal radiation wavelengths cover a range from approximately 0.1 µm to 100 µm, which includes infrared, visible, and a portion of UV in the EM spectrum (Fig. 10.2). Thermal radiation at temperatures above ambient is significantly more important for practical purposes. The heating mechanisms of thermal radiation involve rotation and vibration of molecules due to energy absorption. The wavelength used affects the amount of energy transmitted, temperature, and penetration depth. The shorter the wavelength is, the higher the energy level and greater the penetration depth. Depending on the food product and applied wavelengths, the penetration depth of thermal radiation is generally limited to 0.5 to 5 mm.

As different materials absorb and emit radiation differently, a perfect absorber and emitter (not reflector) called a **blackbody** is used as a standard for the radiative properties of real surfaces. A blackbody is a hypothetical object that ideally absorbs and emits all electromagnetic radiation irrespective of the angle of incidence, frequency, or wavelength and stays in thermal equilibrium. It is called a blackbody because it absorbs all colors of light and thus appears black at low temperature where it is not self-luminous. The quantity of radiation emitted by a blackbody depends on its temperature and is not affected by the object's size or shape. At any given temperature and wavelength, a blackbody emits more energy and more uniformly in all directions than any other surface. The total power or rate of energy emitted by a blackbody (ideal absorber/emitter) surface, known as **emissive power**, is obtained using the equation prescribed by the Stefan-Boltzmann law:

$$E_b = \sigma A T^4 \tag{10.3}$$

where E_b is the rate of heat emission or emissive power (Watts), σ is the Stefan-Boltzmann constant (5.67×10^{-8} W/m^2 K^4), A is the emitting body's surface area

Table 10.1 Emissivity of some common surfaces

Material	Emissivity (ε) at 300 K
Carbon black	0.96
Fireclay brick	0.75
Glass	0.94–0.91
Paper	0.93
Stainless steel, polished	0.07
Water	0.95

Adapted from McAdams [3], Henderson and Perry [4] and others

(m^2), and T is the absolute temperature (K). A real surface at the same temperature, however, will emit power lower than that emitted by a blackbody. The ratio of the emissive power of a real surface (E_{emit}) to that of a blackbody at the same temperature is called **emissivity (ε)**:

$$\varepsilon = \frac{E_{emit}}{E_b} \tag{10.4}$$

As you will recognize, emissivity is dimensionless, and its range is **0 (shiny mirror)$\le \epsilon \le 1$ (black body)**. The emissivity of a surface is not a constant but depends on the temperature and wavelength and the angle between the incident radiation and the normal of the surface. Therefore, for a given temperature, an average value of emissivity estimated over the wavelengths of interest and all directions is used. The emitted or radiated power by a real object is thus:

$$E_{emit} = \sigma \varepsilon A T^4 \tag{10.5}$$

A blackbody has a surface emissivity of unity, and thus, the emissivity value of a real surface is a measure of its proximity to a blackbody and depends on the surface temperature and the surface finish. The emissivity values of some very common surfaces are listed in Table 10.1. A more exhaustive list is available in heat-transfer reference books. Additionally, according to **Wien's law**, at any constant temperature, the maximum energy emission from an object occurs at a certain wavelength (λ_m, in meters), which varies with the inverse of the absolute temperature, as shown below:

$$\lambda_m = \frac{2.9 \times 10^{-3}}{T} \tag{10.6}$$

Due to this phenomenon, when the temperature of an incandescent bulb drops, the wavelenth lengthens and filament looks redder.

Example

1. *Compute the emissive power of (i) a blackbody of one m^2 surface area at 27 °C and (ii) a human body ($\varepsilon = 0.97$) at 37 °C if the surface area is 1.5 m^2.*

Solution: *For unit surface area:*

(i) *For blackbody, Eq. (10.3),* $E_b = \left(5.67 \times \frac{10^{-8}W}{m^2\ K^4} \right) (1\ m^2)(27 + 273\ K)^4 =$ 459.3W.

(ii) *For the human body, Eq. (10.5),*

$$E_{emit} = \left(5.67 \times \frac{10^{-8}W}{m^2\ K^4} \right) (0.97)(1.5\ m^2)(37 + 273\ K)^4 = 785.5W \qquad \blacktriangleleft$$

10.3.1 Absorption, Reflection, and Transmission of Radiation

When radiant electromagnetic energy impinges upon a surface, some of it is absorbed, some reflected, and some transmitted, as shown in Fig. 10.3. The **absorptivity** (α) is the fraction of the incident radiant energy absorbed; the **reflectivity** (ρ) is the fraction reflected; and the **transmissivity** (τ) is the fraction that passes although the object. Thus, it follows that:

$$\alpha + \rho + \tau = 1 \qquad (10.7)$$

As pointed out previously, a blackbody does not reflect or transmit radiation, then:

$$\alpha = 1 \qquad (10.8)$$

In the case of opaque materials, such as most solids, that do not transmit radiation, $\tau = 0$, then:

Fig. 10.3 Radiation on a semi-transparent object

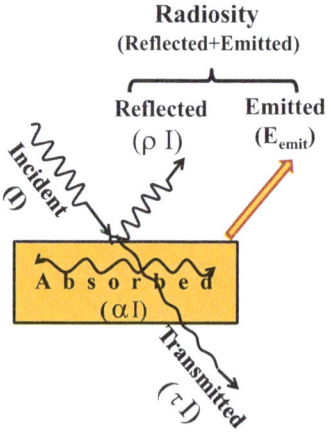

$$\alpha + \rho = 1 \tag{10.9}$$

For materials that do not reflect radiation, such as gases, $\rho = 0$, then:

$$\alpha + \tau = 1 \tag{10.10}$$

In addition to absorption, reflection, and transmission of incident radiation, a material will emit radiation based on its own temperature given by Eq. (10.5). The total sum of the reflected and emitted radiation is known as the **radiosity**, as depicted in Fig. 10.3.

Check Your Understanding

2. Why does a room full of people get warmer than when it is empty? ◀

10.3.2 Kirchhoff's Law

According to Kirchhoff's law, the emissivity (ε) of an object is equal to its absorptivity (α) when it is in equilibrium with its surroundings. This identity can be easily proven by considering a radiating surface of area A of an object within an adiabatic enclosure in equilibrium at a constant temperature (T), as shown in Fig. 10.4. At steady state, the radiation absorbed by the object out of the incident radiation I can be expressed as:

$$E_{abs} = \alpha I = \alpha \sigma A T^4 \tag{10.11}$$

and emitted radiation as:

$$E_{emit} = \sigma \varepsilon A T^4 \tag{10.12}$$

Since the system is in thermal equilibrium:

$$E_{abs} - E_{emit} \tag{10.13}$$

Thus, at any constant temperature:

Fig. 10.4 Isothermal
radiation and Kirchhoff's law

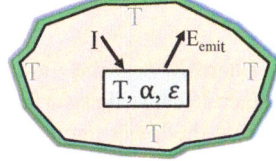

$$\alpha = \varepsilon \tag{10.14}$$

This proves that objects that are good emitters are also good absorbers. Similarly, poor absorbers are poor emitters, very useful information to remember!

10.3.3 Radiation Heat Transfer

The net radiant interchange between two blackbodies of equal areas at temperatures T_1 and T_2 such that $T_1 > T_2$ is given by the Stefan-Boltzmann equation shown below, provided the two bodies see only each other and all energy emitted by one body is absorbed by the second:

$$\dot{q}_{1-2} = \sigma A \left(T_1^4 - T_2^4\right) \tag{10.15}$$

For nonblack body bodies seeing each other only, the emitted radiation is described by the emissivity-modified Stefan-Boltzmann equation given below:

$$\dot{q}_{1-2} = \epsilon \, \sigma A \left(T_1^4 - T_2^4\right) \tag{10.16}$$

During heat transfer by radiation between two real objects some distance apart, in addition to the temperature and properties of the radiating surface, the geometrical orientation of the surfaces relative to each other becomes very important. In such situations, uniform radiation is assumed to occur in all directions. A portion of the emitted radiation from body 1, depending on its geometrical orientation, may strike its own surface, where it may again be absorbed and reflected. Other portions of the radiation from body 1 will strike the surface of body 2, where some of it will be absorbed, some reflected, and the remainder will pass into the surrounding area. This process will continue repeatedly until the entire radiation is dissipated. In such cases where two radiating bodies are exchanging radiation energy only partially, a new factor, called the **view (or shape or form or configuration) factor (F_{12})**, is introduced. Because of the directional nature of radiation, it represents how well the two bodies see each other and depends on the distance between the two bodies and their orientation. The net magnitude of radiation heat transfer is therefore directly proportional to the view factor. Mathematical formulae, graphs, and tables have been developed for determining view factors for many different pairs of surfaces and are given in standard textbooks of radiation heat transfer and may be consulted as needed.

The net radiation energy transfer Eq. (10.16) is then modified to include the view factor as follows:

$$\dot{q}_{1-2} = \epsilon_1 \, \sigma A_1 \, F_{1-2}\left(T_1^4 - T_2^4\right) \tag{10.17}$$

where F_{1-2} is the view factor and indicates the fraction of the radiation from surface 1 that strikes surface 2 directly. It accounts for the radiation absorbed by the body at the lower temperature and ranges between zero and one. In the case of the two surfaces not seeing each other, the shape factor $F_{1-2} = 0$. On the other hand, when surface 2 surrounds surface 1, $F_{1-2} = 1$ and Eq. (10.17) reduces to:

$$\dot{q}_{1-2} = \epsilon_1 \, \sigma A_1 \left(T_1^4 - T_2^4\right) \qquad (10.18)$$

Similarly, view factor F_{1-1} indicates the fraction of radiation from surface 1 striking itself, which is zero for plane or convex surfaces but not for concave and other internal surfaces (Fig. 10.5).

As may be apparent, view factors are complicated to estimate, have been computed for some simple-shaped geometries, and are presented as charts in heat-transfer books. As an example, a chart for obtaining the view factor for two parallel aligned flat plates of equal size is shown in Fig. 10.6. See Example 2 below for an illustration of its use. More sophisticated computations are needed to obtain view factors for many real-life problems.

When the areas of the two radiation emitting surfaces are not equal, i.e., $F_{1-2} \neq F_{2-1}$. In such cases, the view factor **reciprocity rule** applies, and it can be shown that:

$$A_1 F_{1-2} = A_2 F_{2-1} \qquad (10.19)$$

In general, the law of conservation of energy requires that the entire radiation leaving surface 1 of an enclosure is received by the various surfaces within an enclosure such that $\Sigma_{j=1}^{j=n} F1j$.

Example

2. *A plate (2 m wide and 1 m long) at 300 °C is hung parallel to another plate of the same dimensions. If the second plate is at 800 °C and the space between the two plates is 0.5 m, calculate the net heat transfer between the two plates. The emissivity value is assumed to be 0.7.*

* **Solution**: *Since the radiating surface is finite, we need to determine the shape factor using Fig. 10.6: Y/L = 1/0.5 = 2; X/L = 2/0.5 = 4, then F_{1-2} (or F_{i-j}) = 0.5. Using Eq. (10.17), we obtain:*

$$\dot{q}_{1-2} = (0.7)\left(5.67 \times 10^{-8}\,\frac{W}{m^2 K^4}\right)(2\ m^2)(0.5)\left[(1073K)^4 - (573K)^4\right]$$

$$= 48.3\ kW \blacktriangleleft$$

Fig. 10.5 Examples of internal view factor in a radiating body itself

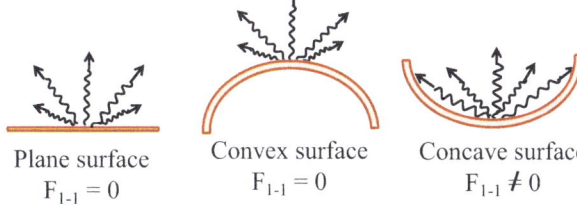

Plane surface
$F_{1-1} = 0$

Convex surface
$F_{1-1} = 0$

Concave surface
$F_{1-1} \neq 0$

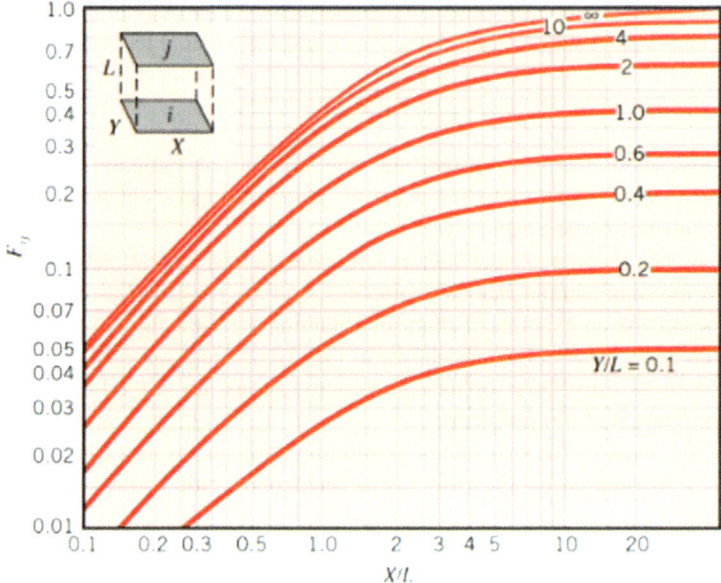

Fig. 10.6 View factor between two aligned parallel rectangles of equal size (from Bergman et al. [1])

When the two bodies have different emissivities (ϵ_1 and ϵ_2) and different areas (A_1 and A_2), then a new factor, Ψ_{1-2}, which accounts for both emissivity and shape, is used, and Eq. (10.17) is modified as follows:

$$\dot{q}_{1-2} = \sigma A_1 \Psi_{1-2} \left(T_1^4 - T_2^4 \right) \tag{10.20}$$

where $\Psi_{1-2} = 1 / \left[\frac{1-\epsilon_1}{\epsilon_1} + \frac{1}{F_{1-2}} + \frac{A_1(1-\epsilon_2)}{A_2\epsilon_2} \right]$ and the net radiation energy transfer equation becomes:

$$\dot{q}_{1-2} = \frac{\sigma \left(T_1^4 - T_2^4 \right)}{\frac{1-\epsilon_1}{A_1\epsilon_1} + \frac{1}{A_1 F_{1-2}} + \frac{1-\epsilon_2}{A_2\epsilon_2}} \tag{10.21}$$

In many real-life thermal processing applications, heat transfer takes place by a combination of the three well-known mechanisms. Hot surfaces lose heat by a combination of convection and radiation. Baking, grilling, toasting, and other similar processes also involve heat transfer by convection and radiation. Heat transfer by radiation between two surfaces at temperatures T_1 and T_2 may also be written in the form of an equation similar to what we have previously studied for heat transfer by convection:

$$\dot{q}_{1-2} = h_{rad} A \left(T_1 - T_2 \right) \tag{10.22}$$

Comparing the above Eqn. with the basic equation of energy transfer by radiation, Eq. (10.16), $\dot{q}_{1-2} = \epsilon\,\sigma\,A\,(T_1^4 - T_2^4)$, a radiative heat-transfer coefficient may be defined as:

$$h_{rad} = \epsilon\,\sigma\,\frac{(T_1^4 - T_2^4)}{(T_1 - T_2)} = \epsilon\,\sigma(T_1 + T_2)(T_1^2 + T_2^2) \tag{10.23}$$

The radiative heat-transfer coefficient is used alone or in combination with the convective heat-transfer coefficient (h_c) for heat-transfer calculations. For the combined effect, an effective heat-transfer coefficient defined below is commonly used.

$$h_{\mathrm{eff}} = h_{\mathrm{rad}} + h_c \tag{10.24}$$

The application of thermal radiation in food processing has not yet achieved widespread applications, although its potential in baking, roasting, broiling, frying, blanching, pasteurization, and drying of foods has been documented. Although quite energy efficient and environmentally benign, one of the limitations of thermal radiation is its penetration depth, which is approximately five mm, lower in comparison with other electromagnetic waves such as microwave and radio frequency. Because of its low penetration characteristics, thermal radiation in the infrared region is generally considered a surface heating technology. It has been shown to be effective in changing the surface color, texture, and consumer acceptability of the product as well as in decontamination of food surfaces and removal of surface moisture. Due to these characteristics, it is used in many food manufacturing operations.

10.4   Dielectric (Microwave and Radio Frequency) Heating

A dielectric material is an electrical insulator that is polarized by an applied electric field. Dielectric heating (DH), also known as high-frequency or capacitive heating, involves heating of a dielectric material (that is nonconducting, nonmetallic and stores and dissipates electric and magnetic energy) by the application of microwaves (MW) or radio frequency (RF) waves of the electromagnetic spectrum. Unlike direct current or alternating current of lower frequencies, the energy in alternating electric current oscillating at radio frequencies gives off electromagnetic waves that can be used as a heating source. MW heating occurs in the frequency range of 300 MHz to 300 GHz (wavelengths between 1 m and 1 mm) and RF from 3 to 300 MHz (wavelengths between 100 m and 1 m). The heating efficiency of DH systems has been estimated to be 50–70% in comparison with approximately 10–20% for conventional ovens.

Both microwave and radio waves are used in communications, mobile phones, radio, TV remote sensing, and cooking and processing of foods. To avoid interference with telecommunications, only certain frequencies are permitted for industrial, scientific, and medical (ISM) uses, including food processing and home cooking, as shown in Table 10.2. Domestic microwave ovens use 2450 MHz, while industrial units are designed to operate at either 896 MHz or 915 MHz. The power rating on these units varies from 0.5 to 1.5 kW.

Table 10.2 Frequencies allocated by the US Federal Communication Commission for ISM use

Frequency (f) (MHz)	Wavelength (λ) (m)	Applications examples
Radio Frequencies (RF)		
13.56 $\pm$ 0.067	22.1	Drying of cereal, crackers, biscuits; eliminates surface cracking by minimizing moisture gradients, defrosting and baking
27.12 $\pm$ 0.160	11.1	
40.68 $\pm$ 0.020	7.4	Pasteurization of dry ingredients, spices, ice cream inclusions; disinfection of produce.
Microwave (MW)		
896 $\pm$ 10	0.335	Tempering of frozen items
915 $\pm$ 13	0.328	Precooking, tempering, pasteurization, sterilization
2375 $\pm$ 50	0.126	Domestic ovens
2450 $\pm$ 50	0.122	Domestic ovens, preheating, pasteurization, sterilization

Check Your Understanding

3. What will be the wavelength of electromagnetic waves produced by a source operating at a frequency of 40 MHz? ◄

Foods act as diamagnetic (nonmagnetic) materials since they contain very trace amounts of magnetic materials such as iron, nickel, and cobalt. They exhibit nearly negligible properties associated with magnetic materials when exposed to electromagnetic fields. Properties that relate to storage and dissipation of electric and magnetic energy, called dielectric properties, are important. The dielectric properties thus control energy coupling and its distribution within a food and determine its thermal response to electromagnetic fields. The dielectric properties are therefore of much interest to us, which in turn depend on the chemical composition as well as the physical structure and geometry of the food.

The dielectric property of a material is given by a **complex relative permittivity (ε_T)**, defined as $\varepsilon_T = \varepsilon/\varepsilon_0$ where ε (Farads/m) is called the material permittivity. It characterizes the interaction between the material and the surrounding electric field, and ε_0 is the permittivity of the free space (vacuum, serving as a reference medium), 8.854×10^{-12} F/m. The complex relative permittivity is defined by both a real and an imaginary component such that

$$\varepsilon_T = \varepsilon' - j\varepsilon'' \qquad (10.25)$$

where ε' is the real portion, called the **dielectric constant,** ε'' is the imaginary component, called the **dielectric loss factor,** and $j = \sqrt{-1}$. Defined relative to the free space, such as permittivity, both ε' and ε'' are also dimensionless. They indicate the ability of a material to interact with the electrical field of the electromagnetic spectrum. The dielectric constant (ε') reflects the ability of a material to be polarized by an electric field and store energy internally. It depends on many factors, such as

Table 10.3 Average dielectric properties of some selected items at 2459 MHz

Item	Temperature (°C)	Dielectric constant (ε')	Dielectric loss factor (ε'')
Apple	20	60	10
Banana	25	62	17
Beef (raw)	25	51	16
Bread	20	5	0.005
Butter (unsalted)	30	24.5	4.2
Cooking oils	25	2.5	0.15
Ice *	0	3.2	0.003
Potato (raw)	25	63	18
Water *	25	78	12.5

Source: * Schiffman [6], Copson [2], and others

temperature, moisture content, and radiation frequency. The dielectric loss factor (ε''), on the other hand, indicates the ability of a material to dissipate and convert electrical energy into thermal energy via frictional dipole rotation and ionic conductivity, thereby increasing the temperature of the material. High dielectric loss materials are strong absorbers of microwaves, low dielectric loss materials are transparent and allow microwaves to pass through them with little attenuation, and opaque materials reflect microwaves.

The ratio of the two components of complex permittivity, ε' and ε'', is known as tan δ or loss tangent. It provides an indication of the relative ability of a material to store electrical energy to how well it is converted into thermal energy.

$$\tan \delta = \varepsilon''/\varepsilon' \tag{10.26}$$

The higher the value of tan δ is, the better the heat generation capacity. Several factors are known to contribute to the loss factor (ε''), including various types of polarizations (millions of times up and down movement per second), such as dipole, electronic, Maxwell–Wagner (charge accumulation at the interface between multi-component systems), and ionic. However, in the permitted range of frequencies used in food applications (Table 10.2), only ionic and dipole polarization mechanisms are the major contributors to the loss factor. Like most biological materials, foods are known as **lossy insulators or nonideal capacitors** since they store as well as dissipate electromagnetic energy passing through them. The loss factor depends on both temperature and frequency. At low frequencies the loss factor has been observed to increase with increasing temperature due to higher ionic conductance while at high frequencies it decreases due to free water dispersion. The dielectric properties of some common items are listed in Table 10.3. Metals reflect microwave and are not heated. Paper, glass, some polymeric films have a low loss factor and do not get heated by microwaves.

Check Your Understanding

4. If all else is held constant, will a dielectric material heat faster when exposed to an MW or RF electromagnetic field? Why? ◄

10.4.1 Fundamentals of Microwave and Radio Frequency Heating

When RF or MW radiation strikes an object, it may be reflected, absorbed, and/or transmitted through the material. As a source of energy, the conversion of EM radiation into heat depends on its interaction with the material. This interaction, as alluded to earlier, is determined by the dielectric properties of each specific material. Described below are the major mechanisms of interaction of foods with RF and MW:

A. Dipole rotation: Water is one of the most polar materials found in nature. As we know, a water molecule has an asymmetric internal distribution of charge because the electrons are closer to the oxygen nucleus than the two hydrogen nuclei. This results in a negatively charged oxygen atom being separated from two positively charged hydrogen atoms. These two centers of equal and opposite charges thus form what is known as an electric dipole. The water molecules are generally randomly oriented and hydrogen bonded. When exposed to an oscillating RF or microwave electric field, the water molecules experience a torque due to the opposite attractive forces and tend to reorient themselves to the direction (or polarity) of the electric field. During the positive half cycle of the propagating waves, the polar molecules align with the field in one direction, and during the negative half cycle, the molecules reorient in the other direction (Fig. 10.7). For example, the electromagnetic field in a microwave oven operating at 2.45 GHz will reverse its polarity 4.9 billion times every second, and this works well since the time needed to flip a water molecule around half cycle is 200 picoseconds. Under the influence of this high-rate and continuously changing polarity of the electric field, water and other polar molecules undergo tremendous amounts of back-and-forth dipolar rotation. Rotating molecules rub, push, and collide with other molecules and atoms in the material and generate localized heat due to friction, such as warming your hands by rubbing them together. Although water molecules are the major contributors to heat generation by dipole rotation in most foods, other molecules, such as some protein, fats, and sugars, may also contribute to the process.

B. Ionic polarization and oscillation: Positively and negatively charged ions or ionic species in a solution are known to move at an accelerated rate toward oppositely charged regions of an electric field and produce electric current. This is known as ionic polarization or ionic conduction, and the changing polarity of microwaves makes the ions oscillate back and forth (Fig. 10.7). When moving

Fig. 10.7 Rotation and polarization of molecules and ions in RF or MW field

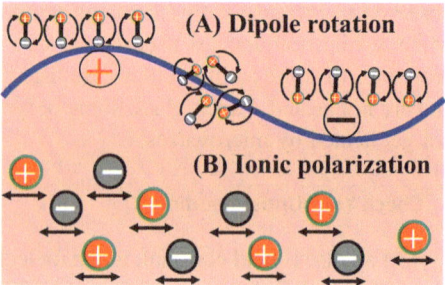

ions face internal resistance due to collisions with neighboring species, their kinetic energy is converted into thermal energy by "ion drag," which is then dissipated within the medium to heat it up. The higher the concentration of charged ions is, the faster the rate of heating.

C. Volumetric and selective heating: Both RF and MW are reflected by metals and travel through glass, paper, parchment, plastics, and ceramics but are absorbed by dielectric materials such as foods and other biological materials. The absorbed energy heats food volumetrically and selectively. It is volumetric because the absorbed energy is converted into heat directly and instantaneously throughout the mass of the material, unlike conventional conduction or convection heating processes. The strength of the field and dielectric properties of the material control the degree of volumetric power absorption and the rate of temperature rise. On the other hand, selective heating of some components in heterogeneous systems such as foods due to differences in their dielectric properties is often advantageous in situations such as drying to remove water with fewer thermal effects on the matrix, creating a thermal gradient to enhance solvent extraction, among others. The fact that the differences in loss factors of various components in a multicomponent system decrease as the frequency of the radiation increases, RF waves with lower frequencies thus heat materials more selectively than microwaves. This selectivity helps in the controlled heating of products with different compositions and properties. These characteristics along with rapid heating times lead to significant advantages in the thermal processing of foods by dielectric heating. Most foods are anisotropic, having different properties in different directions and multicomponent systems. In addition to the frequency of the EM waves employed, the ingredient composition, shape, and size affect the heating rate. As an example, fats and oils in general have low dielectric properties but heat up faster because of their lower specific heat.

10.4.2 Power Generation and Heating Rate

The efficiency of the conversion of electromagnetic radiation to heat is related to the loss factor of the dielectric material and the field strength as well as the frequency of the electromagnetic field. It is given by the following equation:

$$\frac{\dot{q}}{v} = 2\,\pi\,\varepsilon_0 f\,\varepsilon'' E^2 = 55.61 \times 10^{-14}\,f\,\varepsilon'' E^2 \tag{10.27}$$

where $\frac{\dot{q}}{v}$ is the power generated per unit volume (W/cm^3), also called power density, f is the frequency of the field (s^{-1}), and E is the strength of the electric field (V/cm). It is evident from the above equation that the electric field being a square term has the most effect on power density. For a constant f and E, the rate of power density depends on the loss factor, ε''. The greater the loss factor is, the more readily the material will be heated. The name "loss" factor is interesting. Its origin is in electrical devices, where dissipation of electrical energy represented by ε'' is considered an undesirable loss. Heating with RF or microwaves, on the other hand, represents a desirable effect and is a gain factor, but we will continue to use the established nomenclature. The dielectric loss factor of a material is the sum of its loss factors of dipole heating and ionic heating:

$$\varepsilon'' = \varepsilon''_d + \varepsilon''_\sigma = \varepsilon''_d + \frac{\sigma}{2\,\pi\,\varepsilon_0\,\mathrm{f}} \tag{10.28}$$

where ε''_d and ε''_σ are the dipole heating and ionic heating loss factors, respectively, and σ is the electrical conductivity (S/m). The loss factor of foods (Table 10.3) depends on their moisture content, salt content, temperature, and in some cases, structure and geometry. For example, a curved shape can focus short wavelength microwaves better and thus will provide a higher internal heating rate.

Once the electromagnetic energy has been absorbed and generated into thermal energy, its distribution within the food matrix occurs by the conventional methods of conduction and convection. It may thus be helpful to let the heat spread out evenly by leaving the food stand for a while following dielectric heating. Thermal properties such as the specific heat, density, and others also become important, and the rate of temperature rise may be calculated by:

$$\frac{dT}{dt} = \frac{55.61 \times 10^{-14}\, f\, \varepsilon'' E^2}{\rho\, C_p} \tag{10.29}$$

where $\frac{dT}{dt}$ is the time rate of temperature change (°C/s), ρ (kg/m³) is the density, and C_p (J/kg-K) is the specific heat of the dielectric material.

Example

3. A cup of water at 25 °C is heated in a microwave operating at 2450 MHz with a mean field strength of 2 kV/m. Estimate the rate of heating.

   Solution: The dielectric loss factor of water at 25 °C (Table 10.3) is 12.5. Using the values of all the parameters in Eq. (10.29), the rate of heating is obtained as:

$$\frac{dT}{dt} = \frac{55.61 \times 10^{-14}\,\left(2450 \times 10^6\right)(12.5)\left(4 \times 10^2\right)(1000)(kW/m^3)}{(1000\ kg/m^3)(4.18kJ/kg\text{-}K)} = 1.6^0 C/s \ \blacktriangleleft$$

10.4.3 Penetration Depth

The use of EM radiation for heating materials is limited by its penetration depth, which varies with the frequency and field strength. At the same field strength, smaller wavelengths (higher frequencies) are more easily reflected or refracted than longer wavelengths. Thus, RF has deeper penetration power than microwaves, and lower frequencies are more suitable for heating bulk materials more uniformly, while higher frequencies provide more surface heating. The penetration depth (D_P) is defined as the distance at which the incident EM power is reduced to 1/e (approximately 37% since $e = 2.718$) from its value at the entrance. It can be calculated from the dielectric properties of the material and the frequency of the radiation as follows:

$$D_P = \frac{c}{2\pi f} \left[\frac{2}{\varepsilon' \left(\sqrt{1 + (\varepsilon''/\varepsilon')^2} - 1 \right)} \right]^{1/2} \tag{10.30}$$

As may be noted from the above expression, lower frequency radiations such as MW at 915 MHz will penetrate deeper than at 2450 MHz. Similarly, RF in general will exhibit more penetrating depth than MW and may be better suited for heating large volume of products.

10.4.4 Design Features of MW and RF Systems

As discussed earlier, the basic mechanisms involved in the use of RF and MW for heating dielectric materials are similar. Both processes also require high voltage direct current and employ a step-up transformer to transform low voltage alternating current to high voltage current, which is subsequently converted to direct current (~4000 V) to power the system. However, some important differences exist between them in how the two systems are designed and operated. In MW heating, electromagnetic waves, transmitted via a single or multiple waveguides hit the product from all directions. In RF heating, a pair of electrode plates generate the electromagnetic field that hits the product in a unidirectional pattern. For this reason, the design and construction of RF units are generally simpler and less expensive than those of MW systems.

A. Microwave Systems: The major components of a microwave system include a circular symmetric oscillator tube like diode called a magnetron, a waveguide, and a resonance chamber (applicator) for holding and heating the product.

The magnetron is a high-power vacuum tube with resonant cavities that generate electromagnetic radiations by channeling high voltage electrical energy to a hot filament and bouncing electrons around under the influence of a strong magnetic field. Alternatively, a solid-state generator may be used as a source for microwave energy. The generated radiation waves are then transmitted to the chamber with the aid of an antenna at the top of the magnetron and a wave guide. The wave guide consists of a metal tube (rectangular) that reflects and streamlines the energy waves to the inside of the chamber, where they are spread out in three orthogonal directions by reflective fans called stirrers. The radiations then enter the material from all different directions depending on the chamber design and strike the product, Fig. (10.8b). The geometry of the applicator is equally important and is designed to maximize absorption and diffusion of the delivered energy.

In addition to a lower penetration depth, a nonuniform temperature distribution is another limitation of heating with MW. The superposition of a microwave reflected back on itself with the same frequency, amplitude, and wavelength but traveling in opposite directions produces standing (stationary) waves, which are oscillating waves fixed in space, and their energies either get added (most heating, hot spots) or canceled out (i.e., no wave, no heating, cols spot). To mitigate this effect, microwave ovens are designed with a spinning platter (turntable) as well as a stirrer to distribute the microwaves more evenly by breaking up the standing wave pattern.

B. Radio Frequency Systems: An RF unit is made up of two parts, an RF generator and an applicator. The applicator consists of electrodes that pass voltage oscillating at the frequency of the generator through the target material in their proximity in a directional manner. The electrodes are arranged in different configurations to maximize energy delivery to the material without touching it. A common and simple configuration is the parallel plate "through field" electrode system (see Fig. 10.8a). The RF units generate heat more slowly than microwaves but penetrate deeper and are more suitable for materials that require slower heating rates. RF systems are simpler and less expensive than microwave units. The average efficiency of RF heating systems is approximately 70%, compared to 50% for MW. Heating in a batch or continuous mode affects the design and energy utilization efficiency of the setup.

10.5 Ohmic Heating Principle

Ohmic heating, also known as Joule heating, electroconductive heating or electrical resistance heating, is based on an empirical relation known as Ohm's law, first postulated by George Ohm in 1827. It states that the current flowing through a conductor between two points is proportional directly to the difference in voltage and inversely to the resistance of the conductor material. In a conductor, the electrons flow freely from one atom to another, and this flow of electrons constitutes electricity. Mathematically, Ohm's law is written as:

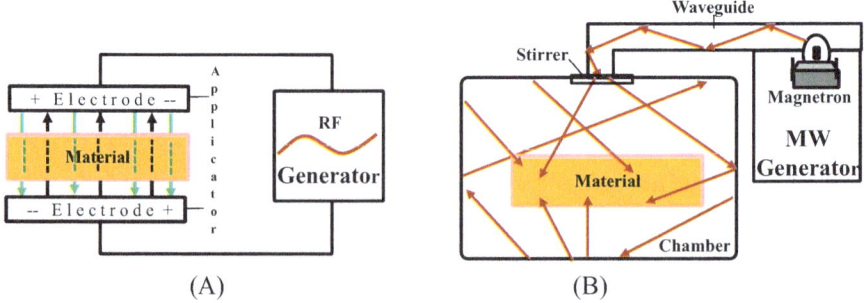

Fig. 10.8 Schematic diagram of RF (**a**) and MW (**b**) heating systems

$$I = \frac{V}{R} \quad \text{or} \quad R = \frac{V}{I} \tag{10.31}$$

where R is the resistance or opposition to the flow of electricity, expressed in units of Ohm, (Ω); V is the voltage, a measure of the strength to move an electrical charge and expressed in volts (V); and I is the current, measured in amperes (A). One ampere is ~6 × 10^{18} electrons flowing per second. In the Ohmic heating of foods, electrical current is passed through the food, which acts as an electrical resistor and, as discussed below, generates heat.

Check Your Understanding

4. If 10 V across a conductor generates a current of 0.5 A, the resistance is 20 Ω. True or False ◄

It is worthwhile to first briefly review the terms resistance, resistivity, conductance, and conductivity that may otherwise create some confusion.

10.5.1 Resistance Vs. Conductance

A conductor's **resistance (R)** is an extrinsic property and indicates the impediment it offers to the flow of electric current. Experimentally, it has been shown that the resistance offered by a conductor depends on its intrinsic property known as electrical **resistivity** (ρ) and its geometry (length and cross-sectional area) (Fig. 10.9). The resistance (R) offered by a conductor in the process is given as follows:

$$R = \rho\left(\frac{L}{A}\right) \tag{10.32}$$

where ρ is the electrical resistivity (Ohm-meters, Ω m) of the conductor, L is the distance between the electrodes, and A is the cross-sectional area of the electrode.

Fig. 10.9 Schematic of an Ohmic heating process

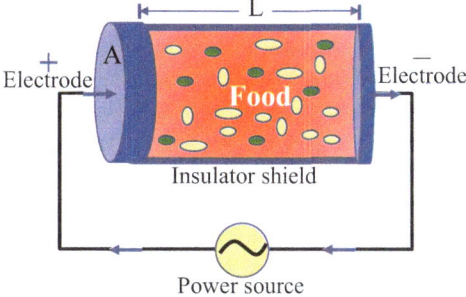

At times, it is more convenient to use the inverse of resistance called **conductance (G)**. It is a more user-friendly term since it tells how good a material is in conducting electricity. A good conductor will have a high conductance. It is written as:

$$G = \frac{1}{R} = \sigma\left(\frac{A}{L}\right) \tag{10.33}$$

The SI unit of conductance (G) is **Siemens, S**, named after German engineer Ernest Siemens. Previously, the unit mho (ohm written backward) was used, and σ is called the electrical **conductivity.** It has the units of S/m and equals $1/\rho$.

Example

4. *A conductor with a radius of 0.5 cm has a resistivity of 0.5 $\Omega.cm$. Determine the resistance and conductance if the conductor is 10 cm long.*
   Solution: Using Eq. (10.32): $R = 0.5\ \Omega.cm \frac{10\ cm}{\pi(0.5cm)^2} = 6.37\ \Omega$.
   Since $1/R$ is conductance, then $G = 1/6.37 = 0.16$ Mho or S ◄

10.5.2 Power Generation

It was not until 1840 that James Prescott Joule first showed that the passage of current through an electrical conductor transforms electrical energy into thermal energy. Thus, the flow of electricity through food also generates heat because of the resistance offered by the food. The applied electrical field ionizes the conductor molecules in foods, which collide with each other and generate energy in proportion to the electrical conductivity of the conductor and the square of the electric field strength. The generated thermal energy then raises the temperature and volumetrically heats the material instantaneously and uniformly, resulting in better quality products. The thermal manifestation of Ohmic heating is thus governed by Joule's law, which states that the rate of energy (i.e., power, P, in Watts) generated is the product of current (I in Amps) and voltage (V) and may be expressed in various forms as indicated below:

$$P = I.V = I^2R = V^2/R \tag{10.34}$$

Substituting for $1/R$ from Eq. (10.33) into Eq. (10.33), we obtain:

$$P = V^2\sigma\left(\frac{A}{L}\right) \tag{10.35}$$

Thus, the power generated is proportional to the square of the voltage, the electrical conductivity (σ) of the material, and the cross-sectional area of the electrode divided by the distance between the electrode terminals.

Table 10.4 Approximate electrical conductivity of selected items

Item (at 25 °C)	Electric Conductivity (S/m)
Deionized water	5.72×10^{-6}
Sea water	4.95
Milk, whole	0.55
Orange juice	0.34
Apple juice	0.26
Apple	0.07
Beef	0.49
Chicken	0.67
Potato	0.04
Strawberry	0.19
Copper	6×10^7
Iron	1.0×10^7
Stainless steel	1.5×10^6
Teflon	1×10^{-24}

The electrical conductivity of foods depends on their composition, ionic strength, and temperature. For good heating, conductivity in the range of 0.1 to 5 S/m is recommended. The electrical conductivity of a few food materials is shown in Table 10.4. Generally, materials with electrical conductivity below 0.01 S/m or above 10 S/m are not amenable to Ohmic heating because they require very high voltages or very large amperes for sufficient heat generation.

The charge in foods is carried by ions that move along the electrical field, and their concentration and mobility dictate the electrical conductivity of foods. Due to the reduction in drag on the movement of the ionized molecules, electrical conductivity increases with temperature and field strength, making Ohmic heating more effective at higher temperatures. Larger solid particles at higher concentrations impede ionic movement and decrease conductivity. The conductivity of liquid foods tends to change linearly with temperature, while the electrical conductivity of cellular foods increases significantly above 70 °C due to electroporation of the cell membrane, breakdown of the cell-wall constituents, and leaching of internal constituents. Electroporation has been found to enhance extraction rates and reduce gelatinization temperature and enthalpy of some foods. Additional factors affecting Ohmic heating include product flow characteristics and viscosity, fouling deposits, and field strength, among others.

The major advantage of Ohmic heating includes four to five times lower energy consumption than conventional heating. In this process, over 95% of the applied electric power is used to heat the product, and heating is far more uniform than heating by microwaves. This helps in maintaining the nutritional and sensory qualities of foods. Other benefits include a shorter heating time, instant shutdown the ability to design heating systems by varying either the electric field strength or product electrical conductivity, and the ability to heat materials of relatively low electrical conductivity by making the electric field strength sufficiently large.

Potential applications include blanching, pasteurization, sterilization, drying, extraction, and fermentation. Some of the limitations of the process are higher cost, inability to heat high fat- and oil-containing foods, and scarcity of data for design and scale-up.

10.5.3 Heat Generation and Temperature

The power (P) dissipated in the conducting material after some time (t) built up heat (Q) such that:

$$Q = P.t \qquad (10.36)$$

If the specific heat (C_p) and mass (m) of the conducting material are known, the temperature rise (ΔT) can be estimated, assuming no loss, with the relation:

$$\Delta T = \frac{Q}{m\ C_p} \qquad (10.37)$$

Example

5. *A 1500 W heater connected to a 220 V electrical system is used to heat 5 kg of milk ($C_p = 4.3$ kJ/kg-K) from 4 °C to 72 °C. Determine the time required if the system efficiency is 80%. How many Amps of current will be used?*
   Solution: *Energy needed, $Q = [(5\ kg)(4.3\ kJ/kg\text{-}{}^0C)(72 \text{ - } 4)^0C]/0.8 = 1827.5\ kJ.$ Heater power output $= 1.5\ kW = 1.5\ kJ/s$, then time required $= 1827.5/1.5\ s = 20\ min.$*
   Additionally, since $1500\ W = I.\ V = I \times 220$, then $I = 1500/220 = 6.82\ A.$ ◄

Problems

10.1 What wavelength range is included in thermal radiation spectrum and what are the characteristics of a blackbody? Does a blackbody actually exist in real life?

10.2 Acting as a blackbody, a furnace wall emits radiation at 1500 K, calculate its emissive power and the wavelength at which emission is maximum.

10.3 Compute the heat loss if the furnace in Problem 10.2 has a 10 cm × 10 cm glass (transmissivity, $\tau = 0.07$) window.

10.4 A layer of food ($\epsilon = 0.75$) at 400 K on a long and wide conveyor belt is exposed to an overhead radiative element ($\epsilon = 0.97$) of the same area. The element is maintained at 1000 K and hung parallel to the conveyor with a small gap between them. Compute the net radiation energy flux.

10.5 Hot oil at 180 °C flows in a 10 m long uninsulated iron pipe with a 10 cm outer diameter. If the ambient temperature is 25 °C, estimate the rate of energy loss from this pipe per day by (a) convection, if the convective heat-transfer coefficient, $h_c = 5$ W/m^2-K, and (b) radiation if the pipe emissivity, $\varepsilon = 0.6$. What percent of the total energy loss is due to radiation?

10.6 Why do fattier foods heat up faster in the microwave and feel hotter than other foods?

10.7 For finish drying of 100 kg/h potatoes from 35 wt.% moisture to 5 wt.% moisture requires 200 kW in a commercial microwave dryer. If the initial temperature of the feed is 25 °C, estimate the efficiency of the unit.

10.8 To pasteurize the apple cider ($C_p = 4.35$ kJ/kg-^{0}C), it takes an electric heater 3 minutes to raise the temperature of one kg of the product from 20 °C to 80 °C. If there is no loss of energy and the available voltage is 250 V, compute the unit's power, resistance, and current.

Bibliography

1. Bergman TL, Lavine AS, Incropera FP, Dewitt DP (2011) Fundamentals of Heat and Mass Transfer.7th edn. Wiley
2. Copson L (1971) Microwave Heating. AVI Publishing Co., Westport, Conn
3. Mc Adams WH (1954) Heat Transmission.3rd edn. McGraw-Hill, NY
4. Henderson SM, Perry RL (1980) Agricultural Process Engineering.3rd edn. AVI Publ. Company, Westport, CT
5. Ryynänen S (1995) The electromagnetic properties of food materials: a review of the basic principles. J Food Eng 26(4):409–429
6. Schiffman RF (1986) Food product development for microwave processing. Food Technol 40: 94–98
7. Smith PG (2011) Introduction to Food Process Engineering.2nd edn. Springer, New York

Mass Transfer: Basic Concepts

11

Mass transfer refers to the net movement of a component (species) in an inhomogeneous system from one location to another in the same or different phase with the aim of making it homogenous and bringing the system closer to equilibrium. Chemical potential difference is the main driving force for mass transfer and depends on various factors like concentration, temperature, pressure, and molecular interactions. Mass transfer occurs by molecular diffusion and bulk flow and does not include movements of materials such as pneumatic conveying of powders, carrying of fruit boxes, or pumping of liquids through pipes. Mass transfer is an important molecular phenomenon that occurs extensively in gases, liquids, and solids in the natural environment as well as in processing and manufacturing industries. It is the basis for numerous biological and chemical processes of direct interest to us, ranging from cooking, digestion, and absorption of foods to separation and purification of products from their matrices.

11.1 Mechanisms of Mass Transfer

Mass transfer occurs by two basic mechanisms, molecular diffusion and convection, analogous to the conduction and convection modes of heat transfer.

Mass transfer by molecular diffusion driven by a concentration gradient results in net transport of molecular mass occurs from a region of high concentration to an area of low concentration by **a random walk process.** It may also occur due to external force fields or other types of gradients, such as thermal and pressure gradients, but their effect is considered negligible in most cases. Molecular diffusion occurs in stagnant fluids or fluids in laminar flow and in solids. Familiar examples include a drop of food coloring spreading over water, salting of cheese, curing of meat, soaking of seeds in water, loss of carbonation in drinks packaged in plastic bottles, etc. A cube of sugar placed in a cup of water begins to dissolve and spread out, albeit very slowly, due to the concentration gradient, as depicted in Fig. 11.1a. It may take a very long time to sweeten the entire cup. Similarly, many high-moisture

© The Author(s), under exclusive license to Springer Nature Switzerland AG 2024
S. S. H. Rizvi, *Food Engineering Principles and Practices*,
https://doi.org/10.1007/978-3-031-34123-6_11

Fig. 11.1 Mass transfer by
diffusion and convection

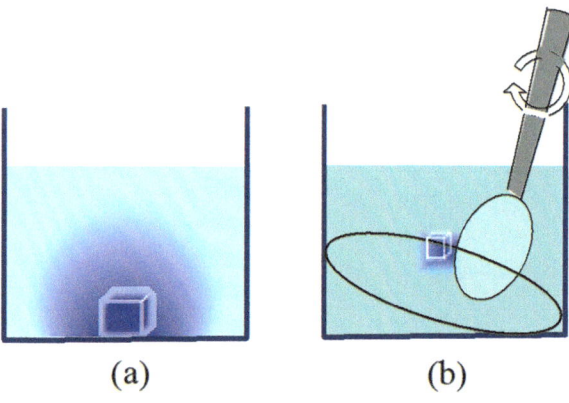

(a)   (b)

foods such as fruits and vegetables left open to the atmosphere will slowly dry out
due to diffusion of water from the surface to the atmosphere. However, the rate of
mass transfer in both these cases can be accelerated significantly by stirring the cup
(Fig. 11.1b) or blowing air past the water-rich foods. This is called **convective mass
transfer.** In James Maxwell (1873), "Mass transfer is due partly to the motion of
translation and partly to that of agitation." In other words, mass transfer occurs
mostly by agitation or by convective currents and eddies of a fluid in turbulent flow
and a small amount by molecular diffusion. In some cases, diffusion may induce
convection. Familiar unit operations such as drying, evaporation, frying, and extraction are examples of convective mass transfer phenomena.

11.1.1 Steady-State Diffusion

Mass transfer due to concentration-based molecular diffusion was first postulated by
Adolf Fick in 1855. He showed that at steady state, the flux of a molecule is
proportional to the concentration gradient, and the proportionality constant is the
so-called binary diffusion coefficient or mass diffusivity. When the bulk fluid is not
moving, a one-dimensional diffusion of species A in species B due to a concentration
gradient is written as follows and is known as **Fick's first law**:

$$J_{Az} = -D_{AB}\left(\frac{dc_A}{dz}\right) \tag{11.1}$$

where J_{Az} is the molar flux of molecule A (kg mole/s.m^2) in the z direction, D_{AB} is
the binary diffusion coefficient (or just the diffusion coefficient) or mass diffusivity
(m^2/s) of molecule A through medium B, c_A is the concentration of A (kg mole/m^3),
and z is the distance in the direction of mass diffusion (m). The negative sign is a
reminder of the fact that mass, like heat and momentum, also flows from a higher
concentration to lower concentration. Mass diffusivity is analogous to the momen-
tum diffusivity (or kinematic viscosity, $\nu = \eta/\rho$) and thermal diffusivity, $\alpha = k/\rho.c_p$,
that were discussed earlier in the fluid mechanics and heat transfer chapters. It is a

measure of a molecule's mobility in a medium. When molecules A and B diffuse at the same rate but in opposite directions to minimize their concentration gradients, it is known as equimolar counter diffusion. The value of mass diffusivity depends on several factors, including the concentration of solute in solution, temperature, and the nature of the fluid and its viscosity.

It is very important to recognize that although c_A is generally expressed in terms of molar concentration, other units, such as mole fraction, mass concentration and mass fraction, are also used. For a gas assumed to behave ideally, molar concentration (c_A) may be expressed in terms of its partial pressure (P_A) and absolute temperature (T) along with the gas constant (R) and $c_A = P_A/RT$, as you will note its utility in later sections. To avoid confusion, it is very important to pay attention to the units involved.

Diffusion is an important but challenging concept. However, by invoking Fick's first law presented above, one can qualitatively explain many real-life phenomena related to cooking and food. For example, to speed up cooking and to extract most flavor from vegetables into the broth, it is recommended to "chop small to chop time." This can be understood by recognizing that small dicing decreases the diffusion distance (dz) through which flavorants and nutrients must travel, thereby increasing dc/dz. and consequently the flux (J).

Equation (11.1) is often expressed in a more fundamental way using mole fraction:

$$J_{Az} = -c\, D_{AB} \frac{dx_A}{dz} \qquad (11.2)$$

where c is the molar concentration of the mixture $(c = c_A + c_B)$ and x_A is the mole fraction of A. It is more useful in certain very special situations, such as the case when the mole fraction of a species is the same at two locations, but its concentration may be different due to, let's say, temperature effects. In such cases, Eq. (11.2) would indicate no diffusion, but Eq. (11.1) would show a finite diffusion.

Diffusion in moving fluids is somewhat more complicated. As may be observed, Eq. (11.1) is analogous to Fourier's law of heat transfer by conduction, but the analogy is slightly simplistic since the flux given by Eq. (11.1) is relative to the average convection flux of the moving fluid and not relative to any fixed axes. Observed from a stationary, fixed point in space, in a binary mixture of A and B, the flux of A is the sum of the average convection flux of A due to the moving fluid and the diffusional flux of A (Fig. 11.2). Mathematically, it may be written as:

$$N_A - N_{AM} + J_A \qquad (11.3)$$

The first term, N_A, is the total flux of A relative to the fixed axes. The second term, N_{AM}, is the average convective flux of A, based on the molar average velocity (v_m), in the bulk fluid relative to the fixed axes. The last term, J_A, represents the diffusion flux relative to the moving fluid. It may be visualized as a school of fish swimming ahead of a flowing stream, the former diffusing in the convective motion of the latter.

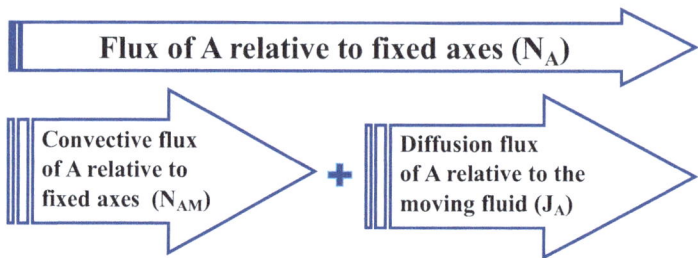

Fig. 11.2 Fluxes in flow relative to fixed axes

In general, the flux is a product of the velocity and concentration of a species; thus, the flux N_{AM} in Eq. (11.3) may be expressed in terms of molar average velocity and concentration as:

$$N_{AM} = c_A v_m \tag{11.4}$$

where $v_m = \sum_{i=1}^{n} x_i v_i$ and x_i is the mole fraction of component i. If the velocities of A and B are v_A and v_B, respectively, then

$$N_{AM} = c_A \left(\frac{c_A}{c} v_A + \frac{c_B}{c} v_B \right) = \frac{c_A}{c} \left(c_A v_A + c_B v_B \right) = \frac{c_A}{c} \left(N_A + N_B \right) \tag{11.5}$$

Substituting Eqs. (11.5) and (11.1) in Eq. (11.3), we obtain

$$N_A = -D_{AB} \frac{dc_A}{dz} + \frac{c_A}{c} \left(N_A + N_B \right) \tag{11.6}$$

The above equation may also be written as

$$N_A = -c D_{AB} \frac{dx_A}{dz} + x_A \left(N_A + N_B \right) \tag{11.7}$$

Equations (11.6) and (11.7) represent the final general equations for diffusion with convection. Diffusion is a microscopic process, while convection is macroscopic and involves movements over much larger distances. Convective movement is generated either by eddies of the turbulent flows or by agitation. As we will see later, in the case of diffusion in stagnant fluids or solids, the convective term is dropped. It should also be noted that in a binary mixture of A and B undergoing equimolar counter diffusion, $N_A = -N_B$, and the convective term drops out.

Check Your Understanding

1. Show that for equimolar counter diffusion in a binary system $D_{AB} = D_{BA}$ ◀

In the case of A and B in the gaseous phase and following the ideal gas law, Eq. (11.7) may be written in terms of pressures as follows:

$$N_A = \frac{-D_{AB}}{RT} \frac{dP_A}{dz} + \frac{P_A}{P_T} (N_A + N_B) \qquad (11.8)$$

where P_T is the total system pressure and P_A is the partial pressure of species A.

Example

1. *Compute the rate of diffusion of water vapor into stagnant ambient air at 1.0 atm. From a tube 2 cm in diameter and 1.5 cm^2 in area and partly filled with water at 25 °C. The diffusion path from the water surface to the top of the tube is 40 cm long. Assume the air is dry and insoluble in water. The diffusion coefficient of water vapor in air is 0.24 cm^2/s.*

   Solution: Using Eq. (11.7) for the molar flux of water (N_A):

$$N_A = - cD_{AB} \frac{dx_A}{dz} + x_A (N_A + N_B)$$

For stagnant air, $N_B = 0$, and the above equation reduces to:

$$N_A(1 - x_A) = - cD_{AB} \frac{dx_A}{dz}$$

Separating the variables and integrating, we obtain:

$$N_A dz = - \frac{cD_{AB}}{1-x_A} dx_A \text{ and } N_A \int_0^z dz = -cD_{AB} \int_{x_{A1}}^{x_{A2}} \frac{dx_A}{1-x_A} \text{ or } N_A z = cD_{AB} \ln \frac{1-x_{A2}}{1-x_{A1}}$$

From Dalton's law of partial pressure, $x_{A1} = P_A/P_T$, where P_T is the atmospheric pressure (1 atm. or 101.3 kPa), and P_A is the partial pressure of water at 25 °C (3.17 kPa from the steam table) = 3.15/101.1 or 0.031 kPa. Additionally, since air is dry, $x_{A2} = 0$, and from the ideal gas law, the molar concentration $c = n/V = P/RT$. On substitution, we obtain:

$$N_A = \frac{P D_{AB}}{R T z} \ln \frac{1-x_{A2}}{1-x_{A1}} \text{ and then}$$

$$N_A = \frac{(1\ atm)\left(\frac{0.24cm^2}{s}\right)}{(82.1\ cm^3.\ atm.\ K^{-1}mol^{-1})(298\ K)(40\ cm)}\ ln\ \frac{1\text{-}0}{1\text{-}0.031}$$
$$= 7.7\ x\ 10^{-9}mol.cm^{-2}.s^{-1}$$

Now the molecular weight of water is 18 g/mol and the given surface area for diffusion is 1.5 cm^2, the diffusion rate of water (R$_A$) becomes:

$$R_A = 7.7\ x\ 10^{-9}mol.\frac{18\ g}{mol}.\frac{1.5\ cm^2}{cm^2.s} = 20.8\ x\ 10^{-8}g/s \quad \blacktriangleleft$$

11.2   Diffusion Coefficient

As a physical constant, the diffusion coefficient indicates how fast a molecule diffuses through a medium, and a higher value means faster diffusion. The diffusion coefficient is inversely correlated with the molecular radius, and high molecular weight molecules have a low diffusion coefficient. For isotropic materials with cubic symmetry, the diffusion coefficient has the same value in all directions.

Diffusion coefficients are used to describe many diffusion-related phenomena and to define dimensionless numbers, which are required in the analysis of various mass transfer operations. The diffusion coefficients are highest in gases (10^{-2}–10^{-1} cm^2/s), followed by liquids (10^{-5}–10^{-2} cm^2/s) and then solids (10^{-10}–10^{-6} cm^2/s). A few typical values of the binary diffusion coefficients of some common molecules in selected matrices are shown in Table 11.1 as examples.

Knowledge of the diffusion coefficient is very useful for estimating the diffusion time in some simple cases. For one-dimensional Brownian motion or random walk processes in gases and liquids, it has been shown (Berg [1]) that the time (t$_d$) needed for a molecule to diffuse a distance (d) is approximated by the following expression:

Table 11.1 Average diffusion coefficients for selected pair of materials

Binary system	Diffusivity (cm^2 s^{-1}) at ~25 °C
Oxygen in air	0.20
Oxygen in water	1.8×10^{-5}
Oxygen in polyethylene	1.0×10^{-6}
Carbon dioxide in air	0.15
Carbon dioxide in water	2.0×10^{-5}
Carbon dioxide in polyethylene	9.0×10^{-6}
Ethanol in air	0.14
Water in air	0.24
Acetic acid in water	0.9×10^{-5}
Salt in cheese	0.3×10^{-4}

$$t_d = \frac{d^2}{2D_{AB}} \qquad (11.9)$$

The diffusion coefficient depends on the size and shape of the molecule as well as on the pressure, temperature, make-up, and physical state (porosity, tortuosity, etc., discussed later) of the system. While gas diffusion coefficients are almost independent of composition, the diffusion coefficients of liquids and solids are concentration dependent. Unlike other transport properties, no standardized methodologies are readily available for either the measurements or a priori determination of diffusion coefficients. Thus, various experimental, theoretical, and estimation approaches have been developed to find the diffusion coefficient values in the gas, liquid, and solid phases. Several experimental approaches have been developed and utilized for the measurement of diffusion coefficients and are available in the literature. Their description and merits are beyond the scope of this chapter. What follows is a concise description of the various theoretical approaches to estimate diffusivities in gases, liquids, and solids.

Example

2. *Estimate the time it would take for oxygen to diffuse 1.0 cm below the surface of lake Cayuga on a calm day.*
   Solution: From Table 11.1, the diffusion coefficient for oxygen in water is 1.8×10^{-5} cm^2/s. Using Eq. (11.9), we obtain:

$$t_d = \frac{1}{2\left(1.8x10^{-5}\right)} = 9 \ x \ 10^4 s. \qquad \blacktriangleleft$$

11.2.1 Diffusion in Gases

Gases diffuse at a fast rate because of their high kinetic energy, and lighter gases do better than heavy gases. Based on the kinetic theory of ideal gases, several equations have been developed to calculate diffusion coefficients for a range of applications. According to this theory, the diffusion coefficient of a gas is directly proportional to its mean free path (λ) and the mean molecular velocity, $<v>$, in a mixture. Accordingly, the binary diffusion coefficient is given as:

$$D_{AB} = \frac{1}{3} \lambda <v> \propto \sqrt{\frac{T}{M}} / \frac{n}{V} \qquad (11.10)$$

where M is the molar mass and n/V is the molar density of the molecules. The above equation predicts that diffusion increases with increasing absolute temperature (T) and decreases with increasing pressure ($D_{AB} \propto 1/P$).

Table 11.2 Atomic and molecular volumes of selected atoms and molecules (Fuller et al. [4])

Atom/Molecule	Diffusion Volume, v (cm³/mol)
C	16.5
H	1.98
O	5.48
N	5.69
O_2	16.6
N_2	17.9
CO_2	26.9
Air	20.1
Water	12.7

A few more semiempirical equations have been proposed, but for engineering work, the equation proposed by Fuller et al. [4] is recommended for gas diffusivity estimation:

$$D_{AB} = \frac{1.0 \times 10^{-3}\, T^{1.75}}{P\left[(\Sigma v)_A^{1/3} + (\Sigma v)_B^{1/3}\right]^2} \left(\frac{1}{M_A} + \frac{1}{M_B}\right)^{1/2} \tag{11.11}$$

where T is the absolute temperature, P is the pressure in atm., Σv is the diffusion volume obtained as the sum of the atomic volumes of all elements in each molecule, and a few selected volumes are given in Table 11.2. When used, the value of D_{AB} is obtained in cm²/s.

Example

3. *Ethanol diffuses from the fermentation broth into a room at 30 °C. Estimate the diffusivity of ethanol in air.*
   Solution: P = 1.0 atm., T = 273 + 30 = 303 K, M_{Etoh} = 46.0, M_{Air} = 29, and from Table 11.2:

$$\Sigma v_{Etoh} = 2(16.5) + 6(1.98) + 1(5.48) = 50.36 \text{ and } \Sigma v_{Air} = 20.1$$

Substituting in Eq. (11.11):

$$D_{Etoh\text{-}Air} = \frac{1.0 \times 10^{-3}\, (303)^{1.75}}{1.0\left[(50.36)_{Etoh}^{1/3} + (20.1)_{Air}^{1/3}\right]^2} \left(\frac{1}{46} + \frac{1}{29}\right)^{1/2}$$

$$= 0.127 \, cm^2/s \qquad \blacktriangleleft$$

11.2.2 Diffusion in Liquids

Diffusion in liquids is approximately four to five orders of magnitude slower than diffusion in gases and often becomes the rate-limiting step in separation and reaction processes. Additionally, the value of the diffusion coefficient of a solute depends on its concentration in the solution, and, unlike gases, the diffusion coefficient D_{AB} is invariably not equal to D_{BA}. Several methods for experimental determination and a few semiempirical approaches for the prediction of diffusion coefficients in liquids are available in the literature (Cussler [3]).

One of the earliest correlations developed for the prediction of liquid diffusivities is the Stokes-Einstein equation, which does not work well for smaller solute molar volumes and has since been modified by others. For most general cases, the Wilke-Chang modification is usually found to be adequate and is given as:

$$D_{AB}\left(cm^2/s\right) = \frac{7.4 \text{ x } 10^{-8} \left(\varphi M_B\right)^{1/2} T\left(K\right)}{\eta_B\left(mPa - s\right) V_A^{0.6}} \tag{11.12}$$

where M_B is the molar mass of solvent, V_A (cm^3/mol) is the volume of the diffusing molecule, and φ is an empirical parameter of the solvent with a value of 2.6 for water, 1.5 for ethanol, and 1.0 for other unassociated solvents.

For **biological solutes** of molar mass greater than 1000 in dilute aqueous solutions, the semiempirical Polson equation shown below is recommended.

$$D_{AB}\left(cm^2/s\right) = \frac{9.4 \text{ x } 10^{-8} T\left(K\right)}{\eta_B\left(mPa - s\right) M_A^{1/3}} \tag{11.13}$$

In the case of multicomponent systems, the diffusion coefficients are determined for each pair of molecules in the system. There are other diffusion coefficients in addition to the binary diffusion coefficients that we discussed above. A **self-diffusion coefficient** is obtained by labeling selected molecules of a pure material and then analyzing their movement in the unlabeled bulk.

11.2.3 Diffusion in Solids

Diffusion is very slow in solids, which include both consolidated materials such as polymers and metals or porous structures of loosely packed particles. The diffusing species could be gases, liquids, or even interdiffusing solids. In the latter, translational molecular movements driven by interstitial point defects, dislocations, and vacancies are very slow but not zero, and thus diffusion occurs over longer time intervals. Alloys such as stainless steel are made by mixing and interdiffusion of solids into solids. Although important, we will limit our further discussion to the diffusion of only gases or vapors in solids.

(A) Diffusion of gases in polymers: Diffusion of gases through consolidated materials such as polymers or metals may be assumed to be a case of a homogenous matrix in which the diffusing gas has a constant diffusivity. Now, we consider a simple case of steady-state diffusion of a gas through a nondiffusing or stationary polymer, as shown in Fig. 11.3. Given that the flux (J) and diffusion coefficient (D) are constant, Eq. (11.1) in generic form can be rearranged and integrated:

$$J \int_{z_1}^{z_2} dz = -D \int_{c_1}^{c_2} dc \tag{11.14}$$

We obtain

$$J = D\left(\frac{c_1 - c_2}{z_2 - z_1}\right) \tag{11.15}$$

The above expression can now be used for steady-state diffusion of molecules through a consolidated material, and we will discuss this next.

Example

4. *Derive an expression for steady-state radial diffusion through a hollow cylinder of inside and outside radii of r_i and r_o, respectively, and length L.*

Solution: In this case, the rate of diffusion ($\dot{J}$), instead of the flux, may be written as:

$$\frac{\dot{J}}{2\pi r L} = -D\frac{dc}{dr}$$

Separating the variables and integrating them with the given limits, we obtain:

$$\dot{J} = -D(c_1 - c_2)\frac{2\pi L}{\ln(r_o/r_i)} \qquad \blacktriangleleft$$

Fig. 11.3 Steady-state diffusion

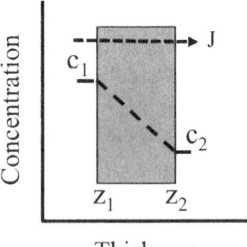

Thickness

Fizzy drinks contained in a polymer (e.g., polyethylene terephthalate, PET) bottle are known to go flat with time due to the diffusion of carbon dioxide through the wall and provide an interesting example of utility to understand and quantify the diffusion process. Carbonated drinks have a higher concentration of carbon dioxide than ambient air, making it possible for carbon dioxide molecules to migrate via diffusion from the beverage into the surrounding air. In such cases, the fundamental transport process of diffusion occurs via the following three basic steps (Fig. 11.4):

- The gas at the high-pressure side adsorbs and solubilizes, following Henry's law, in the solid matrix (wall of the container) with a concentration at the solid interface of c_1
- The gas diffuses via a random walk process toward the low-concentration side according to Fick's law
- Upon reaching the low-pressure side, the gas desorbs to a concentration of c_2 in accordance with Henry's law and dissipates into the ambient air.

However, the solid-phase concentrations c_1 and c_2 are not easy to determine but can be estimated by using more easily measurable parameters P_1 and P_2 and invoking Henry's law for low concentrations. This law says that the concentration of a gas that dissolves in a liquid is directly proportional to the partial pressure of the gas. Approximating Henry's law to the gas–solid system with the proportionality constant representing equilibrium solubility (S) of the gas in the solid matrix at low concentrations, a linear relation is written as:

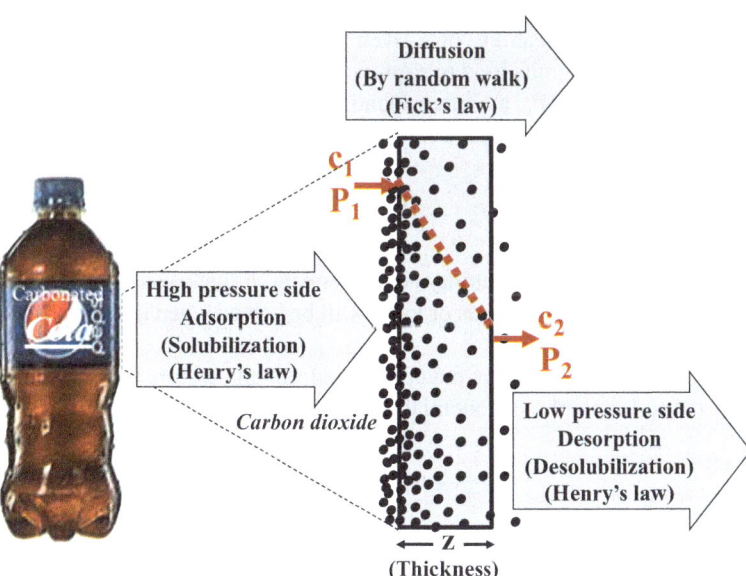

Fig. 11.4 Diffusion through polymeric material

$$c = S P \qquad (11.16)$$

The solubility (S) value is a unique function of a specific pair of gas-polymer systems and temperatures. It may be obtained from the literature or measured experimentally. By substituting Eq. (11.16) into Eq. (11.5), we obtain:

$$J = D\left(\frac{SP_1 - SP_2}{z_2 - z_1}\right) = DS\left(\frac{P_1 - P_2}{z_2 - z_1}\right) \qquad (11.17)$$

The product of the diffusion coefficient (D) and solubility (S) is referred to as permeability (P_m):

$$P_m = DS \qquad (11.18)$$

Equation (11.17) may thus be written as:

$$J = P_m\left(\frac{P_1 - P_2}{z_2 - z_1}\right) \qquad (11.19)$$

Rewriting the above equation in terms of the quantity (Q) instead of molar flux of gas permeating through a polymeric film of area A at time t, we obtain:

$$\frac{Q}{t} = \left[\frac{P_m}{\Delta z}\right] A \Delta P \qquad (11.20)$$

The term $P_m/\Delta z$ is often called the **gas permeance** or **gas transmission rate (GTR),** and it represents the permeability of a given thickness of polymeric material.

Based on the various units used to express concentration, pressure, and thickness, over 30 different units of P_m are often found in the literature. The units of flux will then change accordingly. Under the SI, the units of P_m are:

$$P_m [=] \frac{mL\ (STP).cm}{cm^2.s.Pa} \qquad (11.21)$$

As may be noted, the permeability is independent of thickness, and thus, the rate of transmission (diffusion) of a gas or vapor will be halved when the thickness of the polymeric material is doubled.

The permeability of polymeric materials is indeed temperature dependent and has been reported to follow the exponential Arrhenius relation given below:

$$P_m = P_o\ exp\left(-\frac{E_P}{RT}\right) \qquad (11.22)$$

where P_O is a preexponential factor, E_P is the activation energy for permeation, R is the gas constant, and T is the absolute temperature.

Table 11.3 shows the permeability values for some of the more common gases in several commercial polymers of interest to food packaging.

Table 11.3 Permeability values for a few selected polymers and permeants

| Polymer | P_m at 25 °C, $\frac{mL\ (STP).cm}{cm^2.s.Pa} \cdot 10^{13}$ | | | |
	O_2 (0.35 nm dia.)	CO_2 (0.33 nm dia.)	N_2 (0.36 nm dia.)	H_2O (0.27 nm dia.)
Polyvinylidene chloride (Saran)	0.004	0.02	0.001	7.0
Cellulose hydrate (Cellophane)	0.002	0.189	0.002	18,800
Low density polyethylene (LDPE)	2.20	9.45	0.73	68.2
High density polyethylene (HDPE)	0.30	0.28	0.11	9.0
Polypropylene	1.52	6.00	–	446
Polyethylene terephthalate (PET)	0.014	0.11	0.003	850

Example

5. *Calculate the steady-state diffusion flux of oxygen at 25 °C through LDPE film of 0.2 mm thickness if the partial pressure differential on the two sides of the film is 0.21 atm. Assume negligible surface resistance to diffusion.*

Solution: Using Eq. (11.19) and the permeability value for oxygen in LDPE from Table 11.3, we obtain:

$$J = P_m \left(\frac{\Delta P}{\Delta z}\right) = \frac{2.20 \left(10^{-13}\right)\left(0.21 \times 1.013 \times 10^5\right)}{0.02} = 23.4 \times 10^{-8}\ mL/\left(s.cm^2\right) \blacktriangleleft$$

Frequently, packaging materials consist of multilayered structures of several polymeric films of different thicknesses and permeabilities laminated in series. In such laminates or coated layers, the total permeability may be computed as follows:

$$\frac{1}{P_m} = \frac{\Delta z_1}{\Delta z\ P_{m1}} + \frac{\Delta z_2}{\Delta z\ P_{m2}} + \frac{\Delta z_3}{\Delta z\ P_{m3}} + - - - \tag{11.23a}$$

where $\Delta z = \Delta z_1 + \Delta z_2 + \Delta z_3 + -$ and the subscripts represent the various layers in the laminate. When each layer is of the same thickness, Δz_1, Eq. (11.23a) simplifies to:

$$\frac{1}{P_m} = \frac{\Delta z_1}{\Delta z}\left(\frac{1}{P_{m1}} + \frac{1}{P_{m2}} + \frac{1}{P_{m3}} + - - -\right) \tag{11.23b}$$

(B) Diffusion of gases in porous media: Diffusion of gases through a porous, nonhomogeneous material of porosity or voidage, ε, creates a more tortuous travel path for the molecules, as illustrated in Fig. 11.5. In such cases, diffusion

Fig. 11.5 Diffusion in a
porous medium

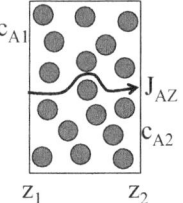

occurs through the pore openings with the permeant remaining in the gas phase,
and no dissolution in the solid matrix is needed. The porosity (ε), which
represents the ratio of open cross-sectional area to total area and may range in
value from zero to one, reduces the diffusion cross-sectional area. Additionally,
in such systems, the porosity creates a longer path and forces the permeant
molecules to travel in a zig-zag manner, which in turn lengthens the pathway
and reduces the diffusion rate. To account for this, a correction factor called
tortuosity, τ, is used. It is generally defined as the squared ratio of the zig-zag
paths followed by a permeant in the porous medium to a straight path and has a
value greater than unity, usually lying in the range of 1. 5 to 5. Both the above
two factors (ε and τ) are therefore incorporated into the diffusion coefficient and
the generic Fick's Eq. 11.1 is modified as follows:

$$J_{Az} = -\frac{\varepsilon D_{AB}}{\tau}\left(\frac{dc_A}{dz}\right) = -D_{eff}\left(\frac{dc_A}{dz}\right) \tag{11.24a}$$

where the effective diffusivity or effective diffusion coefficient, D_{eff}, is defined as
follows:

$$D_{eff} = \frac{\varepsilon D_{AB}}{\tau} \tag{11.24b}$$

Check Your Understanding

2. The diffusion coefficient and effective diffusion coefficient have the same
units. True or false ◀

The diffusion mechanism in porous media discussed to date assumes gaseous
movement in free space such that the gas molecules freely colloid with each other
before hitting the medium through which they are diffusing and follow Fick's laws.
This is known as **Fickian** or **molecular** diffusion, Fig. 11.6a, and occurs only when
the pore diameters are larger than the mean free path, λ, of the diffusing gas,
approximately 10^{-7} m at 1 atm. In many situations, however, this is not the case,
and pore diameters are less than the λ value of the molecules involved and range
from 2 to 50 nm. This leads to the gas molecules colliding with the pore surfaces

Fig. 11.6 Mechanisms of molecular diffusion (**a**) and Knudsen diffusion (**b**)

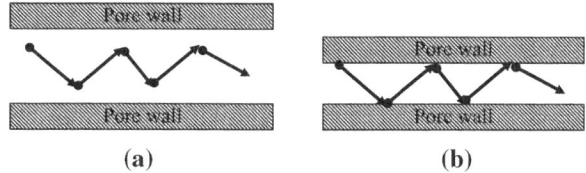

(a)   (b)

more frequently than with each other. The diffusion mechanism then changes to what is known as **Knudsen** diffusion (Fig. 11.6b). In addition to the molecular mass of the permeant and temperature, Knudsen diffusion is also affected by the radius of the pore, and the value of a new Knudsen diffusion coefficient is estimated using the kinetic theory of gases. It is given by the following equation:

$$D_k \left(cm^2/s\right) = 9700 \ r_P \left(\frac{T}{M}\right)^{1/2} \tag{11.25}$$

where r_p is the pore radius (m), T is the absolute temperature (Kelvin), and M is the molar mass of the diffusing molecule. This type of diffusion has been observed to occur in porous catalysts, some soils, stomatal pores in plant leaves, etc.

11.3 Unsteady-State Diffusion

If the concentration of diffusing species at a given point changes with time, then the process becomes a time-dependent or **unsteady state** and is described by Fick's second law in the form of a partial rather than ordinary differential equation. Almost all processes begin with unsteady-state conditions and over time may change to steady state. It is obtained by conducting a material balance along with the application of Eq. (11.1). Figure 11.7 shows a rectangular block of solid B of unit cross-sectional area and small length Δz. If the unsteady-state unidirectional flux of species A into and out of the block is J_{az1} and J_{az2}, respectively, then from mass balance:

$$J_{az1} = J_{az2} - \Delta z \frac{\partial J}{\partial z} \tag{11.26}$$

The net increase in mass in the block becomes:

$$J_{az1} - J_{az2} = - \Delta z \frac{\partial J}{\partial z} \tag{11.27}$$

Now, since the volume of the block is Δz, the above equation may be written as:

Fig. 11.7 Unsteady-state
diffusion

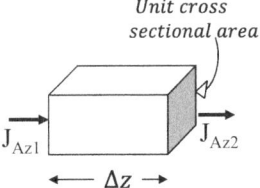

$$\Delta z \frac{\partial c}{\partial t} = -\Delta z \frac{\partial J}{\partial z} \qquad (11.28)$$

Substituting Eq. (11.1) into Eq. (11.28), we obtain:

$$\frac{\partial c}{\partial t} = \frac{\partial}{\partial z}\left(D_{AB}\frac{\partial c}{\partial z}\right) \qquad (11.29)$$

And if D_{AB} is a constant:

$$\frac{\partial c}{\partial t} = D_{AB}\frac{\partial^2 c}{\partial z^2} \qquad (11.30)$$

This is known as **Fick's second law** of diffusion and describes unidirectional unsteady-state mass transfer by diffusion. This shows that the change in the concentration with time is in direct proportion to the concentration gradient, which will disappear with time, the fundamental role of diffusion.

11.3.1 Graphical Solutions

Equation (11.30) describing unsteady-state diffusion is physically distinct but mathematically analogous to unsteady-state heat conduction; thus, the methods used for their solutions are the same. They involve either a trigonometric series that converges for long intervals of time or a series of error functions for small values of time. The analogy also permits the use of Gurney-Lurie charts developed for heat transfer to solve mass transfer problems. To utilize these charts correctly, however, one must remember to use the appropriate definitions and compare them with the heat-transfer parameters, as shown in the table below.

Example

6. *Salting of cheese by immersing in a brine to increase its salt content is an integral part of the cheesemaking process. In a commercial operation, large slabs of 5 cm thick feta cheese are immersed in a dynamic raceway where 3.5*

molar brine at 13 °C is flowing in a serpentine pattern (turbulent flow), making surface resistance to mass transfer negligible (i.e., large convective coefficient). Assuming diffusion is only through the two large parallel flat surfaces 5 cm apart, estimate the salt concentration at the midpoint of the slab after 24 hours.

Solution:

To use the Gurnie-Lurie chart for flat plates, we first calculate the parameters n, m, and X as defined in Table 11.4.

For concentration at the center, n = 0/2.5 cm = 0

Negligible resistance to mass transfer at the surface ($k_c > > D_{AB}$) makes m = 0

The diffusion coefficient of salt in cheese is shown in Table 11.1 as 0.3×10^{-4} cm^2/s and thus:

$$X = \frac{D_{AB}\, t}{x_1^2} = \frac{\left(0.3\, x10^{-4}\, \frac{cm^2}{s}\right)\left(24h.\, \frac{3600\, s}{h}\right)}{(2.5\, cm)^2} = 0.41$$

For X = 0.41, m = 0 and n = 0, from Fig. 9.7 for an infinite plate, the value of Y = 0.4

Therefore, $Y = \frac{c_1 - c}{c_1 - c_0}$ or $0.4 = \frac{c_1 - c}{c_1 - c_0}$ and since c_1 = 3.5 mole/L, c_0 = 0 mol/ L, then: $0.4 = \frac{3.5 - c}{3.5 - 0}$ or c = 2.1 mol/L or 2.1 kmol/m³ ◀

Table 11.4 Unsteady-state heat and mass transfer parameters for graphical solutions

Parameter	Heat Transfer	Mass Transfer
Biot Number (Bi)	$\frac{h_c\, x_1}{k}$	$\frac{k_c\, x_1}{D_{AB}}$
Fourier Number (Fo)	$\frac{\alpha t}{x_1^2}$	$\frac{D_{AB} t}{x_1^2}$
Gurney-Lurie charts	Unaccomplished change: $Y,\ \frac{T_1 - T}{T_1 - T_0}$	$\frac{c_1 - c}{c_1 - c_0}$
	Relative time: $X,\ \frac{\alpha t}{x_1^2}$	$\frac{D_{AB}\, t}{x_1^2}$
	Relative resistance: $m,\ \frac{k}{h_c\, x_1}$	$\frac{D_{AB}}{k_c\, x_1}$
	Relative position: $n,\ \frac{x}{x_1}$	$\frac{x}{x_1}$

where $T = T_0$ and $c = c_0$ at time $t - 0$; T = temperature and c = concentration at $t = t$; T_1 = ambient temperature or c_1 = bulk concentration at $t = 0$; α = thermal diffusivity; k = thermal conductivity; h_c, k_c = convective transfer coefficient; and D_{AB} = mass diffusivity (diffusion coefficient)

Note: the mass transfer equilibrium distribution coefficient (H), discussed later, is assumed to be unity

11.3.2 Solution Based on Average Concentrations

Unlike heat transfer, where we are generally interested in the temperature at the coldest or hottest spot of the food, in mass transfer processes, we are more interested in following the average concentration and time in the object under study. For such cases, integration of the unaccomplished concentration/position lines on the Gurney-Lurie charts provides the average concentration as a function of the Bi and Fo numbers. When Bi is very large, i.e., when the internal resistance controls the process and which is generally a reasonable approximation for most mass transfer processes related to foods, the charts can be replaced with simple shape-specific equations for all except the low values of Fo (<0.2). This comes from the analytical solution of Fick's second law equation and its first-term approximation. The following equations for simple geometries are obtained:

For infinite plates:

$$\frac{c_1 - c_{avg}}{c_1 - c_0} = \frac{8}{\pi^2}\ exp\left(-\frac{\pi^2\,D_{AB}\,t}{4x_1^2}\right) \tag{11.31}$$

For infinite cylinders:

$$\frac{c_1 - c_{avg}}{c_1 - c_0} = 0.692\ exp\left(-\frac{5.78\,D_{AB}\,t}{x_1^2}\right) \tag{11.32}$$

For spheres:

$$\frac{c_1 - c_{avg}}{c_1 - c_0} = \frac{6}{\pi^2}\ exp\left(-\frac{\pi^2\,D_{AB}\,t}{x_1^2}\right) \tag{11.33}$$

where x_1 is half the thickness of the slab or the radius of the cylinder or sphere.

Example

7. *For the brining system described in Example 6 above, determine the time that will be needed for the average salt concentration in the cheese to reach 3.0 mol/L.*

Solution:

For large flat geometry, we will use Eq. (11.31). Therefore, with the initial concentration (c_0) assumed to be zero, the unaccomplished concentration change is:

$$\frac{c_1 - c_{avg}}{c_1 - c_0} = \frac{3.5 - 3.0}{3.5 - 0} = 0.143$$

And then, $\frac{8}{\pi^2}\ exp\left(-\frac{\pi^2\,D_{AB}\,t}{4x_1^2}\right) = 0.143$

Or, $ln\left[(0.143)\left(\frac{\pi^2}{8}\right)\right] = -\frac{\pi^2\,D_{AB}\,t}{4x_1^2}$ *or* $4x_1^2\,\frac{ln\left[(0.143)\left(\frac{\pi^2}{8}\right)\right]}{\pi^2\,D_{AB}} = -t$

$$Or, \; t = 4 \; (2.5 \; cm)^2 \; \frac{1.736}{(3.14)^2 \left(0.3 \; x10^{-4} \frac{cm^2}{s}\right)} = 14.6 \; x \; 10^4 s = 40.76 \; h \; \blacktriangleleft$$

Note that Eqs. (11.31), (11.32), and (11.33) can also be used in processes for the determination of effective diffusivity in porous media and for systems where matrices change during mass transfer, such as in drying. In such cases, D_{AB} is replaced with D_{eff}, and, for example, Eq. (11.33) for a sphere can be written as:

$$\frac{c_1 - c_{avg}}{c_1 - c_0} = \frac{6}{\pi^2} \; exp\left(- \frac{\pi^2 \; D_{eff} \; t}{x_1^2}\right) \tag{11.33a}$$

By transforming the above into a linear equation, we obtain:

$$\ln\left(\frac{c_1 - c_{avg}}{c_1 - c_0}\right) = \ln\left(\frac{6}{\pi^2}\right) - \left(\frac{\pi^2 \; D_{eff} \; t}{x_1^2}\right) \tag{11.33b}$$

A reasonable estimate of D_{eff} can now be obtained from the slope of Eq. (11.33b) by fitting experimental data on the unaccomplished concentration change with time. Additionally, the temperature dependence of D_{eff} is assumed to follow the Arrhenius relationship, and the activation energy (E_a) represents the energy for diffusion.

While practical and useful, the analogy between heat and mass transfer used to solve problems has its limitations, and care must be exercised in its application. Unlike heat transfer, solubilization of the subject matrix may occur during mass transfer processes. This would cause the interfacial boundary between the liquid and solid to move and change the thickness of the hydrodynamic boundary layer. When this happens, the mass transfer coefficients become difficult to control or predict, and the analogy will not hold.

11.4 Mass Transfer by Convection

The unsteady-state diffusion equation developed with the aid of Fig. 11.7 assumed that there was no convective resistance at the surface. However, at the fluid–fluid and fluid–solid interfaces, mass transfer occurs by convection, as in separation processes including drying, concentration, extraction, leaching, etc. In multiphase (solid–liquid–gas) systems, mass transfer becomes more complex. During interphase mass transfer involving gas–liquid (distillation), liquid–liquid (extraction), solid–gas (drying), and solid–liquid (membrane separations such as microfiltration and ultrafiltration) the role of each phase becomes critical. Because of their physical similarities, these processes are amenable to the same mathematical treatments.

Several theories, such as the two-film theory, surface renewal theory, penetration theory, and boundary layer theory, have been proposed to understand and quantify interphase mass transfer processes. The "two-film" or "two-resistance" theory proposed by Whitman in 1923 has been found to be generally applicable to many interphase mass transfer processes and will be discussed next. For others, more advanced mass transfer texts should be consulted.

Whitman theory is based on the assumption that a film of some thickness (z) exists on each side of the interface between two phases, phase-1 and phase-2, as shown in Fig. 11.8. The resistance to mass transfer thus resides within the two films, and molecular diffusion occurs following Fick's law. In Fig. 11.8, the concentration of diffusing species A is represented by its bulk molar concentrations, c_A^1 and c_A^2 in phase 1 and phase 2, respectively, and since $c_A^1 > c_A^2$, mass flux occurs from phase-1 to phase-2. The interfacial concentrations, denoted by c_A^{1i} and c_A^{2i}, also follow the same pattern. The film thicknesses z_1 and z_2 represent the distance from the interface at which the concentration of the diffusing species (A) equals its concentration in the bulk. The shape of the concentration profile and the film thicknesses are both difficult to quantify or predict, and thus Fick's law of diffusion is not useful. Instead, a mass transfer coefficient, k_x, which replaces mass diffusivity and thickness (z), is used. Mass transfer is then expressed in a way similar to Newton's law of cooling in heat transfer by convection. Thus, the molar flux of A (N_A) across the film for phase-1 may be written as:

$$N_A = k_1 \left(c_A^1 - c_A^{1i} \right) \tag{11.34}$$

And for phase-2:

$$N_A = k_2 \left(c_A^{2i} - c_A^2 \right) \tag{11.35}$$

where the coefficients k_1 and k_2 are known as the individual convective film mass transfer coefficients, and depending on the type of phase (liquid or gas) under consideration, they are frequently called the liquid film coefficient (k_L) or gas film coefficient (k_G). The concentration difference indeed provides the driving force, and at steady state, the two fluxes become equal.

Fig. 11.8 Two-film theory for interphase mass transfer

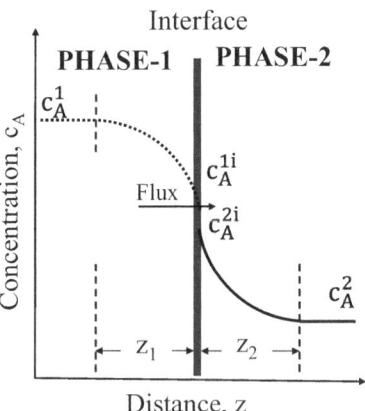

The two-film theory further assumes that the interface offers no resistance to transport and that the concentrations of the diffusing species at the interface are always in equilibrium, as depicted in Fig. 11.9. The equilibrium concentration values in the two phases generally follow some simple proportional relationship, such as Henry's law, when the concentrations are low:

$$c_A^{1i} = H \, c_A^{2i} \qquad\qquad (11.36)$$

where the equilibrium interfacial concentration of species A in phase-1 is c_A^{1i} and c_A^{2i} in phase-2, represented by point I in Fig. 11.9. H is the proportionality constant, referred to as the partition (distribution) coefficient or Henry's law constant. Point B represents the two bulk concentrations of A. The equilibrium concentrations of A corresponding to bulk concentration at point B are c_A^{1*} and c_A^{2*} in phase 1 and phase 2, respectively.

11.4.1 Overall Mass Transfer Coefficient

It is indeed, as noted earlier, not very easy to quantify the concentrations of the diffusing species at the interface (c_A^{1i} and c_A^{2i}), and therefore, Eqs. (11.34) and (11.35) have limited utility. We need to replace them with concentrations that can be measured rather easily. To do this, we proceed as follows and introduce a more manageable mass transfer coefficient, which is based on concentrations and can be determined with accuracy. The overall mass transfer coefficient is based on an overall concentration difference between the bulk compositions of the two phases (see Fig. 11.9). Rearranging Eqs. (11.34) and (11.35) and multiplying Eq. (11.35) by the equilibrium constant H, we obtain:

Fig. 11.9 Equilibrium concentrations of species A between two phases

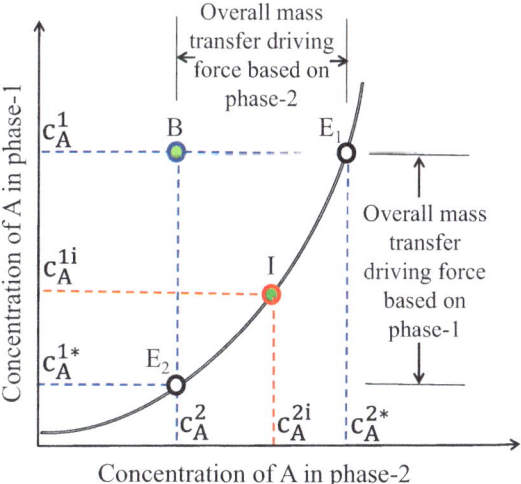

$$\left(c_A^1 - c_A^{1i}\right) = \frac{N_A}{k_1} \tag{11.37}$$

$$\left(Hc_A^{2i} - Hc_A^2\right) = \frac{HN_A}{k_2} \tag{11.38}$$

According to Eq. (11.36), $Hc_A^{2i} = c_A^{1i}$, and thus Eq. (11.38) may be rewritten as:

$$\left(c_A^{1i} - Hc_A^2\right) = \frac{HN_A}{k_2} \tag{11.39}$$

Upon addition of Eqs. (11.37) and (11.39), we obtain:

$$\left(c_A^1 - Hc_A^2\right) = N_A\left(\frac{1}{k_1} + \frac{H}{k_2}\right) \tag{11.40}$$

Again, if the equilibrium relationship shown in Fig. 11.9 holds, then

$$Hc_A^2 = c_A^{1*} \tag{11.41}$$

And Eq. (11.40) may be written as:

$$\left(c_A^1 - c_A^{1*}\right) = N_A\left(\frac{1}{k_1} + \frac{H}{k_2}\right) \tag{11.42}$$

Defining a new **overall mass transfer coefficient, K_1,** such that:

$$\frac{1}{K_1} = \frac{1}{k_1} + \frac{H}{k_2} \tag{11.43}$$

Now, the flux in terms of the new coefficient may be written as:

$$N_A = K_1\left(c_A^1 - c_A^{1*}\right) \tag{11.44}$$

where K_1 is the overall mass transfer coefficient based on the phase-1 driving force, c_A^1 is the bulk phase-1 concentration of A, and c_A^{1*} is the concentration of A in phase-1, which is in equilibrium with the bulk phase-2 concentration of A, c_A^2 (point E_2, Fig. 11.9). As defined before, k_1 and k_2 are individual film mass transfer coefficients.

A similar equation that relates the overall mass transfer coefficient based on phase-2 may be derived to obtain:

$$N_A = K_2\left(c_A^{2*} - c_A^2\right) \tag{11.45}$$

and

$$\frac{1}{K_2} = \frac{1}{Hk_1} + \frac{1}{k_2} \tag{11.46}$$

where K_2 is the overall mass transfer coefficient based on the phase-2 driving force, c_A^{2*} is the concentration of A in phase-2, which is in equilibrium with the bulk phase-1 concentration of A, c_A^1, (point E_1, Fig. 11.9), and c_A^2 is the bulk phase-2-concentration of A.

Check Your Understanding

3. Derive Eq. (11.45) following the steps used to obtain Eq. (11.44). ◀

(A) **Individual film mass transfer versus overall mass transfer resistances:** As may be noted, Eqs. (11.43) and (11.46) are written in terms of the reciprocal of mass transfer coefficients and thus represent resistances and show the relative contribution of the two individual phase resistances. For small values of H, Eq. (11.43) $1/K_1$ approaches $1/k_1$. This would suggest that the major resistance to mass transfer resides in phase-1 or that the process is phase-1 controlled. However, if H is large, with the aid of Eq. (11.46), it can be noted that $1/K_2$ equals $1/k_2$, and thus, phase-2 offers the main resistance to mass transfer and controls the process. Knowledge of the mass transfer rate controlling step is very useful in the design and operation of many industrial processes.

The units of the individual film mass transfer coefficient (k) and the overall mass transfer coefficient (K) depend on the units of the driving force used. Since concentration can be expressed in so many ways, care is required in using the correct units for the mass transfer coefficients. In the above Eq. (11.45), the units used for concentration are moles per volume, and the units of the mass transfer coefficient may thus be written as:

$$K \text{ or } k \left[=\right] \ \frac{kmol. \ m^{-2}. \ s^{-1}}{kmol. \ m^{-3}} = ms^{-1} \tag{11.47}$$

In the case of a gas–liquid system, it is more convenient to use partial pressures in place of concentrations for the gas phase, and Eq. (11.45) may be written as:

$$N_A = K_G \left(p_A^G - p_A^{G*}\right) \tag{11.48}$$

where K_G is the overall mass transfer coefficient based on the gas film, p_A^G is the bulk gas A partial pressure, and p_A^{G*} is the partial pressure of A in equilibrium with the bulk liquid phase concentration of A. In this case, the units of mass transfer coefficients may be derived as:

$$K \text{ or } k \left[=\right] \quad \frac{kmol. \ m^{-2}. \ s^{-1}}{Pa} = \frac{kmol. \ m^{-2}. \ s^{-1}}{N. \ m^{-2}}$$

$$= kmol. \ s^{-1}.N^{-1} \qquad\qquad (11.49)$$

(B) **Estimation of Mass Transfer Coefficients**: The mass transfer coefficient is a function of several variables, including phase behavior and related physico-chemical properties, system configuration, etc., and its measurement is not easy. Empirical correlations in terms of dimensionless numbers have been developed to estimate it for a variety of applications. As we have noted in earlier chapters, these dimensionless numbers are useful tools for engineers to solve practical problems. However, they often appear confusing to scientists who are not familiar with their utility. Generalized correlations for mass transfer that involve several dimensionless numbers are very common. A few simpler correlations that relate the Sherwood number (Sh) to the Reynolds (Re) and Schmidt (Sc) numbers in the following type of equation will be discussed and used next:

$$Sh = a \ Sc^b \ Re^c \qquad\qquad (11.50)$$

where a, b, and c are experimentally determined parameters. Each of the three dimensionless numbers shown in Eq. (11.50) represents the ratio of two transport phenomena and thus their relative importance, as shown below.

The Sherwood number (Sh) describes the ratio of mass transfer rate (K) to mass diffusivity:

$$Sh = \frac{Convective \ mass \ transfer}{Diffusive \ mass \ transfer} = \frac{k_c}{D_{AB}} \qquad\qquad (11.51)$$

where k is the mass transfer coefficient, D_{AB} is the diffusion coefficient of species "A" in the fluid, and L is the characteristic dimension.

The Schmidt number (Sc) describes the ratio of momentum diffusivity to mass diffusivity:

$$Sc = \frac{Momentum \ diffusivity}{Mass \ diffusivity} = \frac{\nu}{D_{AB}} = \frac{\eta}{\rho D_{AB}} \qquad\qquad (11.52)$$

where ν is the momentum diffusivity or kinematic viscosity, η is the viscosity, and ρ is the density. It is analogous to the Prandtl number (Pr), which represents the ratio of momentum diffusivity (ν) to thermal diffusivity (α) and was discussed earlier in the heat transfer chapter.

The Reynolds number (Re) describes the ratio of inertial forces to viscous forces, as discussed in the earlier chapter on fluid mechanics.

Some examples of frequently used correlations of the type represented by Eq. (11.50) for estimation of the mass transfer coefficient under specific situations are shown below:

- Forced convection past a solid sphere:

$$Sh = 2 + 0.6 \ Re^{1/2} Sc^{1/3} \tag{11.53}$$

- Laminar flow ($Re < 5 \times 10^5$) along a flat plate:

$$Sh = 0.32 \ Re^{1/2} Sc^{1/3} \tag{11.54}$$

- Turbulent flow ($Re > 5 \times 10^5$) along a flat plate:

$$Sh = 0.036 \ Re^{0.8} Sc^{1/3} \tag{11.55}$$

- Turbulent flow ($Re > 10^3$) through a circular tube:

$$Sh = 0.026 \ Re^{0.8} Sc^{1/3} \tag{11.56}$$

Notably, however, the accuracy of these correlations is not very high. It ranges from 30 to 50 percent, and experimental verification is often necessary for the design of a specific mass transfer operation.

Example

8. *Washed apples 8 cm in diameter were individually dried in a stream of dry air at 25 °C. If the air velocity is 5 m/s, compute the mass transfer coefficient at the apple surface.*

Solution: Assuming the system to mimic forced convection around a solid sphere, the relevant expression to be used is Eq. (11.53). The Reynolds and Schmidt numbers are calculated using the following properties at 25 °C:

Density, $\rho_{Air} = 1.1 \ kg/m^3$, viscosity, $\eta_{Air} = 1.7 \times 10^{-5} \ kg/m.s$, and D_{AB} for water in air (Table 11 1) $= 0.24 \times 10^{-4} \ m^2/s$

$$Re = \frac{\rho \, D \, v}{\eta} = \frac{(1.1 \ kg/m^3)(0.08 \ m)\left(5 \ \frac{m}{s}\right)}{1.7 \ x \ 10^{-5} \ kg/m.s} = 25{,}882$$

$$Sc = \frac{\eta}{\rho \, D_{AB}} = \frac{1.7 \ x \ 10^{-5} \ kg/m.s}{(1.1 \ kg/m^3)(0.24 \ x \ 10^{-4} \ m^2/s)} = 0.64$$

Using Eq. (11.53), the Sherwood number is obtained as follows:

$$Sh = \frac{kL}{D_{AB}} = 2 + 0.6 \; Re^{1/2} Sc^{1/3} = 2 + 0.6(25,882)^{1/2}(0.64)^{1/3} = 85.18$$

And then, $k = 85.18 \; x \frac{D_{AB}}{L} = 85.18 \; \left(\frac{0.24 \; x \; 10^{-4} \; m^2/s}{0.08 \; m} \right) = 0.026 \; m/s$ ◄

It should be emphasized here that in the interest of simplicity and time, discussion on diffusion and mass transfer has been brief, and many other aspects of the subject have not been addressed. For detailed information, more advanced texts should be consulted.

11.5   Transport Processes Analogy

The subject matters of momentum transfer, heat transfer, and mass transfer are collectively known as transport processes or transport phenomena. The commonality among these processes is that the flow in each process is driven by a concentration gradient in the direction of flow. This leads to mathematical analogy, although the physical mechanisms involved are very different. Let us examine them individually first and then generalize the observation that can be used in practice to simplify and understand processes of interest.

Newton's law of viscosity discussed earlier, Eq. (5.27) also represents the transport of momentum:

$$\tau_{yx} = \frac{F}{A} = -\eta \frac{d(v_x)}{dy} \tag{11.57}$$

where force per unit area (F/A) is the shear stress τ_{yx}. The first subscript denotes the direction of transfer of momentum, while the second subscript refers to the direction of velocity that generates momentum, and the two are in the perpendicular direction. For example, during fluid flow in a pipe, the momentum is transported in the radial direction from the center to the pipe wall. It can be shown that F/A indeed represents momentum per unit area or momentum flux:

$$\frac{F}{A} = \frac{Force}{Area} = \frac{Mass \; x \; Acceleration}{Area} = \frac{Mass \; x \; Velocity}{Area \; x \; Time} = \frac{Momentum}{Area \; x \; Time} \tag{11.58}$$

Now dividing and multiplying Eq. (11.57) by density (ρ)

$$\tau_{yx} = \frac{F}{A} = -\left(\frac{\eta}{\rho} \right) \frac{d(\rho v_x)}{dy} = -v \frac{d(\rho v_x)}{dy} \tag{11.59}$$

The ratio η/ρ or v is called the kinematic viscosity or momentum diffusivity, and the product of density and velocity divided by distance can be shown to be the momentum concentration gradient:

$$\frac{d(\rho v_x)}{dy} = \frac{d(Density \times Velocity)}{dy} = \frac{d[(Mass/Volume) \times Velocity]}{d}$$

$$= \frac{d\left(\frac{Momentum}{Volume}\right)}{d(distance)} = \frac{d(Momentum\ concentration)}{d(distance)} \tag{11.60}$$

$$= Momentum\ Concentration\ Gradient$$

Similarly, starting with Fourier's law for heat conduction presented earlier, Eq. (7.6) and shown below, can also be written in an aternative form as follows:

$$\frac{q_x}{A} = -k\frac{dT}{dx} \tag{11.61}$$

Now, multiplying and dividing the right side of Eq. (11.61) by the product of density (ρ) and specific heat (c_p), we obtain:

$$\frac{q_x}{A} = -\left(\frac{k}{\rho c_p}\right)\frac{d(\rho c_p T)}{dx} = -\alpha\frac{d(\rho c_p T)}{dx} \tag{11.62}$$

where q_x/A is the heat flux in the x direction and α ($=k/\rho c_p$) is the thermal diffusivity. The term $d(\rho c_p T)/dx$ can be shown to represent the energy concentration gradient:

$$\frac{d(\rho c_p T)}{dx} = \frac{d\left[\left(\frac{Mass}{Volume}\right)\left(\frac{Energy}{Mass \times Temp}\right)(Temp)\right]}{d(distance)} = \frac{d\left(\frac{Energy}{Volume}\right)}{d(distance)} \tag{11.63}$$

$$= Energy\ Concentration\ Gradient$$

We now have three analogous Eqs. (11.1), (11.59), and (11.62) for the transport of mass, momentum, and heat. Each relates the flux of a property to its concentration gradient multiplied by a coefficient. All three coefficients D and v and α have the same dimensions of length squared divided by time (e.g., m^2/s) and are referred to as diffusivities, as shown below, with their SI units:

- Mass diffusivity (m^2/s), D

  Momentum diffusivity (m^2/s), $v = \frac{\eta}{\rho}$, and

- Thermal diffusivity (m^2/s), $\alpha = k/\rho c_p$.

The similarity of mass, momentum, and heat equations shown above then leads to the following type of general transport equation of common utility:

$$\psi_y = -\delta \frac{d\Gamma}{dy} \tag{11.64}$$

where ψ_y represents the property flux in the y direction, δ is a proportionality constant called diffusivity, Γ is the property concentration, and y is the distance in the flow direction. The change in concentration with change in distance represents a concentration gradient or may also be considered as a driving force divided by resistance. The above equation may then be interpreted to apply to all the three transport processes with the following generic message:

$$\textit{Rate transport of property} = \left(\frac{\textit{Driving force}}{\textit{Resistance}}\right) \tag{11.65}$$

The above equation resembles Ohm's law in electricity, which we briefly discussed earlier in Chap. 10, Eq. (10.1) and is useful for making general observations regarding process control options.

Problems

11.1 What would you recommend for quick dissolution of a cube of sugar in a cup of hot coffee?

11.2 Examine the permeability values listed in Table 11.3 and explain why carbon dioxide permeability is much higher than nitrogen permeability, although the molecular mass of carbon dioxide is higher than that of nitrogen.

11.3 Explain:
(a) Why is whipped cream made with nitrous oxide and not with nitrogen?
(b) Why do the tanks of deep-sea divers contain a mixture of oxygen and helium and not air?

11.4 A cocci bacterial cell measures 0.2 mm in diameter. If it takes oxygen gas 10 seconds to reach the center of the cell, estimate the diffusion coefficient of oxygen.

11.5 Find the flux of glucose through a 0.1 cm thick, porous membrane. The concentration of glucose was 100 mg/L on one side and 300 mg/L on the other side. The diffusion coefficient of glucose may be assumed to be 0.2×10^{-5} m^2/s.

11.6 A modified atmosphere package of food at 25 °C contains a mixture of N_2 and O_2 at partial pressures of 0.9 and 0.1 atm, respectively. Assuming ideal gas behavior, compute the molar concentration, mole fraction, and mass fraction of each gas.

11.7 Beef strips 1.20 cm thick and 20.0 cm long are dry salted by maintaining the surface at a salt concentration of 0.5 kg/kg salt-free beef. If the initial salt concentration is negligible and the diffusivity of salt in beef is 0.5×10^{-6} cm^2/s, find the mass average salt concentration after 10 hours.

11.8 A long cylindrical extruded bread stick (diameter 1.44 cm) with an initial moisture content (c_o) of 0.0646 g water/g solids was rehydrated by immersion in distilled water in a circulating water bath maintained at 25 °C. At one-minute time intervals, the bread stick was removed from the water bath, blotted, dried with tissue paper to remove superficial water and weighed. The calculated average moisture content with time is shown in the table. If the equilibrium moisture content (c_1) of the bread stick determined by prolonged immersion in water is 0.647 g water/g solids, estimate the effective diffusion coefficient of water in the bread stick.

11.9 A stream of dry air at 25 °C and 20% RH flows over a 25 cm long and 20 cm wide surface of water. If the air velocity is 1 m/s and leaves the surface totally saturated, find the rate of water evaporation. The average kinematic viscosity of air may be taken as 15.0×10^{-6} m^2/s.

11.10 Find the rate of evaporation for the surface described in Problem 11.9 if the air comes in with a relative humidity of 30% and leaves 90% saturated.

Time (min)	g water/g solids (c_{avg})
1	0.4286
2	0.5071
3	0.5456
4	0.5726
5	0.6026

Bibliography

1. Berg HC (1993) Random Walks in Biology. Princeton University Press, Princeton, NJ
2. Crank J (1975) The mathematics of Diffusion.2nd edn. Oxford Science Publication
3. Cussler EL (2009) Diffusion Mass Transfer in Fluid Systems.3rd edn. Cambridge University Press, NY
4. Fuller EN, Schettler PD, Giddings JC (1966) Ind Eng Chem 58:19
5. Geankopolis, C.J. 2003. Transport Processes and Separation Process Principles. 4th edn.
6. Prentice Hall, NJ.
7. Hines, A.L. and Maddox, R.N. 1985. Mass Transfer: Fundamentals and Applications Prentice Hall, NJ

Refrigeration Systems

12

Your kitchen refrigerator (or any other) cools food by removing heat from its interior and dumping it into the kitchen. This means it moves heat from a cold place to a warm place. However, the second law of thermodynamics states that it is impossible for heat to flow from a cold body to a warm body, while the reverse will always happen, and heat (q) will spontaneously flow from a hot reservoir to a cold reservoir, as shown in Fig. 12.1. For heat (q_c) to flow from a cold area to a warmer area, work (w) must be performed such that, according to the first law of thermodynamics, the total heat (q_h) transferred is:

$$q_h = q_c + w \qquad (12.1)$$

The mechanical refrigeration units are designed to do this added work to transfer energy from a lower to a higher temperature and make cooling and freezing of our foods possible, as discussed below.

12.1 Mechanical Refrigeration Systems

Refrigeration units utilize the following familiar principles:

- Heat flows spontaneously from regions of high temperature to regions of low temperature. Work must be performed to make it flow the other way around.
- Isenthalpic (constant enthalpy) expansion of a fluid reduces its temperature (Joule-Thomson effect).
- Fluids can take up and release large amounts of heat with no temperature change when they undergo a phase change.
- Compressing a vapor increases its temperature.

© The Author(s), under exclusive license to Springer Nature Switzerland AG 2024   459
S. S. H. Rizvi, *Food Engineering Principles and Practices*,
https://doi.org/10.1007/978-3-031-34123-6_12

Fig. 12.1 Thermodynamics
and the refrigeration principle

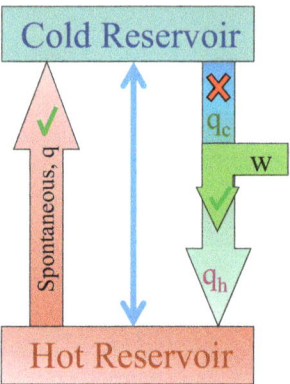

Let us see how these principles are combined in a refrigerator or any refrigeration system that works on the vapor compression principle.

12.1.1 The Vapor Compression Refrigeration Cycle

Figure 12.2 shows the main components of a vapor compression refrigeration system. They form a loop through which a fluid, called a refrigerant, is cycled. As it passes around this loop, the following takes place:

- At point 1 in Fig. 12.2, the refrigerant is a saturated mixture, largely liquid at low pressure and very cold.
- At 2, the refrigerant passes through an evaporator, which is a set of coils inside the refrigerator. It is essentially a heat exchanger in which the cold refrigerant absorbs heat from the less cold air from the surroundings. This absorbed heat vaporizes the refrigerant. Because of the latent heat of this phase change, the refrigerant can store a large amount of energy while remaining cold.
- At 3, the refrigerant is still quite cold but is now mostly or entirely vapor. Even though it is still cold, it carries a great deal of thermal energy.
- At 4, a compressor pumps and compresses the refrigerant. This is where the work is performed, and that work has the effect of increasing the internal energy and hence the temperature of the refrigerant.
- At 5, the refrigerant is at a high pressure and temperature but still vapor.
- At 6, the refrigerant passes through a condenser. This is another set of coils, another heat exchanger, located outside the refrigerator. Since the refrigerant is now hotter than the surroundings, heat radiates into those surroundings. As the refrigerant gives up heat, it condenses to a liquid. Because there is a phase change, a large amount of energy can be released while the refrigerant remains hot.
- At 7, the refrigerant is nearly all liquid and still at a high pressure.
- At 8, the refrigerant is restrained by a small opening called an expansion valve that may be an actual valve but is usually a capillary or a porous block. As the fluid passes through this opening, it undergoes irreversible adiabatic expansion:

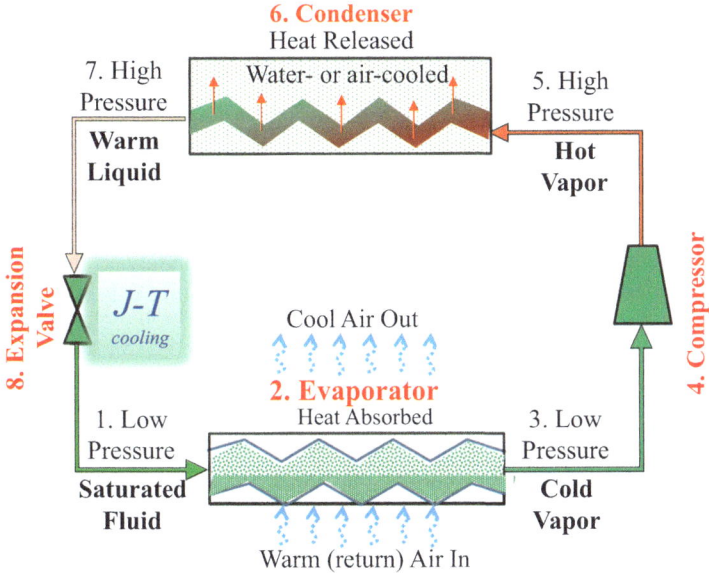

Fig. 12.2 Basics of a vapor compression refrigeration system

loses pressure and volume increase. As it expands, it performs work using its internal energy to do so. This results in cooling of the vapor.
- The expansion valve works as a throttling device to reduce the pressure following the Hagen-Poiseuille equation. The energy balance at this point can be written as:

$$\dot{m}_{ri}\left(h_i + \frac{v_i^2}{2g}\right) = \dot{m}_{ro}\left(h_o + \frac{v_0^2}{2g}\right) \tag{12.2}$$

where $\dot{m}_{ri}$ and $\dot{m}_{ro}$ are the mass flow rates, v_1 and v_o are the velocities, and h_i and h_o are enthalpies of the refrigerant into and out of the expansion valve, respectively. At steady state, the mass flow rate of fluid refrigerant into the expansion valve equals the mass flow rate of fluid refrigerant out. If accounting for all the ways energy enters and leaves the system and small changes in kinetic energy are neglected, the above expression reduces to:

$$h_i = h_o \tag{12.3}$$

where hi is the enthalpy coming into the expansion valve and ho is the enthalpy leaving it. This shows that the expansion process is essentially an **isenthalpic** (i.e., no change in the heat content of the system) process. The temperature change of a real gas or liquid when it expands adiabatically through a valve, as shown above, is described by the **Joule–Thomson coefficient**, μ_j, as $(\Delta T/\Delta P)_h$. For most real gases, μ is positive, and the temperature falls as pressure decreases adiabatically. For gases such as hydrogen and helium, μ is negative, and their temperature rises. For an ideal gas, it is zero. To summarize:

- Refrigerant at a low temperature absorbs heat from inside the refrigerator, and this energy is stored mostly as latent heat.
- Compression raises the temperature of the refrigerant so that it is warmer than the outside surroundings.
- The warm refrigerant delivers latent heat into the surroundings.
- Expansion drops the temperature of the refrigerant so that it is colder than the inside of the refrigerator.

Check Your Understanding

1. If an ideal gas is throttled to a lower pressure, would it also undergo temperature reduction or not? Explain. ◄

12.1.2 Pressure–Volume Changes

There are several ways of graphically tracing the state of the refrigerant. Figure 12.3 traces the pressure–volume changes of the refrigerant as it passes around the loop. The diagram is numbered to correspond to Fig. 12.2.

- From 1 to 3, the refrigerant passes through the evaporator, where it absorbs heat and vaporizes. This results in an increase in volume with no change in pressure, an isobaric process shown as a horizontal line.

Fig. 12.3 Pressure-volume diagram of a refrigerant

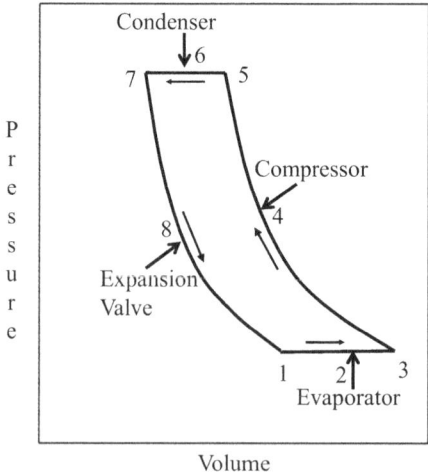

- From 3 to 5, the refrigerant is compressed. This results in a decrease in volume and an increase in pressure.
- From 5 to 7, the refrigerant passes through the condenser, where it releases heat and condenses to a fluid. This results in a decrease in volume with no pressure change, another isobaric process.
- From 7 to 1, the refrigerant passes through the expansion valve. This results in an increase in volume and a decrease in pressure.

12.2   The Mollier Diagram

The refrigeration cycle could also be plotted on a pressure–temperature diagram, but because of phase changes, the isobaric processes are also isothermal, and those parts of the cycle would plot as points rather than lines. The resulting diagram is not very informative. A better plot is done on a pressure-enthalpy diagram. Although only sensible heat produces temperature changes, both latent and sensible heats affect enthalpy, so the resulting plot reveals much about the process.

The Mollier diagram is a pressure-enthalpy diagram on which additional state variables have been drawn. It is constructed as follows:

- Pressure is scaled along the vertical axis (see Fig. 12.4), and states of equal pressure are represented by horizontal lines extending across the diagram.
- The enthalpy is scaled along the horizontal axis (see Fig. 12.5), and states of equal enthalpy are represented by vertical lines extending across the diagram.
- As shown in Fig. 12.6, a dome drawn on the diagram encloses the states where liquid and vapor can coexist. The left side of this dome is the saturated liquid line; to its left are liquid states. The right side of this dome is the saturated vapor line; to its right are vapor states. The critical point is at the top of the dome, and above

Fig. 12.4 Constant pressure lines

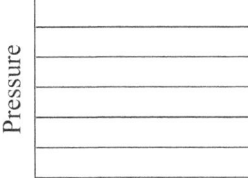

Fig. 12.5 Constant enthalpy lines

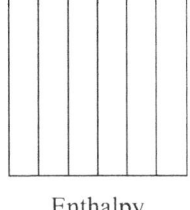

Enthalpy

Fig. 12.6 The two-phase
region

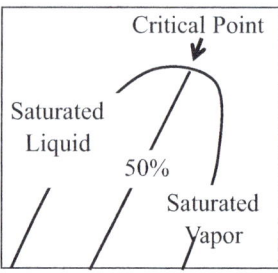

Fig. 12.7 Constant
temperature lines (isotherms)

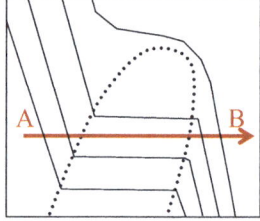

this, the refrigerant exists as a supercritical fluid. At states within the dome,
energy enters and leaves as latent heat. At states outside the dome, energy enters
and leaves as sensible heat.

- Lines are drawn inside the dome to indicate the percent of vapor. Figure 12.6
  shows just the 50% line.
- Figure 12.7 shows constant temperature lines (isotherms) on the diagram. To
  understand the shape of these lines, follow an isobaric process from state A to
  state B. At A, the refrigerant is a cold liquid. As enthalpy is added, the temperature rises, and the process crosses isotherms. When the two-phase region is
  reached, further enthalpy gains result in a phase change rather than a temperature
  rise. For this reason, the isotherms are horizontal and not crossed by the process.
  After leaving the two-phase region, further enthalpy increases the temperature,
  and again, the process crosses isotherms.
- Figure 12.8 shows lines of constant specific volume in the vapor phase where
  large volume changes can occur. Because volume is inversely proportional to
  pressure, the lines higher on the diagram represent lower specific volumes.
  Because volume is proportional to temperature, the lines slope upward.
- Figure 12.9 shows lines of constant entropy in the vapor phase. Entropy is a
  measure of the "disorganization" of a system. In other words, greater entropy
  implies that the positions of the molecules in the system are more random and less
  predictable. For example, the position of any molecule in a crystal is very
  predictable, so a crystal's entropy is low. The position of any molecule in a gas
  is very unpredictable, so the entropy of a gas is high. Decreasing the pressure of
  the gas further reduces predictability, so entropy is lower at the top of the
  diagram. Increasing its temperature reduces predictability, so the lines slope
  upward.

Fig. 12.8 Constant volume lines

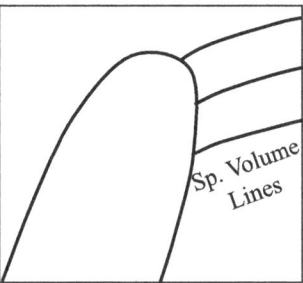

Fig. 12.9 Constant entropy lines

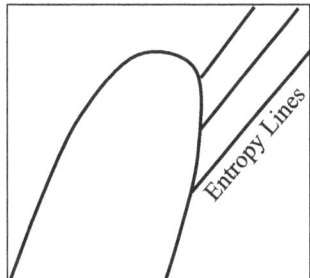

- A pressure-enthalpy diagram of Refrigerant-12 (dichlorodifluoromethane), commonly known as Freon-12 is shown in Appendix A.5.

12.2.1 A Refrigeration Cycle on a Mollier Diagram

Figure 12.10 shows a refrigeration process superimposed on a Mollier diagram. Points on the cyclic process have been numbered to correspond to Figs. 12.2 and 12.3. The following traces the cycle on this diagram:

- From 1 to 3, the refrigerant absorbs heat from the refrigerator and increases in enthalpy. While the phase change is taking place, the curve follows a constant temperature line (isotherm). After vaporization is complete, there will be some increase in temperature between the saturated vapor line and point 3.
- From 3 to 5, the refrigerant undergoes compression. The work performed during compression increases both the pressure and the enthalpy of the system. During this process, the temperature rise increases entropy, while the pressure rise reduces it. These two effects cancel each other, so compression takes place along a line of constant entropy.
- As the refrigerant enters the condenser, it cools until it reaches the saturated vapor line. From point 5 to point 7, it loses enthalpy isothermally as the vapor condenses.

Fig. 12.10 A refrigeration
process on a Mollier diagram

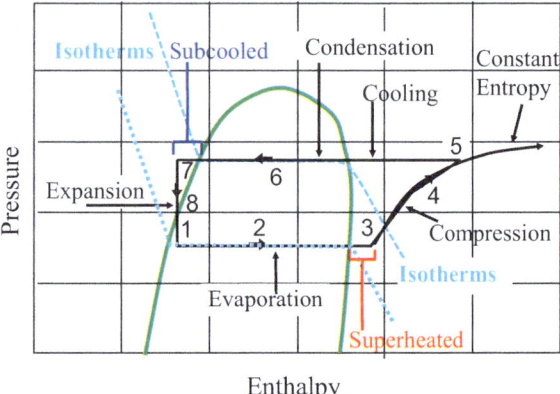

Enthalpy

- From 7 to 1, the refrigerant passes through the expansion valve. Because the valve is rigid, the expansion cannot work on the surroundings. Because it happens rapidly, there is little opportunity for heat to be exchanged with the surroundings. Thus, there is no enthalpy exchanged with the surroundings, and the expansion process must be isenthalpic and plotted as a vertical line on the diagram.
- Since the saturated vapor line slopes, the expansion from 7 to 1 moves into the dome, indicating that some liquid evaporates during expansion. This represents a loss in latent heat that will not be available for cooling. The less vaporization that occurs during expansion, the more efficient the refrigerator will be.

Example

1. *A refrigeration process is schematically shown on a Mollier diagram in Fig. 12.11. The portion of the liquid flashing to vapors for this process at the expansion valve and before entering the evaporator is:*

$$x = \frac{h_1 - h_0}{h_3 - h_0}$$

The saturated mixture then converts to all vapors to state 3 by absorbing energy in the evaporator. ◄

12.3   The Coefficient of Performance (COP)

The cycle in Fig. 12.10 is redrawn in Fig. 12.12 to illustrate the following:

- The horizontal distance from 1 to 3 represents the enthalpy that is removed from the refrigerator (Δh) by the refrigerant to achieve useful cooling:

Fig. 12.11 Vaporization and condensation in a refrigeration process

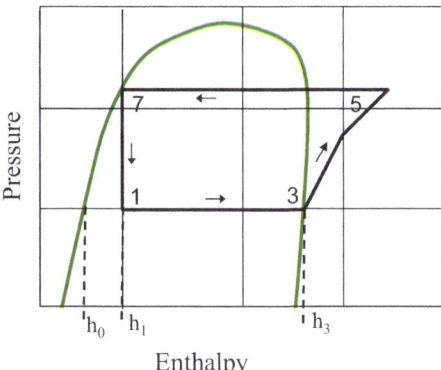

Fig. 12.12 The COP of a refrigeration unit

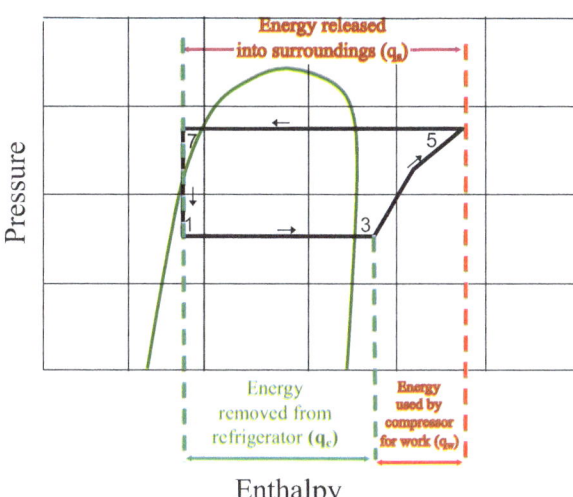

$$\Delta h_{\text{cool}} = \dot{m}_r(h_3 - h_1) \qquad (12.4)$$

where $\dot{m}_r$ is the refrigerant mass flow rate and h_1 and h_3 are enthalpies of refrigerant at states 1 and 3.

- The horizontal distance between 3 and 5 represents the enthalpy added by the work of the compressor and is evaluated as

$$\Delta h_{\text{com}} = \dot{m}_r(h_5 - h_3) \qquad (12.5)$$

where h_3 and h_5 are enthalpies of refrigerant at states 3 and 5.

- The horizontal distance from 5 to 7 represents the enthalpy released into the surroundings and equals:

$$\Delta h_{\text{surr}} = \dot{m}_r(h_5 - h_7) \tag{12.6}$$

As you can see, the enthalpy released to the surroundings will be the sum of the absorbed enthalpy and the work done, so for a constant refrigerant flow rate, we can write the following energy balance equation:

$$\begin{bmatrix} Energy \\ Removed \end{bmatrix} + \begin{bmatrix} Work \\ Done \end{bmatrix} = \begin{bmatrix} Energy \\ Released \end{bmatrix} \text{ or } (h_3 - h_1) + (h_5 - h_3)$$
$$= (h_5 - h_7) \tag{12.7}$$

In an efficient refrigerator, a small amount of work should be required to remove a large amount of enthalpy. The ratio of enthalpy removed to work done is called the **Coefficient of Performance (COP$_R$)** of the refrigeration unit and should be as large as possible, generally greater than unity.

$$COP_R = \frac{Enthalpy\ removed}{Enthalpy\ used\ for\ work} = \frac{(h_3 - h_1)}{(h_5 - h_3)} \tag{12.8}$$

Or more simply, referring to Fig. 12.1, it may be written as:

$$COP_R = \frac{Cooling\ effect}{Energy\ input} = \frac{q_c}{q_w} \tag{12.9}$$

Example

2. If 2 Joules of work is needed to remove 6.5 joules of enthalpy in your kitchen refrigerator, the coefficient of performance is:

$$COP_R = \frac{6.5\ Joules}{2.0\ Joules} = 3.25$$

The energy balance for such a system is: 2 Joules + 6.5 Joules = 8.5 Joules
Therefore, the kitchen receives 8.5 joules of heat for every 6.5 joules removed from the refrigerator. Keep your refrigerator doors closed, particularly in the summer. ◀

12.4   Refrigeration System: Major Components and Their Functions

There are two types of systems in commercial use. The most common type is a direct expansion system schematically illustrated in Fig. 12.13. The other type is a flooded refrigeration system in which a certain level of liquid refrigerant is maintained in the

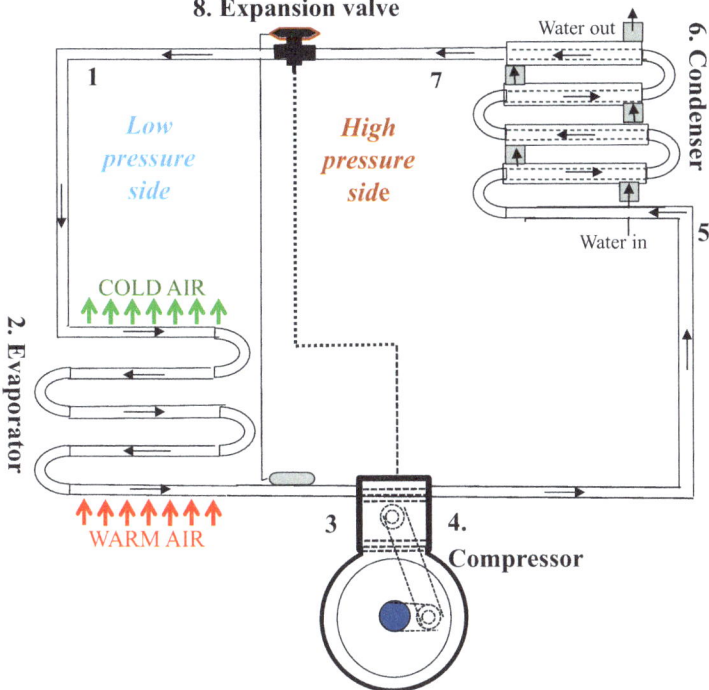

Fig. 12.13 Schematic of a commercial mechanical refrigeration system

evaporator. Flooded systems are economical for large operations because they require lower power input to the compressor. The low temperature attained by the system is determined by the internal evaporator pressure and the type of refrigerant used. In commercial systems, a thermostatic expansion valve controller is used to sense the temperature of the compressor suction line and modulates the expansion valve opening to prevent liquid refrigerant from entering the compressor. The major components and operation of a typical direct expansion system are described below:

12.4.1 Compressor

The compressor takes in low-pressure, low-temperature, and high-volume refrigerant vapor and compresses it adiabatically and reversibly into high-pressure and high-temperature vapor. Their principles of operation are the same as those discussed in Chap. 6. Since energy is added via work done by a reversible adiabatic process, the enthalpy of the system increases at constant entropy. The most common types of compressors in use include the following:

A. Reciprocating or piston-cylinder-type compressors are extensively used. They operate on the same principle that was discussed in Chap. 6 and are used in systems operating at high-pressure differentials.
B. Rotary compressors are employed in small systems such as household refrigerators due to their compactness.
C. Gear compressors are used in more precisely temperature-controlled systems due to their high mechanical efficiency.
D. Centrifugal compressors are used when refrigerants with large specific volumes are used. Unlike the positive displacement compressors mentioned above, throttling the discharge to adjust the volumetric capacity provides flexibility in operation; thus, centrifugal compressors are preferred for large units.

12.4.2 Condenser

The heat absorbed by the refrigerant in the evaporator plus the heat added by the compressor is removed by the condenser, which is basically a heat exchanger and condenses the high-pressure, high-temperature refrigerant vapor back into high-pressure and high-temperature liquid refrigerant via phase change. The process is accompanied by a decrease in both the enthalpy and entropy of the system. The most common types of condensers in general use include the following:

A. Air-cooled condensers consist of a finned or long, horizontal bare pipe that is cooled by ambient air, such as in a domestic refrigerator.
B. Shell-and-tube or tube-in-tube condensers are designed such that cold water runs through the tubes and refrigerant flows over the tubes either contained in a shell or each tube is contained in another tube.
C. Evaporative condensers are used in situations where high ambient temperatures create problems. In this setup, water is circulated over the condenser pipes as a spray or shower, causing part of the water to evaporate from the heat liberated by the condensing refrigerant.

12.4.3 Expansion Valve

The expansion valve is a throttling device and divides the high- and low-pressure sides of a refrigeration system. It is used to control the flow of liquid refrigerant into the evaporator. The following types of such valves are common:

A. Capillary tubes are used in household refrigerators and other small units. These consist of a short tube of very small diameter (0.5 to 2 mm). The pressure drop through the capillary is a function of its diameter and length. Since it is a fixed geometry, it does not respond to heat load change; thus, their actual size is calculated for a given refrigeration effect.

B. Thermal expansion valves can respond to the environment by increasing or decreasing the flow of refrigerant as necessary.
C. Float valves are generally used in flooded systems to maintain a constant level of refrigerant in the evaporator. It opens and closes in response to changes in refrigerant demand.

12.4.4 Evaporator

It is in the evaporator or boiler, as it is often called, that refrigerant converts from liquid to vapor at a constant pressure and picks up heat from the system as enthalpy of vaporization. Thus, both the enthalpy and entropy of the system increase during this process. Like a condenser, it is also basically a heat exchanger, designed differently for different duty. They come in a wide variety of shapes, sizes, and designs. Evaporators are classified based on the method of refrigerant travel, construction design, and type of control. Broadly, they fall into the following two categories:

A. Dry expansion evaporators consist of a set of short pipes (bare or finned) with headers or manifolds in which the flow of the refrigerant is controlled by the expansion valve based on the refrigeration load.
B. Flooded evaporators maintain a constant level of the liquid refrigerant inside by using float control. These are used in large refrigeration systems or in systems where load fluctuation is high. They frequently include an accumulator to further improve the performance.

Check Your Understanding

2. In what state may refrigerant exist in the evaporator? Explain. ◀

12.5   Refrigeration System Rating

Refrigeration systems in the USA have traditionally been rated in "tons." Before the advent of mechanical refrigeration, ice was used as a cooling medium. Like horsepower, the intuitive unit to compare the cooling capacity of newly developed mechanical refrigerators became the melting of a ton of ice. A ton of refrigeration is defined as a cooling capacity equivalent to that provided by ice melting at the rate of 1 ton per 24 hours. Since the latent heat of ice is approximately 144 Btu/lb_m and 1 ton is equal to 2000 lb_m:

$$1 \text{ ton of refrigeration} = 288,000 \text{ Btu/day} = 12,000 \text{ Btu/h} = 200 \text{ Btu/min}$$
$$= 3.517 \text{ kW}$$

Table 12.1 Properties of common refrigerants: in order of boiling point at atm pressure

Refrigerant	Formula	Boiling point (°C)	Freezing point (°C)	Refrigeration Capacity (kJ/kg) h_g $(-15\,°C)-h_f$ (30 °C)	Inflammable Limits, by Volume in Air, %
Water	H_2O	100.0	0.0	2299.4	Nonflammable
R-113	$CCl_2F-CClF_2$	47.6	−35.0	127.3	Nonflammable
R-11	CCl_3F	23.8	−111.0	156.2	Nonflammable
R-114	$CClF_2$	3.8	−94.0	99.2	Nonflammable
Butane	C_4H_{10}	−0.5	−138.0	292.0	1.6–6.5
R-12	CCl_2F_2	−29.8	−158.0	116.6	Nonflammable
Ammonia	NH_3	−33.3	−77.7	1102.2	16–25
R-22	$CHClF_2$	−40.8	−160.0	162.5	Nonflammable
Propane	C_3H_8	−42.1	−188.0	279.9	2.3–7.3
Carbon dioxide	CO_2	−78.4	−56.6	134.2	Nonflammable

In recognition of the fact that cooling capacity is a function of the temperature of the evaporating refrigerant, the following standard operating conditions have been adopted for comparison of different refrigeration systems:

- Refrigerant evaporation temperature: −15 °C.
- Refrigerant condensing temperature: 30 °C.

Based on the above criterion, Table 12.1 shows a list of some refrigerants in decreasing order of their boiling points.

Example

3. *A cold room is maintained at −1 °C by a vapor compression refrigeration unit using R-12* (**Freon-12**). *The evaporator and condenser temperatures are maintained at −10 °C and 30 °C, respectively, and the unit operates under saturated conditions. If the refrigeration load is 20 kW, determine:*

  (a) *High- and low-side pressures,*
  (b) *Refrigeration (cooling) capacity per unit weight of refrigerant,*
  (c) *Coefficient of performance (COP_R),*
  (d) *Refrigerant circulation rate,*
  (e) *Cooling water flow rate if it enters the condenser at 20 °C and leaves at 30 °C.*

   Solution:
   Referring to Fig. 12.14 for R-12, the system operating conditions have been identified on the graph, and their values are used to answer the questions as shown below:

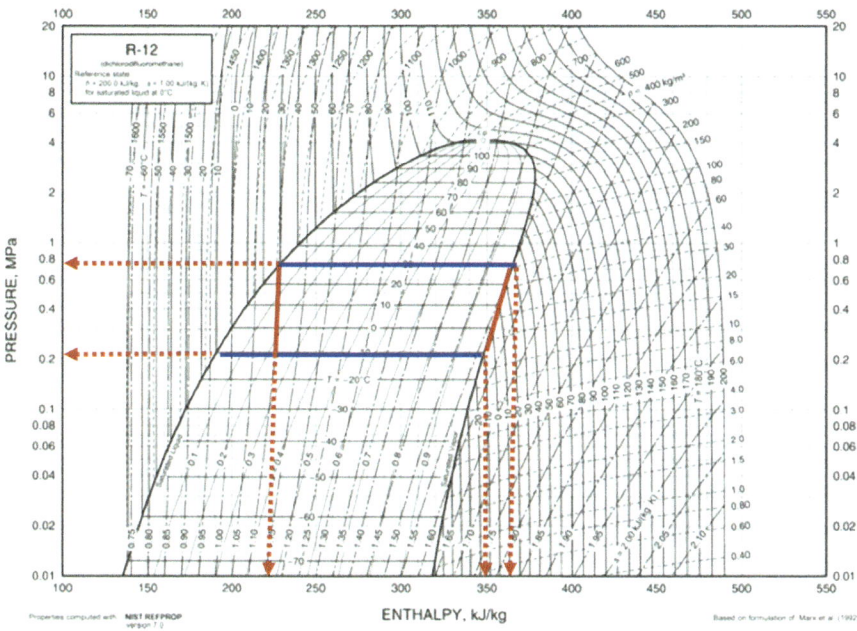

Fig. 12.14   Pressure-enthalpy diagram for Refrigerant-12

(a) *High- and low-side pressures = 0.70 MPa and 0.21 MPa.*

(b) *Refrigeration (cooling) capacity/unit weight of refrigerant = 345–225-*
= 120 kJ/kg.

(c) *Coefficient of performance (COP_R) = enthalpy removed/work done = 120/*
15 = 8.0.

(d) *Refrigerant circulation rate = Load/cooling capacity = (20 kJ/s)/(120 kJ/*
kg) = 0.17 kg/s.

(e) *Cooling water flow rate ($\dot{m}$):*

Total enthalpy to be removed = [(120 + 15) kJ/kg][0.17 kg/s] = 22.95 kJ/s
Then,

$$(\dot{m})\,(4.18\ kJ/kg\text{-}^\circ C)(30\text{-}20)\,^\circ C = 22.95\ kJ/s \quad or \quad \dot{m} = 0.55\ kg/s \quad \blacktriangleleft$$

Check Your Understanding

3. Could a refrigeration system be adopted as a space heating device, commonly
   called a heat pump? Explain how? ◄

12.6   Refrigerant Properties

As discussed above, refrigerants serve as a heat-transfer medium in a refrigeration system by undergoing phase changes. They pick up heat by evaporating at a low pressure and low temperature and give it up on condensing at a higher pressure and temperature. In principle, any fluid can be used as a refrigerant, including water. Although water has a very high enthalpy of vaporization that can absorb more energy per unit mass in the evaporator, its high freezing point makes it unsuitable in most refrigeration applications. In general, a refrigerant with a low boiling point and high enthalpy of vaporization is desired so that it absorbs or discharges heat in a refrigerator more efficiently. Some of the desirable characteristics for refrigerants include the following:

- Boiling and Freezing Points: These determine the service temperature at which the refrigerant can be used, and therefore, $T_{\text{boiling}} < T_{\text{service}}$ and $T_{\text{freezing}} < T_{\text{service}}$.
- Enthalpy of vaporization (kJ/kg): This should be as high as practical, minimizing the rate of refrigerant needed and the size of the unit.
- Evaporating pressure: At any given evaporator temperature, the saturation pressure should not be less than 1 atm to avoid leakage of air, moisture, and other noncondensable gases and for ease of leak detection.
- Critical temperature (T_c): To ensure complete condensation, $T_c > T_{\text{condenser}}$.
- Condensing pressure: It should be as low as possible to minimize the expense of the compressor, condenser, and pipes.
- Viscosity: To minimize the energy requirements for pumping and heat transfer, the viscosity should be small in both liquid and vapor phases.
- Safety: Refrigerants should be nontoxic, nonflammable, and noncorrosive.
- Ozone Depletion Potential: Refrigerant should be nonozone depleting. Since it depends on the presence of chlorine (i.e., CFCs and HCFCs) or bromine in the refrigerant molecules, some of them have been barred under the new regulation. Now, the Total Equivalent Warming Index (TEWI) is used to assess both the ozone depletion potential and the global warming potential of refrigerants, and a low value is indeed preferred.

12.7   Heat Pump vs Refrigeration Unit

Thus far, we have shown that a refrigeration unit removes heat, called the "cooling load," from a low-temperature source and dumps it into a high-temperature sink. On the other hand, a unit working in reverse is known as a heat pump, which is designed to transfer heat, called the "heating load," from a low-temperature to a high-temperature space. Figure 12.15 illustrates the basic principle behind the two operations. Heat pumps are used for space heating by taking heat at lower temperatures from the outside air, bodies of water, etc., and delivering it to heat homes and other spaces. They are an energy-efficient alternative to furnaces and air conditioners for heating and cooling buildings in different seasons. In other

Fig. 12.15 Refrigerator and heat pump operating principles

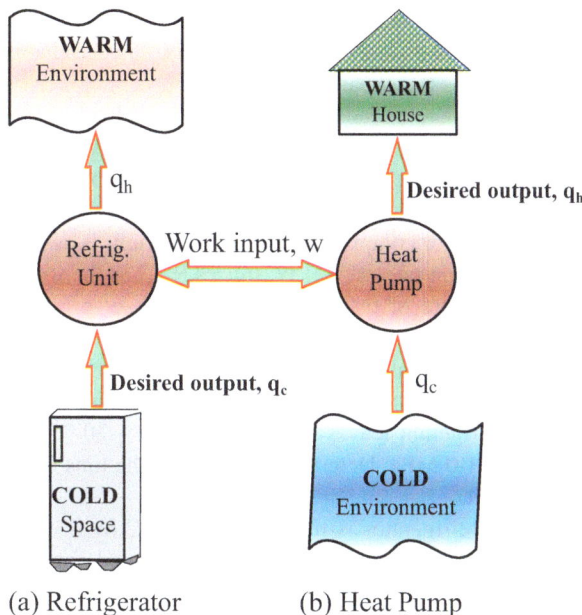

(a) Refrigerator   (b) Heat Pump

applications, energies from both the condenser and evaporators are used simultaneously, such as water heating and space cooling, or vice versa.

Similar to refrigeration systems, the performance of heat pumps is also expressed in terms of the coefficient of performance (COP$_{HP}$). Referring to Fig. 12.1, it may be written as:

$$COP_{Hp} = \frac{Heating\ Effect}{Work\ Input} = \frac{q_h}{w} \qquad (12.10)$$

Problems

12.1 Briefly, answer the following:

  (a) The temperature of walk-in coolers and freezers is generally kept lower than the temperature of reach-in units. Why?

  (b) Will a room with a refrigerator become hotter or cooler if the refrigerator's door is kept open?

  (c) What is a subcooled refrigerant? Is it desirable or undesirable?

  (d) Why and how did the term "ton of refrigeration originate"? Explain.

  (e) While inspecting a refrigerated food storage facility, the health inspectors record the temperature in the warmest part of the unit. Why?

12.2 In an R-12-based refrigeration unit, the condensing temperature is 30 °C, and the evaporator temperature is −5 °C. What is the expected pressure drop across the expansion device?

12.3 Estimate the additional refrigeration load in kWh/day that will be required for storing ten metric ton of apples at 1 °C in an existing cold room operating at the same temperature. The heat of respiration of apples may be assumed to be 1.8 kJ/kg-day.

12.4 A vapor compression refrigeration unit is designed to provide a steady-state cooling rate of 5.0 kW. It uses R-12 as the working fluid with an evaporator pressure of 0.1 MPa and a condenser pressure of 1 MPa. Determine the refrigerant flow rate needed.

12.5 Refrigerant R-32 is used as the working fluid at a mass flow rate of 3 kg/min in a vapor compression unit operating under saturated conditions. The refrigerant leaves the evaporator at −20 °C and has a condenser pressure of 1.0 MPa. Determine COP_R and the tons of refrigeration produced by the unit.

12.6 A refrigerator is maintained at 4 °C by a vapor compression refrigeration system using Freon-12 under saturated condition. The high- and low-side pressure are 0.95 MPa and 0.21 MPa, respectively. The refrigeration load is 30 kW. The cooling water enters the condenser at 20 °C and leaves at 30 °C. Determine:

(f) The temperatures of evaporator and condenser.
(g) Refrigeration (cooling) capacity per unit weight of refrigerant.
(h) Coefficient of performance (COPR).
(i) Refrigerant circulation rate.
(j) Mass flow rate of cooling water.

12.7 A controlled atmosphere storage room is maintained at −1 °C by a vapor compression refrigeration unit using R-12 (Freon-12). The evaporator and condenser temperatures are maintained at −5 and 30 °C, respectively, and the unit operates under saturated conditions. If the refrigeration load is 30 kW, calculate:

(k) Coefficient of performance (COP_R).
(l) Compressor power.
(m) Cooling water flow rate if it enters the condenser at 15 °C and leaves at 40 °C.

12.8 Repeat the above problem if the vapor coming out of the evaporator is superheated by 5 °C, the condenser subcools it by 10 °C and the compressor operates with an efficiency of 80%.

12.9 Two small refrigerators are operating simultaneously to keep products cold in a small shop and the coefficient of performance of refrigerator-1 is known to be 3.6. Collectively, the two refrigerators require 10 Joules to operate and release 48 Joules in the condenser. If refrigerator-1 requires twice as much energy input as refrigerator-2, calculate the coefficient of performance of the latter.

12.10 Show that under the same operating conditions, the COPs of a heat pump and a refrigeration unit are related as $COP_{HP} = COP_R + 1$.

References

1. American Society of Heating, Refrigerating, and Air Conditioning. 2006. Refrigeration Handbook. ASHRAE, New York
2. American Society of Heating, Refrigerating, and Air Conditioning. 1997. Handbook of Fundamentals. ASHRAE, New York
3. Cleland AC (1990) Food Refrigeration Processes Elsevier. , London
4. Dossat RJ (1996) Principle of Refrigeration.4th edn. Prentice Hall
5. Gosney WB (1982) Principles of Refrigeration. Cambridge University Press, Cambridge

Psychrometrics

13

Psychrometrics is concerned with the measurement of moisture in gases, particularly in air. Psychrometrics can be used to answer such questions as:

- What is the relative humidity of the air?
- At what temperature will moisture start to condense out of the air that may cause mold growth?
- How much energy must be expended to dry food?
- Will a food powder cake during storage?
- What is the A_w of food, and what does it mean?
- How can temperature and humidity be controlled for air conditioning?

In this chapter, you will learn how the important physical and thermodynamic properties of gas-vapor mixtures, such as moist air at a constant pressure, are presented in a graphical format called a **psychrometric chart,** and how to use them. First put together by Willis Carrier as a useful tool for designing air conditioning systems, psychrometric charts provide solutions to many practical and important problems related to food processing and preservation. They are very helpful in determining the heating, cooling, humidification, and dehumidification requirements of air for use in processes such as drying. A psychrometric chart is also used for determining what combination of these processes should be used to reach optimal environmental conditions in facilities.

A psychrometric chart shows the following seven properties of moist air:

- dry bulb temperature,
- wet bulb (or saturation) temperature,
- relative humidity,
- moisture content (or humidity ratio),
- dew point,
- specific enthalpy,
- specific volume.

© The Author(s), under exclusive license to Springer Nature Switzerland AG 2024
S. S. H. Rizvi, *Food Engineering Principles and Practices*,
https://doi.org/10.1007/978-3-031-34123-6_13

From knowledge of any two of the above parameters, the values of the other five can be found from the chart, shown in Appendix A.6. These parameters and their uses are explained in more detail below.

13.1 Definitions

This section contains definitions of the parameters that will be read from psychrometric charts.

13.1.1 Partial Pressure of Water in Air

In a mixture of gases and vapors, each component contributes to the total pressure of the mixture in proportion to its mole fraction. As discussed in Chap. 1 (Eq. 1.64), the partial pressure exerted by a component, such as water (P_w), is given by Dalton's law:

$$P_w = y_w P_T \qquad (13.1)$$

where y_w is the mole fraction of water and P_T is the total pressure of the mixture.

Example 1

A certain sample of air contains 2.67 kg of N_2, 0.625 kg of O_2, and 0.278 kg of H_2O at atmospheric pressure. What is the partial pressure of the water vapor?

Solution: *Nitrogen:* $\frac{2.67kg}{28\frac{kg}{kg\text{-}mole}} = 0.095$ *kg-moles*

Oxygen: $\frac{0.625kg}{32\frac{kg}{kg\text{-}mole}} = 0.020$ *kg-moles*

Water: $\frac{0.278kg}{18\frac{kg}{kg\text{-}mole}} = 0.015$ *kg-moles   Mole   fraction   of   water, y_w :*

$\frac{0.015}{0.095+0.020+0.015} = 0.115$ ◀

Partial pressure of water, P_w : (101.3 kPa) (0.115) = 11.65 kPa

13.1.2 Saturation Water Vapor Pressure

In a water–air system at any given temperature, when the rate of evaporation of water equals the rate of its condensation, the system is said to be in thermodynamic equilibrium, and the partial pressure of the water in the air is called the saturation partial pressure. This value is symbolized by P* and can be obtained for a given temperature directly from the saturated steam table. Some empirical relationships are also available in the literature for estimating the saturation pressure of water as a function of temperature.

Example 2

What is the saturation pressure of air at 21 °C? What is the mole fraction of water in air that is saturated at this temperature?

Solution: From the steam table, we read a saturation pressure of water, $P_g = 2.487$ kPa, at 21 °C. Since atmospheric pressure is $P_T = 101.3$ kPa, the mole fraction of water in this air is:

$$y_w = \frac{P_g}{P_T} = \frac{2.487\ kPa}{101.3\ kPa} = 0.025\ or\ 2.5\%\qquad \blacktriangleleft$$

13.1.3 Relative Humidity

Relative humidity as percent (%RH) is 100 times the ratio of the actual partial pressure of water in air to the saturation pressure of water at the same temperature. Instruments used for the measurement of relative humidity are called **hygrometers** and are based on different principles like the measurements wet and dry bulb, electrical resistance, dew point, among others.

Example 3

In a sample of air at 18 °C, water is found to have a partial pressure of 1.5 kPa. What is the relative humidity of this air?

Solution: From the steam table, we find that the saturation vapor pressure of water (P_g) at 18 °C is 2.064 kPa, so

$$RH = 100\ \frac{P_W}{P_g} = 100\left(\frac{1.5\ kPa}{2.064\ kPa}\right) = 73\%\qquad \blacktriangleleft$$

13.1.4 Moisture Content

Moisture content (mc), also called specific humidity or humidity ratio, is the ratio of the mass of water vapor (m_v) in the air to the mass of dry air (m_{da}). The average molecular weight of dry air is 28.94, and the molecular weight of water is 18.0, thus it may be written as:

$$\text{Moisture Content (mc)} = \frac{\text{mass of water}}{\text{mass of dry air}} \frac{(m_w)}{(m_{da})}$$

$$= \frac{(18.0)(\text{moles of water, } n_w)}{(28.94)(\text{moles of dry air, } n_{da})} \qquad (13.2)$$

When the masses of water and dry air are both given in the same units, mc will be a dimensionless number. Usually, however, it is expressed in units as kg/kg, g/kg, or lb_m/lb_m, where the numerator refers to the mass of water and the denominator refers to the mass of dry air.

If the mixture is assumed to follow the ideal gas law, the above equation may be converted to:

$$mc = 0.622 \frac{P_w}{P_a} = 0.622 \frac{P_w}{P_T - P_w} \qquad (13.3)$$

where Pa is the air partial pressure. P_T = Total pressure of moist air.

Example 4

A sample of moist air was dried completely in a desiccator and then 60.0 grams of water was allowed to evaporate into 1.50 kg of the dry air. What is the mass and moisture content of the mixture?

Solution: The mass of the mixture is 1.50 kg + 0.060 kg = 1.56 kg. The moisture content is:

$$mc = \frac{m_w}{m_{da}} = \frac{60 \text{ g}}{1.50 \text{ kg}} = 40.0 \frac{g \text{ water}}{kg \text{ dry air}} \text{ or } 0.040 \frac{kg \text{ water}}{kg \text{ dry air}} \qquad \blacktriangleleft$$

13.1.5 Moisture Content and Relative Humidity

As water makes up a small percentage of air in the mixture, the ratio of the moles of water to the moles of dry air maybe approximated as the mole fraction of water in air, so we can write Eq. (13.2) as:

$$mc = \frac{18.0}{28.94} \text{ (Mole fraction of water in the mixture)} = 0.622 \, y_w \qquad (13.4)$$

Furthermore, if the vapor-air is assumed to behave as an ideal mixture, x_w represents the ratio of the vapor pressure of the water to its saturation vapor pressure. Thus, the moisture content at a given temperature may be written as:

$$mc = 0.622 \left(\frac{P_w}{P_g}\right)_T \tag{13.5}$$

Check Your Understanding

1. At what relative humidity of air the rate of evaporation of water will be zero? ◀

When a material sealed in a container attains equilibrium at a given temperature, the relative humidity it exerts is called the **equilibrium relative humidity (ERH)** or **water activity (A_w)** and is expressed as:

$$A_w = \left(\frac{P_w}{P_g}\right)_T = \frac{\%ERH}{100} \tag{13.6}$$

For an ideal (dilute) solution following Raoult's law, it may be noted that A_w equals the mole fraction of water (x_w).

Water activity is a measure of the energy status of water in a system that is available to support chemical and biological processes. Not all water in food is available to exert vapor pressure and affect its water activity. Depending on the food composition, different amounts of water are energetically associated with food constituents, and thus, the partial pressure exerted by the food at equilibrium changes with its water content. A graph of moisture content against water activity is known as the moisture sorption isotherm. William James Scott was first to suggest in 1953 that microbial growth in food is governed by water activity and not, as was previously thought, by water content. In 1957, he established a limiting water activity level for microbial growth. Food processing and manufacturing protocols now routinely call for the use of water activity to reduce the susceptibility of food to microbial growth and quality loss.

13.1.6 Dry Bulb Temperature

When the temperature of air is measured with an ordinary thermometer, such as the top thermometer in Fig. 13.1, the reading is called the dry bulb temperature (T_{db}). It is the actual temperature of the air.

Fig. 13.1 Wet bulb and dry bulb thermometers

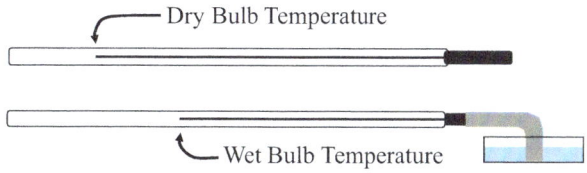

Dry Bulb Temperature

Wet Bulb Temperature

13.1.7 Wet Bulb Temperature

If the bulb on a thermometer is wrapped with a wet cloth like the lower one in Fig. 13.1, water will evaporate into the surrounding air that blows over it and cool the thermometer. The reading obtained under this circumstance is called the wet bulb temperature (T_{wb}), also known as the **saturation temperature**. It is the lowest temperature attained by evaporative cooling of the air. If the air is very dry, the rate of evaporation will be high, and the resulting cooling will result in a wet bulb temperature well below the dry bulb temperature. When the air becomes saturated (100% RH), evaporation will stop, and the wet bulb temperature will equal the dry bulb temperature.

13.1.8 Dew Point

For any given moisture content, cooling will eventually bring the air to a temperature at which the air becomes saturated with moisture. This temperature is called the dew point.

Example 5

A sample of air has a relative humidity of 45% at 24 °C. At what temperature will this air be saturated if the moisture content does not change?

Solution: From the steam table in the Appendix A.4, we find that at 24 °C, the saturation vapor pressure of water is 2.985 kPa. If air is at 45% saturation, it will have water vapor pressure of

$$P_a = \frac{45}{100}(2.985 \ kPa) = 1.343 \ kPa$$

According to the steam table, this is the vapor pressure of water at 11.3 °C. Thus, 11.3 °C is the saturation temperature or the dew point for this air. ◀

Check Your Understanding

2. Under which of the following conditions fog will start to form in ambient air?
(a) Air temp < dew point (b) Air temp = dew point (c) Air temp > dew point. ◀

13.1.9 Specific Volume

Specific volume is the volume occupied by one kg of air and is the reciprocal of its density.

13.1.10 Specific Enthalpy

Dry air has a specific heat of 1.005 kJ/kg-K. To evaporate water into air, enthalpy must be added to the water. Specific enthalpy of moist air is the enthalpy content of a unit mass of dry air plus the enthalpy needed to evaporate water into that air. It is computed by the formula:

$\overline{H}_{va}$ = (Specific enthalpy of air) + (Specific enthalpy of moisture)

$$\overline{H}_{va} = C_{da}(T_{db} - T_0) + \omega\overline{H}_v \tag{13.7}$$

where, $\overline{H}_{wa}$ = specific enthalpy of the vapor-air mixture

C_{da} = specific heat of dry air = 1.005 kJ/kg K
T_{db} = dry bulb temperature of the air
T_0 = reference temperature = 0 °C.
ω = moisture content of the air
$\overline{H}_v$ = specific enthalpy of the vapor in the air. This may be read from a steam table or approximated by the formula: $\overline{H}_v$ = 2501.4 + 1.88 $T_{dew\ point\ (°C)}$.

Example 6

6. *A sample of air at 30 °C has a relative humidity of 56%. What is the enthalpy content of the air?*

Solution: *At 30 °C, the steam table tells us that the saturation vapor pressure of water is 4.246 kPa. At a relative humidity of 56%, the water vapor pressure is:*

$$P_w = (\%RH/100)(P_g) = (0.56)(4.246\ kPa) = 2.335\ kPa$$

From the steam table, we find that at this vapor pressure, the specific enthalpy of vapor is 2538 kJ/kg. As an alternative, we note that the dew point of this air is 19.9 °C, so we can compute the specific enthalpy as:

$$\overline{H}_v = 2501.4 + 1.88(19.9) = 2538.8\ kJ/kg$$

The moisture content of this air is:

$$mc = 0.622\,\frac{P_w}{P_g} = 0.622\,\frac{2.335\ kPa}{4.246\ kPa} = 0.342\,\frac{kg\ moisture}{kg\ dry\ air}$$

The specific enthalpy of the dry air plus the moisture is:

$$\overline{H}_{va} = C_{da}(T_{db} - T_0) + \omega\overline{H}_v$$

$$\overline{H}_{va} = \left(1.005\,\frac{kJ}{kg-K}\right)(30-0)K + \left(0.342\,\frac{kg}{kg}\right)\left(2538.8\,\frac{kJ}{kg}\right) = 898.4\ kJ/kg \quad \blacktriangleleft$$

3. True or False.

- Heating of ambient air results in a decrease in the relative humidity of the air.
- Humidity will not affect the rate at which materials will gain or lose moisture.
- Water droplets in fog freeze on surfaces and turn into frost when surface temperature drops below freezing. ◀

13.2   Psychrometric Charts

A psychrometric chart is a graphical representation of selected thermophysical properties of air-water (gas-vapor) mixtures.

As you can see from the preceding section, there are many parameters that can be used to describe the moisture in air. It should also be clear that these parameters are all interrelated. This makes it possible to combine them all on a single psychrometric chart, which in its overall shape resembles a shoe. Psychrometric charts in SI and Imperial units are given in the Appendix A.6–1 and A.6–2. The following is a description of such charts and how water–air mixture properties are diagrammatically presented and evaluated.

13.2.1 Dry Bulb and Moisture Content Axes

The horizontal axis on a psychrometric chart represents the dry bulb or actual air temperature. The vertical axis represents the moisture content based on dry air (DA). Thus, any point on the chart represents a particular moisture content and temperature. For example, point A in Fig. 13.2 represents air with a 50 g moisture/kg dry air (DA) at 50 °C. All other parameters discussed in the last section can be read from a full psychrometric chart.

13.2.2 Saturation Moisture Content

By consulting a steam table, it is possible to construct a curve on this chart that represents the saturation moisture content as a function of temperature.

Plot a point on the dry bulb and moisture content chart that represents the saturation moisture content at 30 °C.

Fig. 13.2 Psychrometric chart axes

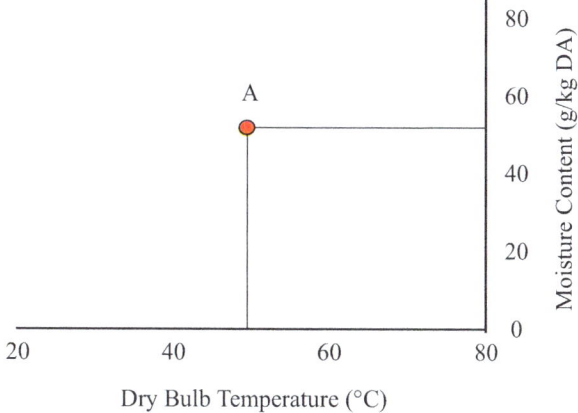

Solution: At 30 °C, the steam table gives a saturation vapor pressure of 4.246 kPa. The moisture content at this vapor pressure is:

$$mc = 0.622\left(\frac{4.246kPa}{101.3kPa}\right) = 0.0261\frac{kg}{kg} \text{ or } 26.1\frac{g}{kg}$$

This point (30 °C, 26.1 g/kg DA) is plotted in Fig. 13.3. ◄

If several calculations of this sort are plotted, the saturation line shown in Fig. 13.3 will result. Since higher moisture contents cannot exist at equilibrium, this curve is the left-most curve on the psychrometric chart.

Example 8

8. *Using the psychrometric chart in the Appendix A.6, determine the saturation moisture content of air at 25 °C.*

Fig. 13.3 Saturation line

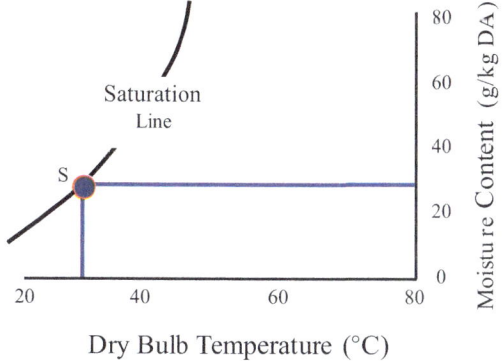

Solution: Find 25 °C on the horizontal axis and go up to the saturation line (the temperature is also recorded here). Go right to the vertical axis and read 20 g/kg DA as the saturation moisture content. ◄

13.2.3 Wet Bulb Temperature

In the last example, we found that dry bulb temperatures are written both on the horizontal axis and on the saturation line. You will also find lines sloping downward to the right, as shown in Fig. 13.4. These lines represent wet bulb temperatures. In Fig. 13.4, a vertical line has been drawn for a dry bulb temperature of 40 °C and a sloping line for a wet bulb temperature of 30 °C. Note that they meet at the saturation line because, in saturated air, no evaporation takes place, and the wet bulb is not cooled below the dry bulb temperature. Be careful not to confuse the wet bulb lines with the more steeply sloping specific volume lines. Wet and dry bulb thermometers can be used together to determine the moisture content of the air.

Example 9

A dry bulb thermometer reads 40 °C, while a wet bulb thermometer reads 30 °C. What is the moisture content of the air?

Solution: A vertical dotted line in Fig. 13.4 indicates the dry bulb temperature of 40 °C. Where it crosses the wet bulb temperature of 30 °C, a dotted horizontal line gives a moisture content of 23 g/kg DA. ◄

13.2.4 Relative Humidity

Figure 13.5 shows a series of curves that are roughly parallel to the saturation curve. These represent relative humidity. Note that if you follow any vertical line, it crosses

Fig. 13.4 Wet and dry bulb temperatures

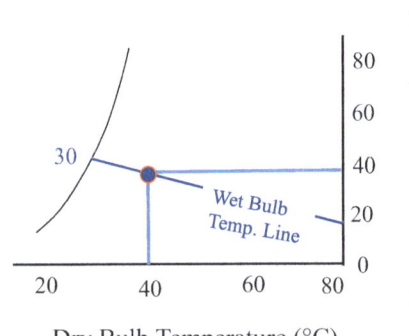

Fig. 13.5 Relative humidity lines

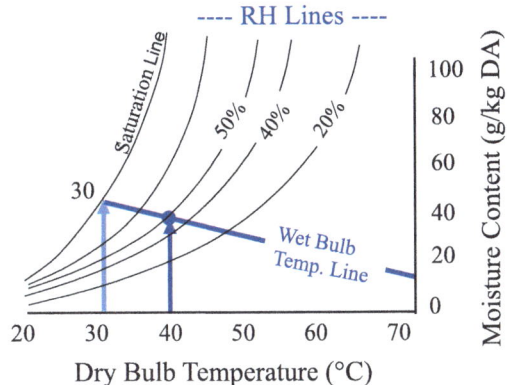

the 20% RH line, for example, 20% of the way between the horizontal axis and the saturation line. Thus, 20% relative humidity occurs when the moisture content is 20% of saturation.

Just as the moisture content can be determined with wet and dry bulb readings, relative humidity can be determined from the same data.

Example 10

What is the relative humidity when the wet bulb reads 30 °C and the dry bulb reads 40 °C?

Solution: The intersection of these temperatures occurs at 50% relative humidity. ◄

13.2.5 Dew Point

The saturation line, also called the dew point line, is the 100% relative humidity line. On this line, the wet and dry bulb temperatures have the same value. For any given moisture content, the dew point is the temperature at which that amount of moisture becomes the saturation moisture content and water will begin to condense out of the moist air. Thus, if you know the current moisture content, simply read across the chart to the saturation line and read the temperature.

Example 11

The current temperature is 53 °C, and the moisture content is 28 gm of moisture per kg of dry air. What is the dew point?

Solution: In Fig. 13.6, point P corresponds to a temperature of 53 °C and a moisture content of 28 gm/kg DA. Reading across to the saturation line, then down to the dry bulb temperature, we find that the dew point *is 30 °C.* ◄

Fig. 13.6 Dew point

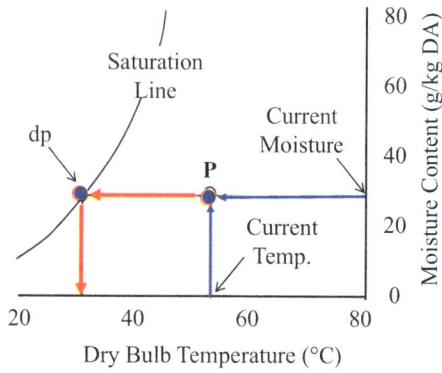

13.2.6 Specific Volume

It is an intensive property, defined as volume per unit mass (inverse of density). When density is very low such as when dealing with air–water mixture or steam, it is easier to work with specific volume numbers and visualize the process.

A second set of lines on the psychrometric charts, steeper than the wet bulb lines, represents the specific volume. These are shown in Fig. 13.7.

Example 12

If the dry and wet bulb temperatures are 58 and 40 °C, respectively, what volume is occupied by a kg of dry air?

Solution: As shown in Fig. 13.7, the intersection of the 40 °C wet bulb and 58 °C dry bulb lines fall on the line for 1.00 m³ per kg of dry air. Notice that, as you would expect, the specific volume increases with temperature. It also increases with moisture content. ◀

Fig. 13.7 Specific volume

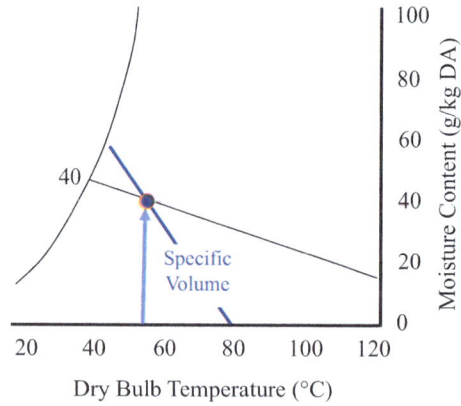

13.2.7 Specific Enthalpy

The specific enthalpy scale is located outside the main body of the psychrometric chart and right above the saturation line. The constant enthalpy lines run diagonally downward, very nearly coinciding with the wet bulb temperature lines. Some charts use the same lines for both the enthalpy and wet bulb temperature but apply a correction factor to the enthalpy. Figure 13.8 shows this feature. Other charts show a separate set of lines for each, such as the charts shown in the Appendix A.6, and no correction to the enthalpy value is needed.

Example 13

A quantity of air at 60 °C has a moisture content of 40 g/kg dry air. What is the enthalpy of this air?

Solution: The intersection of these lines falls on the line for 165.0 kJ/kg dry air. This is the quantity of heat needed to take a kg of dry air from 0 °C to 60 °C and then evaporate 40 gm of water into the air. ◄

13.3   Processes on Psychrometric Charts

Many processes can be traced on a psychrometric chart. Some of the processes of interest like sensible and adiabatic heating/cooling, humidification/dehumidification and mixing of different streams of moist air are described below.

13.3.1 Sensible Heating or Cooling

If air is heated or cooled without a change in moisture content, the process can be represented by a horizontal line, as shown in Fig. 13.9. The change in other parameters can then be read from this line.

Fig. 13.8 Specific enthalpy

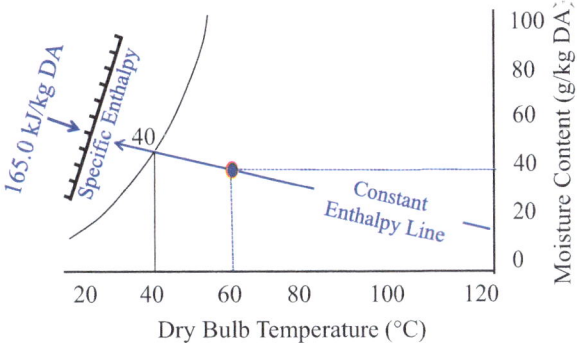

Fig. 13.9 Sensible heating/
cooling

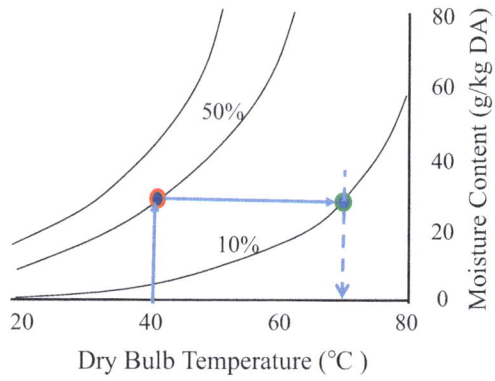

Example 14

Air at 40 °C and 50% RH is heated to 73 °C. What is the relative humidity at the new temperature?

   Solution: *As shown in Fig. 13.9, the point where the vertical line through 40 ° C crosses the 50% relative humidity curve at 24 g moisture/kg DA. Moving horizontally along this line to 73 °C brings you to 10% RH.* ◄

13.3.2 Dehumidification Through Cooling

When air is cooled below its dew point, moisture will condense out on cool surfaces and reduce the moisture content and thus the relative humidity of the air. This is what happens during dehumidification.

Example 15

Air at 53 °C and 60% relative humidity is passed between cold plates that cool it to 25 °C. After reaching equilibrium, the air was removed from the plates and heated to 41 °C. What is the final relative humidity of the air?

   Solution: *As shown in Fig. 13.10, air with 60% RH at 53 °C has a moisture content of 56 g/kg DA. When this is cooled to 25 °C, moisture condenses out until the air has only 19 g/kg DA of moisture. At 41 °C, this moisture content gives 40% RH.* ◄

Fig. 13.10 Dehumidification

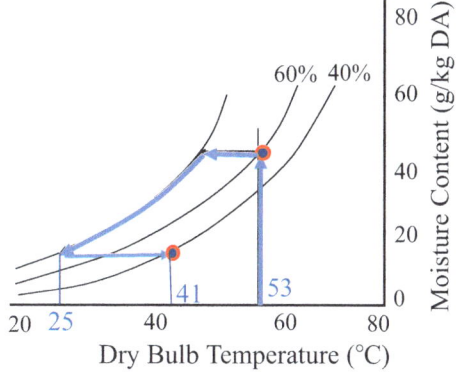

13.3.3 Adiabatic Cooling

If water is allowed to evaporate into air without any heat being exchanged with the surroundings, the enthalpy needed to cause evaporation must come from the air itself. This enthalpy loss will result in cooling of the air. However, this enthalpy will now be possessed by the moisture in the air, so the total enthalpy of the air will not change. Thus, the process will move along a constant enthalpy line, which is practically the same as a wet bulb temperature line.

Example 16

Air at 72 °C has a relative humidity of 15%. By passing over a wet surface, it is cooled adiabatically to 48 °C by flowing over a wet surface. How much moisture must be added per kg of dry air?

Solution: A shown in Fig. 13.11, at 72 °C and 15% RH, the moisture content is 32 g/kg. The enthalpy is 158.0 kJ/kg. At 48 °C, air with the same enthalpy reached ~ 60% RH and a moisture content of 42 g/kg. This is an increase of 42–32 = 10 g water evaporated per kg of dry air. ◄

Fig. 13.11 Adiabatic cooling

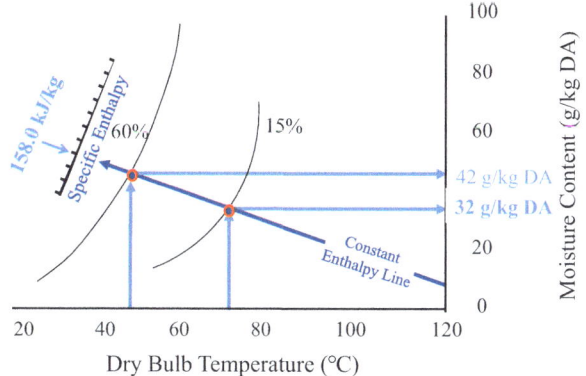

13.3.4 Mixing Two Air Streams

When two air streams are mixed, the moisture content and temperature of the mixture will be a weighted average of these quantities for the two streams. The masses of the streams will be the weighting factors. The equation for this computation is:

$$T_{A+B} = \frac{m_A T_A + m_B T_B}{m_A + m_B} \qquad (13.8)$$

where: m_i = Mass of stream i
T_i = Temperature of stream i

This averaging process can be performed graphically on the psychrometric chart by locating each stream on the chart and joining them with a line. The mixture will be a point (M) on the line between the two points. The distance from A to M will be proportional to the mass of B, while the distance from B to M will be proportional to the mass of A.

Example 17

Air stream A is at 40 °C with a moisture content of 20 g/kg DA and a flow rate of 15 kg/h. Air stream B is at 70 °C with a moisture content of 60 g/kg DA and a flow rate of 5 kg/h. What is the temperature and moisture content of a mixture of the two streams (M)?

Solution: Points A and B are plotted on Fig. 13.12 and joined by a line. The mass flow rates of the two streams are in a ratio of A:B = 15:5 = 3:1. The point M is plotted so that the distance A-M is 1/4 of the length of the line, while the distance M-B is 3/4 of the length. Since A contributes more to the mixture, we ensure that M is closer to A.

From the graph, we read 47.5 °C and 30 g/kg DA. ◀

Fig. 13.12 Mixing of streams

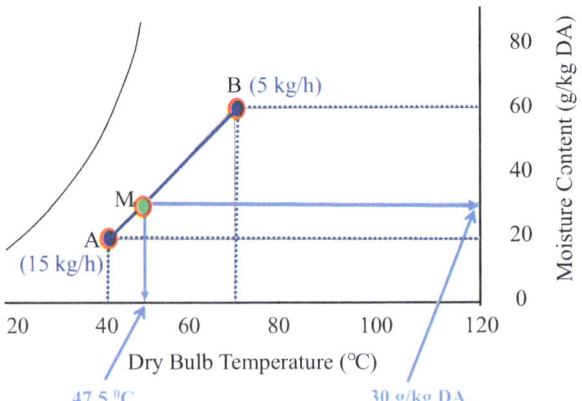

Problems

13.1 Ambient air at 20 °C and 80% relative humidity was heated to 90 °C. What is the specific volume, humidity ratio, specific enthalpy, and wet bulb temperature of the heated air?

13.2 A process requires air at 30 °C at a relative humidity of 40%. The available air is at 35 °C with a relative humidity of 60%. To make the available air suitable for the process, it is passed through a unit that dehumidifies by cooling to the dew point, followed by a reheating unit. The flow rate of air entering the device is 1 m³/s at 1 atm.

    a. What is the mass flow rate of the air?
    b. What is the rate of heat removal in the dehumidifying unit?
    c. What is the rate of moisture removal in the dehumidifying unit?
    d. What is the rate of heat addition in the heating section?
    e. What is the volumetric flow rate of the finished air?

13.3 Air is available at 10 °C and 80% relative humidity. To condition the air to 40 °C and 50% relative humidity, it is first passed at a flow rate of 1 m³/s through a heater and then a humidifier where steam at atmospheric pressure is injected.

    a. At what mass flow rate must steam be injected?
    b. At what temperature must the air leave the heating section?
    c. At what rate must heat energy be added in the heating section?
    d. What is the volumetric flow rate of the air leaving the humidifier?

13.4 Air at 50 °C and 10% RH was used for drying tortilla chips in a dryer. At the dryer exit, the air is found to be totally saturated. Compute the amount of water picked up by the air.

13.5 If air at 50 °C and 20% RH passes through high-moisture food at a flow rate of 10 m³/s, determine the maximum rate of moisture removal that can be achieved.

13.6 Air at 20 °C and 30% RH is heated to 60 °C before being used in a food dryer. During drying of a food material, 32 kg of water is removed every hour and the air leaves the dryer at 45 °C and 70% RH. What is the required flow rate of air?

13.7 An air stream having a volumetric flow rate of 2.5 m³/s, a relative humidity of 60%, and a dew point of 1 °C is electrically heated to 55 °C and used for drying 250 kg of cut fruits with an initial moisture content of 80% on wet basis. If the dryer is operated for one hour and the air exists the dryer saturated, compute the following:

    a. The amount of water removed and the final moisture content of the cut fruits on wet and dry basis.
    b. The electrical cost of heating, assuming electrical energy costs $ 0.08/kW-h.

13.8 On two separate nights, the sky is so clear that maximum radiational cooling can take place. Below are the sundown temperatures and relative humidity for each day. If the dew point is reached above the freezing temperature, dew will form. If the temperature falls below freezing before the dew point is reached, frost will form. Which day is more likely to have frost? At what temperature will the dew or frost start to form in each case?

Day	Temp. (°C)	RH
1	45	50%.
2	40	90%

13.9 To conserve energy, a 10 m³/s flow rate stream of exhaust air at 55 °C and 30% RH is mixed with 20 m³/s of ambient air at 30 °C and 60% RH. Determine the relative humidity, dry bulb temperature, moisture content, and dew point of the air mixture.

13.10 Air stream A flows at 0.65 m³/s at a temperature of 10 °C and 25% RH. Air stream B flows at 1.1 m³/s at a temperature of 20 °C and 50% RH. Both are at 1 atm pressure. The two streams are mixed adiabatically.

c. What is the volumetric flow rate of the mixture?
d. What is the temperature of the mixture?
e. What is the moisture content of the mixture?
f. What is the RH of the mixture?

Bibliography

1. American Society of Heating, Refrigerating and Air-Conditioning Engineers, Inc. 2021 ASHRAE Handbook: Fundamentals. ASHRAE, Atlanta, GA
2. Brooker DB (1967) Mathematical models for psychrometric chart. Trans ASAE 10(4):558
3. Henderson SM, Perry PL, Young JH (1997) Principles of Process Engineering.4th edn. ASAE, St. Joseph, MI

Correction to: Food Engineering Principles and Practices

Correction to: S. S. H. Rizvi, *Food Engineering Principles and Practices*, https://doi.org/10.1007/978-3-031-34123-6

The original version of this book has been revised. The original version of this book was published with the below errors due to an oversight on the part of Springer Nature, and it was corrected as given in the below corrections:

FRONTMATTER:

The following content **"Prefixes in Common Use Before Units"** has been moved from Page ix to xxiv.

Acknowledgement section has been included in Page ix.

Abbreviations:

The following abbreviation has been updated from

m Mass (kg, lb$_m$) to ṁ Mass flow rate (kg/s, lb$_m$/h)

The following abbreviation has been included:

q̇ Heat flow rate (W, BTU/h)

The following abbreviation has been updated from

R Gas constant [kJ/(mol-K)] to

R Gas constant [kJ/(mol-K)] or [l.- atm/mol-K]

Greek Symbols:

The following Greek Symbol has been updated from

α Thermal diffusivity (k/ρC$_p$,m$_2$ /s . ft$_2$ /h), kinetic energy correction factor to

α Thermal diffusivity (k/ρC$_p$,m$_2$ /s . , or ft$_2$ /h), kinetic energy correction factor

---

The updated version of these chapters can be found at
https://doi.org/10.1007/978-3-031-34123-6

Chapter 1:
Equation (1.13) has been updated from

$$\text{Molaity} = \frac{\text{Number solute moles}}{\text{Kilogram of solvent}}$$

to

$$\text{Molality} = \frac{\text{Number of solute moles}}{\text{Kilogram of solvent}}$$

The following sentence in page 14 has been updated from
D. Molar Volume is defined as the volume occupied by 1 M of a substance.
to
D. Molar Volume is defined as the volume occupied by 1 mole of a substance.
The following sentence in page 15 has been updated from
The density of a mixture may be approximated by averaging the densities as shown below:
to
The density of a mixture with no volume change on mixing may be approximated by averaging the densities as shown below:
Page 17: Price density changed to Piece density.
Fig. 1.9 has been updated with the following figure:

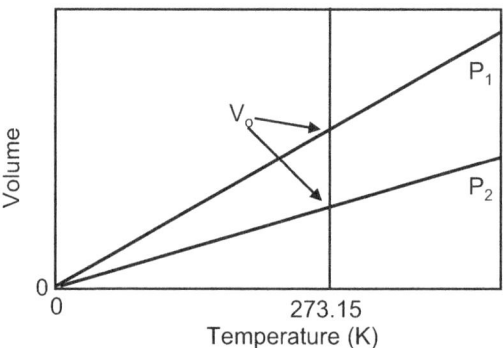

The following equation in Example 12 has been updated from

$$V_2 = \frac{(3\ atm)(50\ ft^3)}{(293.15\ K)}\left(\frac{323.15\ K}{1\ atm}\right) = 163\ ft^3$$

to

$$V_2 = \frac{(3\ atm)(50\ ft^3)}{(293.15\ K)}\left(\frac{323.15\ K}{1\ atm}\right) = 165.35\ ft^3$$

The following equation in Example 12 has been updated from

$$V_2 = \frac{(3\ atm)(50\ ft^3)}{(293.15\ K)}\left(\frac{323.15\ K}{1\ atm}\right) = 163\ ft^3$$

to

$$V_2 = \frac{(3\ atm)(50\ ft^3)}{(293.15\ K)}\left(\frac{323.15\ K}{1\ atm}\right) = 165.35\ ft^3$$

The following equation in Example 17 has been updated from

$$
\begin{aligned}
V &= (z\,n\,RT)/P \\
&= [(0.96)\,(1\ kg/18\ kg/kg\text{-}mole)(8.31\ kJ/kg\text{-}mole\text{-}K)(693.17\ K)]/ \\
&\quad [(30\ bar)]\left[(1\ bar/10^5\ N/m^2)(10^3 N\cdot m)/1\ kJ)\right] = 0.102\ m^3
\end{aligned}
$$

to

$$
\begin{aligned}
V &= (z\,n\,RT)/P \\
&= [(0.96)\,(1\ kg/18\ kg/kg\text{-}mole)(8.31\ kJ/kg\text{-}mole\text{-}K)(653.17\ K)]/ \\
&\quad [(30\ bar)]\left[(1\ bar/10^5\ N/m^2)(10^3 N\cdot m)/1\ kJ)\right] = 0.102\ m^3
\end{aligned}
$$

Fig. 1.17 has been updated with the following figure:

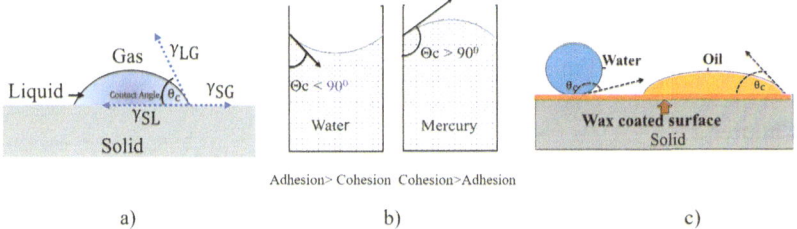

a) b) c)

Equation (1.63a) has been updated from

$$P_1 = x_1 P_g$$

to

$$P_1 = x_1 P*$$

Equation (1.63b) has been updated from

$$P_1 = (1 - x_2)P_g \quad \text{or} \quad \Delta P = P_g - P_1 = x_2\, P_g$$

to

$$P_1 = (1 - x_2)P^* \quad \text{or} \quad \Delta P = P^* - P_1 = x_2\, P^*$$

Equation (1.64) has been updated from

$$P_1 = y_1 P_T = x_1 P_g$$

to

$$P_1 = y_1 P_T = x_1 P^*$$

The following sentence in Problems 1.15 has been updated from

The percentage of bakers is yet another method used to express food formulations and recipes.

to

The baker's percentage is yet another method used to express food formulations and recipes.

The following sentence in Problems 1.39 has been updated from

If 4.0 g of a seasoning (density 900 kg/m^3) is milled to reduce its average size from 5 micrometers to 1 micrometer, compute the percent change in the total surface area of the flour. (Assume spherical shape).

to

If 4.0 g of a seasoning (density 900 kg/m^3) is milled to reduce its average size from 5 micrometers to 1 micrometer, compute the percent change in the total surface area of the seasoning. (Assume spherical shape).

Chapter 2

The following equation in Page 101 has been updated from

$$1 = 0.375 + 0.625 \qquad\qquad => 1 = 1$$
$$0.62 = 0.32(0.375) + 0.80(0.625) \quad => 0.62 = 0.62$$

to

$$1 = 0.375 + 0.625 \qquad\qquad \rightarrow 1 = 1$$
$$0.62 = 0.32(0.375) + 0.80(0.625) \quad \rightarrow 0.62 = 0.62$$

Fig. 2.31 has been updated with the following figure:

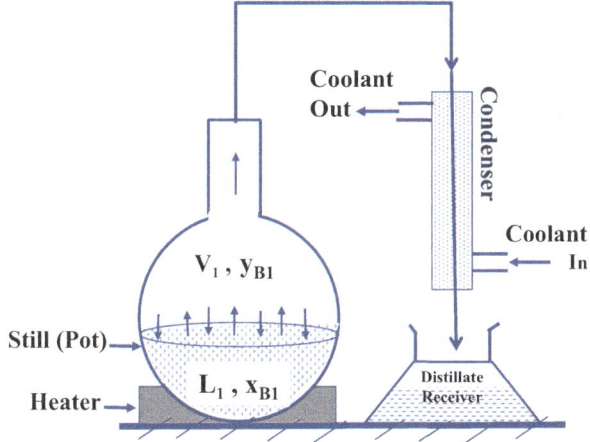

Fig. 2.32 has been updated with the following figure:

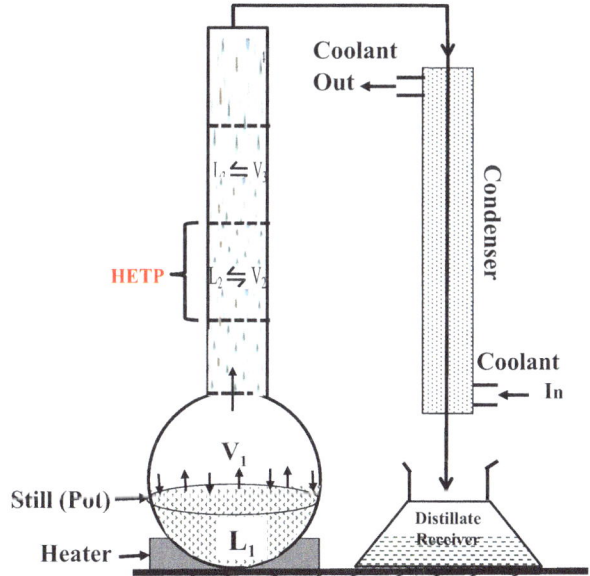

Figure in Page 112 has been updated with the following figure:

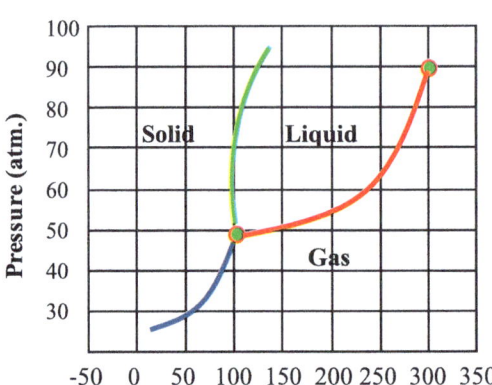

Chapter 3
The following equation in Page 120 has been updated from

$$V = \pi r^2 x = (0.25 \text{ in})^2 (5 \text{ in}) = 0.98 \text{ in}^3$$

$$m = V\rho = 0.98 \text{ in}^3 \left(\frac{62.4 \ lb_m/ft^3}{(12 \ in/ft)^3} \right) = 0.0355 \ lb_m$$

$$E_K = \frac{1}{2} \frac{mv^2}{2g_c} = \frac{1}{2} \frac{(0.355 \ lb_m)(10 \ ft/s)^2}{(32.17 \ ft \cdot lb_m/lb_f \cdot s^2)} = 0.0552 \ ft \cdot lb_f$$

to

$$V = \pi r^2 x = (3.14)(0.25 \text{ in})^2 (5.0 \text{ in}) = 0.98 \text{ in}^3$$

$$m = V\rho = 0.98 \text{ in}^3 \left(\frac{62.4 \ lb_m/ft^3}{(12.0 \ in/ft)^3} \right) = 0.0355 \ lb_m$$

$$E_K = \frac{mv^2}{2g_c} = \frac{1}{2} \frac{(0.0355 \ lb_m)(10 \ ft/s)^2}{(32.17 \ ft \cdot lb_m/lb_f \cdot s^2)} = 0.0552 \ ft \cdot lb_f$$

The following equation in Page 142 has been updated from

$$\Delta h = (4.10) \int_0^{30} dT - (5.30 \times 10^{-3}) \int_0^{30} T dT + (9.95 \times 10^{-4}) \int_0^{30} T^2 dT$$
$$= (4.10)\, T - (5.30 \times 10^{-3})\, (1/2)\, T^2 + (9.95 \times 10^{-4})\, (1/3)\, T^3$$
$$= 4.10\, (30\text{–}0) - (5.30 \times 10^{-3})(30-0)^2/2 + (9.95 \times 10^{-4})(30-0)^3/3$$
$$= 132\ kJ/kg$$

to

$$\Delta \overline{h} = (4.10) \int_0^{30} dT - (5.30 \times 10^{-3}) \int_0^{30} T dT + (9.95 \times 10^{-4}) \int_0^{30} T^2 dT$$
$$= (4.10)\, T - (5.30 \times 10^{-3})\, (1/2)\, T^2 + (9.95 \times 10^{-4})\, (1/3)\, T^3$$
$$= 4.10\, (30\text{–}0) - (5.30 \times 10^{-3})(30-0)^2/2 + (9.95 \times 10^{-4})(30-0)^3/3$$
$$= 132\ kJ/kg$$

The following equation in Page 146 has been updated from

$$\Delta H = m\, C_{pi} \Delta T + m\overline{H}_{sf} + m\, C_{pw} \Delta T$$
$$= (10)\ kg\, [(2.12)(0+20) + (334.1) + (4.18)(80-0)]\ kJ/kg = 7116\ kJ$$

$$\Delta h = [\text{Latent heat}] + [\text{Sensible heat}]$$

to

$$\Delta H = m\, C_{pi} \Delta T + m\overline{H}_{sf} + m\, C_{pw} \Delta T$$
$$= (10)\ kg\, [(2.12)(0+20) + (334.1) + (4.18)(80-0)]\ kJ/kg = 7109.0\ kJ$$

$$\Delta \overline{H} = [\text{Latent heat}] + [\text{Sensible heat}]$$

The following equation in Page 147 has been updated from

$$At - 40\,^{\circ}C : \Delta \overline{H} = -173.7 - 39.14(0.69 + 1.31 + 0.34) = -298.6\ kJ/kg$$

to

$$At - 40\,^{\circ}C : \Delta \overline{H} = -207.14 - 39.14(0.69 + 1.31 + 0.34) = -298.6\ kJ/kg$$

The following equation in Page 148 has been updated from

$$\overline{H} = 298.4 - 184.1 = 114.3 \ kJ/kg$$

$$H = (6 \ kg)114.3\frac{kJ}{kg} = 685.8 \ kJ$$

to

$$\overline{H} = 298.6 - 184.1 = 114.5 \ kJ/kg$$

$$H = (6 \ kg)114.5\frac{kJ}{kg} = 687.0 \ kJ$$

Fig. 3.13 has been updated with the following figure:

A. Specific Heat (kJ/kg-K)

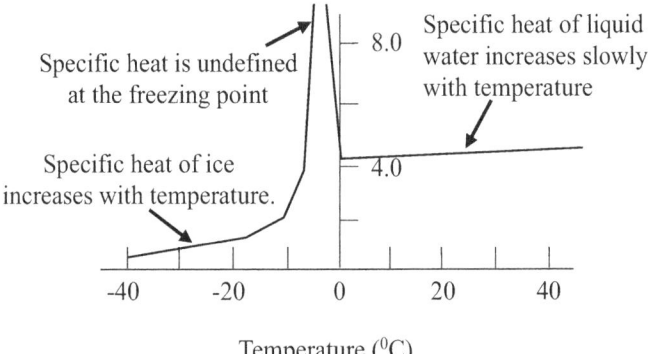

Temperature (^{0}C)

B. Specific Enthalpy (kJ/kg)

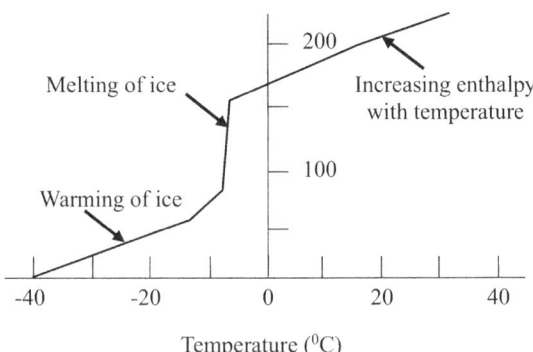

Temperature (^{0}C)

The following equation in Page 150 has been updated from

$$C_p \left(\frac{kJ}{kg}\right) = 9.86 + 0.128T + 0.16T^2 \text{ where } T \text{ is in } (^\circ C)$$

to

$$C_p(kJ/kg\text{-}K) = 9.86 + 0.128T + 0.16T^2 \text{ where } T \text{ is in } (^\circ C)$$

Chapter 5:
Fig. 5.9 has been updated with the following figure:

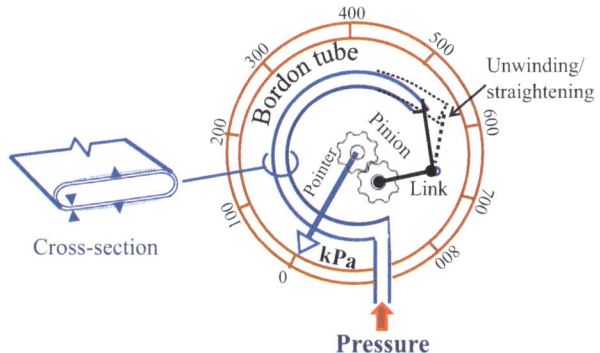

Fig. 5.10 has been updated with the following figure:

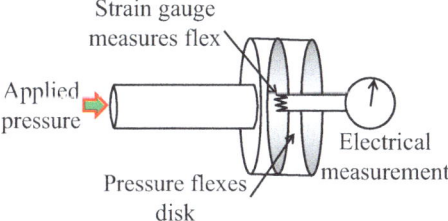

The following equation in Page 191 has been updated from
kP$_a$g and kP$_a$a
to
kPag and kPaa

The following equation in Page 195 has been updated from

$$\text{Volumetric flow rate}\left(\dot{V}\right) = \frac{V}{t} = \frac{Ax}{t} = A \cdot v$$

to

$$\text{Volumetric flow rate}\left(\dot{V}\right) = \frac{V}{t} = \frac{Ax}{t} = A \cdot v$$

The following equation in Page 216 has been updated from

$$\text{Re} = \frac{\rho d <v>}{\eta} = \frac{\left(1000\frac{kg}{m^3}\right)\,(0.032\ m)\,\left(1.04\frac{m}{s}\right)}{\left(1.005 \times 10^{-3}\frac{kg}{ms}\right)} = 3.31 \times 10^{-4}$$

to

$$\text{Re} = \frac{\rho d <v>}{\eta} = \frac{\left(1000\frac{kg}{m^3}\right)\,(0.032\ m)\,\left(1.04\frac{m}{s}\right)}{\left(1.005 \times 10^{-3}\frac{kg}{ms}\right)} = 3.31 \times 10^{4}$$

Fig. 5.42 has been updated with the following figure:

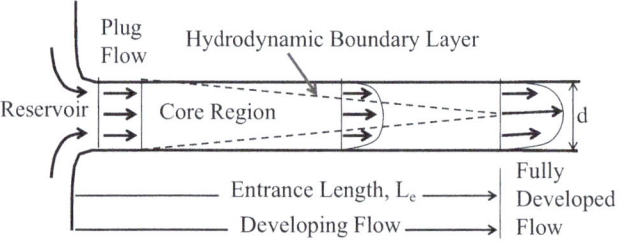

Fig. 5.43 has been updated with the following figure:

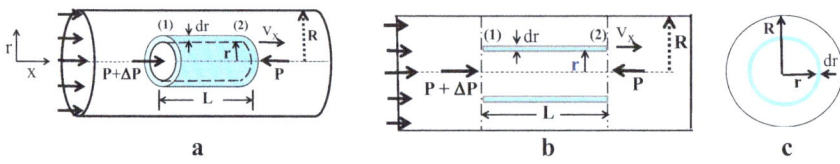

The following equation in Page 238 has been updated from

$$v_{max} = 2 <v> = 2\left(0.085\frac{m}{s}\right) = 0.172 \ m/s$$

$$L = (v_{max})(t) = (0.172)(20 \ s) = 3.44 \ m$$

to

$$v_{max} = 2 <v> = 2\left(0.085\frac{m}{s}\right) = 0.170 \ m/s$$

$$L = (v_{max})(t) = (0.170)(20 \ s) = 3.40 \ m$$

The following equation in Page 243 has been updated from

$$Re = \frac{\rho D <v>}{\eta} = \left(1200 \ \frac{kg}{m^3}\right)(50x \ 10^{-3}m)\left(2.12\frac{m}{s}\right)\left(1.2 \ x \ 10^3 \frac{m.s}{kg}\right)$$

$$= 1.5 \ x \ 10^5 \Rightarrow Turbulent \ flow$$

to

$$Re = \frac{\rho D <v>}{\eta} = \left(1200 \ \frac{kg}{m^3}\right)(50x \ 10^{-3}m)\left(2.12\frac{m}{s}\right)/\left(1.2 \ x \ 10^{-3} \frac{m.s}{kg}\right)$$

$$= 1.5 \ x \ 10^5 \Rightarrow Turbulent \ flow$$

Figure in Page 248 has been updated with the following figure:

1.2 m/s

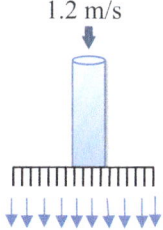

Chapter 6:
The following equation in Page 266 has been updated from

$$P_F = \left(10\frac{L}{s}\right)\left(0.9x\ 1\frac{kg}{L}\right)\left(9.81\frac{m}{s^2}\right)(12.63\ m) = 111.5\ J/s = 111.15\ W$$

$P_B = 111.15/0.69 = 161.1\ W$
to

$$P_F = \left(10\frac{L}{s}\right)\left(0.9x\ 1\frac{kg}{L}\right)\left(9.81\frac{m}{s^2}\right)(12.63\ m) = 111.5\ J/s = 111.15\ W$$

$P_B = 1115.1/0.69 = 1616.1\ W$
Fig. 6.21 has been updated with the following figure:

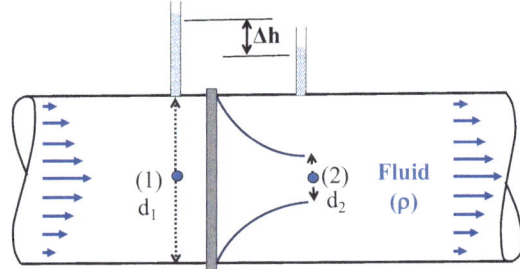

The following equation in Page 276 has been updated from

$$f = \frac{F_D/A}{\frac{\rho <v)^2}{2}}$$

to

$$f = \frac{F_D/A}{\frac{\rho <v>^2}{2}}$$

The following equation in Page 280 has been updated from

$$v = \frac{2}{9}\ r_p^2\ \frac{(\rho_p - \rho)}{\eta}\left(\frac{2\pi N}{60}\right)^2 \rightarrow v = r_p^2\ N^2 R\frac{(\rho_p - \rho)}{410\eta}$$

to

$$v = \frac{2}{9}\ r_p^2\ \frac{(\rho_p - \rho)}{\eta}\left(\frac{2\pi N}{60}\right)^2 R \rightarrow v = r_p^2\ N^2 R\frac{(\rho_p - \rho)}{410\eta}$$

The following equation in Page 285 has been updated from

$$\varepsilon = \frac{\text{volume of voids in bed}}{\text{Total volume of bed (voids + bed)}}$$

to

$$\varepsilon = \frac{\text{volume of voids in bed}}{\text{Total volume of bed (voids + bed material)}}$$

The following equation in Page 293 has been updated from

$$\frac{\eta}{\rho} = \nu = \frac{\pi R^4 g \, h_m}{8 \, L \, \dot{V}}$$

to

$$\frac{\eta}{\rho} = \nu = \frac{\pi R^4 g \, h_m}{8 \, L \, V}$$

Chapter 7:
The following equation in Page 308 has been updated from
C_{fiber} 1:83×10-1 1:76×10-3 T–3:17×10-6 T2
to
$k_{fiber} = 1.83 \times 10^{-1} + 1.76 \times 10^{-3} \, T - 3.17 \times 10^{-6} \, T^2$
Fig. 7.3 has been updated with the following figure:

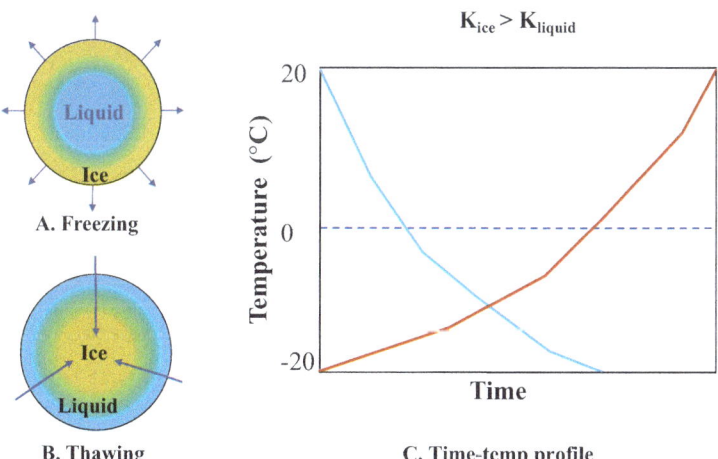

A. Freezing

B. Thawing

C. Time-temp profile

The following equation in Page 314 has been updated from

$$R = \frac{(x_1 - x_2)}{kA} = \frac{\Delta x}{kA} [=] \frac{^\circ c}{w} \text{ or } \frac{^\circ F - h}{B + u}$$

to

$$R = \frac{(x_1 - x_2)}{kA} = \frac{\Delta x}{kA} [=] \frac{^\circ C}{w} \text{ or } \frac{^\circ F - h}{Btu}$$

The following equation in Page 317 has been updated from
0.5" cm
to
0.5 cm
Chapter 8:
Fig. 8.3 has been updated with the following figure:

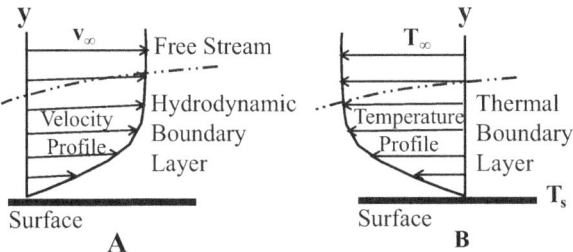

The following equation in Page 353 has been updated from
$(T_{o1} + T_{02})/2,$
to
$(T_{o1} + T_{o2})/2,$
The following equation in Page 357 has been updated from

$$U[=\} \frac{W}{m^2 K} \quad \text{or} \quad U[=\} \frac{Btu}{hr - ft^2 - {^0}R}$$

to

$$U[=] \frac{W}{m^2 K} \quad \text{or} \quad U[=] \frac{Btu}{hr - ft^2 - {^0}R}$$

The following equation in Page 360 has been updated from

$$\frac{1}{U_c} = \frac{1}{1000} = 0.001 \; and \; \frac{1}{U_f} = \frac{1}{100} = 0.01$$

$$R_f = (0.009/0.01)\,(100) = 90\%$$

to

Remove bold for the above two equations

$$\frac{1}{U_c} = \frac{1}{1000} = 0.001 \; and \; \frac{1}{U_f} = \frac{1}{100} = 0.01$$

$$R_f = (0.009/0.01)\,(100) = 90\%$$

Fig. 8.10 has been updated with the following figure:

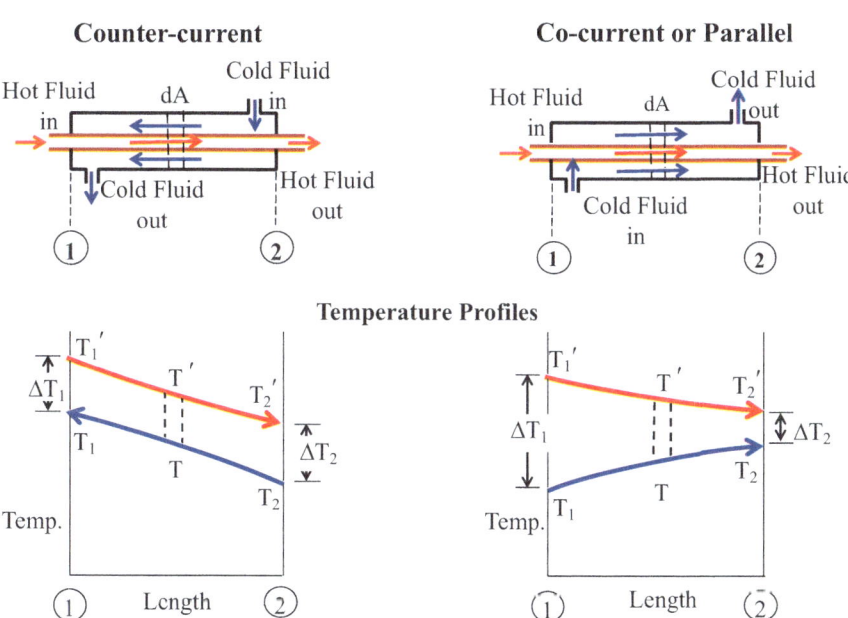

The following equation in Page 363 has been updated from

$$\text{Thus,}\quad \ln\left(\frac{T_2'-T_2}{T_1'-T_1}\right) = -UA\left(\frac{T_1'-T_2'}{\dot{q}} + \frac{T_2-T_1}{\dot{q}}\right) \quad \text{or}\quad \dot{q}$$

$$= \frac{UA\left[-T_1' + T_2'-T_2 + T_1\right]}{\ln\left(\frac{T_2'-T_2}{T_1'-T_1}\right)} = UA\,\frac{\left[(T_2'-T_2)-(T_1'-T_1)\right]}{\ln\left(\frac{T_2'-T_2}{T_1'-T_1}\right)}$$

to

$$\text{Thus,}\quad \ln\left(\frac{T_2'-T_2}{T_1'-T_1}\right) = -UA\left(\frac{T_1'-T_2'}{\dot{q}} + \frac{T_2-T_1}{\dot{q}}\right) \quad \text{or}$$

$$\dot{q} = \frac{UA\left[-T_1' + T_2'-T_2 + T_1\right]}{\ln\left(\frac{T_2'-T_2}{T_1'-T_1}\right)} = UA\,\frac{\left[(T_2'-T_2)-(T_1'-T_1)\right]}{\ln\left(\frac{T_2'-T_2}{T_1'-T_1}\right)}$$

The following equation in Page 375 has been updated from

Convective heat-transfer coefficients: milk = 200 W/m²·K; water = 500 W/m²·K
Specific heats: milk = 4000 J/kg-K; water = 4180 J/kg-K

to

Overall heat transfer coefficients based on the inside area (Ui) = 140.84 W/m²·K.
Specific heats: milk = 4000 J/kg-K; water = 4180 J/kg-K.

Chapter 9:

Fig. 9.6 has been updated with the following figure:

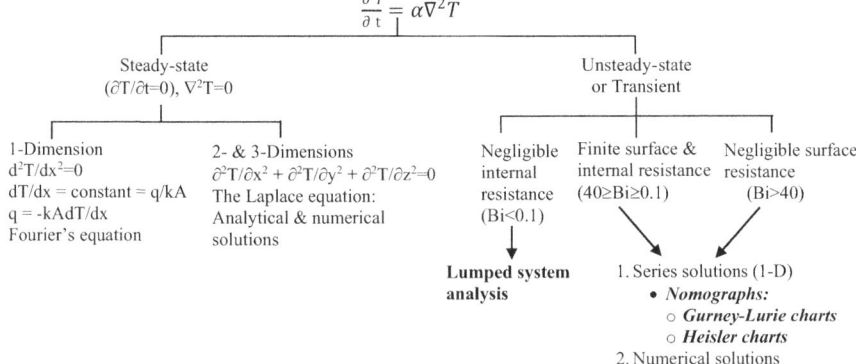

The following equation in Page 392 has been updated from
(a) Bi <0.1, (b) $1 \leq Bi \leq 40$ or (c) Bi >40
to
(a) Bi <0.1, (b) $0.1 \leq Bi \leq 40$ or (c) Bi >40
Fig. 9.9 has been updated with the following figure:

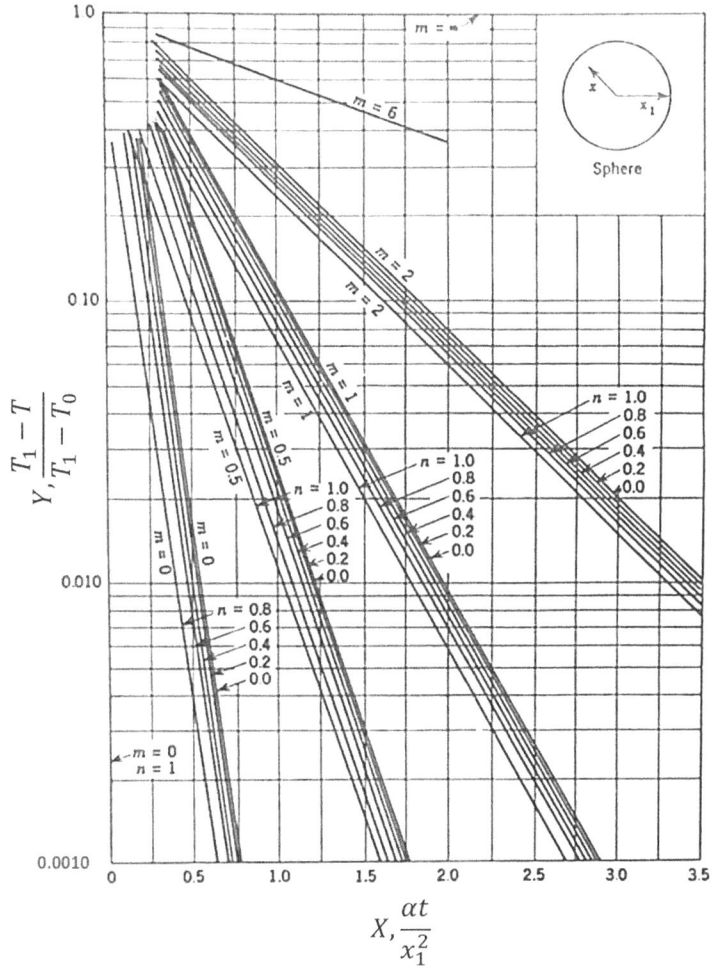

The following equation in Page 395 has been updated from

$$t = \frac{X\rho C_p x_1^2}{k} = \frac{(0.37)\left(800 \frac{kg}{m^3}\right)\left(3500 \frac{J}{kg\,°C}\right)(0.05\ m)^2}{\left(0.30 \frac{W}{m\,°C}\right)} = 8633.3\ s \cong 2.4\ hours$$

to

$$t = \frac{X\rho C_p x_1^2}{k} = \frac{(0.37)\left(900 \frac{kg}{m^3}\right)\left(3500 \frac{J}{kg\,°C}\right)(0.05\ m)^2}{\left(0.30 \frac{W}{m\,°C}\right)} = 8633.3\ s \cong 2.4\ hours$$

The following equation in Page 397 has been updated from
thickness X, Y, and X, so that
to
thickness X, Y, and Z, so that
The following equation in Page 398 has been updated from

$$Y_{XYR} = (Y_X)(Y_Y)(Y_Z) = \left(\frac{T_1 - T}{T_1 - T_0}\right)_{\text{finite slabe}}$$

to

$$Y_{XYZ} = (Y_X)(Y_Y)(Y_Z) = \left(\frac{T_1 - T}{T_1 - T_0}\right)_{\text{finite slab}}$$

Chapter 10:
Table 10.2 has been updated
The following equation in Page 426 has been updated from
Solution: Energy needed, $Q = [(5\ kg)\ (4.3\ kJ/kg\text{-}°C)(72 \text{ - } 4)°C]/0.8 = 1827.5\ kJ$ *Heater power output = 1.5 kW = 1.5 kJ/s, then time required = 1827.5/1.5 s = 20 min.*

to
Solution: Energy needed, $Q = [(5\ kg)\ (4.3\ kJ/kg\text{ - }°C)(72 - 40)°C]/ 0.8 = 1827.5\ kJ$*Heater power output = 1.5 kW = 1.5 kJ/s, then time required = 1827.5/1.5 s = 20 min.*

Chapter 11:

The following equation in Page 430 has been updated from

$$J_{Az} = -D_{AB} \frac{dc_A}{dz}$$

to

$$J_{Az} = -D_{AB} \left(\frac{dc_A}{dz} \right)$$

The following equation in Page 432 has been updated from

$$N_{AM} = c_A \left(\frac{c_A}{C} v_A + \frac{c_B}{C} v_B \right) = \frac{c_A}{c} \left(c_A v_A + c_B \; v_B \right) = \frac{c_A}{c} \left(N_A + N_B \right)$$

to

$$N_{AM} = c_A \left(\frac{c_A}{c} v_A + \frac{c_B}{c} v_B \right) = \frac{c_A}{c} \left(c_A v_A + c_B \; v_B \right) = \frac{c_A}{c} \left(N_A + N_B \right)$$

The following equation in Page 436 has been updated from

$$D_{Etoh\text{-}Air} = \frac{1.0 \times 10^{-3} \, (273)^{1.75}}{1.0 \left[(50.36)_{Etoh}^{1/3} + (20.1)_{Air}^{1/3} \right]^2} \left(\frac{1}{46} + \frac{1}{29} \right)^{1/2}$$
$$= 9.23 \times 10^{-4} \text{cm}^2/\text{s}$$

to

$$D_{Etoh\text{-}Air} = \frac{1.0 \times 10^{-3} \, (303)^{1.75}}{1.0 \left[(50.36)_{Etoh}^{1/3} + (20.1)_{Air}^{1/3} \right]^2} \left(\frac{1}{46} + \frac{1}{29} \right)^{1/2}$$
$$= 0.127 \text{ cm}^2/\text{s}$$

The following equation in Page 441 has been updated from

$$J = P_m \left(\frac{\Delta P}{\Delta z} \right) = \frac{2.20 \, \left(10^{-13} \right) (0.21)}{0.02} = 2.31 \times 10^{-12} \text{ mL}/\left(\text{s.m}^2 \right)$$

to

$$J = P_m \left(\frac{\Delta P}{\Delta z} \right) = \frac{2.20 \, \left(10^{-13} \right) \left(0.21 \times 1.013 \times 10^5 \right)}{0.02} = 23.4 \times 10^{-8} \text{ mL}/\left(\text{s.cm}^2 \right)$$

The following equation in Page 452 has been updated from

$$Sh = \frac{\text{Mass tranfer coefficient}}{\text{Mass diffusivity}} = \frac{kL}{D_{AB}}$$

to

$$Sh = \frac{\text{Convective mass transfer}}{\text{Diffusive mass transport}} = \frac{k_c}{D_{AB}}$$

Chapter 13:
The following equation in Page 481 has been updated from

$$RH = 100 \frac{P_W}{P_g} = 100 \frac{1.5 \text{ kPa}}{2.064 \text{ kPa}} = 73\%$$

to

$$RH = 100 \frac{P_W}{P_g} = 100 \left(\frac{1.5 \text{ kPa}}{2.064 \text{ kPa}} \right) = 73\%$$

The following equation in Page 482 has been updated from

$$mc = \frac{18.0}{28.94} \ (\text{Mole fraction of water in the mixture}) = 0.622 \ x_w$$

to

$$mc = \frac{18.0}{28.94} \ (\text{Mole fraction of water in the mixture}) = 0.622 \ y_w$$

The following equation in Page 483 has been updated from

$$mc = 0.622 \frac{P_w}{P_g} = 0.622 \ (\%RH)/100$$

to

$$mc = 0.622 \left(\frac{P_w}{P_g} \right) T$$

The following equation in Page 484 has been updated from

$$P_g = \frac{45}{100} (2.985 \text{ kPa}) = 1.343 \text{ kPa}$$

to

$$P_a = \frac{45}{100} (2.985 \text{ kPa}) = 1.343 \text{ kPa}$$

The following equation in Page 485 has been updated from

$$\overline{H}_{va} = C_{da}(T_{db} - T_0) + \omega \overline{H}_v$$

to

$$\overline{H}_{va} = C_{da}(T_{db} - T_0) + \omega \overline{H}_v$$

The following equation in Page 486 has been updated from

$$mc = 0.6219\left(\frac{4.246\text{kPa}}{101.3\text{kPa}}\right) = 0.0261\frac{\text{kg}}{\text{kg}} \text{ or } 26.1\frac{\text{g}}{\text{kg}}$$

to

$$mc = 0.622\left(\frac{4.246\text{kPa}}{101.3\text{kPa}}\right) = 0.0261\frac{\text{kg}}{\text{kg}} \text{ or } 26.1\frac{\text{g}}{\text{kg}}$$

Fig. 13.5 has been updated with the following figure:

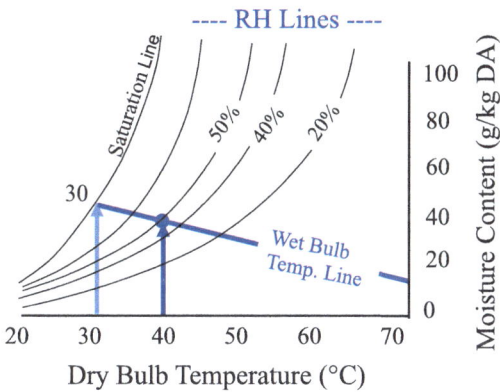

BACKMATTER:

Appendices:

The following equation in Page 508 has been updated from

1mm = 10-6 m

1 psia = 2.0360 in. Hg = 51.715 mm Hg = 6.89476 dyne/cm^2 = 6894.76 N/m^2

to

1mm = 10-3 m

1 psia = 2.0360 in. Hg = 51.715 mm Hg = 6.89476 dyne/cm^2 = 68947.6 N/m^2

The following equation in Page 509 has been updated from

$$\log (a^n) = n \log a \quad \log (ab) = \log a + \log b \quad \log(a/b) = \log a$$
$$- \log b \quad \ln(a^n) = n \ln a$$

to

$$\log (a^n) = n \log a \quad \log (ab) = \log a + \log b \quad \log(a/b) = \log a$$
$$- \log b \quad \ln(a^n) = n \ln a$$

Glossary

- **Integer:** Any number that can be expressed without the use of decimals or fractions ($\ldots -2, -1, 0, 1, 2 \ldots$).
- **Rational number:** Any number that can be written as a fraction m/n, where m (the **numerator**) is an integer and n (the **denominator**) is any non-zero integer.
  - **Proper fraction:** A fraction m/n where $m < n$.
  - **Improper fraction:** A fraction m/n where $m > n$.
- **Irrational number:** Any number that cannot be expressed as a ratio of integers (e.g., π, $\sqrt{2}$, etc.)
- **Imaginary number (i):** The even root (square, fourth, sixth, etc.) of a negative number. Imaginary numbers are expressed as multiples of i, where $i = \sqrt{-1}$. A **complex number** is the sum of a real number and an imaginary number.
- **Real number:** Any non-imaginary number (integer, rational, or irrational).
- **Prime number:** A positive integer that is only divisible by itself and 1 (e.g., 2, 3, 5).
- **Infinity (∞):** A conceptual "number," used to represent a quantity larger than any real number.
- **Zero (0):** The real integer used to define positive and negative numbers. Numbers are positive if they are greater than zero, and negative if they are less than zero. The product of zero and any other number is zero; the result of dividing any other number by zero is infinity.
- **Slope:** The ratio $\frac{\Delta y}{\Delta x}$ of any straight line that can be plotted using Cartesian coordinates.
- **Instantaneous slope:** The slope of a straight-line tangent to a curve that can be plotted using Cartesian coordinates. The slope of the line is said to be the curve's instantaneous slope at the point of tangency.
- **Vector:** A numerical expression of a quantity with both magnitude and direction. A quantity that only has a magnitude is known as a **scalar**.
- **Length:** A measure of distance between two points.
- **Mass:** A measure of the quantity of matter in a sample.
- **Time:** A measure of the ordering of events relative to regular periodic phenomena (the Earth's orbit around the Sun, the Earth's rotation about its axis, etc.).

S. S. H. Rizvi, *Food Engineering Principles and Practices*,
https://doi.org/10.1007/978-3-031-34123-6

- **Temperature:** A measure of the average kinetic energy of the molecules in a sample.
- **Mole:** A measure of a unit amount of a substance, defined as 6.022×10^{23} entities (atoms, molecules, particles, etc.). The number 6.022×10^{23} is known as **Avogadro's number**.
- **Electric current:** The flow of charged particles through a conductor, measured by the number of charges passing through a set point per unit time.
- **Luminous intensity:** A measure of the brightness of light.
- **Force:** A push or pull acting on a body.
- **Weight:** The downward force exerted on an object by a gravitational field.
- **Pressure:** The force per unit area exerted on an object by a fluid (gas or liquid). For solids, this quantity is referred to as **normal stress**.
  - **Absolute pressure:** The exact force per unit area exerted by a substance on a surface.
  - **Gauge pressure:** The difference between absolute pressure and atmospheric pressure.
  - **Vacuum:** Any absolute pressure that is less than atmospheric pressure. Can be expressed as a negative gauge pressure, or as the negative of gauge pressure (e.g., 10 psi of vacuum $= -10$ psig $= (14.7 - 10) \rightarrow 4.7$ psia).
- **Energy:** The ability to do work or transfer heat to or from an object.
- **Work:** The energy required to displace an object by a measurable distance. If a force is applied to an object, but there is no displacement (movement), then no work is done; similarly, if an object is displaced but no force is applied, there is still no work done.
- **Power:** The rate at which work is done or energy is transferred over time.
- **Concentration:** A measure of the amount of a chosen substance in a mixture or solution.
  - **Mass fraction:** The mass of a substance divided by the mass of the total mixture. Can be measured on either a **wet ("as-is") basis** (total mass includes mass of all components) or a **dry basis** (total mass does not include mass of water). May also be expressed as a percentage of the total mass (**mass percent**).
  - **Parts per million (ppm):** The mass of a component per million mass units of mixture or solution, generally used for very low concentrations.
  - **Mole fraction:** The number of moles of a substance divided by the total number of moles of all components in the solution or mixture. May also be expressed as a percentage of the total number of moles (**mole percent**).
  - **Molarity (M):** The number of moles of solute dissolved in 1 liter of solution.
  - **Molality (m):** The number of moles of solute dissolved in 1 kilogram of solvent (not solution!).
  - **Molar volume:** The volume occupied by one mole of a substance.
- **Density:** The mass per unit volume of a substance.
  - **Solids density:** The true density of a substance, excluding all voids, air pockets, etc.

– **Particle/piece density:** The apparent density of a porous particle.
– **Bulk density:** The mass of particles occupying a unit volume, including voids between particles. May vary depending on processing, particle size, etc.
– **Specific volume:** The volume occupied by a unit mass of a substance, i.e. the reciprocal of density.
– **Specific gravity:** The ratio of a material's density to the density of a reference material (usually water for solids and liquids, or air for gases).
• **Ideal gas:** A gas in which the volume occupied by gas molecules/particles is negligible relative to the space between particles, and there are no significant intermolecular forces (attraction or repulsion). Gases at high temperatures and low pressures can generally be considered ideal.
• **Partial pressure:** The pressure exerted by a component gas within a mixture of gases.
• **Partial volume:** The volume occupied by a component gas within a mixture of gases.
• **Critical temperature:** The temperature above which a gas cannot be liquefied (i.e., the liquid and gas phases of a fluid become indistinguishable).
– **Critical pressure:** The pressure required to liquefy a gas at its critical temperature.
– **Critical volume:** The volume occupied by a gas at its critical temperature and pressure.
• **Surface tension:** The work required per unit area to break the surface of a liquid.
• **Colligative property:** A property of a solution that is dependent on the number of solute particles present in solution, and independent of the type of solute.
• **Osmotic pressure:** The pressure caused by a difference in concentration of two solutions separated by a semi-permeable membrane. Can also be defined as the pressure required to prevent osmosis between two solutions of varying concentrations, separated by a semi-permeable membrane.
– **Osmosis:** The flow of solvent through a semi-permeable membrane, from a region of low solute concentration to a region of high solute concentration.
– **Reverse osmosis:** The use of high pressure to force solvent to flow through a semi-permeable membrane, from a region of high solute concentration to a region of low solute concentration (i.e., in the opposite direction of normal osmosis).
• **Stress:** An applied force per unit area that causes deformation of a solid body. Pressure is a type of stress.
– **Strain:** Deformation of a solid body caused by stress.
– **Tensile stress:** Stress caused by a force perpendicular to the surface of a body, which causes the body to elongate. A form of **normal stress**.
– **Compressive stress:** Stress caused by a force perpendicular to the surface of a body, which causes the body to be shortened. A form of **normal stress**.
– **Shear stress:** Stress caused by a force parallel to a surface.
– **Plastic deformation:** Permanent deformation caused by high stress on an elastic material.

- **System:** The part of the universe being analyzed in a given problem.
  - **Surroundings:** All parts of the universe that are not part of the system. Typically, only the system's immediate surroundings are of interest.
  - **Boundary:** The "surface" separating a system from its surroundings. May be physical, imaginary, or both, but must be clearly defined.
  - **Isolated system:** A system that is impermeable to both matter and energy (heat and work).
  - **Closed system:** A system that is permeable to energy, but not matter.
  - **Open system:** A system that is permeable to both matter and energy.
  - **Adiabatic system:** A closed system that is permeable to work, but not heat.
- **Intensive property:** A property of a system that is independent of the system's mass (e.g., density, temperature, pressure).
- **Extensive property:** A property of a system that is dependent on the system's mass (e.g., volume, weight, energy).
- **Isothermal process:** A process that runs at constant temperature.
- **Isobaric process:** A process that runs at constant pressure.
- **Isochoric (isovolumic) process:** A process that runs at constant volume.
- **Isentropic process:** A process that runs at constant entropy. Isentropic processes must be adiabatic and have no energy loss due to friction.
- **State function:** A relationship between two or more system variables that can be used to determine the value of one variable if the rest are known. An equation describing such a relationship is called an **equation of state**.
- **Equilibrium:** A state in which a defined system will undergo no changes unless mater or energy is exchanged with its surroundings.
  - **True equilibrium:** An equilibrium state which can be reached from any direction.
  - **Metastable equilibrium:** An equilibrium state which can only be reached from certain directions. A system may remain in metastable equilibrium for extended periods of time, but will leave this state if disturbed.
- **Steady state:** A state in which a system is not in equilibrium, but the system variables do not change with time.
- **Batch process:** A process which begins at a fixed initial state and progresses to a final state.
- **Continuous process:** A process in which materials are continuously fed through an input stream and product is continuously removed through an output stream.
- **Phase:** A homogenous, physically distinct portion of a system that can be mechanically separated from the rest of the system.
- **Saturated steam:** Steam that is at its vapor pressure. Saturated steam may be present alongside vapor water.
- **Steam quality:** A measure of the quantity of steam in a mixture of steam and liquid water. Defined as the mass (weight) percent of vapor in the total mixture.
- **Superheated steam:** Dry steam that is at a temperature higher than its saturation temperature (at a given pressure). The difference between the steam's current temperature and the saturation temperature is called the **degree of superheat**.

- **Triple point:** The temperature and pressure at which all three phases of water can coexist at equilibrium.
- **Supercritical fluid:** A substance that has been heated/pressurized to above its critical temperature and pressure. Supercritical fluids have properties of both liquids (high density and solvation ability) and gases (high compressibility, variable density, and gas-like diffusivity and viscosity).
- **Internal energy:** Energy associated with the molecules within a system.
  - **Nuclear energy:** Energy associated with the binding of particles within atomic nuclei (protons and neutrons).
  - **Chemical energy:** Energy associated with the binding of atoms to make molecules.
  - **Thermal energy:** Energy associated with the random motion of molecules and electron shifts within the molecule.
  - **Molecular energy:** Energy associated with the attractions and repulsions between molecules.
- **Kinetic energy:** Energy associated with motion of a system (not molecular motion).
- **Potential energy:** Energy associated with the elevation of a system relative to some reference level.
- **Heat:** Energy that moves between a system and its surroundings due to a difference in temperature.
- **Point function:** Any variable that is a function of the state of a system, independent of how the system reached the aforementioned state.
- **Path function:** Any variable that is a function of the path taken by a system in order to reach its current state, rather than the current state itself.
- **Sensible heat:** Heat used to increase the temperature of a system.
- **Latent heat:** Heat used to change the phase of a system.
- **Heat capacity (specific heat):** The amount of heat required to increase the temperature of one gram of a material by one degree Celsius.
- **Fluid:** A substance that does not permanently resist deformation and assumes the shape of the container in which it is placed.
  - **Liquid:** A generally-incompressible fluid with definite volume. If placed in a container larger than its volume, a liquid will form a level top surface.
  - **Gas:** A fluid with no definite volume, which will expand to fill any container with no top surface. Gases are compressible.
- **Hydrostatic pressure:** The pressure exerted by a static fluid above a measurement point due to the force of gravity.
- **Pressure head:** A means of reporting pressure. It is the height of a specified fluid that would produce the equivalent hydrostatic pressure (e.g., atmospheric pressure has a head of 760 millimeters of mercury, or mmHg).
- **Gradient:** The rate at which a quantity changes over distance.
- **Local velocity:** The velocity of a fluid at an exact point within a fluid flow.
- **Viscosity:** A measure of a fluid's resistance to flow (shear). Also known as **dynamic viscosity**.

- **Shear rate:** The rate at which a fluid's velocity changes with respect to distance from a shear force. Also known as **velocity gradient**.
- **Kinematic viscosity:** The ratio of a fluid's viscosity to its density.
- **Newtonian fluid:** A fluid that obeys Newton's law of viscosity.
  - **Pseudoplastic (shear-thinning) fluid:** A non-Newtonian fluid whose viscosity decreases as shear rate increases.
  - **Dilatant (shear-thickening) fluid:** A non-Newtonian fluid whose viscosity increases as shear rate increases.
- **Bingham plastic:** A material that does not deform until a threshold shear stress (**yield stress**) is met or exceeded. A Bingham plastic behaves as a solid for any stress less than its yield stress, and a Newtonian fluid for stresses greater than its yield stress.
- **Herschel-Bulkley fluid:** A material with a positive yield stress, which behaves as a non-Newtonian fluid above its yield stress.
- **Thixotropic fluid:** A fluid with time-dependent viscosity. When subjected to a constant shear rate, a thixotropic fluid becomes less viscous over time.
- **Rheoplastic fluid:** A fluid with time-dependent viscosity. When subjected to a constant shear rate, a rheoplastic fluid becomes more viscous over time.
- **Ideal fluid:** A material that undergoes permanent shear when subjected to shear stress. The fluid retains its current shape when the stress is removed.
- **Ideal solid:** A material that is completely elastic and follows Hooke's law. When subjected to shear stress, the material deforms, but returns to its original shape and size when the stress is removed.
- **Viscoelastic material:** A material with behaviors of both an ideal fluid and an ideal solid. The material will deform when stressed and return to nearly its original shape if the stress is removed quickly, but continued stress may cause permanent deformation.
- **Laminar flow:** A flow pattern in which layers of fluid slide over each other without significant mixing.
- **Turbulent flow:** A flow pattern in which the fluid moves chaotically, with significant mixing.
- **Plug flow:** A flow pattern in which the fluid velocity is nearly uniform.
- **Hydraulic radius:** The ratio of a flow's cross-sectional area to the wetted perimeter of the pipe/channel, etc. through which the fluid is flowing.
- **Boundary layer:** The region of slower-moving fluid in a fluid flow near a wall.
- **Entrance length:** The distance from the entry to a pipe/channel, etc. to where a fluid flow is fully developed. A flow is said to be fully developed when the shape of its velocity profile becomes constant at all cross sections.
- **Fluid power:** Energy added to a fluid flow system by a pump.
- **Brake (actual) power:** Energy produced (work done) by a pump. Brake power multiplied by pump efficiency equals fluid power.
- **Wall drag (skin drag):** Drag force exerted by a fluid on a solid due to transfer of momentum perpendicular to the direction of flow. Causes shear stress at pipe walls, etc.

- **Form drag:** Drag caused by changes in the direction of fluid flow around an immersed object.
- **Food contact surface (FDA definition):** Any surface in "direct contact with food residue, or where food residue can drip, drain, diffuse, or be drawn."
- **Conduction:** Transfer of heat from molecule to molecule within a material. Does not require movement of the material as a whole.
- **Convection:** Transfer of heat by mass movement of a material (fluid).
  - **Natural convection:** Convection caused by uneven heating of a fluid.
  - **Forced convection:** Convection caused by forced movement of a fluid, via use of pumps, fans, etc.
- **Radiation:** Transfer of heat between objects via electromagnetic radiation, without heating the intervening material. Does not require a transfer medium.
- **Thermal diffusivity:** The rate at which temperature change travels (diffuses) through an object.
- **Coefficient of performance:** A measure of a refrigeration system's efficiency. Defined as the ratio of enthalpy removed to work done.
- **Relative humidity:** The ratio of the partial pressure of water in air to the saturation partial pressure, expressed as a percentage.
- **Moisture content (specific humidity, humidity ratio):** The mass ratio of water vapor in air to dry air.
- **Dry bulb temperature:** The actual temperature of the air, as measured with an ordinary thermometer.
- **Wet bulb temperature:** The saturation temperature of the air, measured using a thermometer with the bulb wrapped in a wet cloth.
- **Dew point:** The temperature at which air is saturated, given a specified moisture content.

Laws, Theorems, Etc.

- **Pascal's Law:** Pressure applied to/by an incompressible, enclosed fluid is transmitted undiminished to every part of the fluid and the walls of the container.
- **Avogadro's Law:** For a constant temperature and pressure, equal **volumes** of any ideal gas must contain the same number of molecules.
- **Dalton's Law of Partial Pressures:** The total pressure exerted by a gaseous mixture is equal to the sum of the component gases' partial pressures.
- **Amagat's Law of Partial Volumes:** The total volume occupied by a gaseous mixture is equal to the sum of the component gases' partial volumes.
- **Hooke's Law:** Applied stress is linearly related to the strain experienced by a material. Materials that obey Hooke's law are said to be ideal.
- **First Law of Thermodynamics:** Energy cannot be created or destroyed, only transferred between objects or changed from one form to another.
- **Second Law of Thermodynamics:** Natural processes will spontaneously occur if they lead to an increase in the system's entropy (disorder). Universal entropy will

always tend towards a maximum; no process may decrease the total entropy of the universe.

- **Third Law of Thermodynamics:** A perfect crystalline solid has zero entropy at absolute zero. Absolute zero is unattainable.
- **Newton's Law of Viscosity:** Viscosity is a proportionality constant relating shear stress to shear rate. Fluids that obey this law are called **Newtonian fluids**.
- **Reynolds number:** $Re = \dfrac{\rho \bar{v} d}{\mu} = \dfrac{\text{molecular momentum transport}}{\text{viscous momentum transport}} = \dfrac{\text{kinetic/inertial forces}}{\text{viscous forces}}$
- **Nusselt number:** $Nu = \dfrac{h_c d}{k} = \dfrac{\text{heat transfer by convection}}{\text{heat transfer by conduction}}$
- **Prandtl number:** $Pr = \dfrac{C_p \mu}{k} = \dfrac{\text{momentum diffusivity}}{\text{thermal diffusivity}}$
- **Brinkman number:** $\dfrac{\mu \bar{v}^2}{k \Delta T} = \dfrac{\text{heat generated by viscous dissipation}}{\text{heat transfered by conduction}}$
- **Grashof number:** $Gr = \dfrac{g \rho_b B (\Delta T) x^3}{\mu^2}, \; B = \dfrac{1}{\rho_b}\left(\dfrac{d\rho}{dT}\right) = \dfrac{1}{\rho_b}\left(\dfrac{\rho_b - \rho_i}{\Delta T}\right)$
- **Biot number:** $Bi = \dfrac{h_c x_i}{k_{\text{solid}}} = \dfrac{\text{internal resistance to heat transfer}}{\text{external or boundary resistance to heat flow}}$
- **Fourier number:** $Fo = \dfrac{\alpha t}{x_1^2}$

Useful Equations

- **Conversion of mass fraction from wet to dry basis:**

$$\text{MF}_{\text{db}} = \frac{m_k}{m_1 + m_2 + \ldots + m_{n-1}} = \frac{m_k}{\displaystyle\sum_{i=1}^{n-1} m_i}$$

- **Boyle's Law:** For constant T, $V \propto \frac{1}{P} \rightarrow \frac{P_1}{P_2} = \frac{V_2}{V_1}$
- **Charles's Law:** For constant P, $\frac{dV/V}{dT} = \beta = 2.1 * 10^{-4} \; \frac{\text{cm}^3/\text{cm}^3}{^\circ\text{C}}$
- **Gay-Lussac's Law:** For all gases at constant P, $V = V_0 + V_0 \beta T = V_0 + \frac{V_0}{273.15} T$
- **Ideal Gas Law:** $PV = nRT$
- **Compressibility factor for non-ideal gases:** $z = \dfrac{(PV)_{\text{actual}}}{nRT}$
- **Henry's Law:** $P_i = k_H C = k_H x_i = \frac{x_i}{k_H'}$
- **Young's modulus:** $E = \frac{\sigma}{\varepsilon}$
- **Poisson's ratio:** $\nu = -\dfrac{\varepsilon_{\text{trans}}}{\varepsilon_{\text{longitudinal}}} = \dfrac{d - d_0/d_0}{l - l_0/l_0}$
- **Shear modulus (modulus of rigidity):** $G = \frac{\tau}{\gamma}$
- **Bulk modulus:** $B = -\frac{\Delta P}{\Delta V/V}$

- **Gibb's phase rule:** $df = 2 + C - P$
- **Kinetic energy:** $\text{KE} = \frac{1}{2}mv^2$
- **Potential energy:** $\text{PE} = mgh = \dfrac{mgh}{g_c}$
- **Enthalpy:** $H = E + PV \rightarrow \Delta H = \Delta E - P\Delta V$
- **Heat capacity:** $C = \dfrac{q}{m\Delta T} \rightarrow \dfrac{dq}{dT}$
  - **Constant pressure:** $C_P = \left(\dfrac{dq}{dT}\right)_P = C_V + R$
  - **Constant volume:** $C_V = \left(\dfrac{dq}{dT}\right)_V = C_P - R$
- **Hydrostatic pressure:** $\Delta P = \rho g \Delta h$
- **Power law:** $\tau = K\dot{\gamma}^n$
- **Power law for plastics:** $\tau = \tau_0 + K\dot{\gamma}$
- **Continuity equation:** $\rho_1 \langle v_1 \rangle A_1 = \rho_2 \langle v_2 \rangle A_2$
- **Overall energy balance:** $\Delta H + \frac{1}{2\alpha}\Delta \langle v \rangle^2 + g\Delta h = q_T - W_s$
- **Mechanical energy balance:** $\frac{1}{2\alpha}\langle v_2 \rangle^2 + gh_2 + \frac{P_2}{\rho} + \sum F + W_s = \frac{1}{2\alpha}\langle v_1 \rangle^2 + gh_1 + \frac{P_1}{\rho}$
- **Bernoulli equation:** $\frac{1}{2}\langle v_2 \rangle^2 + gh_2 + \frac{P_2}{\rho} = \frac{1}{2}\langle v_1 \rangle^2 + gh_1 + \frac{P_1}{\rho}$
- **Torricelli equation:** $\langle v \rangle = \sqrt{2gh}$
- **Hagen-Poiseuille equation:** $\langle v \rangle = \dfrac{\Delta P r_w^2}{8\eta l} = \dfrac{v_{max}}{2}$
- **Fanning equation:** $F_f = \dfrac{\Delta P}{\rho} = 4f\left(\dfrac{l}{d}\right)\left(\dfrac{\langle v \rangle^2}{2}\right) = 4f\left(\dfrac{l}{d}\right)\left(\dfrac{\langle v \rangle^2}{2g_c}\right)$
- **Expansion losses:** $h_{ex} = K_{ex}\dfrac{\langle v_1 \rangle^2}{2\alpha} = \left(1 - \dfrac{A_1}{A_2}\right)^2 \dfrac{\langle v_1 \rangle^2}{2\alpha}$
- **Contraction losses:** $h_c = K_c \dfrac{\langle v_2 \rangle^2}{2\alpha} = 0.55\left(1 - \dfrac{A_2}{A_1}\right)\dfrac{\langle v_2 \rangle^2}{2\alpha}$
- **Valves and fittings losses:** $h_f = K_f \dfrac{\langle v_1 \rangle^2}{2}$
- **Drag coefficient:** $C_D = \dfrac{F_D/A_p}{\rho \frac{v_\infty^2}{2}}$
- **Stokes' Law:** $v = \dfrac{2}{9}r_p^2 \dfrac{(\rho_p - \rho)g}{\eta}$
- **Stokes' Law for centrifugal separation:** $v = \dfrac{r_p^2 N^2 R(\rho_p - \rho)}{410\eta}$
- **Centrifugal force:** $F_c = \omega^2 R = \left(\dfrac{2\pi N}{60}\right)^2 R$
- **Centrifugal number (relative centrifugal force):** $N_c = \dfrac{\omega^2 R}{g} = \left(\dfrac{2\pi N}{60}\right)^2 \dfrac{R}{g}$
- **Fourier's First Law of Conduction:** $\dot{q} = -kA\frac{dT}{dx}$

- **Log-mean area:** $A_{\mathrm{LM}} = \dfrac{A_0 - A_i}{\ln\left(\frac{A_0}{A_i}\right)}$

- **Newton's Law of Cooling:** $\dot{q} = hA\Delta T = hA\left(T_w - T_f\right)$

- **Overall heat transfer coefficient:** $U = \dfrac{1}{A\sum R} = \dfrac{1}{\frac{1}{h_i} + \frac{\Delta x}{k_A} + \frac{q}{h_o}}$

- **Log-mean temperature difference:** $(\Delta T)_{\mathrm{LM}} = \dfrac{\Delta T_1 - \Delta T_2}{\ln\left(\frac{\Delta T_1}{\Delta T_2}\right)}$

- **Fourier's Second Law:** $\dfrac{\partial T}{\partial t} = \alpha \nabla^2 T$

  - **Cartesian coordinates:** $\dfrac{\partial T}{\partial t} = -\alpha\left[\dfrac{\partial^2 T}{\partial x^2} + \dfrac{\partial^2 T}{\partial y^2} + \dfrac{\partial^2 T}{\partial z^2}\right]$

  - **Cylindrical coordinates:** $\dfrac{\partial T}{\partial t} = -\alpha\left[\dfrac{\partial^2 T}{\partial r^2} + \dfrac{1}{r}\left(\dfrac{\partial T}{\partial r}\right) + \dfrac{1}{r^2}\left(\dfrac{\partial^2 T}{\partial \theta^2}\right) + \dfrac{\partial^2 T}{\partial z^2}\right]$

- **Thermal diffusivity:** $\alpha = \dfrac{k}{\rho C_p}$

- **Lumped parameter analysis:** $\dfrac{T_1 - T}{T_1 - T_0} = \exp[-(Bi)(Fo)]$

- **Fick's First Law of Diffusion:** $J_{A,z} = -D_{AB}\dfrac{dc_A}{dz}$

- **Fick's Second Law:** $\dfrac{\partial C}{\partial t} = D\dfrac{\partial^2 C}{\partial x^2}$

Appendices

Appendix A.1: Useful Constants and Formulae

Constants and Conversion Factors

Acceleration due to Gravity

$g = 9.80665$ m/s^2 = 980.665 cm/s^2 = 32.174 ft/s^2

g_c (gravitational constant) = 32.174 lb$_m$-ft/lb$_f$-s^2 = 980.665 g-cm/g$_f$-s^2

Density

1 g/cm^3 = 62.43 lb$_m$/ft^3 = 1000 kg/m^3

1 g/cm^3 = 8.345 lb$_m$/U.S. gallon

1 lb$_m$/ft^3 = 16.018 kg/m^3

Force

1 kg-m/s^2 = 1 Newton (N)

1 lb$_f$ = 4.4482 N

1 g-cm/s^2 (dyne) = 10^{-5} kg-m/s^2 = 10^{-5} N = 2.2481 × 10^{-6} lb$_f$

1 kg force = 9.806 N

Gas law constant (R)

8.314462 J/mol-K

8314.462 J/kmole-K

8.20573 × 10^{-5} m^3-atm/mol-K

0.082057 L-atm/mol-K

1.9872 cal/mol-K

1.9858 Btu/lbmol-R

Heat Capacity, Enthalpy

1 Btu/lb$_m$-°F = 4.1868 kJ/kg-°K = 1.0 cal/g-°C

1 Btu/lb$_m$ = 2326.0 J/kg

1 ft-lb$_f$/lb$_m$ = 2.989 J/kg

1 kcal/gmol = 4.184 × 103 kJ/kg mol

Heat Flux

1 Btu/h-ft^2 = 3.1546 W/m^2

1 Btu/h = 0.29307 W

1 cal/h = 1.1622 × 10^{-3} W

© The Author(s), under exclusive license to Springer Nature Switzerland AG 2024
S. S. H. Rizvi, *Food Engineering Principles and Practices*,
https://doi.org/10.1007/978-3-031-34123-6

Heat Transfer Coefficient

1 kcal/h-m^2-°F = 0.2048 Btu/h-ft^2-°F

1 Btu/h-ft^2-°F = 5.6783 × 10^{-4} W/cm^2-°C

1 Btu/h-ft^2-°F = 5.6783 W/m^2-°C

Length

1 inch = 2.54 cm = 25.4 mm

1 mm = 10^{-3} m

1 m = 3.2808 ft = 39.37 inches

1 mile = 1760 yard = 5280 ft

1 foot = 0.3048 m

Mass

1 lb$_m$ = 453.59 g = 16 oz

1 kg = 1000 g = 2.2046 lb$_m$

1 metric ton = 1000 kg

1 short ton = 2000 lb$_m$

1 long ton = 2240 lb$_m$

1 oz = 28.349 g

Power

1 hp = 745.7 Watts = 0.7457 kW = 550 ft-lb$_f$/s = 0.7068 Btu/s

1 Watt = 14.34 cal/min = 1 J/s

Pressure

1 bar = 1 × 10^5 Pa = 1 × 10^5 N/m^2

1 N/m^2 = 10 dynes cm^{-2} = 9.8692 × 10^{-6} atm

1 atm = 14.696 psia = 101.325 kPa = 1.01325 bar = 760 mm Hg = 29.921 in. Hg

1 atm = 33.90 ft water at 4 °C

1 psia = 2.0360 in. Hg = 51.715 mm Hg = 68947.6 dyne/cm^2 = 6894.76 N/m^2

1 mm Hg or 1 torr (0 °C) = 133.322 Pa = 0.13322 kPa

Rotational Speed

1cycle = 1 revolution = 2 pi radians (rad)

1 Hertz (Hz) = 1 cycle/s

1 revolution per minute (RPM) = 0.1047 rad/s

1 revolution per second = 1 Hz = 6.283 rad/s

Temperature

$T_{Celsius} = (T_{Fahrenheit} - 32)/1.8$

$T_{Kelvin} = T_{Celsius} + 273.15$

$T_{Kelvin} = (T_{Fahrenheit} + 459.67)/1.8$

Torque

1 lb$_f$-in= 0.113 N-m

1 lb$_f$-ft= 1.356 N-m

1 N-m = 10^7 dyne·cm = 0.73756 ft-lb$_f$ = 9.4783 × 10^{-4} Btu = 1 J

Viscosity, Dynamic

1 cP = 10^{-2} g/cm-s (Poise) = 2.4191 lb$_m$/ft-h = 6.7197 × 10^{-4} lb$_m$/ft-h

1 cP = 2.0886 × 10^{-5} lb$_f$-s/ft^2

1 cP = 10^{-3} Pa-s = 10^{-3} kg/m-s = 10^{-3} N-s/m^2

1 Pa-s $= 1$ N-s/m$^2 = 1$ kg/m-s $= 1000$ cP

Viscosity, Kinematic

1 Stokes $=$ cm^2/s $= 100$ mm^2/s $= 10^{-4}$ m^2/s $= 3.875$ ft^2/h

1 cS $= 10^{-2}$ cm^2/s

Volume

1 liter $= 1000$ cm^3

1 ft$^3 = 28.317$ liter $= 0.028317$ m$^3 = 7.481$ U.S. gallons

1 m$^3 = 1000$ liter $= 264.17$ U.S. gallons

1 U.S. gallon $= 4$ quarts $= 3.7854$ liters

1 British gallon $= 1.20094$ U.S. gallons

Work, Energy, and Heat

1 J $= 1$ N-m $= 1$ kg-m^2/s$^2 = 10^7$ g $\times$ cm^2/s^2 (erg)

1 Btu $= 1055.06$ J $= 1.05506$ kJ $= 252.16$ cal $= 778.17$ ft-lb$_f$

1 cal $= 4.1868$ J

1 hp $\times$ h $= 0.7457$ kW-h

1 Btu/ft$^3 = 37.25895$ kJ/m^3

1 ft$\cdot$lb$_f$/lb$_m$ $= 2.989$ J/kg

Mathematical Relations

Algebra

$$\frac{1}{a^x} = a^{-x} \quad a^x a^y = a^{(x+y)} \quad \frac{a^x}{a^y} = a^{(x-y)}$$

Logarithms

If $a = 10^x$, then $\log a = x$ and If $a = e^x$, then $\ln a = x$

$$\log (a^n) = n \log a \quad \log (ab) = \log a + \log b \quad \log(a/b) = \log a - \log b \quad \ln(a^n) = n \ln a$$

$$\ln (ab) = \ln a + \ln b \quad \ln (a/b) = \ln a - \ln b$$

Trigonometry

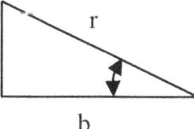

Right Triangle

$A = bh/2$

$r^2 = h^2 + b^2$

$\sin \theta = h/r$

$\cos \theta = b/r$

$\tan \theta = h/b$

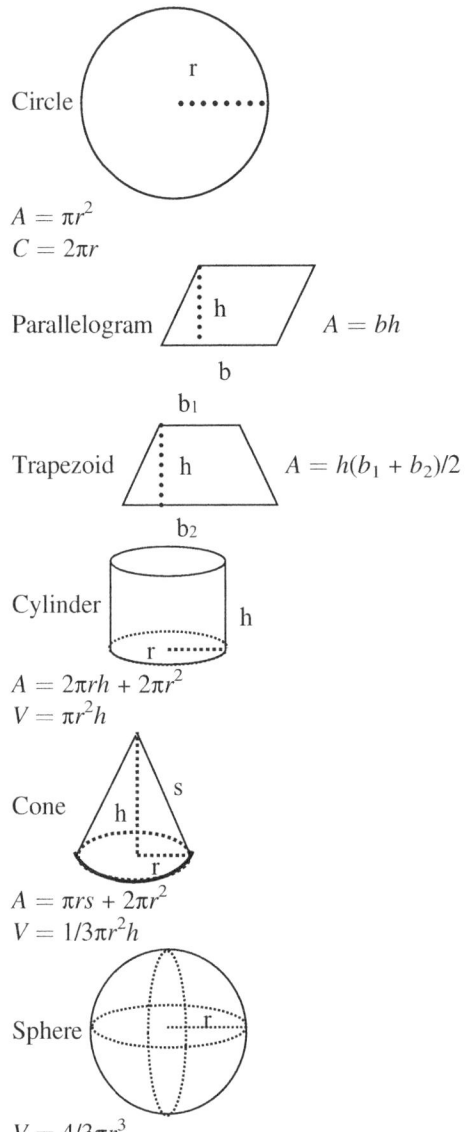

Circle

$A = \pi r^2$
$C = 2\pi r$

Parallelogram $A = bh$

Trapezoid $A = h(b_1 + b_2)/2$

Cylinder

$A = 2\pi rh + 2\pi r^2$
$V = \pi r^2 h$

Cone

$A = \pi rs + 2\pi r^2$
$V = 1/3\pi r^2 h$

Sphere

$V = 4/3\pi r^3$
$A = 4\pi r^2$

Appendix A.2: Thermophysical Properties of Selected Gases, Liquids, and Solids

Substance	Temperature (°C)	Density (kg/m^3)	Specific heat (J/kg-K)	Conductivity (W/m-K)	Diffusivity (m^2/s)
Gases					
Air	0	1.29	1005	0.024	18.6E−06
	20	1.20	1006	0.026	21.2E−06
	100	0.95	1012	0.031	33.0E−06
Water vapor	100	0.60	2030	0.024	19.0E−06
	200	0.47	1960	0.030	32.6E−06
Liquids					
Water	0	999.8	4210	0.553	13.5E−08
	20	998.2	4180	0.599	14.3E−08
	40	992.2	4180	0.629	15.1E−08
	60	983.2	4180	0.652	16.1E−08
	80	971.8	4190	0.670	16.4E−08
	100	958.4	4210	0.684	17.1E−08
Milk	20	1032.0	4000	0.0	13.5E−08
Ethanol	20	790.0	2440	0.167	8.7E−08
Solids					
Ice	0	917	2100	2.22	1.1E−06
	−10	933	2040	2.32	1.2E−06
	−20	948	1950	2.43	1.3E−06
Aluminum	20	2700	900	230.0	95.0E−06
Stainless steel (18/8)	20	7950	480	15.0	4.0E−06
Glass	20	2700	800	0.75	0.4E−06
Polyethylene	20	930	2300	0.32	0.15E−06
Tallow (Beef Fat)	20	950	1675	0.18	9.0E−08

Appendix A.3: Sieves for Particle Size Classification

US sieve size	Tyler mesh size	Aperture Inches	[a]Microns
No. 3	3 Mesh	0.2650	6730
No. 4	4 Mesh	0.1870	4760
No. 5	5 Mesh	0.1570	4000
No. 6	6 Mesh	0.1320	3360
No. 7	7 Mesh	0.1110	2830
No. 8	8 Mesh	0.0937	2380
No. 10	9 Mesh	0.0787	2000

(continued)

(continued)

US sieve size	Tyler mesh size	Aperture	
		Inches	[a]Microns
No. 12	10 Mesh	0.0661	1680
No. 14	12 Mesh	0.0555	1410
No. 16	14 Mesh	0.0469	1190
No. 18	16 Mesh	0.0394	1000
No. 20	20 Mesh	0.0331	841
No. 25	24 Mesh	0.0280	707
No. 30	28 Mesh	0.0232	595
No. 35	32 Mesh	0.0197	500
No. 40	35 Mesh	0.0165	420
No. 45	42 Mesh	0.0138	354
No. 50	48 Mesh	0.0117	297
No. 60	60 Mesh	0.0098	250
No. 70	65 Mesh	0.0083	210
No. 80	80 Mesh	0.0070	177
No. 100	100 Mesh	0.0059	149
No. 400	400 Mesh	0.0015	37

[a]International Standard Organization (ISO) standards

Appendix A.4: Steam Tables

Saturated Steam Table (SI Units)

Temper-ature (°C)	Vapor Pressure (kPa)	Specific Volume (m^3/kg)		Enthalpy (kJ/kg)		Entropy ($kJ/kg \cdot K$)	
		Liquid	Sat'd Vapor	Liquid	Sat'd Vapor	Liquid	Sat'd Vapor
0.01	0.6113	0.0010002	206.136	0.00	2501.4	0.0000	9.1562
3	0.7577	0.0010001	168.132	12.57	2506.9	0.0457	9.0773
6	0.9349	0.0010001	137.734	25.20	2512.4	0.0912	9.0003
9	1.1477	0.0010003	113.386	37.80	2517.9	0.1362	8.9253
12	1.4022	0.0010005	93.784	50.41	2523.4	0.1806	8.8524
15	1.7051	0.0010009	77.926	62.99	2528.9	0.2245	8.7814
18	2.0640	0.0010014	65.038	75.58	2534.4	0.2679	8.7123
21	2.487	0.0010020	54.514	88.14	2539.9	0.3109	8.6450
24	2.985	0.0010027	45.883	100.70	2545.4	0.3534	8.5794
25	3.169	0.0010029	43.360	104.89	2547.2	0.3674	8.5580
27	3.567	0.0010035	38.774	113.25	2550.8	0.3954	8.5156
30	4.246	0.0010043	32.894	125.79	2556.3	0.4369	8.4533
33	5.034	0.0010053	28.011	138.33	2561.7	0.4781	8.3927
36	5.947	0.0010063	23.940	150.86	2567.1	0.5188	8.3336
40	7.384	0.0010078	19.523	167.57	2574.3	0.5725	8.2570
45	9.593	0.0010099	15.258	188.45	2583.2	0.6387	8.1648
50	12.349	0.0010121	12.032	209.33	2592.1	0.7038	8.0763
55	15.758	0.0010146	9.568	230.23	2600.9	0.7679	7.9913
60	19.940	0.0010172	7.671	251.13	2609.6	0.8312	7.9096
65	25.03	0.0010199	6.197	272.06	2618.3	0.8935	7.8310
70	31.19	0.0010228	5.042	292.98	2626.8	0.9549	7.7553
75	38.58	0.0010259	4.131	313.93	2635.3	1.0155	7.6824
80	47.39	0.0010291	3.407	334.91	2643.7	1.0753	7.6122
85	57.83	0.0010325	2.828	355.90	2651.9	1.1343	7.5445
90	70.14	0.0010360	2.361	376.92	2660.1	1.1925	7.4791
95	84.55	0.0010397	1.9819	397.96	2668.1	1.2500	7.4159
100	101.35	0.0010435	1.6729	419.04	2676.1	1.3069	7.3549
105	120.82	0.0010475	1.4194	440.15	2683.8	1.3630	7.2958
110	143.27	0.0010516	1.2102	461.30	2691.5	1.4185	7.2387
115	169.06	0.0010559	1.0366	482.48	2699.0	1.4734	7.1833
120	198.53	0.0010603	0.8919	503.71	2706.3	1.5276	7.1296
125	232.1	0.0010649	0.7706	524.99	2713.5	1.5813	7.0775
130	270.1	0.0010697	0.6685	546.31	2720.5	1.6344	7.0269
135	313.0	0.0010746	0.5822	567.69	2727.3	1.6870	6.9777
140	316.3	0.0010797	0.5089	589.13	2733.9	1.7391	6.9299
145	415.4	0.0010850	0.4463	610.63	2740.3	1.7907	6.8833
150	475.8	0.0010905	0.3928	632.20	2746.5	1.8418	6.8379
155	543.1	0.0010961	0.3468	653.84	2752.4	1.8925	6.7935
160	617.8	0.0011020	0.3071	675.55	2758.1	1.9427	6.7502
165	700.5	0.0011080	0.2727	697.34	2763.5	1.9925	6.7078
170	791.7	0.0011143	0.2428	719.21	2768.7	2.0419	6.6663
175	892.0	0.0011207	0.2168	741.17	2773.6	2.0909	6.6256
180	1002.1	0.0011274	0.19405	763.22	2778.2	2.1396	6.5857
190	1254.4	0.0011414	0.15654	807.62	2786.4	2.2359	6.5079
200	1553.8	0.0011565	0.12736	852.45	2793.2	2.3309	6.4323
225	2548	0.0011992	0.07849	966.78	2803.3	2.5639	6.2503
250	3973	0.0012512	0.05013	1085.36	2801.5	2.7927	6.0730
275	5942	0.0013168	0.03279	1210.07	2785.0	3.0208	5.8938
300	8581	0.0010436	0.02167	1344.0	2749.0	3.2534	5.7045

From Keenan et al. (1969), with permission from John Wiley & Sons, Inc.

Saturated Steam Table (Imperial Units)

Temperature ($°F$)	Vapor Pressure (psia)	Specific Volume (ft^3/lb_m)		Enthalpy (btu/lb_m)		Entropy ($btu/lb_m \cdot °F$)	
		Liquid	Sat'd Vapor	Liquid	Sat'd Vapor	Liquid	Sat'd Vapoi
32.02	0.08866	0.016022	3302	0.00	1075.4	0.000	2.1869
35	0.09992	0.016021	2948	3.00	1076.7	0.00607	2.1764
40	0.12166	0.016020	2445	8.02	1078.9	0.01617	2.1592
45	0.14748	0.016021	2037	13.04	1081.1	0.02618	2.1423
50	0.17803	0.016024	1704.2	18.06	1083.3	0.03607	2.1259
55	0.2140	0.016029	1431.4	23.07	1085.5	0.04586	2.1099
60	0.2563	0.016035	1206.9	28.08	1087.7	0.05555	2.0943
65	0.3057	0.016042	1021.5	33.09	1089.9	0.06514	2.0791
70	0.3622	0.016051	867.7	38.09	1092.0	0.07463	2.0642
75	0.4300	0.016061	739.7	43.09	1094.2	0.08402	2.0497
80	0.5073	0.016073	632.8	48.09	1096.4	0.09332	2.0356
85	0.5964	0.016085	543.1	53.08	1098.6	0.10252	2.0218
90	0.6988	0.016099	467.7	58.07	1100.7	0.11165	2.0083
95	0.8162	0.016114	404.0	63.06	1102.9	0.12068	1.9951
100	0.9503	0.016130	350.0	68.05	1105.0	0.12963	1.9822
110	1.2763	0.016166	265.1	78.02	1109.3	0.14730	1.9574
120	1.6945	0.016205	203.0	88.00	1113.5	0.16465	1.9336
130	2.225	0.016247	157.17	97.98	1117.8	0.18172	1.9109
140	2.892	0.016293	122.88	107.96	1121.9	0.19851	1.8892
150	3.722	0.016343	96.99	117.96	1126.1	0.21503	1.8684
160	4.745	0.016395	77.23	127.96	1130.1	0.23130	1.8484
170	5.996	0.016450	62.02	137.97	1134.2	0.24732	1.8293
180	7.515	0.016509	50.20	147.99	1138.2	0.26311	1.8109
190	9.343	0.016570	40.95	158.03	1142.1	0.27866	1.7932
200	11.529	0.016634	33.63	168.07	1145.9	0.29400	1.7762
210	14.125	0.016702	27.82	178.14	1149.7	0.30913	1.7599
212	14.698	0.016716	26.80	180.16	1150.5	0.31213	1.7567
220	17.188	0.016772	23.15	188.22	1153.5	0.32406	1.7441
230	20.78	0.016845	19.386	198.32	1157.1	0.33880	1.7289
240	24.97	0.016922	16.327	208.44	1160.7	0.35335	1.7143
250	29.82	0.017001	13.826	218.59	1164.2	0.36772	1.7001
260	35.42	0.017084	11.768	228.76	1167.6	0.38193	1.6864
270	41.85	0.017170	10.066	238.95	1170.9	0.39597	1.6731
280	49.18	0.017259	8.650	249.18	1174.1	0.40986	1.6602
290	57.33	0.017352	7.467	259.44	1177.2	0.42360	1.6477
300	66.98	0.017448	6.472	269.73	1180.2	0.43720	1.6356
310	77.64	0.017548	5.632	280.06	1183.0	0.45067	1.6238
320	89.60	0.017652	4.919	290.43	1185.8	0.46400	1.6123
330	103.00	0.017760	4.312	300.84	1188.4	0.47722	1.6010
340	117.93	0.017872	3.792	311.30	1190.8	0.49031	1.5901
350	134.53	0.017988	3.346	321.80	1193.1	0.50329	1.5793
360	152.92	0.018108	2.961	332.35	1195.2	0.51617	1.5688
370	173.23	0.018233	2.628	342.96	1197.2	0.52894	1.5585
380	195.60	0.018363	2.339	353.62	1199.0	0.54163	1.5483
390	220.2	0.018498	2.087	364.34	1200.6	0.55422	1.5383
400	247.1	0.018638	1.8661	375.12	1202.0	0.56672	1.5284
410	276.5	0.018784	1.6726	385.97	1203.1	0.57916	1.5187
450	422.1	0.019433	1.1011	430.2	1205.6	0.6282	1.4806

From Keenan et al. (1969), with permission from John Wiley & Sons, Inc

Superheated Steam Table (SI Units)

Absolute Pressure, kPa (Sat. Temp., °C)		Temperature (°C)							
		100	150	200	250	300	360	420	500
10 (45.81)	v	17.196	19.512	21.825	24.136	26.445	29.216	31.986	35.679
	H	2687.5	2783.0	2879.5	2977.3	3076.5	3197.6	3320.9	3489.1
	s	8.4479	8.6882	8.9038	9.1002	9.2813	9.4821	9.6682	9.8978
50 (81.33)	v	3.418	3.889	4.356	4.820	5.284	5.839	6.394	7.134
	H	2682.5	2780.1	2877.7	2976.0	3075.5	3196.8	3320.4	3488.7
	s	7.6947	7.9401	8.1580	8.3556	8.5373	8.7385	8.9249	9.1546
75 (91.78)	v	2.270	2.587	2.900	3.211	3.520	3.891	4.262	4.755
	H	2679.4	2778.2	2876.5	2975.2	3074.9	3196.4	3320.0	3488.4
	s	7.5009	7.7496	7.9690	8.1673	8.3493	8.5508	8.7374	8.9672
100 (99.63)	v	1.6958	1.9364	2.172	2.406	2.639	2.917	3.195	3.565
	H	2672.2	2776.4	2875.3	2974.3	3074.3	3195.9	3319.6	3488.1
	s	7.3614	7.6134	7.8343	8.0333	8.2158	8.4175	8.6042	8.8342
150 (111.37)	v		1.2853	1.4443	1.6012	1.7570	1.9432	2.129	2.376
	H		2772.6	2872.9	2972.7	3073.1	3195.0	3318.9	3487.6
	s		7.4193	7.6433	7.8438	8.0720	8.2293	8.4163	8.6466
400 (143.63)	v		0.4708	0.5342	0.5951	0.6548	0.7257	0.7960	0.8893
	H		2752.8	2860.5	2964.2	3066.8	3190.3	3315.3	3484.9
	s		6.9299	7.1706	7.3789	7.5662	7.7712	7.9598	8.1913
700 (164.97)	v			0.2999	0.3363	0.3714	0.4126	0.4533	0.5070
	H			2844.8	2953.6	3059.1	3184.7	3310.9	3481.7
	s			6.8865	7.1053	7.2979	7.5063	7.6968	7.9299
1000 (179.91)	v			0.2060	0.2327	0.2579	0.2873	0.3162	0.3541
	H			2827.9	2942.6	3051.2	3178.9	3306.5	3478.5
	s			6.6940	6.9247	7.1229	7.3349	7.5275	7.7622
1500 (198.32)	v			0.13248	0.15195	0.16966	0.18988	0.2095	0.2352
	H			2796.8	2923.3	3037.6	3.1692	3299.1	3473.1
	s			6.4546	6.7090	6.9179	7.1363	7.3323	7.5698
2000 (212.42)	v				0.11144	0.12547	0.14113	0.15616	0.17568
	H				2902.5	3023.5	3159.3	3291.6	3467.6
	s				6.5453	6.7664	6.9917	7.1915	7.4317
2500 (223.99)	v				0.08700	0.09890	0.11186	0.12414	0.13998
	H				2880.1	3008.8	3149.1	3284.0	3462.1
	s				6.4085	6.6438	6.8767	7.0803	7.3234
3000 (233.90)	v				0.07058	0.08114	0.09233	0.10279	0.11619
	H				2855.8	2993.5	3138.7	3276.3	3456.5
	s				6.2872	6.5390	6.7801	6.9878	7.2338

From Keenan et al. (1969), with permission from John Wiley & Sons, Inc
v = specific volume (m^3/kg), H = enthalpy (kJ/kg); s = entropy (kJ/kg·K)

Superheated Steam Table (*Imperial Units*)

Absolute Pressure, psia (Sat. Temp., °F)		Temperature (°F)								
		200	300	400	500	600	700	800	900	1000
1.0 (101.70)	v	392.5	452.3	511.9	571.5	631.1	690.7	750.3	809.9	869.5
	H	1150.1	1195.7	1241.8	1288.5	1336.1	1384.5	1433.7	1483.8	1534.8
	s	2.0508	2.1150	2.1720	2.2235	2.2706	2.3142	2.3550	2.3932	2.4294
5.0 (162.21)	v	78.15	90.24	102.24	114.20	126.15	138.08	150.01	161.94	173.86
	H	1148.6	1194.8	1241.2	1288.2	1335.8	1384.3	1433.5	1483.7	1534.7
	s	1.8715	1.9367	1.9941	2.0458	2.0930	2.1367	2.1775	2.2158	2.2520
10.0 (193.19)	v	38.85	44.99	51.03	57.04	63.03	69.01	74.98	80.95	86.91
	H	1146.6	1193.7	1240.5	1287.7	1335.5	1384.0	1433.3	1483.5	1534.6
	s	1.7927	1.8592	1.9171	1.9690	2.0164	2.0601	2.1009	2.1393	2.1755
14.696 (211.99)	v		30.52	34.67	38.77	42.86	46.93	51.00	55.07	59.13
	H		1192.6	1239.9	1287.3	1335.2	1383.8	1433.1	1483.4	1534.5
	s		1.8157	1.8741	1.9263	1.9737	2.0175	2.0584	2.0967	2.1330
20.0 (227.96)	v		22.36	25.43	28.46	31.47	34.77	37.46	40.45	43.44
	H		1191.5	1239.2	1286.8	1334.8	1383.5	1432.9	1483.2	1534.3
	s		1.7805	1.8395	1.8919	1.9395	1.9834	2.0243	2.0627	2.0989
60.0 (292.73)	v		7.260	8.353	9.399	10.425	11.440	12.448	13.452	14.454
	H		1181.9	1233.5	1283.0	1332.1	1381.4	1431.2	1481.8	1533.2
	s		1.6496	1.7134	1.7678	1.8165	1.8609	1.9022	1.9408	1.9773
100.0 (327.86)	v			4.934	5.587	6.216	6.834	7.445	8.053	8.657
	H			1227.5	1279.1	1329.3	1379.2	1429.6	1480.5	1532.1
	s			1.6517	1.7085	1.7582	1.8033	1.8449	1.8838	1.9204
150.0 (358.48)	v			3.221	3.679	4.111	4.531	4.944	5.353	5.759
	H			1219.5	1274.1	1325.7	1376.6	1427.5	1478.8	1530.7
	s			1.5997	1.6598	1.7110	1.7568	1.7989	1.8381	1.8750
200.0 (381.86)	v			2.361	2.724	3.058	3.379	3.693	4.003	4.310
	H			1210.8	1268.8	1322.1	1373.8	1425.3	1477.1	1529.3
	s			1.5600	1.6239	1.6767	1.7234	1.7660	1.8055	1.8425
250.0 (401.04)	v				2.150	2.426	2.688	2.943	3.193	3.440
	H				1263.3	1318.3	1371.1	1423.2	1475.3	1527.9
	s				1.5948	1.6494	1.6970	1.7401	1.7799	1.8172
300.0 (417.43)	v				1.766	2.004	2.227	2.442	2.653	2.860
	H				1257.5	1314.5	1368.3	1421.0	1473.6	1526.5
	s				1.5701	1.6266	1.6751	1.7187	1.7589	1.7964
400 (444.70)	v				1.2843	1.4760	1.6503	1.8163	1.9776	2.136
	H				1245.2	1306.6	1362.5	1416.6	1470.1	1523.6
	s				1.5282	1.5892	1.6397	1.6884	1.7252	1.7632

From Keenan et al. (1969), with permission from John Wiley & Sons, Inc
  v = specific volume (ft³/lbm), H = enthalpy (btu/lbm); s = entropy (btu/lbm·°F)

Appendix A.5: Pressure-Enthalpy (Mollier) Diagrams

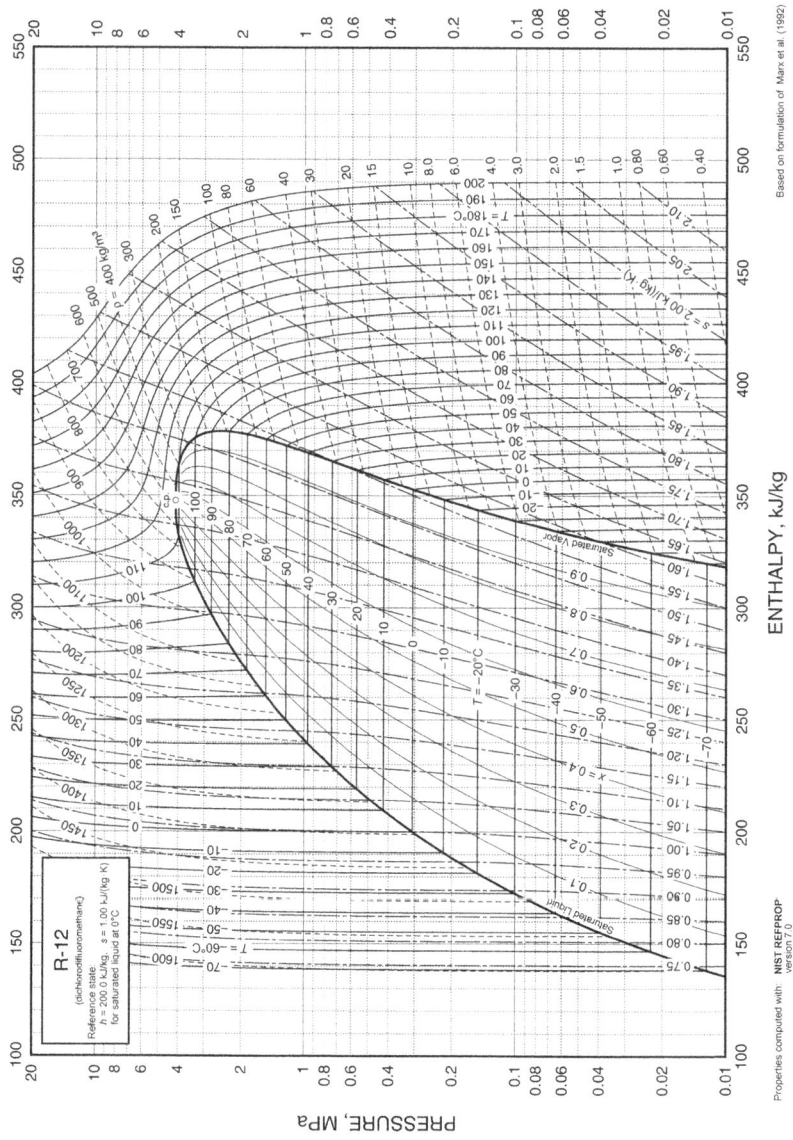

Fig. A.1 Pressure-enthalpy diagram for Refrigerant 12. ©ASHRAE, www.ashrae.org. (2009) ASHRAE Handbook-(Fundamentals: 30.2), with permission

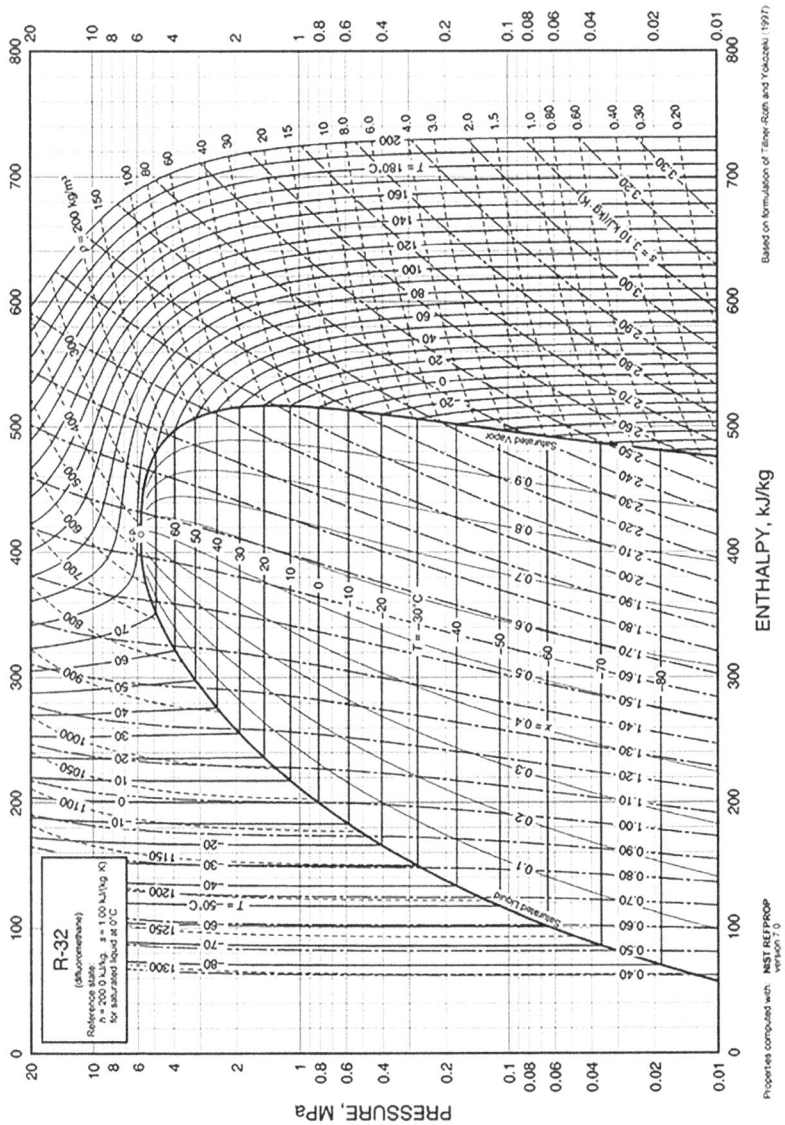

Fig. A.2 Pressure-enthalpy diagram for Refrigerant 32. ©ASHRAE, www.ashrae.org. (2009) ASHRAE Handbook-(Fundamentals: 30.8), with permission

Appendix A.6: Psychrometric Charts

Appendix A.6.1: SI Units

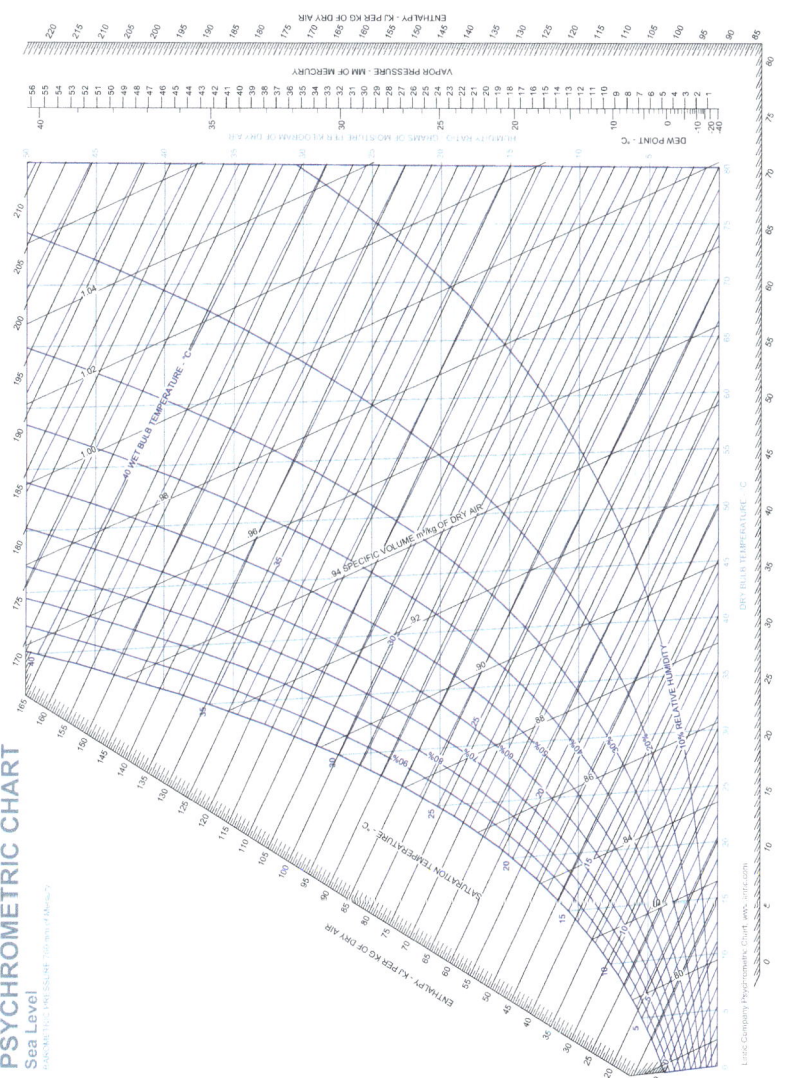

Fig. A.3 Psychrometric chart—SI units

Appendix A.6.2: Imperial Units

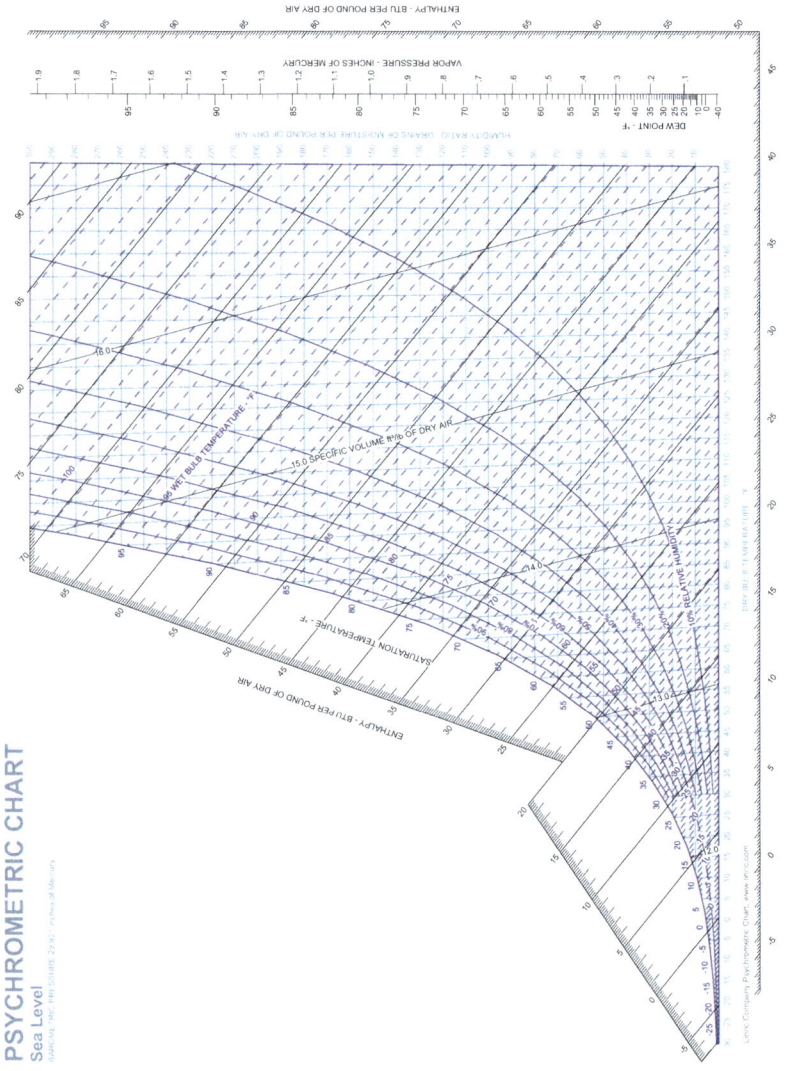

Fig. A.4 Psychrometric chart—imperial units

Appendix A.7: Food Engineering Reference Books

1. Batty JC, Folkman SL (1983) Food engineering fundamentals. Wiley, New York
2. Blakebrough N (ed) (1968) Biochemical and biological engineering science, vol 2. Academic Press, London
3. Brennan JG, Butler JR, Cowell ND, Lilly AEV (1969) Food engineering operations. Elsevier, Barking, Essex, UK
4. Broker DB, Bakker-Arkema FW, Hall CW (1974) Drying cereal grains. AVI Publishing, Westport, CO
5. Bruin S (1988) Preconcentration and drying of food materials. Elsevier, Barking, Essex, UK
6. Charm SE (1983) Fundamentals of food engineering.2nd edn. Allyn & Bacon, Boston, MA
7. Cheryan M (1986) Ultrafiltration handbook. Technomic Publishing, Lancaster, PA
8. Clarke RJ (1957) Process engineering in the food industries. Philosophical Library, New York, NY
9. Decareau RV, Peterson RA (1986) Microwave processing and engineering. Ellis Horwood, Chichester, UK
10. Earle RL (1966) Unit operation in food processing. Pergamon Press, Oxford, UK
11. Farrall AW (1973) Engineering for dairy and food products. Robert E. Krieger Publishing, Huntington, NY
12. Farrall AW (ed) (1976–1979) Food engineering systems. AVI Publishing, Westport, CO
13. Field RW, Howell JA (1989) Process engineering in food industry. Elsevier, Barking, Essex, UK
14. Heldman DR, Singh RP (1981) Food process engineering. AVI Publishing, Westport, CO
15. Hall CW, Farrall AW, Rippen AL (1986) Encyclopedia of food engineering. AVI Publishing, Westport, CO
16. Hallstrom B, Skjoldebrand C, Trfarth C (1988) Heat transfer and food products. Elsevier Applied Science, London, UK
17. Harper JW, Hall CW (1976) Dairy technology and engineering. AVI Publishing, Westport, CO
18. Harper JM (1981) Extrusion of foods. CRC Press, Boca Raton, FL
19. Henderson SM, Perry RL, Young JH (1997) Principles of process engineering. ASAE, St. Joseph, MI
20. Imholte TJ (1984) Engineering for food safety and sanitation. Technical Institute of Food Safety, Crystal, MN
21. Jowitt R, Escher F, Hallstrom B, Meffert AFT, Spiess WEL, Voss G (1983) Physical properties of foods, I. Elsevier Applied Science, London
22. Jowitt R, Escher F, Kent M, McKenna B, Roques M (1987) Physical properties of food, II. Elsevier Applied Science, London
23. LeMaguer B (1984) Food engineering and process applications. Elsevier Applied Science, London

© The Author(s), under exclusive license to Springer Nature Switzerland AG 2024

S. S. H. Rizvi, *Food Engineering Principles and Practices*,

https://doi.org/10.1007/978-3-031-34123-6

24. Kessler HG (1981) Food engineering and dairy technology. V. A. Kessler Publisher, Freising, Germany
25. Leninger HA, Beverloo WA (1975) Food process engineering. D. Riedel Publishing, Dordrecht, The Netherlands
26. Linko P, Makki Y, Olkku J, Larinka J (1980) Food process engineering. Elsevier Applied Science, London
27. Merkel JA (1974) Basic engineering principles. AVI Publishing, Westport, CO
28. McKenna B (1984) Engineering and food industry. Elsevier Applied Science, London
29. Mohsenin NN (1980) Thermal properties of foods and agricultural products. Gordon and Breach Science, Reading, MA
30. Mohsenin NN (1984) Electromagnetic radiation properties of foods and agricultural products. Gordon Breach Science, New York, NY
31. Okos MA (1986) Physical and chemical properties of foods. Marcel Dekker, New York, NY
32. Rizvi SSH, Mittal GS (1992) Experimental methods in food engineering. Van Nostrand Reinhold, New York, NY
33. Rao MA, Rizvi SSH (2014) Engineering properties of foods.4th edn. CRC Press, Boca Raton, FL
34. Sharma SK, Mulvaney SJ, Rizvi SSH (2000) Food process engineering: theory and laboratory experiments. Wiley-Interscience, New York, NY
35. Singh RP, Heldman DR (2014) Introduction to food engineering.5th edn. Academic Press, Orlando, FL
36. Singh RP (1986) Energy in food processing. Elsevier Science Publishers, Amsterdam, The Netherlands
37. Singh RP, Medina A (1989) Food properties and computer aided engineering of food processing systems. D. Riedel Publishing, Dordrecht, The Netherlands
38. Smith PG (2011) Introduction to food process engineering.2nd edn. Springer, New York, NY
39. Thorne S (ed) (1981–1989) Developments in food preservation. Elsevier, London
40. Toledo RT, Singh RK, Kong F (2018) Fundamentals of food process engineering.4th edn. Springer, Cham, Switzerland
41. Watson E, Harper JC (1988) Elements of food engineering.2nd edn. Van Nostrand Reinhold, New York, NY

Index

523

The manufacturer's authorised representative in the EU is Springer
Nature Customer Service Centre GmbH, Europaplatz 3, 69115 Heidelberg,
Germany. If you have any concerns regarding our products, please
contact ProductSafety@springernature.com

Printed and bound by CPI Group (UK) Ltd, Croydon, CR0 4YY

23/04/2026

02095594-0013